Lecture Notes in Computer Science 8055

Commenced Publication in 1973
Founding and Former Series Editors:
Gerhard Goos, Juris Hartmanis, and Jan van Leeuwen

Hendrik Decker   Lenka Lhotská
Sebastian Link   Josef Basl   A Min Tjoa (Eds.)

Database and Expert Systems Applications

24th International Conference, DEXA 2013
Prague, Czech Republic, August 26-29, 2013
Proceedings, Part I

 Springer

Volume Editors

Hendrik Decker
Instituto Tecnológico de Informática, Valencia, Spain
E-mail: hendrik@iti.es

Lenka Lhotská
Czech Technical University in Prague, Czech Republic
E-mail: lhotska@fel.cvut.cz

Sebastian Link
The University of Auckland, New Zealand
E-mail: s.link@auckland.ac.nz

Josef Basl
University of Economics, Prague, Czech Republic
E-mail: basl@vse.cz

A Min Tjoa
Vienna University of Technology, Austria
E-mail: amin@ifs.tuwien.ac.at

ISSN 0302-9743   e-ISSN 1611-3349
ISBN 978-3-642-40284-5   e-ISBN 978-3-642-40285-2
DOI 10.1007/978-3-642-40285-2
Springer Heidelberg Dordrecht London New York

Library of Congress Control Number: 2013944804

CR Subject Classification (1998): H.2, H.3, H.4, I.2, H.5, J.1, C.2

LNCS Sublibrary: SL 3 – Information Systems and Application,
incl. Internet/Web and HCI

Typesetting: Camera-ready by author, data conversion by Scientific Publishing Services, Chennai, India

Printed on acid-free paper

Springer is part of Springer Science+Business Media (www.springer.com)

Preface

The book you are reading comprises the research articles as well as the abstracts of invited talks presented at DEXA 2013, the 24th International Conference on Database and Expert Systems Applications. The conference was held in Prague, the lovely Czech capital, where DEXA already took place in 1993 and 2003. The presented papers show that DEXA has successfully stayed true to a core of themes in the areas of databases, intelligent systems, and related applications, but also that DEXA promotes changing paradigms, new developments, and emerging trends. For the 2013 edition, we had called for novel results or qualified surveys in a wide range of topics, including:

* Acquisition, Modeling, Management and Processing of Knowledge
* Authenticity, Consistency, Integrity, Privacy, Quality, Security of Data
* Availability
* Constraint Modeling and Processing
* Database Federation and Integration, Interoperability, Multi-Databases
* Data and Information Networks
* Data and Information Semantics
* Data and Information Streams
* Data Provenance
* Data Structures and Data Management Algorithms
* Database and Information System Architecture and Performance
* Data Mining and Data Warehousing
* Datalog 2.0
* Decision Support Systems and Their Applications
* Dependability, Reliability and Fault Tolerance
* Digital Libraries
* Distributed, Parallel, P2P, Grid, and Cloud Databases
* Incomplete and Uncertain Data
* Inconsistency Tolerance
* Information Retrieval
* Information and Database Systems and Their Applications
* Metadata Management
* Mobile, Pervasive and Ubiquitous Data
* Modeling, Automation and Optimization of Processes
* Multimedia Databases
* NoSQL and NewSQL Databases
* Object, Object-Relational, and Deductive Databases
* Provenance of Data and Information
* Replicated Databases
* Semantic Web and Ontologies
* Sensor Data Management

* Statistical and Scientific Databases
* Temporal, Spatial, and High-Dimensional Databases
* User Interfaces to Databases and Information Systems
* WWW and Databases, Web Services
* Workflow Management and Databases
* XML and Semi-structured Data

In response to this call, we received 174 submissions from all over the world, of which 43 are included in these proceedings as accepted full papers, and 33 as short papers. We are grateful to the many authors who submitted their work to DEXA. Decisions on acceptance or rejection were based on at least three reviews for each submission. Most of the reviews were detailed and provided constructive feedback to the authors. We owe our deepest thanks to all members of the Program Committee and to the external reviewers who invested their expertise, interest, and time in their reviews.

The program of DEXA 2013 was enriched by three exceptional invited keynote speeches, presented by distinguished colleagues:

- Trevor Bench-Capon: "Structuring E-Participation in Policy Making Through Argumentation"
- Tova Milo: "Making Collective Wisdom Wiser"
- Klaus-Dieter Schewe: "Horizontal and Vertical (Business Process) Model Integration"

In addition to the main conference track, DEXA 2013 also featured seven workshops that explored a wide spectrum of specialized topics of growing general importance. The organization of the workshops was chaired by Franck Morvan, A. Min Tjoa, and Roland R. Wagner, to whom we say "many thanks indeed" for their smooth and effective work.

Special thanks go to the host of DEXA 2013, the Prague University of Economics, where, under the admirable guidance of the DEXA 2013 General Chairs Josef Basl and A. Min Tjoa, an excellent working atmosphere was provided.

Last, but not at all least, we express our heartfelt gratitude to Gabriela Wagner. Her professional attention to detail, skillful management of the DEXA event as well as her preparation of the proceedings volumes are greatly appreciated.

August 2013

Hendrik Decker
Lenka Lhotská
Sebastian Link

Organization

Honorary Chair

Makoto Takizawa   Seikei University, Japan

General Chairs

Josef Basl   University of Economics, Prague,
     Czech Republic
A. Min Tjoa   Technical University of Vienna, Austria

Conference Program Chairs

. Hendrik Decker   Instituto Tecnológico de Informática, Valencia,
     Spain
Lenka Lhotská   Czech Technical University, Czech Republic
Sebastian Link   The University of Auckland, New Zealand

Publication Chair

Vladimir Marik   Czech Technical University, Czech Republic

Workshop Chairs

Franck Morvan   IRIT, Paul Sabatier University, Toulouse,
     France
A. Min Tjoa   Technical University of Vienna, Austria
Roland R. Wagner   FAW, University of Linz, Austria

Program Committee

Slim Abdennadher   German University, Cairo, Egypt
Witold Abramowicz   The Poznan University of Economics, Poland
Hamideh Afsarmanesh   University of Amsterdam, The Netherlands
Riccardo Albertoni   Italian National Council of Research, Italy
Rachid Anane   Coventry University, UK
Annalisa Appice   Università degli Studi di Bari, Italy
Mustafa Atay   Winston-Salem State University, USA
Spiridon Bakiras   City University of New York, USA

Ying Ding	Indiana University, USA
Zhiming Ding	Institute of Software, Chinese Academy of Sciences, China
Gillian Dobbie	University of Auckland, New Zealand
Peter Dolog	Aalborg University, Denmark
Dejing Dou	University of Oregon, USA
Cedric du Mouza	CNAM, France
Johann Eder	University of Klagenfurt, Austria
Suzanne Embury	The University of Manchester, UK
Bettina Fazzinga	University of Calabria, Italy
Leonidas Fegaras	The University of Texas at Arlington, USA
Victor Felea	"Al.I. Cuza" University of Iasi, Romania
Stefano Ferilli	University of Bari, Italy
Flavio Ferrarotti	Victoria University of Wellington, New Zealand
Filomena Ferrucci	Università di Salerno, Italy
Flavius Frasincar	Erasmus University Rotterdam, The Netherlands
Bernhard Freudenthaler	Software Competence Center Hagenberg GmbH, Austria
Hiroaki Fukuda	Shibaura Institute of Technology, Japan
Steven Furnell	Plymouth University, UK
Aryya Gangopadhyay	University of Maryland Baltimore County, USA
Yunjun Gao	Zhejiang University, China
Manolis Gergatsoulis	Ionian University, Greece
Bernard Grabot	LGP-ENIT, France
Fabio Grandi	University of Bologna, Italy
Carmine Gravino	University of Salerno, Italy
Sven Groppe	Lübeck University, Germany
William Grosky	University of Michigan, USA
Jerzy Grzymala-Busse	University of Kansas, USA
Francesco Guerra	Università degli Studi Di Modena e Reggio Emilia, Italy
Giovanna Guerrini	University of Genova, Italy
Antonella Guzzo	University of Calabria, Italy
Abdelkader Hameurlain	Paul Sabatier University, France
Ibrahim Hamidah	Universiti Putra Malaysia, Malaysia
Wook-Shin Han	Kyungpook National University, Republic of Korea
Takahiro Hara	Osaka University, Japan
André Hernich	Humboldt-Universität zu Berlin, Germany
Francisco Herrera	University of Granada, Spain
Steven Hoi	Nanyang Technological University, Singapore
Estevam Rafael Hruschka Jr.	Federal University of Sao Carlos, Brazil
Wynne Hsu	National University of Singapore, Singapore

Yu Hua   Huazhong University of Science and
                            Technology, China
Jimmy Huang   York University, Canada
Xiaoyu Huang   South China University of Technology, China
Michal Huptych   Czech Technical University in Prague,
                            Czech Republic
San-Yih Hwang   National Sun Yat-Sen University, China
                            (Taiwan Province)
Theo Härder   TU Kaiserslautern, Germany
Ionut Emil Iacob   Georgia Southern University, USA
Sergio Ilarri   University of Zaragoza, Spain
Abdessamad Imine   University of Nancy, France
Yasunori Ishihara   Osaka University, Japan
Adam Jatowt   Kyoto University, Japan
Peiquan Jin   University of Science and Technology of China,
                            China
Anne Kao   Boeing, USA
Dimitris Karagiannis   University of Vienna, Austria
Stefan Katzenbeisser   Technische Universität Darmstadt, Germany
Sang-Wook Kim   Hanyang University, Republic of Korea
Benny Kimelfeld   IBM Almaden, USA
Hiroyuki Kitagawa   University of Tsukuba, Japan
Carsten Kleiner   University of Applied Sciences and Arts
                            Hannover, Germany
Solmaz Kolahi   Oracle, USA
Ibrahim Korpeoglu   Bilkent University, Turkey
Harald Kosch   University of Passau, Germany
Michal Krátký   Technical University of Ostrava,
                            Czech Republic
Petr Kremen   Czech Technical University in Prague,
                            Czech Republic
Arun Kumar   IBM Research - India, India
Ashish Kundu   IBM T J Watson Research Center, Yorktown
                            Heights, USA
Josef Küng   University of Linz, Austria
Kwok-Wa Lam   University of Hong Kong, Hong Kong
Nadira Lammari   CNAM, France
Gianfranco Lamperti   University of Brescia, Italy
Anne Laurent   LIRMM, University of Montpellier 2, France
Mong Li Lee   National University of Singapore, Singapore
Alain Léger   FT R&D Orange Labs Rennes, France
Daniel Lemire   LICEF, Université du Québec, Canada
Lenka Lhotská   Czech Technical University, Czech Republic
Wenxin Liang   Dalian University of Technology, China
Stephen W. Liddle   Brigham Young University, USA
Lipyeow Lim   Unversity of Hawaii at Manoa, USA

Tok Wang Ling	National University of Singapore, Singapore
Sebastian Link	The University of Auckland, New Zealand
Volker Linnemann	University of Lübeck, Germany
Chengfei Liu	Swinburne University of Technology, Australia
Chuan-Ming Liu	National Taipei University of Technology, China (Taiwan Province)
Fuyu Liu	Microsoft Corporation, USA
Hong-Cheu Liu	University of South Australia, Australia
Hua Liu	Xerox Research Labs at Webster, USA
Jorge Lloret Gazo	University of Zaragoza, Spain
Peri Loucopoulos	Harokopio University of Athens, Greece
Jiaheng Lu	Renmin University of China, China
Jianguo Lu	University of Windsor, Canada
Alessandra Lumini	University of Bologna, Italy
Hui Ma	Victoria University of Wellington, New Zealand
Qiang Ma	Kyoto University, Japan
Stéphane Maag	TELECOM SudParis, France
Elio Masciari	ICAR-CNR, Università della Calabria, Italy
Norman May	SAP AG, Germany
Jose-Norberto Mazón	University of Alicante, Spain
Dennis McLeod	University of Southern California, USA
Brahim Medjahed	University of Michigan - Dearborn, USA
Alok Mishra	Atilim University, Ankara, Turkey
Harekrishna Mishra	Institute of Rural Management Anand, India
Sanjay Misra	University of Technology, Minna, Nigeria
Jose Mocito	INESC-ID/FCUL, Portugal
Lars Moench	University of Hagen, Germany
Riad Mokadem	IRIT, Paul Sabatier University, France
Anirban Mondal	University of Tokyo, Japan
Yang-Sae Moon	Kangwon National University, Republic of Korea
Reagan Moore	University of North Carolina at Chapel Hill, USA
Franck Morvan	IRIT, Paul Sabatier University, France
Mirco Musolesi	University of Birmingham, UK
Tadashi Nakano	University of California, Irvine, USA
Ullas Nambiar	IBM Research
Ismael Navas-Delgado	University of Málaga, Spain
Martin Necasky	Charles University in Prague, Czech Republic
Wilfred Ng	University of Science and Technology, Hong Kong
Javier Nieves Acedo	University of Deusto, Spain

Levent V. Orman	Cornell University, Ithaca, New York, USA
Mourad Oussalah	University of Nantes, France
Gultekin Ozsoyoglu	Case Western Reserve University, USA
George Pallis	University of Cyprus, Cyprus
Christos Papatheodorou	Ionian University and "Athena" Research Centre, Greece
Marcin Paprzycki	Polish Academy of Sciences, Warsaw Management Academy, Poland
Óscar Pastor López	Universidad Politécnica de Valencia, Spain
Dhaval Patel	National University of Singapore, Singapore, Singapore
Jovan Pehcevski	European University, Macedonia, Former Yugoslav Republic
Jorge Perez	Universidad de Chile, Chile
Reinhard Pichler	Technische Universität Wien, Austria
Olivier Pivert	Ecole Nationale Supérieure des Sciences Appliquées et de Technologie, France
Clara Pizzuti	Institute for High Performance Computing and Networking (ICAR)-National Research Council (CNR), Italy
Jaroslav Pokorny	Charles University in Prague, Czech Republic
Pascal Poncelet	LIRMM, France
Elaheh Pourabbas	National Research Council, Italy
Xiaojun Qi	Utah State University, USA
Fausto Rabitti	ISTI, CNR Pisa, Italy
Claudia Raibulet	Università degli Studi di Milano-Bicocca, Italy
Isidro Ramos	Technical University of Valencia, Spain
Praveen Rao	University of Missouri-Kansas City, USA
Rodolfo F. Resende	Federal University of Minas Gerais, Brazil
Claudia Roncancio	Grenoble University/LIG, France
Edna Ruckhaus	Universidad Simon Bolivar, Venezuela
Massimo Ruffolo	ICAR-CNR, Italy
Igor Ruiz-Agundez	University of Deusto, Spain
Giovanni Maria Sacco	University of Turin, Italy
Shazia Sadiq	The University of Queensland, Australia
Simonas Saltenis	Aalborg University, Denmark
Carlo Sansone	Università di Napoli "Federico II", Italy
Igor Santos Grueiro	Deusto University, Spain
Ismael Sanz	Universitat Jaume I, Spain
N.L. Sarda	I.I.T. Bombay, India
Marinette Savonnet	University of Burgundy, France
Raimondo Schettini	Università degli Studi di Milano-Bicocca, Italy
Peter Scheuermann	Northwestern University, USA

Klaus-Dieter Schewe   Software Competence Centre Hagenberg,
                               Austria
Erich Schweighofer   University of Vienna, Austria
Florence Sedes   IRIT, Paul Sabatier University, Toulouse,
                               France
Nazha Selmaoui   University of New Caledonia, New Caledonia
Patrick Siarry   Université Paris 12 (LiSSi), France
Gheorghe Cosmin Silaghi   Babes-Bolyai University of Cluj-Napoca,
                               Romania
Leonid Sokolinsky   South Ural State University, Russian
                               Federation
Bala Srinivasan   Monash University, Australia
Umberto Straccia   Italian National Research Council, Italy
Darijus Strasunskas   DS Applied Science, Norway
Lena Strömbäck   Swedish Meteorological and Hydrological
                               Institute, Sweden
Aixin Sun   Nanyang Technological University, Singapore
Raj Sunderraman   Georgia State University, USA
David Taniar   Monash University, Australia
Cui Tao   Mayo Clinic, USA
Maguelonne Teisseire   Irstea - TETIS, France
Sergio Tessaris   Free University of Bozen-Bolzano, Italy
Olivier Teste   IRIT, University of Toulouse, France
Stephanie Teufel   University of Fribourg, Switzerland
Jukka Teuhola   University of Turku, Finland
Taro Tezuka   University of Tsukuba, Japan
Bernhard Thalheim   Christian Albrechts Universität Kiel, Germany
Jean-Marc Thevenin   University of Toulouse 1 Capitole, France
Helmut Thoma   Thoma SW-Engineering, Basel, Switzerland
A Min Tjoa   Vienna University of Technology, Austria
Vicenc Torra   IIIA-CSIC, Spain
Traian Marius Truta   Northern Kentucky University, USA
Vassileios Tsetsos   National and Kapodistrian University of
                               Athens, Greece
Theodoros Tzouramanis   University of the Aegean, Greece
Maria Vargas-Vera   Universidad Adolfo Ibanez, Chile
Krishnamurthy Vidyasankar   Memorial University of Newfoundland, Canada
Marco Vieira   University of Coimbra, Portugal
Jianyong Wang   Tsinghua University, China
Junhu Wang   Griffith University, Brisbane, Australia
Qing Wang   The Australian National University, Australia
Wei Wang   University of New South Wales, Sydney,
                               Australia
Wendy Hui Wang   Stevens Institute of Technology, USA
Gerald Weber   The University of Auckland, New Zealand
Jef Wijsen   Université de Mons, Belgium

Andreas Wombacher	University Twente, The Netherlands
Lai Xu	Bournemouth University, UK
Ming Hour Yang	Chung Yuan Christian University, China (Taiwan Province)
Xiaochun Yang	Northeastern University, China
Haruo Yokota	Tokyo Institute of Technology, Japan
Zhiwen Yu	Northwestern Polytechnical University, China
Xiao-Jun Zeng	University of Manchester, UK
Zhigang Zeng	Huazhong University of Science and Technology, China
Xiuzhen (Jenny) Zhang	RMIT University, Australia
Yanchang Zhao	RDataMining.com, Australia
Yu Zheng	Microsoft Research Asia
Xiaofang Zhou	University of Queensland, Australia
Qiang Zhu	The University of Michigan, USA
Yan Zhu	Southwest Jiaotong University, China

External Reviewers

Giuseppe Amato	ISTI-CNR, Italy
Abdelkrim Amirat	University of Nantes, France
Edimilson Batista dos Santos	Federal University of Sao Joao Del Rei, Brazil
Souad Boukhadouma	University of Nantes, France
Sahin Buyrukbilen	City University of New York, USA
Changqing Chen	Yahoo, USA
Jimmy Ka Ho Chiu	La Trobe University, Australia
Camelia Constantin	UPMC (university Pierre and Marie Curie), Paris, France
Matthew Damigos	NTUA, Greece
Andrea Esuli	ISTI-CNR, Italy
Fabrizio Falchi	ISTI-CNR, Italy
Ming Fang	Georgia State University, USA
Nikolaos Fousteris	Ionian University, Greece
Maria Jesús García Godoy	Universidad de Málaga, Spain
Di Jiang	Hong Kong University of Science and Technology, Hong Kong
Christos Kalyvas	University of the Aegean, Greece
Anas Katib	University of Missouri-Kansas City, USA
Julius Köpke	University of Klagenfurt, Austria
Christian Koncilia	University of Klagenfurt, Austria
Janani Krishnamani	Georgia State University, USA
Meriem Laifa	University of Bordj Bouarreridj, Algeria
Szymon Łazaruk	Poznan University of Economics, Poland
Chien-Hsian Lee	National Sun Yat-sen University, Taiwan
Fabio Leuzzi	University of Bari, Italy
Dingcheng Li	Mayo Clinic, USA

Table of Contents – Part I

Parallel Processing

XML and RDF

Enterprise Models

Query Evaluation and Optimization

Semantic Web

Sampling

Industrial Applications

Communities

Table of Contents – Part II

Queries, Streams, and Uncertainty

Storage and Compression

Query Processing

Security

Distributed Data Processing

Metadata Modeling and Maintenance

Pricing and Recommending

Security and Semantics

Horizontal and Vertical Business Process Model Integration[*]
(Abstract)

Klaus-Dieter Schewe[1,2]

[1] Software Competence Center Hagenberg, Austria
kd.schewe@scch.at
[2] Johannes-Kepler-University Linz, Austria
kd.schewe@cdcc.faw.jku.at

Modelling information systems in general is a complex endeavour, as systems comprise many different aspects such as the data, functionality, interaction, distribution, context, etc., which all require different models. In addition, models are usually built on different levels of abstraction and the switch from one of these levels to another one may cause mismatches. Horizontal model integration refers to the creation of system models by successive enlargement, whereas vertical model integration refers to the systematic, seamless refinement process of high-level abstract (conceptual) models down to running systems. Our research on horizontal and vertical model integration has concentated on business process models. The results will be reported in the monograph [5].

With respect to horizontal model integration several submodels have to be defined and integrated. The common model to start with addresses the *control flow model*, i.e. a business process is decsribed in an abstract way by a set of activities and gateways, the latter ones for splitting and synchronisation, plus start and termination events. Depending on whether one, all or an arbitrary selection of (outgoing) paths are enabled in splitting gateways, we adopt the common distinction between XOR-, AND- and OR-gateways with an analogous distinction for the synchronisation gateways. However, this terminology is in a sense misleading, as there need not be a well-nested structure, in which a splitting-gateway corresponds to exactly one synchronisation gateway. This is one of the reasons, why we formalise the semantics of each of the constructs by means of Abstract State Machines (ASMs, [2]). As a state-based rigorous method, ASMs support the unambiguous capture of the semantics of OR-synchronisation [1]. Furthermore, on grounds of ASMs necessary subtle distinctions and extensions to the control flow model such as counters, priorities, freezing, etc. can be easily integrated in a smooth way. All constructs found in a control flow model are supposed to be exceuted in parallel for all process instances.

The control flow model is then extended by a *message model* and an *event model*. For this refinement in ASMs – mainly conservative extensions – are

---

[*] The research reported in this paper was supported by the European Fund for Regional Development as well as the State of Upper Austria for the project *Vertical Model Integration* within the program "Regionale Wettbewerbsfähigkeit Oberösterreich 2007-2013".

H. Decker et al. (Eds.): DEXA 2013, Part I, LNCS 8055, pp. 1–3, 2013.

exploited. In particular, the ground specification of firing conditions that depend on the state of the control flow, data, events and resources and actions that update this state [3] requires that only conditions and actions are refined. While messages are easily captured by means of specifications of sender and receiver, it becomes more subtle to define details such as synchronised vs. asynchronised messaging, delivery failure, rejection, message box overflows, etc. In H-BPM the ASM-based specification of messaging from S-BPM [4] has been adopted. For the event model it is necessary and sufficient to specify what kind of events are to be observed, which can be captured on the grounds of monitored locations in ASMs, and which event conditions are to be integrated into the model.

The next horizontal extensions concern the *actor model*, i.e. the specification of responsibilities for the execution of activities (roles), as well as rues governing rights and obligations. This leads to the integration of deontic constraints [6], some of which can be exploited to simplify the control flow [7]. In this way subtle distinctions regarding decision-making responsibilities in BPM can be captured. Horizontal model integration through refinement is then extended towards an *interaction model* and a *data model*. For this, an abstract dialogue model is adopted [8] capturing interaction by means of operations on views that are defined on top of a database schema. In this way the data model results from view integration, but global consistency has to be addressed, as a global database infers dependencies between activities that are not visible on the control flow level.

Finally, an *exception handling model* has to be integrated to complete the horizontal integration picture. This is still in a preliminary state in H-BPM. Overall, the general idea is that an exception is a disruptive event that requires partial rollback and depending on the state the continuation with a different subprocess.

Vertical integration is achieved by further refining the involved ASMs in a development process that is targeting the executable specification of a workflow engine that is enriched with components for data and dialogue handling and exception processing. Throughout the process rigorous quality assurance methods have to be applied.

References

1. Börger, E., Sörensen, O., Thalheim, B.: On defining the behavior of OR-joins in business process models. Journal of Universal Computer Science 15(1), 3–32 (2009)
2. Börger, E., Stärk, R.: Abstract State Machines. Springer, Heidelberg (2003)
3. Börger, E., Thalheim, B.: Modeling workflows, interaction patterns, web services and business processes: The ASM-based approach. In: Börger, E., Butler, M., Bowen, J.P., Boca, P. (eds.) ABZ 2008. LNCS, vol. 5238, pp. 24–38. Springer, Heidelberg (2008)
4. Fleischmann, A., et al.: Subject-Oriented Business Process Management. Springer, Heidelberg (2012)
5. Kossak, F., et al.: The Hagenberg Business Process Modelling Method H-BPM (2014) (forthcoming)

6. Natschläger, C., Kossak, F., Schewe, K.D.: BPMN to Deontic BPMN: A trusted model transformation. Journal of Software and Systems Modelling (to appear, 2013)
7. Natschläger, C., Schewe, K.D.: A flattening approach for attributed type graphs with inheritance in algebraic graph transformation. Electronic Communications of the EASST 47, 160–173 (2012)
8. Schewe, K.D., Schewe, B.: Integrating database and dialogue design. Knowledge and Information Systems 2(1), 1–12 (2000)

Structuring E-Participation in Policy Making through Argumentation

Trevor Bench-Capon

Department of Computer Science, University of Liverpool,
England
tbc@liverpool.ac.uk

An important feature of democracies is that citizens can engage their Governments in dialogues about policies. They tend to do so in one of three ways: they may seek a justification of some policy or action; they may object to all or some aspects of a policy; or they may make policy proposals of their own.

For the first, the reply need only to state a justification. For the second, having offered the justification, the respondent needs first to understand what the citizen objects to, and then to give an answer to the specific points. For the third, first a well formulated proposal must be elicited from the citizen, and then that proposal can then be critiqued from the standpoint of the government's own beliefs and values. Current e-participation systems too often lack structure. Most commonly they take the form of petitions or threaded discussions. Petitions allow the expression of general feelings, but they are unable to express objections with precision. Too often they are ill expressed and conflate a variety of different arguments, so that it is not clear what people are subscribing too. Threaded discussions allow people to feel that they have expressed their views, but they lack structure. Thus arguments are typically ill-formed, and the lack of structure also makes comparison, aggregation and assimilation difficult. In consequence Government replies are often general, bland and superficial and do not address the particular objections of the citizens. To address these issues, we need tools that are firrmly grounded on a well defined model of argument.

Common to all three scenarios is the notion of justifying an action. Justifying actions is a form of practical reasoning, and has traditionally made use of the practical syllogism of Aristotle. For the purposes of computational modelling, the traditional syllogism has been re-expressed in the form of an argumentation scheme. As stated in [1], the scheme brings together knowledge of the current circumstances, the effects of actions, the goals being pursued and the values which will be promoted if the goals are attained:

- In the current circumstances (R), action ac should be performed by agent ag, since this will bring about a new set of circumstances (S), which will realise a goal (G). Realising G in R will promote social value (V).

Following the notion of argumentation schemes in [3], an argument made using an argumentation scheme can be challenged using characteristic *critical questions*. Seventeen such critical questions are given in [1], covering the formulation of the

H. Decker et al. (Eds.): DEXA 2013, Part I, LNCS 8055, pp. 4–6, 2013.

problem (what is considered relevant, the causal relations in the domain etc), current beliefs (what is true now, how will other agents respond if ag does ac) and the evaluation of the actions (does realising G promote V, is there a better way to promote V, etc). This scheme and the critical questions can be used to structure justification of policy, and critiques of such justifications.

Note that this scheme requires knowledge of several sorts: knowledge of what can be considered relevant to the question, knowledge of what actions are available, knowledge of what is the case, knowledge of the consequences of these actions, knowledge of other agents who can influence the results of the actions, knowledge of what is desirable, and knowledge of preferences between values. Such knowledge can be captured in the form of an Action-based Alternating Transition System (AATS) [4], augmented to label the transitions with the values promoted and demoted. The scheme and the critical questions are given in terms of an AATS extended with such labels in [1].

This scheme, and its underlying AATS model, can be used as the basis of tools to support e-participation. First the domain is modelled as an AATS. An example of such a model can be found in [7], where the domain related to the formulation of policy on the introduction of speed cameras to reduce traffic accidents was modelled. Such a model can then be used to support several policy related tasks. The task of selecting a policy from among the several available is considered in [7]. We have also developed two interactive web tools to support the second and third tasks mentioned above.

For the second task, where the policy-maker presents a policy to citizens and solicits their points of agreement and disagreement, we provide the *Structured Consultation Tool* (SCT), written in PhP and accessing a MySQL database. The user is presented with five screens, one each for an introduction, circumstances, consequences, values, and a summary page. These screens explore the various elements of the argumentation scheme of [1], and ask the user a series of yes/no questions, the responses to which can be interpreted by someone familiar with the scheme as posing particular critical questions relating to the AATS model. This structures the interaction in terms of the model and the scheme, but does not require the user to be aware of this, and so the tool remains simple to use. In this way a fine grained response can be obtained, and assimilated with other responses: this is not possible with free text as found in threaded discussions, which lack the required unifying structure. More details can be found in [5].

The third task is supported by the *Critique Tool* (CT), based on the same database, argumentation scheme, and also implemented using MySQL and PHP. Rather than the policy-maker presenting a policy for critique, the user is able to create her own policy proposal interactively by selecting from a menu of choices relating to circumstances, actions, consequences and values. Internally this is structured using the argumentation scheme and then critiqued from the basis of the AATS model and preferences of the Government. The justification is again structured using the argumentation scheme, and the critique again takes the form of a range of appropriate critical questions, which are generated automatically from the model. Thus, the citizen can proactively engage with with policy-making rather than simply reacting to a given policy proposal. Again the structure is exploited without requiring the user to be aware of the structure, allowing the arguments and the criticisms to be well formed and precise without compromising usability. More details of this tool can be found in [6].

Both tools also provide access to additional supporting information through links to other web sites, including external sites. These may offer independent support for the views of the Government, or may set out the pros and cons for the citizen to consider. The tools are (June 2013) available at

- http://impact.uid.com:8080/impact/ and
- http://cgi.csc.liv.ac.uk/ maya/ACT/

A major problem with current e-participation systems is organising the replies for comparison, aggregation and assimilation. One answer to this is to make use of a well defined argumentation structure to organise policy justifications and critiques of these justifications. I have described:

- An argumentation scheme to structure justification and critiques;
- A semantical structure for models to underpin this scheme
- A tool to facilitate a precise critique of the scheme
- A tool to elicit a well frormed justification and generate an automatic critique.

Acknowledgements. This work was partially supported by the FP7-ICT-2009-4 Programme, IMPACT Project, Grant Number 247228. The views are those of the author. I would especially like to thank my colleagues Adam Wyner, Katie Atkinson and Maya Wardeh. The work described here has its origins in [2], also presented in Prague.

References

1. Atkinson, K., Bench-Capon, T.J.M.: Practical reasoning as presumptive argumentation using action based alternating transition systems. Artif. Intell. 171(10-15), 855–874 (2007)
2. Greenwood, K., Bench-Capon, T.J.M., McBurney, P.: Structuring dialogue between the people and their representatives. In: Traunmüller, R. (ed.) EGOV 2003. LNCS, vol. 2739, pp. 55–62. Springer, Heidelberg (2003)
3. Walton, D.: Argumentation Schemes for Presumptive Reasoning. Lawrence Erlbaum Associates, Mahwah (1996)
4. Wooldridge, M., van der Hoek, W.: On obligations and normative ability: Towards a logical analysis of the social contract. J. Applied Logic 3(3-4), 396–420 (2005)
5. Wyner, A., Atkinson, K., Bench-Capon, T.: Towards a structured online consultation tool. In: Tambouris, E., Macintosh, A., de Bruijn, H. (eds.) ePart 2011. LNCS, vol. 6847, pp. 286–297. Springer, Heidelberg (2011)
6. Wyner, A.Z., Atkinson, K., Bench-Capon, T.: Model based critique of policy proposals. In: Tambouris, E., Macintosh, A., Sæbø, Ø. (eds.) ePart 2012. LNCS, vol. 7444, pp. 120–131. Springer, Heidelberg (2012)
7. Wyner, A.Z., Bench-Capon, T.J.M., Atkinson, K.: Towards formalising argumentation about legal cases. In: Ashley, K.D., van Engers, T.M. (eds.) ICAIL, pp. 1–10. ACM (2011)

Making Collective Wisdom Wiser

Tova Milo

School of Computer Science,
Tel Aviv University, Israel
milo@post.tau.ac.il

Many popular sites, such as Wikipedia and Tripadvisor, rely on public participation to gather information – a process known as crowd data sourcing. While this kind of collective intelligence is extremely valuable, it is also fallible, and policing such sites for inaccuracies or missing material is a costly undertaking. In this talk we will overview the MoDaS project that investigates how database technology can be put to work to effectively gather information from the public, efficiently moderate the process, and identify questionable input with minimal human interaction [1–4, 7]. We will consider the logical, algorithmic, and methodological foundations for the management of large scale crowd-sourced data as well as the development of applications over such information.

The goal of the project is to develop solid scientific foundations for Web-scale data sourcing. We believe that such a principled approach is essential to obtain knowledge of superior quality, to realize the task more effectively and automatically, be able to reuse solutions, and thereby to accelerate the pace of the adoption of this new technology that is revolutionizing our life. This requires the development of formal models capturing all the diverse facets of crowd-sourced data. This also encompasses developing the necessary reasoning capabilities for managing and controlling data sourcing, cleaning, verification, integration, sharing, querying and updating, in a dynamic Web environment [5, 6, 8–12]. Such a technological breakthrough will open the way for developing a new and otherwise unattainable universe of knowledge in a wide range of applications, from scientific fields to social and economical ones.

Acknowledgment. This work has been partially funded by the European Research Council under the European Community's Seventh Framework Programme (FP7/2007-2013) / ERC grant MoDaS, agreement 291071, by the Israel Ministry of Science, and by the US-Israel Bi national Science foundation.

References

1. Amsterdamer, Y., Grossman, Y., Milo, T., Senellart, P.: Crowd mining. In: SIGMOD (2013)
2. Boim, R., Greenshpan, O., Milo, T., Novgorodov, S., Polyzotis, N., Tan, W.-C.: Asking the right questions in crowd data sourcing. In: ICDE, pp. 1261–1264 (2012)
3. Davidson, S., Khanna, S., Milo, T., Roy, S.: Using the crowd for top-k and group-by queries. In: ICDT (2013)

H. Decker et al. (Eds.): DEXA 2013, Part I, LNCS 8055, pp. 7–8, 2013.

4. Deutch, D., Greenshpan, O., Kostenko, B., Milo, T.: Declarative platform for data sourcing games. In: WWW, pp. 779–788 (2012)
5. Franklin, M.J., Kossmann, D., Kraska, T., Ramesh, S., Xin, R.: Crowddb: answering queries with crowdsourcing. In: SIGMOD (2011)
6. Guo, S., Parameswaran, A.G., Garcia-Molina, H.: So who won?: dynamic max discovery with the crowd. In: SIGMOD Conference, pp. 385–396 (2012)
7. Kaplan, H., Lotosh, I., Milo, T., Novgorodov, S.: Answering planning queries with the crowd. In: VLDB (2013)
8. Liu, X., Lu, M., Ooi, B.C., Shen, Y., Wu, S., Zhang, M.: Cdas: A crowdsourcing data analytics system. PVLDB 5(10), 1040–1051 (2012)
9. Marcus, A., Wu, E., Madden, S., Miller, R.C.: Crowdsourced databases: Query processing with people. In: CIDR, pp. 211–214 (2011)
10. Park, H., Pang, R., Parameswaran, A.G., Garcia-Molina, H., Polyzotis, N., Widom, J.: Deco: A system for declarative crowdsourcing. PVLDB 5(12), 1990–1993 (2012)
11. Selke, J., Lofi, C., Balke, W.-T.: Pushing the boundaries of crowd-enabled databases with query-driven schema expansion. PVLDB 5(6), 538–549 (2012)
12. Wang, J., Kraska, T., Franklin, M.J., Feng, J.: Crowder: Crowdsourcing entity resolution. PVLDB 5(11), 1483–1494 (2012)

Preferences Chain Guided Search and Ranking Refinement

Yann Loyer, Isma Sadoun, and Karine Zeitouni

PRiSM, CNRS UMR 8144
Université de Versailles Saint Quentin, France

Abstract. Preference queries aim at increasing personalized pertinence of a selection. The most famous ones are the skyline queries based on the concept of dominance introduced by Pareto. Many other dominances have been proposed. In particular, many weaker forms of dominance aim at reducing the size of the answer of the skyline query. In most cases, applying just one dominance is not satisfying as it is hard to conciliate high pertinence, i.e. a strong dominance, and reasonable size of the selection. We propose to allow the user to decide what dominances are reliable, and what priorities between those dominances should be respected. This can be done by defining a sequence, eventually transfinite, of dominances. According to that sequence, we propose operators that compute progressively the ranking of a dataset by successive application of the dominances without introducing inconsistencies. The principle of progressive refinement provides a great flexibility to the user that can not only dynamically decide to stop the process whenever the results satisfies his/her wishes, but can also navigates in the different levels of ranking and be aware of the level of reliability of each successive refinement. We also define maximal selection and top-k methods, and discuss some experimentations of those operators.

1 Introduction

Considerable attention has recently been paid to preference queries. Those queries aim to improve the pertinence of information retrieval that may be different from one user to another. They take into account user's preferences and have been studied following two different ways [11]. The first approach personalizes a given query by expanding it to include preferences. The second approach uses explicit preference operators in the query, such that the *skyline operator* [2] which is based on the concept of *dominance* or *efficiency* introduced by Pareto.

Considering a set of alternatives that can be compared with respect to a finite set of criteria, Pareto defined an alternative A to be more efficient than another one B (or to dominate B) if there is at least one criterion that suggests to prefer A to B while there exists none that suggests the contrary. The set of optimal alternative, i.e. those that are not dominated by any other one, is called the frontier of Pareto. The skyline operator computes that frontier.

For example, consider a relational schema R of basketball players with five attributes *(points, rebounds, blocks, steals, played games)*. Suppose a coach wants to recruit a player. He will search for the best player, i.e., according to the skyline

H. Decker et al. (Eds.): DEXA 2013, Part I, LNCS 8055, pp. 9–24, 2013.

Table 1. Table of basket-ball players

	points	rebounds	blocks	steals	games
t_1	20	20	20	20	100
t_2	22	20	19.5	15	80
t_3	16	24	16	15	100
t_4	15	10	18.5	10	105
t_5	10	12	10	10	105
t_6	10	8	10	12	105
t_7	5	5	5	5	110
t_8	10	10	0	0	110
t_9	8	5	0	0	100

approach, the player who scores the most points, who makes the most rebounds, blocks, steals and whose experience (number of played games) is the greatest. Of course such a player may not exist! Consider the set of players given in table 1. The player t_1 scores more points, makes more rebounds, blocks and steals than player t_4. But player t_4 has played more games than player t_1. That means t_1 is preferable to t_4 with respect to some criteria, while t_4 is preferable to t_1 with respect to others. It follows that those two points are incomparable with respect to the traditional dominance. The reader will easily verify that, in this set, the only player that is dominated is t_9 (for instance by t_1). All the other players are in the answer of the skyline query. Obviously, that query will be useless for the coach to select the best player.

In the context of high dimensional databases, skyline queries alone do not provide an efficient decision support. It is therefore necessary to refine the selection. For example, the user may consider that the difference on the number of *played games* is not big enough to refute the domination of t_1 over t_4. Making such an assumption, one can assert that t_1 dominates t_4 with respect to a weaker form of dominance. Different approaches have been proposed to overcome that limitation. The main idea consists in introducing more comparability by defining other, mostly weaker, dominance relations. To name a few : ϵ-dominance [13], K-dominance [3], dominance-back [14] and quasi-dominance [6]. The problem of selecting data points with some designated dominances has been investigated in [4, 8–12]. The relevance of the different dominance relations is obviously disputable and depends on the context and/or the user. More generally, the dominance relation could even be defined by the user itself or be obtained from queries to experts, communities of users or web services. The dominance could even be obtained by the integration of information coming from different, eventually inconsistent, sources, using different combinations relying on operators such as set operators or other specific operators (e.g., see [11]). In fact, a dominance relation should simply be defined as a binary relation over the set of tuples.

It follows that the user is confronted with the choice of a dominance relation. Indeed, given a set of dominance relations, the user shall consider as more appropriate or relevant a dominance than another one. Once his choice done, he can refine the skyline set with respect to the chosen dominance using the operator, known as *winnow* [5] or *Best* [11], that returns the maximal elements – or their iterated versions that rank the whole dataset. We propose an alternative version of *Winnow* that does not systematically remove tuples involved in cycles, contrarily to usual approaches allowing cyclic dominance relations. We believe that eliminating from the result the tuples belonging to a cycle, whatever their relations to the other tuples, may cause some relevant results to be missed. However, the result provided may still not satisfy the user as it may still contains too many incomparable tuples. Thus the user could decide to apply another dominance on the previous result in order to refine it even more. Of course, the application of a second dominance should not introduce inconsistencies with the first one, nor lead to inverse the ranking of two tuples. Pushing that idea further, the user shall assign some priorities to the dominance relationships. We propose to generalize that reasoning to *any given set of binary relations* over a same set of tuples. The user decides of (a) a selection, eventually transfinite, in the set of dominance relations of those he wants to use, (b) values for the parameters of each dominance relation that requires some, and (c) a strict total order, called *preferences chain*, over the set of relations he selected. Guided by that chain, we apply successively the dominance relations to refine progressively the answer set. Each step may allow new comparisons between tuples, but only between tuples that were considered incomparable and ranked at the same level by the precedent step.

In this paper, we define an operator that computes *preferences chain guided rankings* of a set of tuples (that can eventually be the skyline set). It applies successive dominance relations in a way that each application of one more dominance refines the ranking provided by the precedent one. We also define two variants of that operator: an operator that computes *preferences chains guided searches*, by reducing the answer set as much as possible to obtain only the very best tuples according to user's preferences, and a top-k operator. The principle of progressive refinement provides a great flexibility to the user that can not only determine priorities between dominances he decides to rely on, but can also dynamically decide to stop the process whenever the results satisfies his wishes, or even navigates in the different levels of ranking, being aware of the level of reliability of each successive refinement. Finally, we provide experimental results that show the effectiveness and the efficiency of our algorithms.

2   Preliminaries

Let $R(d_1, ..., d_n)$ denotes a database relation schema with n attributes where each attribute d_i takes values from a numerical domain $dom(d_i)$. Let $d = \{d_1, ..., d_n\}$ be the set of attributes of R and $dom(d)$ be the domain of d, defined by $dom(d) = dom(d_1) \times ..... \times dom(d_n)$. We use t to denote a tuple $(u_1, u_2, ..., u_n) \in dom(d)$ of R, and r to denote a relation or dataset on R, i.e. a set of tuples in $dom(d)$. Let R^* be the set of datasets on R.

Definition 1 (Dominance relation) *A dominance relation over a dataset r is a binary relation over $r \times r$.*

Dominance relations are also called qualitative preference relations. The dominance relation that leads to the skyline set is called traditional dominance.

Definition 2 (Traditional dominance relation) *Let t, t' be two tuples in a dataset r on $R(d_1, ..., d_n)$. t dominates t' for $d' \subseteq d$, denoted by $TD_{d'}(t, t')$, iff $\forall d_i \in d'$ $(t[d_i] \geq t'[d_i])$ $\wedge$ $\exists d_j \in d'$ $(t[d_j] > t'[d_j])$.*

Intuitively, a tuple t dominates another tuple t', or is preferable to t', if and only if it is better on at least one dimension, and it is not worse on any other dimension. A tuple t is maximal with respect to TD if and only if there exists no tuple t' in r such that $TD(t', t)$. The *answer to the skyline query* over a dataset r is the set of maximal elements of r with respect to TD. Without loss of generality, we will assume that $d' = d$ and use $TD(t, t')$ instead of $TD_{d'}(t, t')$. Note that we consider that a value is better than another one if it is greater. We choose to privilege maximization. The definition can be obviously modified to privilege minimisation by replacing $>$ (resp. $\geq$) by $<$ (resp. $\leq$).

Example 1 (running example) *We have $TD(t_1, t_9)$. As t_9 is dominated, it is not a skyline tuple, it shall be deleted. Note that if we remove t_9, there is no other possible comparison with respect to TD between the remaining players.*

When the number of dimensions increases, the number of tuples that are not comparable w.r.t. the traditional dominance relation increases in such proportions that the result is often useless. A common approach to overcome that limitation consists in ranking or shrinking the skyline set by relaxing the dominance relation, i.e. in defining some *weaker* dominance relationships. Due to lack of space, we recall formal definitions for just a few dominance relationsthat will be used for illustrating and experimenting our approach.

Example 2 (running example) *Let r be a dataset on $R(d_1, ..., d_n)$. Let t and t' be two tuples in r. We can use the two dominance relation below.*

Quasi-dominance QD [6]. *Let $q = (q_1, ..., q_n) \in \mathbb{R}^n$ be the indifference threshold on R. (t, t') is in the quasi-dominance relation w.r.t. q, denoted $QD(q)$ (QD for short), iff*
$|\{i \mid t'[d_i] - t[d_i] > q_i\}| = 0$ $\wedge$ $|\{i \mid t'[d_i] - t[d_i] > 0\ \}| < |\{i | t[d_i] - t'[d_i] > 0\}|$.
A tuple t quasi-dominates a tuple t' even if t' is better than t on some dimensions, but only with differences that do not exceed the indifference thresholds and on a number of dimensions that is strictly smaller than the number of dimensions where t is better than t'. Quasi-dominance can be seen as a relaxation of the traditional dominance. Some pairs of tuples that were incomparable with respect to TD may now be compared with respect to QD. Of course QD may be considered as less reliable. However it provides us some useful information to select – or to rank, tuples in a set such as the skyline set. But one can decide to use an even weaker relation if QD returns too many tuples.

k-dominance relation [3]. *Let k be an integer. (t, t') is in the k-dominance relation w.r.t. k, denoted $KD(k)$, iff*
$\exists d' \subseteq d$ $(|d'| = k$ $\wedge$ $\forall d_i \in d'$ $(t[d_i] \geq t'[d_i])$ $\wedge$ $\exists d_i \in d'$ $(t[d_i] > t'[d_i]))$.

In other words, t k-dominates t' iff $TD_{d'}(t,t')$ for at least one subset d' of d such that $|d'| \geq k$.

The following table summarizes all the relationships between the remaining players (Skylines) for five dominance relations. Each line gives the sets of players dominated by each player in the column w.r.t. a given dominance. The indifference threshold q is instantiated to $q = (1,1,1,1,10)$.

dominance	t_1	t_2	t_3	t_4	t_5	t_6	t_7	t_8
$QD(q)$	$t_4,...,t_8$		$t_5,...,t_8$	t_7,t_8	t_7,t_8	t_7		
$QD(q*2)$	$t_2,t_4,...,t_8$		$t_5,...,t_8$	$t_5,...,t_8$	t_7,t_8	t_7		
$QD(q*4)$	$t_2,...,t_8$	$t_4,...,t_8$	$t_4,...,t_8$	$t_5,...,t_8$	t_7,t_8	t_7		
$KD(4)$	$t_2,...,t_8$	$t_4,...,t_8$	$t_5,...,t_8$	$t_5,...,t_8$	t_6,t_7,t_8	t_5,t_7		
$KD(3)$	$t_2,...,t_8$	$t_3,...,t_8$	$t_2,t_4,...,t_8$	$t_5,...,t_8$	t_4,t_6,t_7,t_8	t_5,t_7,t_8	t_8	t_6,t_7

Note that increasing the value of the indifference threshold (resp. decreasing the value of k) used by QD (resp. KD) relaxes the dominance in the sense that more comparisons between players can be done.

3   Maximality-Based Selection

For a given dominance relation θ, similarly to the skyline operator (equivalent to *Winnow* and *Best* operators with respect to the TD), we propose to select as "best" tuples with respect to θ those that are not dominated by any other tuple with respect to θ. But we consider a different definition of *maximality*. The usual definition asserts that a tuple is maximal if it is not dominated by any other one. If this is acceptable for pre-orders such as TD, it is not appropriate anymore for cyclic relations. For instance, suppose that a small subset s of r form a cycle w.r.t. a dominance relation and that no tuples in s is dominated by a tuple in $r \setminus s$ while any tuple in $r \setminus s$ is dominated by a tuple in r. In that case, there is no maximal tuple. We believe that cycles should be seen as set of equivalent tuples, i.e. elements that can not be preferred to each other (in our example, s should be the set of maximal elements). To this end, we will use the classical notion of transitive closure in order to derive a pre-order from each dominance relationships. Note that we restrict the transitive closure to a subset of tuples.

Definition 3 (Partial transitive closure θ^+) *Let θ be a dominance relation over a dataset r. The partial transitive closure of θ over $s \subset r$, denoted θ_s^+, is the binary relation over s such that $\forall(t,t') \in s^2$, $\theta_s^+(t,t')$ iff $\exists(t_1,...,t_v) \in s^v(t_1 = t \wedge t_v = t' \wedge \theta(t_1,t_2) \wedge ... \wedge \theta(t_{v-1},t_v)))$.*

Relying on the transitive closure of a relation, we propose the definition of a new algebraic operator that can be seen as an alternative to the *Winnow* operator.

Definition 4 (max_θ: maximality-based selection) *Let θ be a dominance relation over a dataset r. An element t in r is said to be maximal w.r.t. θ iff $\forall t' \in r$ $(\neg(\theta_r^+(t',t)) \vee \theta_r^+(t,t'))$. The maximality-based selection w.r.t. θ in r is the set of maximal elements of r w.r.t. θ is denoted $max_\theta(r)$.*

A tuple is maximal with respect to a dominance relation θ if and only if it dominates all the tuples that dominate it, i.e. iff $\forall t' \in r \ (\theta_r^+(t',t) \Rightarrow \theta_r^+(t,t'))$. Of course, if it is not dominated by any tuple, then it is maximal. It is immediate that max_θ is a filter that selects some elements in r.

Theorem 1 *Let θ be a dominance relation over a dataset r. $max_\theta(r) \subseteq r$ holds.*

Now consider the relation $\rightleftharpoons_\theta$ over r defined by: $\forall t, t' \in r$, $t \rightleftharpoons_\theta t'$ iff $(t = t') \vee (\theta_r^+(t',t) \wedge \theta_r^+(t,t'))$. It is immediate that $\rightleftharpoons_\theta$ is reflexive, symmetric and transitive, thus an equivalence relation over r. Let $r/\rightleftharpoons_\theta$ be the quotient set of r w.r.t. the equivalence relation $\rightleftharpoons_\theta$, and $\theta_\rightleftharpoons$ be the dominance obtained by replacing each tuple in θ by the representant of its equivalence class in $r/\rightleftharpoons_\theta$. The following property holds.

Theorem 2 $max_\theta(r) = Winnow_{\rightleftharpoons_\theta}(r/\rightleftharpoons_\theta)$.

Example 3 *Consider another set of players r' that contains the players given in the following table and $q = (1,1,1,1,10)$.*

	points	rebounds	blocks	steals	games	dominated players / QD(q)
t_{10}	20	20	20	20	200	$t_{11}, t_{13}, t_{14}, t_{15}$
t_{11}	19.5	19.5	20	20	202	$t_{12}, t_{13}, t_{14}, t_{15}$
t_{12}	20.5	20.5	19.5	19.5	201	$t_{10}, t_{13}, t_{14}, t_{15}$
t_{13}	1	1	1	1	203	t_{14}, t_{15}
t_{14}	0.5	0.5	0.5	0.5	204	t_{15}
t_{15}	0	0	0	0	205	

All the players are dominated so $Winnow_{QD(q)}(r') = \emptyset$ while our approach does not eliminate all the players. Players t_{10}, t_{11} and t_{12} seem to be largely better than the three other players. They form a cycle in the dominance relation. They are only dominated by players that they also dominate via transitive closure. We consider those three players are equivalent. We have $max_{QD(q)}(r') = \{t_{10}, t_{11}, t_{12}\}$.

4   Preferences Chain Guided Selection

As explained above, there exist many dominance relations that one can choose to reduce the size of the answer set. However, pairs of incomparable tuples may still be too numerous. In such a case, the relaxation may be insufficient to satisfy the user. In order to compare those tuples, one should have the possibility to successively use some other forms of dominance to refine progressively the answer set. Let recall that an ordinal i is said to be a successor ordinal if there exists an ordinal j such that $i = j + 1$, whereas an ordinal α is said to be a limit ordinal if there does not exist any ordinal j such that $\alpha = j + 1$. Ordinals where defined by Cantor in order to deal with *transfinite sequence*, i.e. infinite sequences that are not limited to the first limit ordinal ω. The sequence of ordinals can be represented by $0, 1, \ldots, i-1, i, \ldots, \omega, \omega + 1, \ldots, 2\omega, 2\omega + 1 \ldots \omega^{\omega^\omega}, \ldots$

Definition 5 (Preferences chain) *Let r be a dataset on $R(d_1, ..., d_n)$. Let θ_i be a dominance relation over a dataset r for all ordinal i. A preferences chain Θ is defined as a transfinite sequence of dominance relations $\theta_1, \ldots, \theta_i, \theta_{i+1}, \ldots, \theta_\lambda, \ldots$ where i denotes a successor ordinal and λ a limit ordinal.*

Note that with Θ, we denote a chain of dominance relations, whereas with θ, we denote a dominance relation.

Example 4 (running example) *The following sequences are examples of preferences chains:*
- $\langle KD(n), KD(n-1), \ldots, KD(1)\rangle$ as a progressive filtering relying on KD;
- $\langle TD, QD(q), QD(q*2), QD(q*3), \ldots, QD(\omega)\rangle$ that could be used to recursively refine the skyline set by applying the QD with infinitely increasing indifference threshold until a fixed point. In the rest of the paper, we will refer to the first chain as KD_n, and to the second as QD_n.

The intuition is that we first consider the more pertinent dominance and select the tuples we consider to be the best with respect to that dominance. Then, in (and only in) that selection, we apply the same process with respect to the partial transitive closure of the next dominance relation. Then we continue to apply iteratively the process with the next dominance relations.

Definition 6 (Preferences chain guided filtering) *Let Θ be a preferences chain over a dataset r. The preferences chain guided filtering of r w.r.t. Θ is defined as the sequence $\langle sky_n^\Theta \rangle$, where $sky_0^\Theta = r$, $sky_n^\Theta = max_{\theta_n}(sky_{n-1})$, for any successor ordinal n, and $sky_\lambda^\Theta = \bigcap_{i<\lambda} sky_i$ for any limit ordinal λ.*

Relying on the lattice structure of $\mathcal{P}(r)$ and Theorem 1, we can assert that preferences chain guided filtering is a well-defined concept.

Theorem 3 *Let Θ be a preferences chain over a dataset r. The preferences chain guided filtering $\langle sky_n^\Theta \rangle$ of r w.r.t. Θ is a non-increasing sequence with respect to set inclusion that reaches its limit in a finite number of steps.*

We can now define a new operator that computes a progressive filtering.

Definition 7 (Preferences chain guided search operator SKY_Θ) *Let Θ be a preferences chain over $dom(d)$. The preferences chain guided search operator $SKY_\Theta : R^* \to R^*$ associates to a dataset r the limit of the preferences chain guided filtering $\langle sky_n^\Theta \rangle$ of r w.r.t. Θ.*

Note that our operator SKY_θ can be used to refine the skyline set by simply relying on a sequence of relations that begins with $\theta_1 = TD$. If the sequence contains only the dominance TD, then SKY_{TD} returns the skyline set. Note also that our approach is inspired by outranking methods proposed in ELECTRE IV [6] but does not rely on the same selection of tuples. Given a dominance relation θ and a tuple t, the qualification of t, denoted $qualif_\theta(t)$ is defined as the difference between the number of tuples that t dominates and the number of tuples that dominate t. More formally, $qualif_\theta(t) = |\{t'|t' \in r \wedge \theta(t, t')\}| - |\{t'|t' \in r \wedge \theta(t', t)\}|$.

We believe that the selection method w.r.t. qualification is not appropriate to the context of skyline queries where the number of tuples may be very big and the distribution may contain concentrated zones in the multi-dimensional space that may give unexpected results. However, our approach captures the selection w.r.t. qualification as a particular case. Indeed, given a dominance relation θ, one can easily infer another dominance relation θ^q defined by $\theta^q(t, t')$ iff $qualif_\theta(t) > qualif_\theta(t')$ and use that new relation instead of θ. Moreover, ELECTRE IV applies successively a finite pre-defined relaxing sequence of dominance relations, where a relaxing sequence is such that for all i, $\theta_i(t, t') \Rightarrow \theta_{i+1}(t, t').$, i.e. that each relation θ_i is included into the next one θ_{i+1} which weaken its conditions of satisfaction. In our approach, we generalize the outranking concept to any sequence of dominance relations.

Example 5 (running example) *The following table shows the progressive filtering of the chains QD_n and KD_n after each successive dominance θ_i, until their respective limits SKY_{QD_n} and SKY_{KD_n}.*

dominances	QD_n	KD_n
θ_1	$t_1,t_2,t_3,t_4,t_5,t_6,t_7,t_8$	$t_1,t_2,t_3,t_4,t_5,t_6,t_7,t_8$
θ_2	t_1,t_2,t_3	t_1
θ_3	t_1,t_3	
θ_4	t_1	

Now suppose the computation does not provide enough answers for the user, we propose an alternative approach that provides a ranking of the set. The user will have the possibility to choose either a complete ranking or a partial one (as in a top-k approach).

5 Preferences Chain Guided Ranking Refinement

First, we need to define the ranking with respect to a given dominance relation. That ranking is defined as an *ordered partition* of a dataset r, i.e. a *list* of *disjoint* subsets of r whose union is equal to r.

Definition 8 (θ-decomposition operator Γ_θ) *Let θ be a dominance relation over a dataset r. The decomposition of r w.r.t. θ, denoted $\Gamma_\theta(r)$, is defined as the ordered partition $\langle \gamma_0, \ldots, \gamma_p \rangle$ of r, where*

$$\gamma_0 = max_\theta(r), \ \gamma_i = max_\theta(r \setminus \bigcup_{j=0}^{i-1} \gamma_j) \ for \ 1 \leq i, and \ p = max\{i \mid \gamma_i \neq \emptyset\}$$

That operator first computes the set γ_0 of maximal tuples with respect to θ. That set is the first set of the ordered partition. It represents the "first choice tuples" with respect to θ. Then it removes those selected tuples from the original set r. It computes the set γ_1 of maximal tuples in the remaining set r with respect to the partial transitive closure over that set. The resulting set is the second set of the partition and represents the "second choice tuples". The computation is iterated until there is no more tuples into the original set. Note that this operator is similar to the successive iterations of the *Winnow* and *Best* operators called respectively *Winnow iterated* and *Best**, but relies on a different notion of maximality.

Theorem 4 $\Gamma_\theta(r) = iterated\_Winnow_{\rightleftharpoons_\theta}(r/\Rightarrow_\theta)$.

Example 6 (running example) *The first step of the computation of $\Gamma_{QD}(r)$ is $\gamma_0 = max_{QD}(r) = \{t_1, t_2, t_3\}$. Then we remove those players from r and compute the maximal players with respect to QD in the remaining set. We obtain $\gamma_1 = max_{QD}(r \setminus \{t_1, t_2, t_3\}) = \{t_4, t_5, t_6\}$. Finally, the computation ends with $\gamma_2 = max_{QD}(r \setminus \{t_1, t_2, t_3, t_4, t_5, t_6\}) = \{t_7, t_8\}$. Thus $\Gamma_{QD}(r) = \langle\{t_1, t_2, t_3\}, \{t_4, t_5, t_6\}, \{t_7, t_8\}, \{t_9\}\rangle$. We consider that the best players are t_1, t_2 and t_3. If we need more players or can not afford one of those, then we will choose between t_4, t_5 and t_6. If necessary, we will choose between t_7, t_8. The worst choice of our set is t_9.*

Similarly to the SKY_θ operator, the idea is to refine progressively the ranking of the set of skyline tuples by successively applying the relations of a preferences chain. Thus, once the decomposition of r with respect to a dominance θ_i as been computed, we need to apply the same method with the next dominance θ_{i+1}. As the dataset r as already be pre-sorted, θ_{i+1} should not be apply over the entire set r but only inside the different subsets of r in order to refine the ranking, i.e. to refine the ordered partition of r.

Definition 9 (generalized θ-decomposition operator $\hat{\Gamma}_\theta$) *Let $\langle r_0, \ldots, r_m \rangle$ be an ordered partition of a dataset r. Let θ be a dominance relation over r. The decomposition of $\langle r_0, \ldots, r_m \rangle$ w.r.t. θ, denoted $\hat{\Gamma}_\theta(r_0, \ldots, r_m)$, is the ordered partition of r defined by $\hat{\Gamma}_\theta(r_0, \ldots, r_m) = \langle \Gamma_\theta(r_0), \ldots, \Gamma_\theta(r_m) \rangle$.*

Let $\mathcal{O}(S)$ be the set of *lists of disjoint subsets* of a set S. For instance, in our running example, $\langle\{t_1, t_2, t_3\}, \{t_4, t_5, t_6\}\rangle$ and $\langle\{t_1, t_2, t_3\}, \{t_4, t_5, t_6\}, \{t_7, t_8\}, \{t_9\}\rangle$ are elements of $\mathcal{O}(r)$. Note that if $\langle x_1, \ldots, x_n \rangle$ in $\mathcal{O}(S)$ is such that $\bigcup_{1 \leq i \leq n} x_i = S$, then $\langle x_1, \ldots, x_n \rangle$ is an ordered partition of S. Let the order $\preceq$ over $\tilde{\mathcal{O}}(S) \times \mathcal{O}(S)$ be defined by $\langle x_1, \ldots, x_n \rangle \preceq \langle y_1, \ldots, y_m \rangle$, read $\langle y_1, \ldots, y_m \rangle$ is *finer* than $\langle x_1, \ldots, x_n \rangle$, iff for all y_i, y_j in $\langle y_1, \ldots, y_m \rangle$ such that $i \leq j$, there exist x_j, $x_{j'}$ in $\langle x_1, \ldots, x_n \rangle$ such that $y_i \subseteq x_j$ and $y_{i'} \subseteq x_{j'}$ and $i' \leq j'$. The following result asserts that the application of the operator $\hat{\Gamma}_\theta$ on an ordered partition is a refinement of that partition.

Theorem 5 *Let $\langle r_0, \ldots, r_m \rangle$ be an ordered partition of a dataset r. Let θ be a dominance relation over r. Then $\langle r_0, \ldots, r_m \rangle \preceq \hat{\Gamma}_\theta(r_0, \ldots, r_m)$ holds.*

Example 7 *Let γ be the decomposition of r of example 5 and $\theta = QD(2 * q)$.*

$$\hat{\Gamma}_\theta(\gamma) = \hat{\Gamma}_\theta(\{t_1, t_2, t_3\}, \{t_4, t_5, t_6\}, \{t_7, t_8\}, \{t_9\})$$
$$= \langle \Gamma_\theta(\{t_1, t_2, t_3\}), \Gamma_\theta(\{t_4, t_5, t_6\}), \Gamma_\theta(\{t_7, t_8\}), \Gamma_\theta(\{t_9\}) \rangle$$
$$= \langle\{t_1, t_3\}, \{t_2\}, \{t_4\}, \{t_5, t_6\}, \{t_7, t_8\}, \{t_9\}\rangle$$

In the set $\{t_1, t_2, t_3\}$, the only possible comparison with respect to $QD(q*2)$ is $QD(q*2)(t_1, t_2)$. Thus the decomposition of $\{t_1, t_2, t_3\}$ with respect to $QD(q*2)$ is $(\{t_1, t_3\}, \{t_2\})$. Similarly, in $\{t_4, t_5, t_6\}$, we have $QD(q*2)(t_4, t_5)$ and $QD(q*2)(t_4, t_6)$ thus the decomposition is $(\{t_4\}, \{t_5, t_6\})$. Finally t_7 and t_8 are incomparable with respect to $QD(q*2)$. It appears clearly that the computed decomposition preserves the dominance with respect to $QD(q*2)$, but provides a finer ordered partition of r by allowing less strict comparison conditions.

Finally, applying successively the decomposition operator will provide the user a global ranking as refined as possible with respect to the sequence of dominance relations that he selected and ordered.

Definition 10 (Preferences chain guided ranking) *Let Θ be a preferences chain over a dataset r. The* preferences chain guided ranking *of r w.r.t. Θ is defined as the sequence $\langle rank_n^\Theta \rangle$, where $rank_0^\Theta = r$, $rank_{n+1}^\Theta = \hat{\Gamma}_{\theta_{n+1}}(rank_n^\Theta)$ for any successor ordinal n, and $rank_\alpha^\Theta = max_{\preceq}\{rank_n^\Theta, n < \alpha\}$ for any limit ordinal α.*

Relying on Theorem 5, the above ranking is a well-defined concept.

Theorem 6 *Let Θ be a preference chain over a dataset r. The preferences chain guided ranking $\langle rank_n^\Theta \rangle$ of r w.r.t. Θ is a non-decreasing sequence w.r.t. $\preceq$ that reaches its limit, which is an ordered partition of r, in a finite number of steps.*

We can now define a new operator that computes an ordered partition by progressive refinement.

Definition 11 (Preferences chain guided ranking operator $Rank_\Theta$) *Let Θ be a preferences chain over $dom(d)$. The Θ-ranking operator $Rank_\Theta$ associates to the relation r in R^* the limit of the preferences chain guided ranking $\langle rank_n^\Theta \rangle$ of r w.r.t. Θ.*

Example 8 *In example 7, we have computed $\hat{\Gamma}_{QD(q*2)}(\hat{\Gamma}_{QD}(r))$. As far as the only supplementary comparison provided by $QD(q*4)$ is $QD(q*4)(t_1, t_3)$, applying on that partition the decomposition with respect to $QD(q*4)$ will lead to the decompose $\{t_1, t_3\}$ into $\langle\{t_1\}, \{t_3\}\rangle$. The following table presents rankings w.r.t. the chains QD_n and KD_n. We obtain $\langle\{t_1\}, \{t_3\}, \{t_2\}, \{t_4\}, \{t_5, t_6\}, \{t_7, t_8\}, \{t_9\}\rangle$ with QD_n and $\langle\{t_1\}, \{t_2, t_3\}, \{t_4\}, \{t_5, t_6\}, \{t_7, t_8\}, \{t_9\}\rangle$ with KD_n.*

θ	QD_n	KD_n
θ_1	$\{t_1, t_2, t_3, t_4, t_5, t_6, t_7, t_8\}, \{t_9\}$	$\{t_1, t_2, t_3, t_4, t_5, t_6, t_7, t_8\}, \{t_9\}$
θ_2	$\{t_1, t_2, t_3\}, \{t_4, t_5, t_6\}, \{t_7, t_8\}, \{t_9\}$	$\{t_1\}, \{t_2, t_3\}, \{t_4\}, \{t_5, t_6\}, \{t_7, t_8\}, \{t_9\}$
θ_3	$\{t_1, t_3\}, \{t_2\}, \{t_4\}, \{t_5, t_6\}, \{t_7, t_8\}, \{t_9\}$	$\{t_1\}, \{t_2, t_3\}, \{t_4\}, \{t_5, t_6\}, \{t_7, t_8\}, \{t_9\}$
θ_4	$\{t_1\}, \{t_3\}, \{t_2\}, \{t_4\}, \{t_5, t_6\}, \{t_7, t_8\}, \{t_9\}$	$\{t_1\}, \{t_2, t_3\}, \{t_4\}, \{t_5, t_6\}, \{t_7, t_8\}, \{t_9\}$

6 Preferences Chain Guided Top-k Approximation

Now suppose the user does not desire a complete ranking but a sub-selection of the k "best" skyline tuples. We propose an alternative approach that provides a top-k selection into the relation. The approach is very similar as it can be seen as a particular case of global ranking. Of course, one could directly use the global ranking to get a top-k selection. However we propose a variation of the above definition to compute more efficiently the top-k query by renouncing to a result as refined as possible.

Definition 12 (top-k θ-decomposition operator Γ_θ^k) *Let k be an integer such that $k < |r|$. Let θ be a dominance dataset over r. The k-decomposition of r w.r.t. θ, denoted $\Gamma_\theta^k(r)$, is defined as the list $\langle \gamma_0^k, \ldots, \gamma_p^k \rangle$ of subsets of r, where*

$$\gamma_0^k = max_\theta(r), \ \gamma_i^k = max_\theta(r \setminus \bigcup_{j=0}^{i-1} \gamma_j) \text{ for } 1 \leq i \text{ and } p = min\{i \mid | \bigcup_{j=0}^p \gamma_j^k | \geq k\}.$$

That operator is very similar to the θ-decomposition operator Γ_θ^k of definition 8, but instead of computing a complete partition of r, we stop the computation as soon as the number of selected tuples exceeds k.

Example 9 (running example) *Consider $k = 4$. Then*

$$\Gamma_{QD}^k(r) = \langle \{t_1, t_2, t_3\}, \{t_4, t_5, t_6\} \rangle$$

The first subset is not sufficient to get four tuples. Adding the second one already gives too many answers. We stop the computation. The remaining tuples are removed.

In that list of subsets, if we have to remove some more tuples, then we apply the next relaxing dominance, but only on the last subset of the sequence which contains the least preferable tuples of the selection.

Definition 13 (generalized top-k θ-decomposition $\hat{\Gamma}_\theta^k$) *Let $\langle r_0, \ldots, r_m \rangle$ be a list of disjoint subsets of a dataset r. Let θ be a dominance relation over r. The top-k decomposition of $\langle r_0, \ldots, r_m \rangle$ w.r.t. θ, denoted $\hat{\Gamma}_\theta^k(\langle r_0, \ldots, r_m \rangle)$, is the list of subsets of r defined as follows:*

$$\hat{\Gamma}_\theta^k(\langle r_0, \ldots, r_m \rangle) = \begin{cases} \langle r_0, \ldots, r_m \rangle & \text{if } k \geq | \bigcup_{j=0}^m r_j| \\ \langle r_0, \ldots, r_{m'} \rangle & \text{if } \exists m'(m' < m \wedge k = | \bigcup_{j=0}^{m'} r_j|) \\ \langle r_0, \ldots, r_{m''}, \Gamma_\theta^{k'}(r_{m''+1}) \rangle & \text{otherwise} \end{cases}$$

where $m'' = max\{i \mid | \bigcup_{j=0}^{i-1} r_j| < k\}$ and $k' = k - | \bigcup_{j=0}^{m''} r_j|$.

Theorem 7 *Let $\langle r_0, \ldots, r_m \rangle$ be a list of subsets of a dataset r. Let θ be a dominance relation over r. Then $\langle r_0, \ldots, r_m \rangle \preceq \hat{\Gamma}_\theta(r_0, \ldots, r_m)$ holds.*

Definition 14 (Preferences chain guided top-k approximation) *Let Θ be a preferences chain over a dataset r. The preferences chain guided top-k approximation of r w.r.t. Θ is defined as the sequence $\langle topk_n^\Theta \rangle$, where $topk_0^\Theta = r$, $topk_{n+1}^\Theta = \hat{\Gamma}_{\theta_{n+1}}(topk_n^\Theta)$ for any successor ordinal n, and $topk_\alpha^\Theta = max_\preceq \{topk_n^\Theta, n < \alpha\}$ for any limit ordinal α.*

Relying on Theorem 7, we can assert that preferences chain guided top-k approximation is a well-defined concept.

Theorem 8 *Let Θ be a preferences chain over a dataset r. The preferences chain guided top-k approximation $\langle topk_n^\Theta \rangle$ of r w.r.t. Θ is a non-decreasing sequence w.r.t. $\preceq$ that reaches its limit in a finite number of steps.*

We can now define a new operator that computes an partially ranked top-k approximation.

Definition 15 (Preferences chain guided top-k operator $Topk_\Theta$) *Let Θ be a preferences chain over $dom(d)$. The preferences chain guided top-k operator $Topk_\Theta$ associates to the dataset r in R^* the limit of the preferences chain guided top-k approximation $\langle topk_n^\Theta \rangle$ of r w.r.t. Θ.*

Example 10 *Suppose $k = 2$. The chain QD_n leads us to a selection of exactly 2 players, while the chain KD_n provides an approximation of k. Note also that the $Topk_\Theta$ result is not as finely ranked as as the result of $Rank_\Theta$.*

dominances	QD_n	KD_n
θ_1	$\{t_1, t_2, t_3, t_4, t_5, t_6, t_7, t_8\}, \{t_9\}$	$\{t_1, t_2, t_3, t_4, t_5, t_6, t_7, t_8\}, \{t_9\}$
θ_2	$\{t_1, t_2, t_3\}$	$\{t_1\}, \{t_2, t_3\}$
θ_3	$\{t_1, t_3\}$	$\{t_1\}, \{t_2, t_3\}$

Note that if we apply the $Topk_\Theta$ operator with $k = 1$, then we obtain the set of SKY_θ of definition 7.

7   Experiments

Algorithms. Due to lack of space in the short version of this paper, we omitted the algorithms that are direct implementations of the definitions given in the precedent sections. The algorithms are provided in [7]. We only explain here how maximality with respect to the partial transitive closure of a given dominance θ over a relation r can be tested. As already suggested above by Theorem 2, we compute the quotient set $r/ \rightleftharpoons_\theta$ of r w.r.t. the equivalence relation $\rightleftharpoons_\theta$. Algorithm 1 contracts all the cycles in the dominance θ and returns an acyclic dominance denoted $\theta\_contraction$. Each contracted cycle is represented by one of its elements, and the equivalence classes are stored in the structure $\mathcal{C}(S)$. To avoid the complete computation of the transitive cloture of r, we first transfer recursively to $\theta\_contraction$ all the lines in θ that do obviously not belong to any cycle (lines such that the first tuple is never dominated or such that the second one does not dominate any tuple). In the remaining lines, we select one, (t, t'), and search all the tuples that can be dominated by t through transitve closure, contracting cycles on the fly. Once t has been completely processed and all cycles containing t contracted, we transfer all the lines containing t in $\theta\_contraction$. Then we repeat recursively the whole process until θ is empty.

```
Data: a dominance relation θ    Result: θ_contraction
while θ ≠ ∅ do
    move recursively to θ_contraction all lines in θ with a point /*not in
    cycle*/ in (Π_dominant(θ) \ Π_dominated(θ)) ∪ (Π_dominated(θ) \ Π_dominant(θ)) ;
    (t,t'):= first(θ);
    foreach (t₁, t₂) in θ do
        if t₁ = t then
            if (t₂, t₁) exists in θ then // cycle
                remove (t₁, t₂) and (t₂, t₁) from θ;
                replace t₂ by t₁ in θ;
                add contents(t₂) to contents(t₁) in C(S);
                remove from C(S) the line identified by t₂;
                (t₁,t₂):= first(θ);
            else
                foreach (t₃, t₄) in θ do
                    if t₃ = t₂ then  insert (t₁, t₄) into θ ; // transit.

    move to θ_contraction all lines in θ containing t;
update θ_contraction w.r.t. C(S);
```

Algorithm 1. Contractθ

Experimental Settings. Experimentations were carried out in Java under Windows 7 with a capacity of 1GB memory, an Intel Core 2 Duo and an Oracle 11g DBMS, on real and data. This data set is the NBA data set [1] that contains 21671 records of players' season statistics over 17 attributes from the first season of NBA in 1946 to the season 2009. We evaluate and compare the quality and cost of different operators SKY_Θ, $Rank_\Theta$, $TopK_\theta$. In our test, the indifference threshold q used in $QD(q)$ is the tuple $(0.1 * \sigma(d_1), \ldots, 0.1 * \sigma(d_n))$ where $\sigma(d_i)$ is the standard deviation of $\Pi_{d_i}(r)$. We will use the two preferences chains KD_n and QD_n used in our running example.

Progressive Filtering and Ranking Capacity. Figure 1(a) and (b) show the size of the selection after each step, until the limit $SKY_\Theta(r)$, of the filtering of the NBA data set guided respectively by the preferences chains KD_n and QD_n. The result illustrates the progressive nature of the preferences chain guided filtering. Similarly, Figure 1(c) represents the decomposition of r after each step of the ranking guided by the preferences chain QD_n of a set of 28 skyline tuples computed on a sample of 100 tuples in the NBA database. We can observe the progressive refinement of the ordered partitioning of the data set. Each line, from TD to QD(q*8), shows the more refined partition of the set obtained after the application of one more dominance relation. The user can stop at any step or continue the ranking. In this figure, we chose to stop at QD(q*8) only for the space capacity and the clarity of the ranking representation.

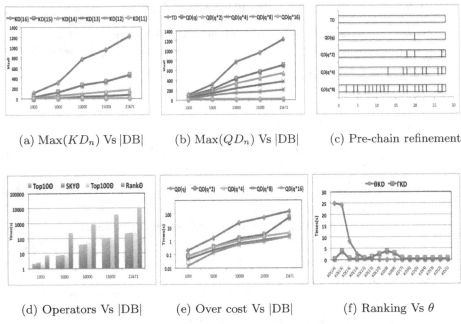

(a) Max(KD_n) Vs |DB| (b) Max(QD_n) Vs |DB| (c) Pre-chain refinement

(d) Operators Vs |DB| (e) Over cost Vs |DB| (f) Ranking Vs θ

Fig. 1. Experiments: Results size (Top), Runtime (bottom)

Performance. Figure 1(d) compares the performance of the three mains algorithms or operators SKY_Θ, $Rank_\Theta$, $TopK_\theta$ (for $K = 10$ and $K = 100$) for $\Theta = QD_n$. As expected, the complete ranking of the set is very more expensive than the selection and topK which refine, at each step, only one subset of tuples obtained at the precedent step, while the ranking apply the computation in all the subsets. Note that the topK computation may be, in particular cases, more efficient then SKY_Θ computation, e.g. when the cardinality of the union of n first sets provided by the QD is equal to k (in that case, the algorithm stops without computing the other relations, while the SKY_Θ algorithm will continue the decomposition until there remains one tuples or the limit is reached). Figure 1(e) represents the over cost of each step of the computation of the selection guided by the preferences chain QD_n. Theorem 3 asserts that each step takes in input a smaller set than the precedent step, thus there are less tuples to compare. Effectively the runtime of each successive dominance of the chain is smaller than the precedent one. Figure 1(f) concerns the ranking guided by the preferences chain KD_n. The highest curve represents the cost of the total runtime of computation of each dominance in all subsets of the ordered partition. Similarly, the lowest curve represents the cost of the total runtime for ranking all the subsets once the dominance is computed. As expected from Theorem 6, the over cost of each supplementary dominance converges to 0 as the ordered partition becomes progressively finer and the number of comparisons to be tested smaller. Note that the most expensive part of the computation is for the first dominances of the chain. Consequently, a guideline for the user, from a performance point of view, is to begin its chain with dominances that can be optimized. This is the case of those satisfying transitivity or anti-monotonicity properties among which Pareto dominance.

8 Conclusion

We propose a formal framework for progressive filtering of skylines sets with respect to users' preferences. Our approach, inspired by multi-criteria analysis methods, is very flexible as it allows the user *(a)* to define different thresholds of preference that represent different levels of relaxation of dominance conditions, *(b)* to choose between a global ranking, a (partially or totally ranked) top-k query and a restricted set of "best" elements, and *(c)* to decide to stop the progressive filtering as soon as the result satisfies him (or to navigate between the results of the different steps of filtering in order to choose the one that fits his will). We provide not only the formal framework but also algorithms and experiments. Generalization and optimization will be topics for future research.

References

1. NBA basketball statistic, `http://databasebasketball.com/stats.download`
2. Börzsönyi, S., Kossmann, D., Stocker, K.: The skyline operator. In: Proceedings of the IEEE 17th International Conference on Data Engineering (ICDE 2001), pp. 421–430 (2001)
3. Chan, C.Y., Jagadish, H.V., Tan, K.-L., Tung, A.K.H., Zhang, Z.: Finding k-dominant skylines in high dimensional space. In: Proceedings of the ACM SIGMOD International Conference on Management of Data (SIGMOD 2006), pp. 503–514 (2006)
4. Chan, C.-Y., Jagadish, H.V., Tan, K.-L., Tung, A.K.H., Zhang, Z.: On High Dimensional Skylines. In: Ioannidis, Y., Scholl, M.H., Schmidt, J.W., Matthes, F., Hatzopoulos, M., Böhm, K., Kemper, A., Grust, T., Böhm, C. (eds.) EDBT 2006. LNCS, vol. 3896, pp. 478–495. Springer, Heidelberg (2006)
5. Chomicki, J.: Preference formulas in relational queries, vol. 28, pp. 427–466. ACM, New York (2003)
6. Figueira, J., Mousseau, V., Roy, B.: Electre methods. In: Figueira, J., Greco, S., Ehrgott, M. (eds.) Multiple Criteria Decision Analysis: State of the Art Surveys, pp. 133–162. Springer (2005)
7. Pre-Chain, `http://www.prism.uvsq.fr/users/zeitouni/papers/DEXA2903.pdf`
8. Lee, J., You, G.-W., Hwang, S.-W.: Telescope: Zooming to interesting skylines. In: Kotagiri, R., Radha Krishna, P., Mohania, M., Nantajeewarawat, E. (eds.) DASFAA 2007. LNCS, vol. 4443, pp. 539–550. Springer, Heidelberg (2007)
9. Lin, X., Yuan, Y., Zhang, Q., Zhang, Y.: Selecting stars: The k most representative skyline operator. In: Proceedings of the 23rd International Conference on Data Engineering (ICDE 2007), pp. 86–95 (2007)
10. Papadias, D., Tao, Y., Fu, G., Seeger, B.: An optimal and progressive algorithm for skyline queries. In: Proceedings of the 2003 ACM SIGMOD International Conference on Management of Data, SIGMOD 2003, pp. 467–478. ACM, New York (2003)

11. Stefanidis, K., Koutrika, G., Pitoura, E.: A survey on representation, composition and application of preferences in database systems, vol. 36, pp. 19:1–19:45. ACM, New York (2011)
12. Vlachou, A., Vazirgiannis, M.: Ranking the sky: Discovering the importance of skyline points through subspace dominance relationships, vol. 69, pp. 943–964. Elsevier Science Publishers B. V., Amsterdam (2010)
13. Xia, T., Zhang, D., Tao, Y.: On skylining with flexible dominance relation. In: Proceedings of the IEEE 24th International Conference on Data Engineering (ICDE 2008), pp. 1397–1399 (2008)
14. Yang, J., Fung, G.P., Lu, W., Zhou, X., Chen, H., Du, X.: Finding superior skyline points for multidimensional recommendation applications, vol. 15, pp. 33–60. Kluwer Academic Publishers, Hingham

Efficient XML Keyword Search:
From Graph Model to Tree Model

Yong Zeng[1], Zhifeng Bao[1], Tok Wang Ling[1], and Guoliang Li[2]

[1] National University of Singapore
[2] Tsinghua University

Abstract. Keyword search, as opposed to traditional structured query, has been becoming more and more popular on querying XML data in recent years. XML documents usually contain some ID nodes and IDREF nodes to represent reference relationships among the data. An XML document with ID/IDREF is modeled as a graph by existing works, where the keyword query results are computed by graph traversal. As a comparison, if ID/IDREF is not considered, an XML document can be modeled as a tree. Keyword search on XML tree can be much more efficient using tree-based labeling techniques. A nature question is whether we need to abandon the efficient XML tree search methods and invent new, but less efficient search methods for XML graph. To address this problem, we propose a novel method to transform an XML graph to a tree model such that we can exploit existing XML tree search methods. The experimental results show that our solution can outperform the traditional XML graph search methods by orders of magnitude in efficiency while generating a similar set of results as existing XML graph search methods.

1 Introduction

Keyword search, as opposed to traditional structured query, has been becoming more and more popular on querying XML data in recent years. Users are free from learning the query language and XML schema by simply using some keywords to query the XML data. It attracts a lot of research efforts recently [9,7,13,14,2]. For example, Figure 1 shows an XML document about a company with *project*, *part* and *supplier*. Each node is assigned a unique Dewey label [12]. To know the price of a part $p1$, users can simply issue a keyword query Q=``p1 price'' without knowing the schema or any query language (e.g., XPath).

XML documents usually contain some ID nodes and IDREF nodes to represent reference relationships among the data. For example, in Figure 1, every *part* used by each *project* has a reference indicating its *supplier*. An XML document with ID/IDREF is usually modeled as a digraph, where the keyword query results are usually computed by graph traversal [9,5,10,8]. Then the problem is reduced to the problem of finding Minimal Steiner Tree (MST) or its variants in a graph, where an MST is defined as a minimal subtree containing all query keywords in either its leaves or root. Since this problem is NP-complete [6], a lot of works are interested in finding the "best" answers of all possible MSTs, i.e. finding top-K results according to some criteria, like subtree size, etc.

H. Decker et al. (Eds.): DEXA 2013, Part I, LNCS 8055, pp. 25–39, 2013.
© Springer-Verlag Berlin Heidelberg 2013

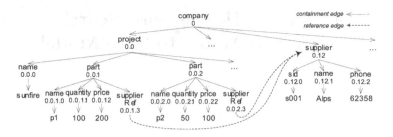

Fig. 1. An Example XML Document (with Dewey Labels)

As a comparison, if we do not consider ID/IDREF, an XML document can be modeled as a tree. Keyword search on an XML tree can be much more efficient based on the tree structure. The results are defined as minimal subtrees containing all query keywords, which is actually a variant of MST adapted to XML tree. Because in a tree, finding an MST for a set of nodes reduces to finding the lowest common ancestor (LCA) of that set of nodes, which can be efficiently addressed by node label computation. For example, if we do not consider the ID/IDREF in Figure 1, given a query Q="p1 price", a node labeled 0.0.1.0 matches keyword "p1" and another node labeled 0.0.1.2 matches keyword "price", then the MST connecting these two nodes is actually the subtree rooted at their lowest common ancestor (LCA), i.e. node 0.0.1. Calculating the LCA simply requires calculating the common label prefix of those two nodes, i.e. 0.0.1 is the prefix of 0.0.1.0 and 0.0.1.2. It is very efficient and does not need any graph traversal.

There are abundant efficient XML tree search methods available, which efficiently calculate the query results based on node labels rather than graph traversal. They build inverted lists of query keywords, in the form of ($keyword$: $dewey_1, dewey_2, dewey_3, ...$), where $dewey_i$ represents the label of a node containing the $keyword$.

XML graph is indeed a tree with a portion of reference edges. This observation offers a great opportunity to accelerate keyword search on XML graph. In this paper, instead of adopting traditional graph search methods, we propose a novel approach to transform an XML graph to a tree model, such that we can exploit XML tree search methods to accelerate query evaluation. The rest of the paper is organized as follows. We present the related work in Sec. 2. Preliminaries are in Sec. 3. We discuss how to transform an XML graph to a tree model for efficient query evaluation in Sec. 4 and how it works on complex reference patterns in Sec. 5. Further extension of our approach is in Sec. 6. The algorithm is presented in Sec. 7. Experiments are in Sec. 8 and we conclude in Sec. 9.

2 Related Work

XML Graph Keyword Search. An XML graph can be modeled as a digraph [9]. Keyword search on a graph is usually reduced to the Steiner Tree problem or its variants: given a graph $G = (V, E)$, where V is a set of nodes and E is a set of edges, a keyword query result is defined as a minimal subtree T in G such that the leaves or the root of T contain all keywords in the query. The Steiner tree problem is NP-complete [6], and many works are interested in finding the "best" answers of all possible Steiner trees, i.e. finding top-K results according to some criteria, like subtree size, etc. Backward expanding strategy is used by BANKS [4] to search for Steiner trees in a graph. To improve the efficiency, BANKS-II [10] proposed a bidirectional search strategy to reduce the search space, which searches as small portion of graph as possible. Later [5] designed a dynamic programming approach (DPBF) to identify the optimal Steiner trees containing all query keywords. BLINKS [8] proposes a bi-level index and a partition-based method to prune and accelerate searching for top-k results in graphs. XKeyword [9] presented a method to optimized the query evaluation by making use of the graph's schema in XML. [3] proposed an object-level matching semantics on XML graph based on the assumption that the object information is given.

XML Tree Keyword Search. In XML tree, LCA (lowest common ancestor) semantics is first proposed and studied in [7] to find XML nodes, each of which contains all query keywords within its subtree. For a given query $Q = \{k_1,...,k_n\}$ and an XML document D, L_i denotes the inverted list of k_i. Then the LCAs of Q on D are defined as $LCA(Q)=\{v \mid v = lca(m_1,...,m_n), v_i \in L_i (1 \leq i \leq n)\}$. Subsequently, SLCA (smallest LCA [13]) is proposed, which is indeed a subset of $LCA(Q)$, of which no LCA in the subset is the ancestor of any other LCA. ELCA [7,14], which is also a widely adopted subset of $LCA(Q)$, is defined as: a node v is an ELCA node of Q if the subtree T_v rooted at v contains at least one occurrence of all query keywords, after excluding the occurrences of keywords in each subtree $T_{v'}$ rooted at v's descendant node v' and already contains all query keywords. [2] proposed a statistical way to identify the search target candidates. Recently, more efficient algorithms have been proposed for SLCA and ELCA computation based on hash index [16] and set intersection operation[15].

3 Preliminaries

3.1 Data Model

We model an XML graph as a digraph, where each node of the graph corresponds to an element of the XML data, with a tag name and (optionally) some value. Each containment relationship between a parent node a and a child node b in the XML data corresponds to a containment edge in the digraph, represented as $a \rightarrow b$. Each ID/IDREF reference in the XML data corresponds to a reference edge in the digraph, represented as $a \dashrightarrow b$, where a is the IDREF node and b is the ID node. Thus an XML graph is denoted as $G = (V, E, R)$, where V is a set of nodes, E is a set of containment edges and R is a set of reference edges.

We use T_n to denote the query result rooted at node n. A node n is usually represented by its label or *tag:label*, where tag is the tag name of n. To accelerate the keyword query processing on XML tree model, existing works adopt the dewey labeling scheme [12], as shown in Figure 1.

3.2 Reference Types

If the reference edges are not considered, an XML graph will reduce to an XML tree. There are three types of reference edges in an XML graph: *basic references* (as mentioned in our data model), *sequential references* and *cyclic references*. When an object a references an object b, while b also references a third object c, sequential references occur. Cyclic references happen when containment edges and reference edges form a cycle in an XML graph.

4 Transforming Query Processing over XML Graph to XML Tree

In order to fully exploit the power of tree search methods over the XML digraph, we realize two challenges to tackle: (1) how to transform an XML graph to a proper tree model, which can work with different reference patterns; (2) how to apply existing tree search techniques onto such a tree model. We start addressing these challenges by focusing on the case of *basic references* first, then discuss how the proposed solution can handle sequential and cyclic cases in Sec. 5.

4.1 Real Replication

As shown in Figure 1, every IDREF node in an XML graph points to a particular object with a unique ID value. An object is in the form of a subtree. Therefore, a straightforward yet naive transformation is to just to make a real replication of all such subtrees being referenced. For every reference edge $a \dashrightarrow b$ in the XML graph, we make a replication of the subtree T_b rooted at b and put it under a. Figure 2 shows a transformed XML tree based on the XML graph in Figure 1, where the subtrees in dotted circles are the replication of the subtree $T_{0.12}$.

The transformed XML tree is identical to the original XML graph in the sense that they can infer each other. As a result, any existing keyword search method designed for XML tree can be applied on it without any change.

However, even though the real replication approach can work well for the case of basic references, it is **not usable in practice** because:

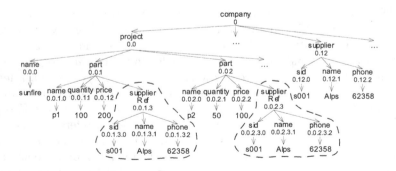

Fig. 2. Naive Method: Real Replication

- The number of nodes will increase due to the replication of subtrees. We will show in Sec. 5 that, in the worse case, the number of nodes will increase exponentially for the case of sequential references and cyclic references. The space cost is unacceptable in practice.
- Some duplicate results may be generated (as shown in Example 1).

Example 1. If we issue a query $Q=$ "Alps phone" to find the phone number of supplier *Alps* in Figure 1, the real replication method will get the transformed XML tree in Figure 2 and do the keyword search on it. By ELCA search method, we get three results: $T_{supplier:0.12}$, $T_{supplierRef:0.0.1.3}$ and $T_{supplierRef:0.0.2.3}$ respectively. The last two results, which are the same as the first one, are actually redundant. Because they are found within the replicated subtrees, while the same results should have already been found in the original subtree. □

4.2   Virtual Replication

As discussed in the previous section, real replication is not usable in practice. From Example 1 we observe that, a result is **redundant** if it is found within the replicated subtrees, because it must have been found in the original subtree as well. Thus, a result is **non-redundant** only if the root of the result is not within any replicated subtree. Based on this observation, we find that the cost of replicating the subtrees is not necessary because we do not need to search within any replicated subtree.

Instead, we propose to use a special node, i.e. the IDREF node, to virtually represent the whole replicated subtree (without inducing any new node), which is able to find the same set of non-redundant results as the real replication method. This is what we call **virtual replication**. For instance, Figure 3(a) shows the idea of using one node to represent the whole replicated subtree. As compared to Figure 2 of real replication, here we use node supplierRef:0.0.1.3 in Figure 3(a) to represent the whole replicated subtree under it.

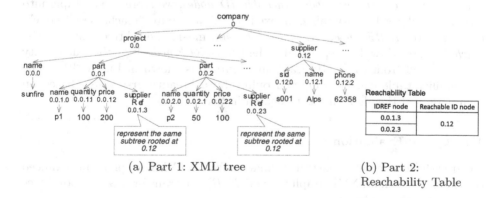

(a) Part 1: XML tree

(b) Part 2: Reachability Table

Fig. 3. Advanced Method: Virtual Replication (Two Parts)

Example 2. For a query Q="Alps part" in Figure 2, the real replication method will get the following results: $T_{part:0.0.1}$ and $T_{part:0.0.2}$. These two results are non-redundant because their roots, part:0.0.1 and part:0.0.2, are not within any replicated subtree.

Now by virtual replication, keyword "Alps" will no longer match the node 0.0.1.3.1 and 0.0.2.3.1 in Figure 2. Instead, it will match two IDREF nodes 0.0.1.3 and node 0.0.2.3 in Figure 3(a), because we use these two IDREF nodes to represent the whole replicated subtrees. But the final results are still the same: (1) $T_{part:0.0.1}$, which is computed from node 0.0.1.3 (matching keyword "Alps") and node 0.0.1 (matching keyword "part") in Figure 3(a); (2) $T_{part:0.0.2}$, which is computed from node 0.0.2.3 (matching keyword "Alps") and 0.0.2 (matching keyword "part"). □

In this manner, we do not induce any new node while it is able to get the same set of non-redundant results as the real replication method. A proof can be seen at Appendix (section 10).

In order to know which IDREF node represents which subtree, we need a data structure to keep track of the information that which subtree will be replicated under which IDREF node. For such a purpose, we maintain a table called **reachability table**, as shown in Figure 3(b). The table is based on a concept called **reachable**.

Definition 1. *Reachable.* *Given an IDREF node n, if there is a directed path from n to a node m in the XML graph, where the last edge of the path is an reference edge, then we say m is a reachable ID node of n.*

Example 3. Given the XML graph in Figure 1, we can find that from the IDREF node 0.0.1.3, there is a directed path from it to node 0.12, where the path ends with a reference edge. So node 0.12 is a reachable ID node for node 0.0.1.3. Similarly, node 0.12 is a reachable ID node for node 0.0.2.3. □

For every pair of *(IDREF node, reachable ID node)*, we store it as a tuple into a table called reachability table, indexed by the attribute "reachable ID node". Every pair of *(IDREF node, reachable ID node)* means the subtree rooted at the *reachable ID node* will be replicated under the *IDREF node*. E.g., the reachability table inferred from the XML graph in Figure 1 is shown in Figure 3(b). The reachability table can be computed offline by a breadth-first search based on each node and the algorithm is presented in Sec. 7.

4.3   Query Evaluation

So far we have completed the transformation from an XML graph to an advanced tree model. Given an XML graph $G = (V, E, R)$, we transform G to a novel tree model consisting of two parts:

1. An XML tree $GT = (V, E, \emptyset)$, which is exactly the same as G with all the reference edges dropped.

2. A reachability table *table*, which maintains the information of which subtree will be virtually replicated under which IDREF node.

Now, we will present how to make an efficient keyword query evaluation based on our transformed tree model.

As discussed in Sec. 1, existing keyword search methods on XML tree do not traverse the tree to search query results. Instead, they compute results based on nodes' labels, e.g., the dewey label. Such labels are stored in an inverted list index in form of $keyword : dewey_1, dewey_2, dewey_3, ...$, where $dewey_i$ represents a node containing the *keyword*. Any LCA-based keyword search method for XML tree will build such an index. Given a keyword query $Q = \{k_1, k_2, ..., k_n\}$, they will retrieve an inverted list for each keyword k_i, and then compute the results based on the inverted lists.

Similarly, after we transform an XML graph to tree model in *virtual replication*, we will also build such an inverted list index. Our tree model consists of an XML tree and a reachability table. The inverted list index will be built on the XML tree, while later the reachability table will help to expand the inverted lists to handle ID/IDREF.

With the index ready, we exploit the existing XML tree keyword search methods and evaluate a keyword query on our tree model in three steps:

1. Retrieve the inverted lists for each keyword in a query.
2. Expand the inverted lists retrieved in step 1.
3. Apply an existing XML tree keyword search method to the expanded inverted lists.

Step 1. Given a query $Q = $ "$k_1 k_2...k_n$", one inverted list will be retrieved from the index for each keyword. E.g., given a query $Q=$"Alps part", based on our tree model in Figure 3, we will first retrieve the inverted lists as follows:
"Alps" : 0.12.1
"part" : 0.0.1, 0.0.2, ...

Take note that the keyword "Alps" only matches one node, i.e. 0.12.1, because the inverted list is built on the XML tree in Figure 3(a). So in this step, we only find out the nodes in the XML tree matching the keywords before replication.

Step 2. With the help of the reachability table, we will try to find out whether there is any node in the replicated subtree matching the keywords as well. We can do it in the following way: for each dewey label retrieved in step 1, we check each of its ancestors to see whether the ancestor appears in the *reachable ID node* column of the reachability table. If yes, we add the corresponding IDREF nodes to the inverted list.

E.g., for the dewey label 0.12.1 retrieved in step 1 in the above example, 0.12.1 has two ancestor (prefix): 0.12 and 0. We can find that its ancestor 0.12 appears in the *Reachable ID node* column of the reachability table in Figure 3(b). This means the subtree $T_{0.12}$ is reachable by some IDREF nodes and it should be replicated under those IDREF nodes. So the keyword should match those IDREF nodes as well. Then we add the corresponding IDREF nodes to the inverted list. But its ancestor 0 does not appear in the *Reachable ID node*

column. After that, the expanded inverted list will be:
"Alps" : 0.0.1.3, 0.0.2.3, 0.12.1

After we do the same thing to the "part" inverted list, it will become:
"part" : 0.0.1, 0.0.2, ...

The reachability table is organized in a B+ tree and indexed by the column *Reachable ID node*. So given the dewey label of a reachable ID node, the corresponding IDREF nodes can be retrieved efficiently.

Step 3. After step 2, the final inverted lists are ready. Now we can apply any existing keyword search methods designed for XML tree, like SLCA, ELCA, etc., as all of them operate on the inverted lists for result computation.

5   Sequential References and Cyclic References

Section 4 presents our solution on transformation and query evaluation to basic references case, and in this section we would like to discuss how they are capable of handling the cases of sequential references and cyclic references as well.

5.1   Sequential References

In this case, e.g. in Figure 4(a), to make a complete replication, the subtree rooted at employee:0.88 should be replicated not only under managerRef:0.12.2, but also supplierRef:0.0.1.3. Therefore, if we adopt the real replication approach in Sec. 4.1, the number of nodes may increase exponentially because each object in the sequential references could have multiple references to some other objects.

For the virtual replication in Sec. 4.2, we do not need to induce any new nodes into the XML graph. For the XML graph in Figure 4(a), according to Definition 1, we just construct a reachability table as shown in Figure 4(b). E.g., there is a directed path from supplier:0.0.1.3 to employee:0.88, where the path ends with a reference edge. So node 0.88 is a reachable ID node for node 0.0.1.3.

5.2   Cyclic References

In the case of cyclic references, our reachability concept in Definition 1 is still able to handle it. E.g., in Figure 5(a), there is a directed path from node authorRef:0.0.3 to node author:0.12, where the path ends with a reference edge.

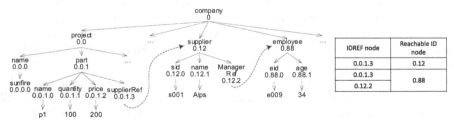

(a) XML Graph with Sequential References   (b) Its Reachability Table

Fig. 4. Constructing Reachability Table for Sequential References

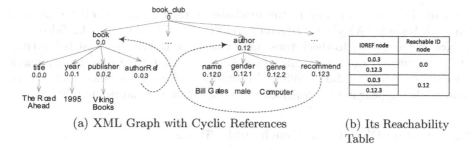

(a) XML Graph with Cyclic References   (b) Its Reachability Table

Fig. 5. Constructing Reachability Table for Cyclic References

So node 0.12 is a reachable ID node for node 0.0.3. There is also a path from node authorRef:0.0.3 to node book:0.0, where the path ends with a reference edge. So node 0.0 is also a reachable ID node for node 0.0.3. So we will have a reachability table shown in Figure 5(b), and we can find that every ID node in a cycle is reachable by all the IDREF nodes in that cycle.

5.3   Reachability Table Space Complexity

Let the number of IDREF nodes in an XML graph be L, where each IDREF node corresponds to one reference edge, then there could be at most L different ID nodes which are referenced by a reference edge. In the worst case, if every IDREF node can reach all these L ID nodes, then the space complexity is $O(L^2)$ in the worst case. The worst case only happens when all the ID/IDREF nodes forms a big cycle. Furthermore, the L IDREF nodes only occupy a small portion of all nodes in an XML graph in practice (around 5% in real-life data set in our experiments in Sec. 8). This is because every IDREF node must belong to a particular object in the XML, and the attribute information of an object, like ID, name, etc., can only be described by non-IDREF nodes.

6   Further Extension and Optimization

6.1   Removing Unnecessary Checking of the Reachability Table

For query evaluation on our transformed XML tree model, we need to expand the inverted lists by checking the reachability table. However, we find that many of the checking is unnecessary. E.g., given the reachability table in Figure 3(b) and the following inverted lists to be expanded: "Alps" : 0.12.1 and "part" : 0.0.1, 0.0.2, ...

As discussed in Sec. 4.3, in step 2 we need to check the ancestor of each dewey label to see whether their ancestors appear in the *Reachable ID node* column of the reachability table. But the ancestors of 0.0.1, 0.0.2, ... do not appear in that column, thus the checking is in vain. To avoid unnecessary checking, we can add a check bit to each dewey label in the inverted list index, indicating whether we need to check such a dewey in the reachability table. E.g., the above inverted lists will become "Alps" : 0.12.1(true) and "part" : 0.0.1(false), 0.0.2(false), ...

Now we only need to check those dewey labels with the check bit being true. Only the ancestors of 0.12.1 will be checked in reachability table. Such a check bit can be computed during offline by checking whether the ancestors of each dewey label appear in the *Reachable ID node* column of the reachability table.

6.2   Adding Distance to Reachability Table

Some of the XML tree keyword search methods need to rank the query results by some criteria. For example, one of the common criteria is the size of the results. It is usually measured by the sum of path length from the result root to each match node. To meet such a need, we can extend our virtual replication method (in Sec. 4.2) by adding a column called *distance* to the reachability table. The *distance* value records the distance from the IDREF node to the reachable ID node. If an IDREF node can reach a reachable ID node by more than one paths, we record the distance of the shortest one. Because substructure with minimal size is in favor in both XML tree search and XML graph search.

Take the reachability table in Figure 4(b) as an example. We can extend the table with a *distance* column and the distance values for the the three tuples are 1, 3 and 1 respectively. For the second tuple, the distance value is 3 because the IDREF node supplierRef:0.0.1.3 need to go through a path $0.0.1.3 \dashrightarrow 0.12 \rightarrow 0.12.2 \dashrightarrow 0.88$ to the reachable ID node employee:0.88. The length of such a path is 3.

When we measure the path length from a result root to a match node of a keyword, it consists of three parts: (1) distance from the result root to the IDREF node; (2) distance from the IDREF node to the reachable ID node; (3) distance from the reachable ID node to the match node. The first and the third part can be found in the XML tree, the second part can be found in the reachability table *distance* column. E.g., given a result root part:0.0.1 and a match node eid:0.88.0, the distance from the result root to the match node is the sum of three parts: (1) distance from part:0.0.1 to supplierRef:0.0.1.3 is 1; (2) distance from supplierRef:0.0.1.3 to employee:0.88 is 3, which can be found in the reachability table *distance* column; (3) distance from employee:0.88 to eid:0.88.0 is 1. Therefore the total path length is 5.

7   Algorithms

In this section, we present Algorithm 1 to transform an XML graph to our tree model, which consists of an XML tree and a reachability table.

Given an XML graph, the XML tree part can be easily generated by removing all the reference edges (line 17). The main task here is to generate the reachability table. We will first assign a dewey label to each node in the XML graph (line 3). Then based on each IDREF node n in the XML graph (line 4), we do a breadth-first search to explore the reachable ID nodes until no more new ID node can be further explored (line 5-14). The first node to be explored is the ID node being referenced by n and it will be pushed to a queue (line 5-6). The ID nodes which have been visited will be stored in a set (line 7). Each time we will take a node from the queue to explore until there is no more node in the queue (line 8-9). If the node taken from the queue is not visited before (line 10), we will visit it and mark it as explored (line 11). Then we will further explore within the node. For each IDREF node within it (line 12), we will add the corresponding ID node to the queue (line 13-14), which stores the nodes waiting to be explored. This process will terminate until no more node to explore (line 8). Finally, it will add all reachable ID nodes to the reachability table (line 15-16).

For the algorithm of doing query evaluation based on our tree model, it is similar to the 3 steps discussed in Sec. 4.3 and existing XML tree search algorithms can be easily found in the literature [13,14]. So the pseudo code will be omitted here.

Algorithm 1. transformXMLGraphToTree(XT)

---

    input : XML Graph XG
    output: Transformed XML Tree XT and reachability table RT
1 // Construct reachability table
2 Table RT;
3 assignDeweyLabel(XG); //regardless of reference edges
4 **foreach** *IDREF node* $n \in XG$ **do**
5   np = the ID node which n references to;
6   Queue *nodesToExplore*={np};
7   Set *exploredNodes* = {$\emptyset$};
8   **while** *nodesToExplore* $\neq \emptyset$ **do**
9   v = *nodesToExplore*.removeFirst();
10   **if** *exploredNodes.notContains(v)* **then**
11   *exploredNodes* = *exploredNodes* $\cup v$;
12   **foreach** *IDREF node* $m \in$ *the subtree rooted at v* **do**
13   mp = the node which m references to;
14   *nodesToExplore*.add(mp);
15   **foreach** *node* $r \in$ *exploredNodes* **do**
16   RT.addTuple(n.dewey, r.dewey);
17 XT = removeAllReferenceEdges(XG); // Generate the XML tree
18 **return** XT and RT;

---

8 Experiments

In this section, we present the experimental results comparing our approach with two graph-search-based methods. One is XKeyword [9], which is dedicated for XML graph by making use of the XML schema. Another one is BLINKS [8], which is one of the most efficient pure graph search method so far by building a bi-level index.

Experimental Settings. All algorithms are implemented in Java. The experiments were performed on a 2.83GHz Core 2 Quad machine with 3GB RAM running 32-bit windows 7. Berkeley DB Java Edition [1] is used to organize our reachability table in a B+ tree and store the inverted lists. MySQL [11] is used to support XKeyword. BLINKS does not need any database support since it is an in-memory approach.

Data Set. To test the real impact of the keyword search methods, we use a 200MB subset of real-life XML data set with ID/IDREF, ACMDL [1] , in our experiments. It contains publications from 1990 to 2001 indexed by the ACM Digital Library. There are 38K publications and 253K citation (as IDREF) among the publications. Totally 4.5M XML nodes and 4.8M XML edges are included. We can see that IDREF nodes (253K) are 5% of all XML nodes.

8.1   Comparing the Results

There are abundant search methods available designed for XML tree. Here we adopt one of them, i.e. ELCA [14], to work on our transformed tree model. Most of the XML tree methods focus on finding a meaningful subset of all possible results with regard to users' search intention. However, studying whether these subset of results are meaningful regarding users' search intention is not the main focus of this paper. So here we study the similarity between the subset found by tree methods and the subset found by graph methods in terms of result overlap rather than users' search intention.

We generate 100 random queries with two keywords, three keywords, four keywords and five keywords respectively. For each group of queries, we compare the top-20 results found by XKeyword and BLINKS on the original XML graph, to the top-20 results found by ELCA on our transformed tree model. For a fair comparison, all results are ranked by the size of the corresponding Minimal Steiner Tree, i.e. the sum of the path length.

Table 1 shows the average result overlap between our approach and the graph methods, which is calculated as (*# of same results in top-20*)/20. Two results are the same only if the root and each match node are the same.

Table 1. Result Overlap Between Our Approach and Graph Methods

Graph Methods	XKeyword				BLINKS			
# keywords	2	3	4	5	2	3	4	5
Average Result Overlap	77.9%	83.0%	85.4%	82.9%	92.1%	89.0%	90.5%	91.2%

As we can tell from Table 1, averagely 16 out of top-20 results are the same between our approach and XKeyword, while averagely 18 out of top-20 results are the same between our approach and BLINKS. Because XKeyword sets a maximum result size to constrain the search space, sometimes it finds less than 20 results. Therefore the result overlap is smaller.

---

[1] Thanks to Craig Rodkin at ACM Headquarters for providing the ACMDL dataset.

8.2 Performance

Next we will study the performance of our approach with the transformed tree model. XML tree search methods can be very efficient. E.g., ELCA can compute the results by linearly scanning the inverted lists. Here we will compare our approach with two graph search methods, XKeyword and BLINKS. However, BLINKS is an in-memory approach, which throws out-of-memory errors when handling the ACMDL date set. In order to be able to compare the performance of these three approaches, we have to downgrade the data set size to 45MB, which is the maximum data size BLINKS can handle on our machine. Later we will compare on the full data set with only our approach and XKeyword.

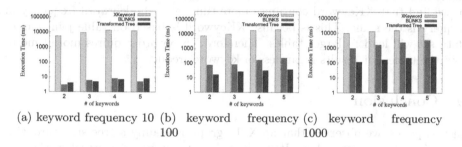

(a) keyword frequency 10 (b) keyword frequency 100 (c) keyword frequency 1000

Fig. 6. Query Execution Time (45MB data Size)

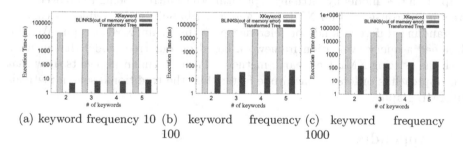

(a) keyword frequency 10 (b) keyword frequency 100 (c) keyword frequency 1000

Fig. 7. Query Execution Time (200MB Data Size)

Figure 6 shows the execution time of the three approaches. Our approach performs a full ELCA computation while XKeyword and BLINKS perform a top-20 results computation. We generate 100 random queries for each combination of *keyword frequency* and *# of keywords*. We can see that our approach outperforms XKeyword by orders of magnitude. This is because XKeyword stores the node information in relational tables to accommodate very large graphs, then the results computation is based on table join. Although schema information can help prune some search space, it is still not efficient.

For BLINKS, our approach is faster than it by an order of magnitude when keyword frequency is 100 or 1000. But our approach runs neck and neck with BLINKS when the frequency is around 10. We find that this is because BLINKS is an in-memory approach, which loads the whole graph into memory and does not need to access disk during the whole query evaluation. Yet it is not scalable to large data set. With 1.5 GB heap size assigned to JVM on our machine, 45MB is the maximum data size it can handle without out of memory. For our approach, we store the inverted lists and reachability table in database, so the disk access dominates the query evaluation time when keyword frequency is low.

Now we will compare XKeyword and our approach on the full data set. Figure 7 shows the experiment results. As we can see, our approach is still orders of maganitude faster than XKeyword on full data set. Comparing Figure 7 to Figure 6, we find that XKeyword consumes more time even the keyword frequency in a query is the same. This is because XKeyword is based on table join. Larger data set will lead to larger tables. Therefore XKeyword requires more time to join the tables for results regardless of keyword frequency.

9 Conclusion

In this paper, we observed that an XML graph is mainly a tree structure with a portion of reference edges. It motivated us to proposed a novel method to transform an XML graph with ID/IDREF to a tree model, such that we can exploit abundant efficient XML tree search methods. We transform an XML graph to a tree model by virtually replicating the subtrees being referenced. Our tree model consists of an XML tree and a reachability table, which is capable of handling different kinds of reference patterns in an XML graph. We also designed a query evaluation framework based on our tree model. It can work with the existing XML tree search methods. The experimental results show that our approach is orders of magnitude faster than the traditional search methods on XML graph.

10 Appendix

We declare: *Virtual Replication* will find the same set of non-redundant results as *Real Replication*.

Proof. ($a \prec b$ denotes that node a is an ancestor of b. $a \preceq b$ denotes that $a \prec b$ or $a = b$.) Step 1: to prove that every non-redundant result found by *real replication* can also be found by the *virtual replication*. Let any non-redundant result found by the *real replication* be T_r, which is a subtree rooted at node r. It should be an LCA of a set of nodes $M_{real} = \{n_1, n_2, ..., n_k, \hat{n_1}, , \hat{n_2}, ..., \hat{n_l}\}$ matching the query keywords, where $\hat{n_j}$ is a match node appearing in a replicated subtree and n_i is a match node not in any replicated subtree. Each match node corresponds to a keyword in the query. The LCA relationship can be represented as two properties: ① $r \preceq n_i (1 \leq i \leq k)$, $r \preceq \hat{n_j} (1 \leq j \leq l)$; ② $\nexists r' \prec r$ s.t. $r' \preceq n_i (1 \leq i \leq k)$

and $r' \preceq \hat{n}_j (1 \leq j \leq l)$. In the *virtual replication* method, suppose we use an IDREF node $\hat{N}_j$ to represent the replicated subtree which $\hat{n}_j$ is in, we have ③ $\hat{N}_j \preceq \hat{n}_j$. Then we can prove that the same result T_r can also be calculated based on the following set of match nodes $M_{virtual} = \{n_1, n_2, ..., n_k, \hat{N}_1, , \hat{N}_2, ..., \hat{N}_l\}$. Formally, we need to prove r is the LCA of $M_{virtual}$. Since T_r is a non-redundant result, we have ④ $r \prec \hat{N}_j (1 \leq j \leq l)$. So from ① and ④, we have ⑤ $r \preceq n_i (1 \leq i \leq k)$, $r \preceq \hat{N}_j (1 \leq j \leq l)$. Next we need to prove ⑥ $\nexists r' \prec r$ s.t. $r' \preceq n_i (1 \leq i \leq k)$ and $r' \preceq \hat{N}_j (1 \leq j \leq l)$ by contradiction. If ⑥ is not true, with ③ we can infer that $\exists r' \prec r$ s.t. $r' \preceq n_i (1 \leq i \leq k)$ and $r' \preceq \hat{N}_j \preceq \hat{n}_j (1 \leq j \leq l)$, which contradicts with ②. So with ⑤ and ⑥ being true, r is the LCA of $M_{virtual}$ as well. Step 1 is finished. Step 2: to prove that every non-redundant result found by *virtual replication* can also be found by *real replication*. The proof is similar to step 1, which is omitted here due to space limitation.   □

References

1. Berkeley, D.B.: `http://www.sleepycat.com`
2. Bao, Z., Ling, T.W., Chen, B., Lu, J.: Effective XML keyword search with relevance oriented ranking. In: ICDE (2009)
3. Bao, Z., Lu, J., Ling, T.W., Xu, L., Wu, H.: An effective object-level XML keyword search. In: Kitagawa, H., Ishikawa, Y., Li, Q., Watanabe, C. (eds.) DASFAA 2010. LNCS, vol. 5981, pp. 93–109. Springer, Heidelberg (2010)
4. Bhalotia, G., Hulgeri, A., Nakhe, C., Chakrabarti, S., Sudarshan, S.: Keyword searching and browsing in databases using banks. In: ICDE (2002)
5. Ding, B., Yu, J.X., Wang, S., Qin, L., Zhang, X., Lin, X.: Finding top-k min-cost connected trees in databases. In: ICDE (2007)
6. Dreyfus, S.E., Wagner, R.A.: The steiner problem in graphs. Networks (1971)
7. Guo, L., Shao, F., Botev, C., Shanmugasundaram, J.: XRANK: Ranked keyword search over XML documents. In: SIGMOD (2003)
8. He, H., Wang, H., Yang, J., Yu, P.S.: Blinks: ranked keyword searches on graphs. In: SIGMOD (2007)
9. Hristidis, V., Papakonstantinou, Y., Balmin, A.: Keyword proximity search on XML graphs. In: ICDE 2003 (2003)
10. Kacholia, V., Pandit, S., Chakrabarti, S., Sudarshan, S., Desai, R., Karambelkar, H.: Bidirectional expansion for keyword search on graph databases. In: VLDB (2005)
11. MySQL, `http://www.mysql.com`
12. Vesper, V., `http://www.mtsu.edu/vvesper/dewey.html`
13. Xu, Y., Papakonstantinou, Y.: Efficient keyword search for smallest LCAs in XML databases. In: SIGMOD (2005)
14. Xu, Y., Papakonstantinou, Y.: Efficient LCA based keyword search in XML data. In: EDBT (2008)
15. Zhou, J., Bao, Z., Wang, W., Ling, T.W., Chen, Z., Lin, X., Guo, J.: Fast SLCA and ELCA computation for XML keyword queries based on set intersection. In: ICDE (2012)
16. Zhou, R., Liu, C., Li, J.: Fast ELCA computation for keyword queries on XML data. In: EDBT (2010)

Permutation-Based Pruning
for Approximate K-NN Search

Hisham Mohamed and Stéphane Marchand-Maillet

Université de Genève, Geneva, Switzerland
{hisham.mohamed,stephane.marchand-maillet}@unige.ch

Abstract. In this paper, we propose an effective indexing and search algorithms for approximate K-NN based on an enhanced implementation of the Metric Suffix Array and Permutation-Based Indexing. Our main contribution is to propose a sound scalable strategy to prune objects based on the location of the reference objects in the query ordered lists. We study the performance and efficiency of our algorithms on large-scale dataset of millions of documents. Experimental results show a decrease of computational time while preserving the quality of the results.

Keywords: Metric Suffix Array (MSA), Permutation-Based Indexing, Approximate Similarity Search, Large-Scale Multimedia Indexing.

1 Introduction

Searching for similar objects in a database is a fundamental problem for many applications, such as information retrieval, visualization, machine learning and data mining.

In metric spaces [1], several techniques have been developed for improving the performance of searching, by decreasing the number of direct distance calculations [1]. One of the recent techniques is the *permutation-based indexing* [2,3]. The idea behind it is to represent each object by a list of permutations of selected neighboring items (reference points). The similarity between any two objects is then derived by comparing the two corresponding permutation lists. In this work, we propose an enhanced implementation of the Metric Suffix Array (MSA) proposed in [4] which is one of the recent data structure for permutation based-indexing. In [4], regardless of the number of permutations, the number of objects, and the number of K-NN which need to be retrieved, the complete MSA has to be scanned in order to retrieve the most promising results. Here, we propose an enhanced implementation of the MSA to avoid scanning the complete MSA. The main idea is to prune cells representing objects that have a high difference in their permutations ordering. Hence, only a small part of the array is scanned, which improves the running time. In addition, we propose different strategies for selecting the reference points. To validate our claims, we test the enhanced MSA and the selection strategies on a high dimensional large dataset containing several millions of objects.

The rest of the paper is organized as follows. Section 2 proposes a review of the related work. In sections 3 and 4, we introduce our modeling for permutation-based indexing and a formal justification for our proposed indexing and searching procedures. Finally, we present our results in section 5 and conclude in section 6.

H. Decker et al. (Eds.): DEXA 2013, Part I, LNCS 8055, pp. 40–47, 2013.

2   Prior Work

The idea of *permutation-based indexing* was first proposed in [3,2]. Amato and Savino [3] introduced the metric inverted files (MIF) to store the permutations in an inverted file. Then, Mohamed and Marchand-Maillet [5] proposed three distributed implementation of the MIF. In [6], authors proposed the *brief permutation index*. The main idea is to encode the permutation as a binary vector and to compare these vectors using the Hamming distance. In [7], authors proposed the *prefix permutation index* (PP-Index). PP-Index stores the prefix of the permutations only and the similarity between objects is measured based on the length of its shared prefix. Furthermore, in [4] authors proposed the MSA, which is a fast and an effective data structure for storing the permutations, by saving half of the processing memory which is needed in [3,5,7]. The work presented in this paper is based on [4] considered as current state of the art. We provide an enhanced implementation of the MSA for fast and effective retrieval and we test it against [4].

3   Indexing Model

Permutation-based indexes aim to predict the proximity between elements according to how they order their distances towards a set of reference objects [2,3].

Definition 1. *Given a set of N objects o_i, $D = \{o_1, \ldots, o_N\}$, a set of reference objects $R = \{r_1, \ldots, r_n\} \subset D$, and a distance function which follows the metric space postulates, we define the ordered list of R relative to $o \in D$, L_o, as the ordering of elements in R with respect to their increasing distance from o:*

$$L_o = \{r_{i_1}, \ldots, r_{i_n}\} \text{ such that } d(o, r_{i_j}) \leq d(o, r_{i_{j+1}}) \ \forall j = 1, \ldots, n-1$$

Then, for any $r \in R$, $P(L_o, r)$ indicates the position of r in L_o. In other words, $P(L_o, r) = j$ such that $r_{i_j} = r$. Further, given $\bar{n} > 0$, $\bar{L}_o$ is the pruned ordered list of the $\bar{n}$ first elements of L_o.

Figure 1(b) gives the pruned ordered lists $\bar{L}_{o_i}$, where $\bar{n} = 2$, for D and R illustrated in Figure 1(a).

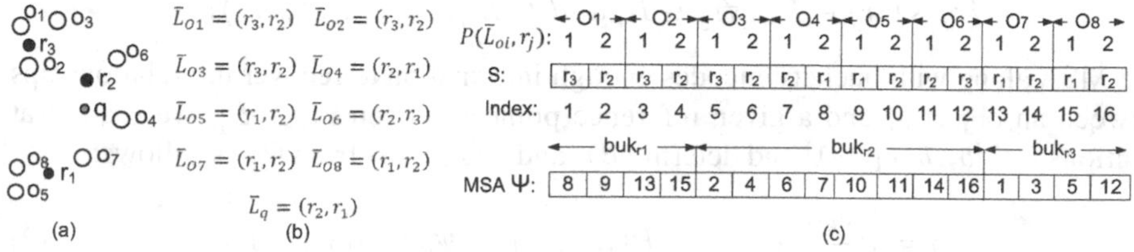

Fig. 1. a) White circles are data objects o_i; black circles are reference objects r_j; the gray circle is the query object q b) Pruned ordered lists $\bar{L}_{o_i}$, $\bar{n} = 2$. c) Example of Metric Suffix Array and buckets $b_j = \text{buk}_{r_j}$

In K-NN similarity queries, we are interested in ranking objects (to extract the K first elements) and not so much in the actual inter-object distance values. Permutation-based indexing relaxes distance calculations by assuming that they will be approximated in terms of their ordering when comparing the ordered lists of objects. Here, we consider the Spearman Footrule Distance (d_{SFD}) between ordered lists. Formally,

$$d(q, o_i) \overset{\text{rank}}{\simeq} d_{\text{SFD}}(q, o_i) = \sum_{\substack{r_j \in L_q \\ r_j \in L_{o_i}}} |P(\bar{L}_q, r_j) - P(\bar{L}_{o_i}, r_j)| \tag{1}$$

To efficiently answer users queries, the Metric Suffix Array (MSA) was proposed in [4]. Independent of the number of reference points, objects or K-NN, the complete MSA should be scanned in order to retrieve the most similar objects to the query. Here, we use our rank-based approximation to postulate that objects having similar ordered lists as the query (based on d_{SFD}) are candidates similar objects, which we could filter by direct distance calculation. Hence, as per Eq.(1), for a query q, if we organise the MSA cells to characterise objects such that $P(\bar{L}_q, r_j) = P(\bar{L}_{o_i}, r_j)$, we can avoid accessing the complete MSA to obtain the list C of candidate similar objects. We therefore propose below an enhanced structure for the MSA to help reducing the searching time. We first recall formally the construction of the MSA [4] .

Given all pruned ordered lists $\bar{L}_{o_i}$, we construct $S = \bigcup_{i=1}^{N} \bar{L}_{o_i} = \{r_{i_1}, \ldots, r_{i_M}\}$, where $M = \bar{n}.N$. The set S can then be seen as a string of length M on the alphabet R. A Metric Suffix Array Ψ acts like a Suffix Array [8,9,10] over S. More specifically, Ψ is a set of M integers corresponding to the permutation induced by the lexical ordering of all M suffixes in S ($\{r_{i_k}, \ldots, r_{i_M}\} \; \forall k \leq M$). In [4], the MSA is sorted into buckets.

Definition 2. *A bucket for reference point r_j is a subset of the MSA Ψ from position b_j. The bucket is identified to its position in Ψ so that $b_j \equiv \Psi_{[b_j, b_{j+1}-1]}$.*

Bucket b_j contains the positions of all the suffixes of S of the form $\{r_j, \ldots, r_{i_M}\}$, i.e. where reference point r_j appears first.

For example, in Figure 1(c), the string S corresponding to the objects and reference points shown in Figure 1(a) is given. The MSA Ψ is also shown. The bucket b_2 for reference point r_2 (buk$_{r_2}$) contains the positions in S of suffixes starting by r_2.

At query time, $\bar{L}_q$ will be computed. From Eq.(1), we therefore need to characterise the objects

$$\{o_i \text{ s.t } r_j \in \bar{L}_{o_i}, \forall r_j \in \bar{L}_q \text{ and } P(\bar{L}_{o_i}, r_j) = P(\bar{L}_q, r_j)\}$$

The MSA along with buckets encodes enough information to recover the relationships between an object o_i and a given reference point r_j. Given $r_j \in \bar{L}_q$, we scan Ψ at positions $k \in [b_j, b_{j+1} - 1]$ and determine i, and $P(\bar{L}_{o_i}, r_j)$ from Ψ_k as follows:

$$i = \left\lfloor \frac{\Psi_k}{\bar{n}} \right\rfloor + 1 \qquad P(\bar{L}_{o_i}, r_j) = (\Psi_k \bmod \bar{n}) + 1 \tag{2}$$

Within each bucket b_j, $P(\bar{L}_{o_i}, r_j) \leq \bar{n}$. In order to speed up further the scanning of buckets, we sort them according to the value of $P(\bar{L}_{o_i}, r_j)$. A sub-bucket $b_j^{(l)}$ points to objects o_i such that $P(\bar{L}_{o_i}, r_j) = l$ (see Figure 2(b)).

4 Practical Setup

4.1 Indexing

Algorithm 1 details the indexing process. Line 1 builds the MSA using pruned lists $\bar{L}_{o_i}$. In lines 2-3, buckets are sorted based on $P(\bar{L}_{o_i}, r_j)$ (Eq.(2)) using the *quicksort* algorithm. All the suffixes representing the objects that have the same $P(\bar{L}_{o_i}, r_j)$ are located next to each other, which makes it easy to divide each bucket into $\bar{n}$ sub-buckets $b_j^{(l)}$ (Line 4).

Algorithm 1

IN: Domain D of N , R of n ,$\bar{n}$
OUT: The MSA Ψ with sub-buckets $b_j^{(l)}$
1.   *Build the MSA Ψ and buckets b_j*
2.   *For each $r_j \in R$*
3.   quickSort$(\Psi, b_j, b_{j+1} - 1)$
4.   *Define the sub-buckets $b_j^{(l)}$ within each bucket b_j*

Theoretically, the indexing complexity is $O(\bar{n}\eta(1 + log\eta))$, where η is the average size of the sub-buckets ($\eta \leq N$). The average memory usage is $O(M + (\bar{n} \times n))$.

4.2 Searching

Equation (1) measures the discrepancy in ranking from common reference objects between each object and the query. In practice, it can be simplified by counting the co-occurrences of each object with the query in the same (or adjacent) sub-buckets. That is, each object o_i scores

$$s_i = \left| \left\{ r_j \in \bar{L}_q \text{ such that } (r_j \in \bar{L}_{o_i} \text{ and } |P(\bar{L}_{o_i}, r_j) - P(\bar{L}_q, r_j)) \leq 1 \right\} \right| \quad (3)$$

Objects o_i are then sorted according to their decreasing s_i scores. This sorted candidate list provides an approximate ranking of the database objects relative to the submitted query. This approximate ranking can be improved by direct distance calculation (DDC). For a K-NN query, we apply DDC on the $K_c = \Delta \times K$ first objects in our sorted candidate list and call $\Delta > 1$ the *DDC factor*. The effect of Δ is explored in our performance evaluation (section 5). The search procedure is described in Algorithm 2.

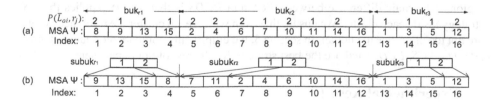

Fig. 2. a) MSA sorted by bucket b) MSA sorted by sub-buckets

First $\bar{L}_q$ is computed (line 1). For each $r_j \in \bar{L}_q$, active sub-buckets are identified (line 3). For each object o_i pointed, a count $s[i]$ of co-occurrences is computed using (Eq.3) (lines 4-5). Line 6 sorts the candidate list in decreasing order of their score (counters s). Final DDC filtering is performed in lines 7-9.

Algorithm 2

IN: Query: q, R of n, MSA, $b_j^{(l)}$, s[0 . . . N]
OUT: Sorted Objects list: out
1.   *Create the query ordered list $\bar{L}_q$*
2.   *For $r_j \in \bar{L}_q, l = P(\bar{L}_q, r_j)$*
3.   *For $k = b_j^{(l)}$ to $b_j^{(l+1)} - 1$*
4. $i = \left\lfloor \frac{\Psi_k}{\bar{n}} \right\rfloor + 1$
5. $s[i] = s[i] + 1$
6.   *sort(s)*
7. $K_c = K_{NN} \times \Delta$
8. $C \leftarrow s[K_c]$
9. $out = \text{calc\_distance}(C, K_c, q)$
10.   *sort(out)*

Theoretically, the computational complexity to retrieve the K_c is $O(2\bar{n}\eta)$.

4.3 Reference Points Selection

In [3,7], the selection of R is done randomly, we name this strategy *Random Selection* (RS). We propose three alternative strategies for selecting the reference points. The first strategy is the *distributed selection (DS)*. In *DS*, close reference points are neglected based on a certain threshold value. Hence, if one of the new selected points is close to an already selected point, the selection is ignored. Using this technique, we ensure that the points are well-distributed over the database. That leads to a relevant encoding of each object using the permutations. The second and the third strategies are based on the k-mean algorithm for clustering. We call them *post-clustering selection (PCS)* and *post-clustering distributed selection (PCDS)*. The main idea behind the two strategies is that the dataset is divided into a number of clusters to support the selection of the reference points. For instance, if we need 1,000 reference points and we create 5 clusters, 200 reference points are selected from each cluster. We thus ensure that the objects located in the same cluster have the same reference points as the primary items in their order lists. This helps to improve the identification of the objects for eliminating unwanted regions. The main difference between *PCS* and *PCDS* is that, *PCDS*, applies the *DS* strategy inside each cluster. We ensure that even within each cluster the reference points are not too close one to another. In section 5, we empirically compare the four strategies *(RS, DS, PCS, PCDS)* and use the best strategy for the rest of our experiments.

5 Experimental Results

The average recall (RE), average position error (PE) [1], and average indexing and running time are measured and compared with that listed in [4]. All the experiments were done using 250 different queries that were selected randomly from the dataset. The sub-bucket algorithm was implemented in C++. The experiments were done on a 2.70GHz processor, with 128Gb of memory and 512GB storage capacity. Our conducted experiments are based on a dataset of 5 millions visual shape features (21-dimensional), which were extracted from the 12-million ImageNet corpus [11].

Selecting the reference points (R): Figures 3a and 3b show the average *RE* and *PE* for the reference point selecting strategies (section 4.3) for 10 K-NN using MSA-Full [4].

The *RS* technique gives the lowest RE and the highest PE values. Using RS, there is a high probability that some reference points are close one to another. That makes them at the same distance from the objects, leading to inefficient encoding. *DS* gives a higher RE and lower PE compared to the *RS*, as the objects are identified using equally distributed reference points. For the *PCS* and the *PCDS* techniques, the dataset is clustered into 5 and 10 clusters. From the figures, we see that the RE and PE augment with the number of clusters. This is based on the dataset and the number of clusters that we can get out of the dataset. Also, we can see that the *PCDS* technique gives better RE and PE values than the *PCS*. The reason is that the reference points which were selected from each cluster are well-spread within the clusters.

Comparing the four strategies, when the number of reference points increases the RE increases and PE decreases until a certain limit. The *DS* and the *PCDS* give the best RE and PE because these techniques ensure a good distribution of the reference points. On the other hand, *DS* is better than the *PCDS* in terms of time consumption. There is no time used to cluster the data before selecting the reference points. We therefore apply *DS* in the rest of our experiments.

Sub-buckets and DDC Factor: Figure 4a shows the average RE and PE for different numbers of reference points using MSA-Full, MSA-NN [4] and the sub-bucket implementations ($\Delta = 100$) for 10 K-NN. From the figure, for the three algorithms, when the number of reference points increases, RE increases and PE decreases. For the sub-bucket algorithm, even with small number of reference points, we are able to achieve higher RE and lower PE than scanning the complete MSA using high number of

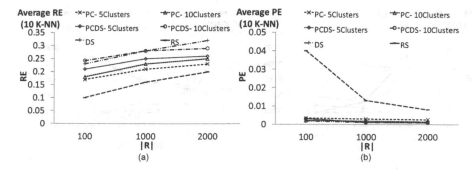

Fig. 3. Average *RE*(a) and *PE*(b) (250 queries) for top 10-NN using MSA-Full [4]

reference points for MSA-Full and MSA-NN. For instance, using $n = 100$ and $\Delta = 100$, for 10 K-NN, our algorithm is able to achieve higher RE than using $n = 2000$ for MSA-Full and MSA-NN, with a faster response time (next paragraph).

Figure 4b shows the RE and PE for different numbers of reference points using different values of Δ for 10 K-NN. As a reminder, for 10 K-NN and $\Delta = 40$, the algorithm calculates the distance between the query and the top $K_c = 400$ candidates objects. As we can see, when Δ increases, RE increases and PE decreases, as the number of the objects that are compared to the query increases.

Comparing MSA-Full, MSA-NN and MSA-sub buckets Algorithms: Table 1 shows the average indexing and searching time (in seconds) for the sub-bucket implementation (including scanning the MSA and accessing the hard-disk) compared to MSA-Full and MSA-NN proposed in [4]. For indexing, we can see that the indexing time for the three algorithms increases with the increase in the number of reference points.

Table 1. Indexing and searching time(in seconds) for sub-bucket (Δ=100) compared to [4]

$|R|$	Index-Full	Search-Full	Index-NN	Search-NN	Index-Subbuckets	Search-Subbucket
100	45	4	86	1.5	113	**0.35**
1000	517	46	735	12	1622	**0.45**
2000	1081	94	1651	25	3523	**0.76**

In addition, the indexing time for the sub-buckets algorithm is higher than that of the other algorithms. This is due to the sorting and definition of the sub-buckets after building the MSA. However, since the indexing process is an off-line process, this increase is accepted.

For searching, for different n, using $\Delta = 100$, it appears clearly from Table 1 that the sub-bucket technique is faster than [4], as the algorithm does not need to scan all the MSA cells nor all the database.

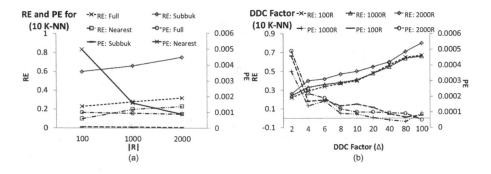

Fig. 4. Using 250 queries a) Comparing sub-bucket to [4]. b) Effect of Δ.

6   Conclusion

We have presented an enhanced indexing technique based on the Metric Suffix Array (MSA), representing the current state of the art implementation for *permutation-based indexing*. Our main idea is to prune the MSA cells which represent objects that have high difference in their permutations ordering. Hence, only a small part of the array is scanned. With a combination of direct distance calculations, we showed through an experimental analyses that our algorithm gives better results in terms of time and precision compared to that proposed in [4]. In addition, we empirically showed how the selection of the reference points can affect the performance. There is much to improve on this selection for *permutation-based indexing*. This is the subject of our future work.

Acknowledgment. This work is jointly supported by the Swiss National Science Foundation (SNSF) via the Swiss National Center of Competence in Research (NCCR) on Interactive Multimodal Information Management (IM2) and the European COST Action on Multilingual and Multifaceted Interactive Information Access (MUMIA) via the Swiss State Secretariat for Education and Research (SER).

References

1. Zezula, P., Amato, G., Dohnal, V., Batko, M.: Similarity Search: The Metric Space Approach. Advances in Database Systems, vol. 32. Springer (2006)
2. Gonzalez, E., Figueroa, K., Navarro, G.: Effective proximity retrieval by ordering permutations. IEEE Transactions on Pattern Analysis and Machine Intelligence 30(9) (September 2008)
3. Amato, G., Savino, P.: Approximate similarity search in metric spaces using inverted files. In: International Conference on Scalable Information Systems, pp. 28:1–28:10 (2008)
4. Mohamed, H., Marchand-Maillet, S.: Metric suffix array for large-scale similarity search. In: ACM WSDM 2013 Workshop on Large Scale and Distributed Systems for Information Retrieval, Rome, IT (February 2013)
5. Mohamed, H., Marchand-Maillet, S.: Parallel approaches to permutation-based indexing using inverted files. In: 5th International Conference on Similarity Search and Applications (SISAP), Toronto, CA (August 2012)
6. Téllez, E.S., Chávez, E., Camarena-Ibarrola, A.: A brief index for proximity searching. In: Bayro-Corrochano, E., Eklundh, J.-O. (eds.) CIARP 2009. LNCS, vol. 5856, pp. 529–536. Springer, Heidelberg (2009)
7. Esuli, A.: Pp-index: Using permutation prefixes for efficient and scalable approximate similarity search. In: Proceedings of LSDSIR 2009, vol. i, pp. 1–48 (July 2009)
8. Manber, U., Myers, E.W.: Suffix arrays: A new method for on-line string searches. SIAM J. Comput. 22(5), 935–948 (1993)
9. Schürmann, K.B., Stoye, J.: An incomplex algorithm for fast suffix array construction. Softw., Pract. Exper. 37(3), 309–329 (2007)
10. Mohamed, H., Abouelhoda, M.: Parallel suffix sorting based on bucket pointer refinement. In: 5th Cairo International Biomedical Engineering Conference (CIBEC), pp. 98–102 (2010)
11. Deng, J., Dong, W., Socher, R., Li, L.J., Li, K., Fei-Fei, L.: ImageNet: A Large-Scale Hierarchical Image Database. In: CVPR 2009 (2009)

Dynamic Multi-probe LSH: An I/O Efficient Index Structure for Approximate Nearest Neighbor Search

Shaoyi Yin[*], Mehdi Badr, and Dan Vodislav

ETIS, Univ. of Cergy-Pontoise / CNRS, France
yinshaoyi@gmail.com, {mehdi.badr,dan.vodislav}@u-cergy.fr

Abstract. Locality-Sensitive Hashing (LSH) is widely used to solve approximate nearest neighbor search problems in high-dimensional spaces. The basic idea is to map the "nearby" objects into a same hash bucket with high probability. A significant drawback is that LSH requires a large number of hash tables to achieve good search quality. Multi-probe LSH was proposed to reduce the number of hash tables by looking up multiple buckets in each table. While optimized for a main memory database, it is not optimal when multi-dimensional vectors are stored in a secondary storage, because the probed buckets may be randomly distributed in different physical pages. In order to optimize the I/O efficiency, we propose a new method called Dynamic Multi-probe LSH which groups small hash buckets into a single bucket by dynamically increasing the number of hash functions during the index construction. Experimental results show that our method is significantly more I/O efficient.

Keywords: Locality sensitive hashing, indexing, high-dimensional database, approximate nearest neighbor search.

1   Introduction

Nearest neighbor search (NNS), also known as similarity search, consists in finding, for a given point in a high-dimensional space, the closest points from a given set. The nearest neighbor search problem arises in many application fields, such as pattern recognition, computer vision, multimedia databases (e.g. content-based image retrieval), recommendation systems and DNA sequencing. Various indexing structures have been proposed to speed up the nearest neighbor search. Early proposed tree-based indexing methods such as R-tree [10], K-D-tree [2], SR-tree [14], X-tree [3] and M-tree [5] return exact query results, but they all suffer the "curse of dimensionality": it has been shown in [19] that they exhibit linear complexity at high dimensionality, and that they are outperformed on average by a simple sequential scan of the database if the number of dimensions exceeds even moderate values, e.g. around 10.

---

[*] Shaoyi YIN is currently an associate professor at IRIT Laboratory, Paul Sabatier University, France.

H. Decker et al. (Eds.): DEXA 2013, Part I, LNCS 8055, pp. 48–62, 2013.
© Springer-Verlag Berlin Heidelberg 2013

In fact, for most of the applications, exact nearest neighbors are not more meaningful than the so-called ε-approximate nearest neighbors, because the feature vectors used to represent the objects are usually already imprecise. This phenomenon, called semantic gap is inherent e.g. in content-based image retrieval (CBIR), where feature vectors expressing low-level image properties are used to answer high-level user queries. Formally, the goal of the ε-approximate NNS is to find the data object within the distance $\varepsilon \times R$ from a query object, where R is the distance from the query point to its exact nearest neighbor. The most well-known methods for solving the ε-approximate NNS problem in high-dimensional spaces are Locality Sensitive Hashing (LSH) [9], [12] and its variants [1], [15].

The basic LSH method uses a family of locality-sensitive hash functions to hash nearby objects into the same bucket with a high probability. Several hash functions are combined to produce a compound hash signature corresponding to finer hash buckets. For a given query object, the indexing method hashes this object into a bucket, takes the objects in the same bucket as ε-approximate NNS candidates, and then ranks the candidates according to their distances to the query object. A side effect of combining several hash functions is that some "nearby" points may be hashed into different buckets. In order to increase the probability of finding all the nearest neighbors, the LSH method usually creates multiple hash tables and each hash table is built by using independent locality-sensitive hash functions. The number of hash tables is usually over a hundred [9] and sometimes several hundred [4]. This becomes a problem in terms of space consumption. To reduce the number of hash tables needed, the Multi-probe LSH method [15] has been proposed by Lv et al.

The main idea of Multi-probe LSH is to build on the basic LSH indexing method, but to use a carefully derived probing sequence to look up multiple buckets that have a high probability of containing the nearest neighbors of a query object. By probing multiple buckets in each hash table, the method requires far fewer hash tables than the previously proposed LSH methods. As an in-memory algorithm, Multi-probe LSH method is very efficient; however, if the feature vectors cannot be stored in main memory, the query cost becomes rather high, because the probed buckets may be randomly distributed in different disk pages. In this paper, we consider the case where the feature vectors are stored in a secondary storage, even though we suppose that the index structure itself is still in main memory.

In order to improve the query efficiency of the Multi-probe LSH, we present in this paper a new method called Dynamic Multi-probe LSH (DMLSH). The main modification to Multi-probe LSH is that we dynamically adapt the number of hash functions for each bucket in order to produce buckets whose size fits a disk page. With the same parameter setting, our method always requires less I/O cost and provides higher query accuracy than Multi-probe LSH. The experimental results have shown that the gain is significant.

The rest of this paper is organized as follows. We first review the background knowledge and the related work in Section 2, and then present our DMLSH method in Section 3. We describe experimental studies in Section 4 and conclude in Section 5.

2   Background and Related Work

Approximate Nearest Neighbor Search. Definition (ε-Nearest Neighbor Search (ε-NNS)) [9]. Let us consider the normed space l_p^d representing the d-dimensional Euclidian space R^d with the l_p norm and the distance $D(\cdot,\cdot)$ induced by this norm. Given a set P of points in l_p^d, solving the ε-NNS problem in P consists in preprocessing P so as to efficiently return a point $p \in P$ for any given query point q, such that $D(p,q) \leq (1+\varepsilon) D(q,P)$, where $D(q,P)$ is the distance of q to its closest point in P.

Note that the above definition generalizes to any metric space. It also generalizes naturally to finding $K>1$ approximate nearest neighbors. In the approximate K-NNS problem, we wish to find K points $p_1, \ldots, p_k$, such that the distance of p_i to the query point q is at most $(1+\varepsilon)$ times the distance from q to the i-th nearest point in P.

Locality Sensitive Hashing. The basic idea of LSH is to use hash functions that map similar objects into the same hash bucket with high probability. Performing a similarity search query on an LSH index consists of two steps: 1) using LSH functions to select "candidate" objects for a given query q, and 2) ranking the candidate objects according to their distance to q.

Definition (Locality-Sensitive Hash Family) [9], [12]. A family $H = \{h: S \rightarrow U\}$ is called $(r, \varepsilon, p_1, p_2)$-sensitive, with $p_1 > p_2 > 0$, $\varepsilon > 0$, if for any $p, q \in S$, the following conditions hold:

- If $D(p, q) \leq r$ then $Pr_H[h(p)=h(q)] \geq p_1$;

- If $D(p, q) > (1+\varepsilon)r$ then $Pr_H[h(p)=h(q)] \leq p_2$.

Here S is a set of objects and $D(\cdot,\cdot)$ is the distance function of elements in the set S.

Different LSH families can be used for different distance functions. Families for Jaccard distance, Hamming distance, l_1 and l_2 distances are known. The most widely used one is the LSH family for Euclidean distance proposed by Datar et al.[7]. Each function is defined on R^d as follows: $h_{a,b}(v) = \lfloor (a \cdot v + b)/W \rfloor$, where a is a random d-dimensional vector and b is a real number chosen uniformly from the range $[0, W]$. $a \cdot v$ is the dot product of vectors a and v. Each hash function maps a d-dimensional vector v onto into an integer value.

Given a locality-sensitive hash family $H = \{h: S \rightarrow U\}$, an LSH index is constructed as follows. (1) For an integer $M > 0$, define a family $G = \{g: S \rightarrow U^M\}$ of compound hash functions; for any $g \in G$, $g(v) = (h_1(v), h_2(v), \ldots, h_M(v))$, where $h_j \in H$ for $1 \leq j \leq M$. (2) For an integer $L>0$, choose $g_1, g_2, \ldots g_L$ from G, independently and uniformly at random. Each of the L functions g_i ($1 \leq i \leq L$) is used to construct one hash table, resulting in L hash tables. (3) Insert each vector v into the hash bucket to which $g_i(v)$ points to, for $i = 1, \ldots, L$. A K-NNS query for vector q is processed in two steps. (1) Compute the hash value $g_i(q)$ and retrieve all the vectors in bucket $g_i(q)$ for $i = 1, \ldots, L$ as candidates. (2) Rank the candidates according to their distances to the query object q, and then return the top K objects. Note that compound hash functions reduce the probability that distant vectors belong to a same bucket, but increase the risk of nearby points separated into different buckets. Merging candidates from several hash tables reduces the risk of missing close objects.

Fig. 1 presents an example of 2-NNS, where $L = 3$. Objects $p_1, p_2, ..., p_5$ are inserted into three hash tables corresponding to hash functions g_1, g_2, g_3. For a given query q, by checking the three buckets $g_1(q), g_2(q), g_3(q)$, we get the candidates p_1, p_3, p_4, p_5. By ranking their distances to q, we return p_1 and p_3 as the 2-NN objects.

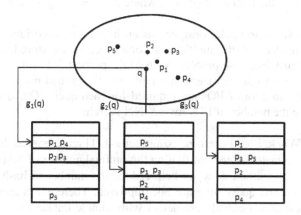

Fig. 1. LSH index structure

The main drawback of the above index structure is that it may require a large number of hash tables to cover most nearest neighbors. For example, over 100 hash tables are needed to achieve 1.1-approximation in [9], and as many as 583 hash tables are used in [4]. Thus, the whole data structure takes too much space. In order to reduce the number of hash tables while keeping a good approximation ratio, Multi-probe LSH [15] was proposed.

Multi-probe LSH. The main idea of the Multi-probe LSH method is to use a carefully derived probing sequence to check multiple buckets that are likely to contain the nearest neighbors of a query object. Since each hash table provides more candidates, the number of needed hash tables could be reduced.

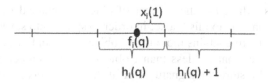

Fig. 2. Distance from q to the boundary of its neighbor bucket

Multi-probe LSH is based on the family of locality-sensitive functions of Datar et al.[7] described above, returning integer values. To derive the probing sequence, the authors first define a hash perturbation vector $\Delta = \{\delta_1, ..., \delta_M\}$ where $\delta_i \in \{-1, 0, 1\}$ and M is the number of hash functions for each hash table. Given a query q, the basic LSH method checks the hash bucket $g(q) = (h_1(q), h_2(q), ... , h_M(q))$, while

Multi-probe LSH checks also the buckets $g(q) + \Delta_1$, $g(q) + \Delta_2$, ..., $g(q) + \Delta_T$. These buckets are ordered according to their "success probability" which is a score estimated using formula score(Δ) $= \sum_{i=1}^{M} x_i (\delta_i)^2$, where $x_i(\delta_i)$ is the distance from q to the boundary of the bucket $h_i(q) + \delta_i$. For example, in Fig. 2, $x_i(1)$ is the distance from q to the boundary of the bucket $h_i(q) + 1$, where $h_i(q) = (a_i \cdot q + b_i)/W$ and $f_i(q) = a_i \cdot q + b_i$.

Multi-probe LSH is originally designed as an in-memory algorithm. In this paper, however, we consider that the multidimensional vectors are stored in a secondary storage such as hard disk. The problem with Multi-probe LSH in this new context is that many probes are needed for each hash table and the probed buckets are randomly stored in the disk, so a lot of I/Os are required for each query. Our objective in this paper is to reduce the number of I/Os for a K-NN search.

Other Related Work. LSH Forest indexing method [1] represents each hash table by a prefix tree to eliminate the need of finding the optimal number of hash functions per table. However, this method does not help reduce the number of hash tables, so the space consumption and query time are not improved. There exists some other work which tends to estimate optimal parameters with sample datasets [8], use improved hash functions [11], [13], [17], [18] or divide the dataset into clusters before building LSH indexes [16]. All these methods are complementary to the Multi-probe LSH and our improved structure and could be combined with our method in order to achieve better performance and quality.

3   Dynamic Multi-probe LSH

3.1   Overview

The main idea of DMLSH is to dynamically vary the granularity of buckets in order to adapt the number of objects they contain to the size of a disk page. We use the same locality-sensitive hash functions and the same probing sequence as Multi-probe LSH. More precisely: 1) Instead of directly building a hash table by using all M functions, we first build a hash table by using only one LSH function. If a bucket contains more than l objects (where l is the number of objects contained in a disk page), we add a second LSH function to this bucket in order to split it into several small buckets. If some small bucket still contains more than l objects, we continue adding LSH functions until each bucket contains less than l objects or the number of functions used becomes to be M. We store the signatures of all these buckets in to a B+ tree. Note that these signatures have different lengths, so the keys in the B+ tree have variable size. 2) We use the sequence probing algorithm of Multi-probe LSH to generate the signatures of the buckets to be probed. If the bucket signature exists in the B+ tree index, we will take the objects in the corresponding bucket as candidates; otherwise, we will check the bucket whose signature is a prefix of the generated signature.

Let us explain these principles through an example. In Fig. 3(a), a basic LSH table has been built using 2 hash functions h_1 and h_2. For a given query q, if we use Multi-probe algorithm to generate 6 probes, the buckets chosen are those of signatures 11,

01, 10, 00, 21 and 20. It means that, we need to access 6 pages on the disk to get 7 candidates. In Fig. 3(b), instead of using directly 2 hash functions, we use one hash function h_1 first, then only if the number of objects in a bucket exceeds a threshold l (in this example $l=2$), we add a second hash function for this bucket. For the same query q, we use the algorithm of the Multi-probe LSH to generate the same probing sequence, i.e. 11, 01, 10, 00, 21 and 20. For a signature that corresponds to no bucket (e.g. 11), we take the bucket whose signature is its prefix (i.e. bucket 1 for signature 11). Thus, the probing sequence becomes 1, 01, 00 and 2. Only 4 disk pages need to be accessed.

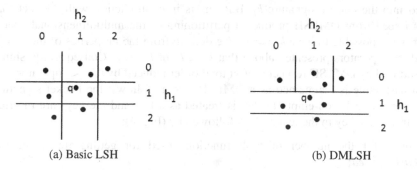

(a) Basic LSH   (b) DMLSH

Fig. 3. Dynamically adding hash functions

3.2   Index Construction

As in the Multi-probe LSH, we use L hash tables, and for each hash table we randomly generate M LSH functions. The difference between our method and Multi-probe LSH is that not every inserted object needs to be hashed with all the M functions, and the hash results are stored in a B+ tree rather than a normal hash table. However, the search that we do in the B+ tree is not an exact match. Instead, we may return a prefix of the searched key, for example, when searching for 001010, we may return 001 as a result. Thus, we have slightly modified the search algorithm of the traditional B+ tree. Note that signatures (keys) stored in the B+ tree are sequences of "digits", where a digit is an integer value returned by an LSH function. For simplicity, our examples only consider binary digits.

DMLSH Tree. We call DMLSH tree the specific (in memory) B+ tree that stores the signatures of the buckets produced by the DMLSH method, together with some extra information about these buckets. Note that only *non-empty buckets* are stored in a DMLSH tree.

The only modification that we have made to the B+ tree algorithm described in [6] is that we overloaded the comparison operators of the keys. The new definitions of the operators "==", "<" and ">" are as follows:

- Given two keys k_1 and k_2, we say that $k_1 == k_2$ if one of k_1 or k_2 is a prefix of the other one (including the case where $k_1 = k_2$ as sequences).

- If $k_1 == k_2$ is not true, it is easy to show that k_1 and k_2 have the form $k_1 = k + d_1 + s_1$ and $k_2 = k + d_2 + s_2$, where k, s_1, s_2 are digit sequences (possibly empty) and d_1, d_2 are single digits, with $d_1 \mathrel{!=} d_2$. We say that $k_1 < k_2$ (respectively $k_1 > k_2$) if $d_1 < d_2$ (respectively $d_1 > d_2$).

For example, we have $001 == 00101$, $0101 < 011$ and $10110 > 100010$, etc.

The following property holds for DMLSH bucket signatures: *For any two distinct signatures k_1 and k_2 produced by DMLSH, we have either $k_1 < k_2$ or $k_1 > k_2$.*

Proof. Let us suppose that for two distinct signatures produced by DMLSH we have $k_1 == k_2$ with e.g. k_1 being a prefix of k_2. In this case, the bucket of signature k_2 is included into the one of signature k_1. But this is in contradiction with the fact that buckets issued from DMLSH produce a partitioning of the multidimensional space. Since it is not possible to have $k_1 == k_2$, we deduce from the properties of the "$==$", "$<$" and "$>$" operators presented above that $k_1 < k_2$ or $k_1 > k_2$. Consequently, signatures produced by DMLSH respect a strict total order induced by the "$<$" operator.

Each hash table is maintained as a DMLSH tree as follows: the bucket signature (hash value) of each non-empty bucket is treated as a key and the keys are inserted into the tree. Each key in the leaf node is followed by (Fig. 4):

1. a counter for the number of hash functions used for getting this signature, *nb_hashes_used*;
2. a counter for the number of vectors in the corresponding hash bucket, *nb_vectors*;
3. the address of the page containing these vectors, @;
4. a set *HV* containing the hash values of these vectors computed by using all the M functions.

| Bucket number | nb_hashes_used | nb_vectors | @ | HV |

Fig. 4. Data structure of each item in the leaf nodes

Construction of the DMLSH Index. Algorithm 1 shows the process of the index construction. For each of the L hash tables, we generate M random LSH functions and an initial empty DMLSH tree. Then, for each vector v in the database S, we insert v in each of the DMLSH trees with *InsertVector* (Algorithm 2).

Algorithm 1. Construction of the index structure
---
for i = 1 to L **do**
    for j = 1 to M **do**
        Generate a random LSH function $h_{i,j}$
    end for
    Create an empty DMLSH tree Tree$_i$
end for
for each v $\in$ S **do**
    InsertVector(v) (**Algorithm 2**)
end for

As shown below, vector insertion respects the DMLSH strategy to generate new buckets that use as few hash functions as possible. Only when the size of a bucket exceeds a threshold, we add a new hash function for this bucket and distribute the objects into smaller buckets. When the number of hash functions used becomes to be M, we stop adding new hash functions and we will store the newly inserted objects in the overflow pages of the full bucket.

Object Insertion. As shown in Algorithm 2, when inserting a new vector v, we insert it into each of the L hash tables. For each hash table, we compute the hash value $g(v)$ of vector v, using the M functions, and search the key $g(v)$ in the corresponding B+ tree. If a prefix of $g(v)$ exists in the tree, the function $Tree_i.find(g(v))$ will return the item in the leaf node which corresponds to the prefix key. Note that a prefix may not exist in the tree, because empty buckets are not stored in the B+ tree. If there is no prefix of the searched key, item is NULL and we will add into the tree a new item with the shortest prefix of $g(v)$ which is not a prefix of any other existing key. This is done by function $Tree_i.insert(g'(v), item)$. The length of this prefix is computed by function Nb_hash (Algorithm 3). Finally, we insert the vector into the bucket page linked to the found (or inserted) prefix key and update the other fields of the leaf item with $AddVectorToItem$ (see Algorithm 4).

Algorithm 2. *InsertVector(v)*: insert a vector v

for i = 1 to L **do**
    $g(v) = (h_{i,1}(v), h_{i,2}(v), ..., h_{i,M}(v))$
    item = $Tree_i.find(g(v))$
    if item == NULL **then**
        item = *New_leaf_item()*
        nb_hashes_used = $Nb\_hash(v, i)$ (**Algorithm 3**)
        $g'(v) = (h_{i,1}(v), h_{i,2}(v), ..., h_{i,nb\_hashes\_used}(v))$
        $Tree_i.insert(g'(v), item)$
    AddVectorToItem(v, item, i) (**Algorithm 4**)
end for

Algorithm 3. *Nb_hash(v, i)*: determine nb_hashes_used for vector v in $Tree_i$

$g(v) = (h_{i,1}(v), h_{i,2}(v), ..., h_{i,M}(v))$
pred = $Tree_i.find\_pred(g(v))$
succ = $Tree_i.find\_succ(g(v))$
lcc_pred = length(longest_common_prefix(pred, g(v)))
lcc_succ = length(longest_common_prefix(succ, g(v)))
lcc = max(lcc_pred, lcc_succ)
return lcc+1;

Algorithm 3 shows how to determine the number of hash functions to use for a newly inserted vector whose complete hash value does not have a prefix key in the tree. We first compute the key $g(v)$ using all M hash functions, then we search in the tree for *pred* which is the greatest key smaller than $g(v)$ and *succ* which is the smallest key larger than $g(v)$. The next step is to compute *lcc* which is the maximum length of the longest common prefix between $g(v)$ and *pred/succ*. Note that if *pred* or *succ* do not exist, the corresponding length of the common prefix is 0. At the end, we return *lcc*+1 as the number of hash functions to be used.

Algorithm 4 shows how to add a vector v into an item. Insertion is possible only if the counter *nb_vectors* is below the threshold l ($l = B/sizeof(v)$, where B is the size of a disk page), or if the maximum number M of hash functions is reached. Insertion adds v into the page @ (or into an overflow page) and the full signature $g(v)$ to *HV*. If insertion is not possible, the bucket is "split" as follows: its item is removed from the B+ tree and all its vectors are reinserted in buckets using one more hash function (Algorithm 5). If the bucket containing the reinserted vector already exists in the tree, the vector is directly inserted; otherwise, a new bucket is created.

Algorithm 4. *AddVectorToItem*(v, item, i): add a vector v into a leaf entry item of Tree$_i$

---

if item.nb_vectors < l **or** item.nb_hashes_used==M **then**
    item.nb_vectors++
    $g(v) = (h_{i,1}(v), h_{i,2}(v), ..., h_{i,M}(v))$
    Add g(v) into item.HV
    Add v into the page at address item.@ or into an overflow page
else
    Tree$_i$.remove(item)
    for each v$_j$ ∈ item **do**
        ReinsertVector(v$_j$, item.nb_hashes_used+1, i)
    end for

Algorithm 5. *ReinsertVector*(v, k, i): reinsert vector v into Tree$_i$ with k hash functions

---

$g(v) = (h_{i,1}(v), h_{i,2}(v), ..., h_{i,k}(v))$
item = Tree$_i$.find(g(v))
if item == NULL **then**
    item = *New_leaf_item*()
    item.nb_hashes_used = k
    Tree$_i$.insert(g(v), item)
end if
AddVectorToItem(v, item, i)   (**Algorithm 4**)

Example. Since the insertion algorithm is the same for all the hash tables, we only consider one hash table as an example. For simplicity, we assume the threshold $l = 2$ and the maximum number of hash functions $M = 2$. Initially, we have four objects p_1, p_2, p_3 and p_4, with $h_1(p_1) = 0$, $h_1(p_2) = 1$, $h_1(p_3) = 1$ and $h_1(p_4) = 0$. Their complete hash values with function $g = (h_1, h_2)$ are: $g(p_1) = 00$, $g(p_4) = 01$, $g(p_2) = g(p_3) = 11$. These values are stored in the set *HV* following the prefix keys. The index is shown in Fig. 5(a). The format of the element in the leaf nodes is defined by Fig. 4. When we insert object p_5 with $h_1(p_5) = 1$, $g(p_5) = 10$, the counter *nb_vectors* for the bucket 1 becomes 3 which is larger than *l*, so we need to split this hash bucket (i.e. add one hash function h_2). We reinsert all the objects of bucket 1. Fig. 5(b) shows the result: new entries with keys 10 and 11 are inserted in the B+ tree and the old entry with key 1 is deleted; for key 0, the bucket is not split.

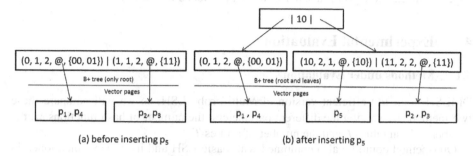

| 10 |

(0, 1, 2, @, {00, 01}) | (1, 1, 2, @, {11})   (0, 1, 2, @, {00, 01})   (10, 2, 1, @, {10}) | (11, 2, 2, @, {11})

B+ tree (only root)   B+ tree (root and leaves)

Vector pages   Vector pages

p_1, p_4 p_2, p_3 p_1, p_4 p_5 p_2, p_3

(a) before inserting p_5 (b) after inserting p_5

Fig. 5. Example of a DMLSH index

3.3   Approximate K Nearest Neighbor Search

Algorithm. For a given query *q*, we repeat the following process for each of the hash tables. We build an empty B+ tree *Probed_Tree* in memory, used to memorize the signatures of the accessed buckets. This tree has the same properties as a DMLSH tree (Section 3.2.1), but its items only contain a bucket signature. We compute the hash value of *q* using all *M* hash functions, i.e. $g(q) = (h_1(q), ..., h_M(q))$ and generate the probing sequence for $g(q)$. The algorithm of generating the probing sequence can be found in [15]. We note *T* the number of probes.

For each probe, we search the probed key in *Probed_Tree*. If its prefix is found, it means that the bucket has already been checked, so we skip this probe. Otherwise, we search the probed key in the DMLSH tree. If a prefix key p_k is found and $g(q)$ exists in the set *HV*, this means that the bucket of signature $g(q)$ is not empty and is included into the "physical" bucket of signature p_k. Therefore, we add the vectors in the page(s) linked to p_k into the candidate set and insert p_k into *Probed_Tree*.

After retrieving all the candidates, we compute the distances from them to the query *q*, rank them in increasing order of their distances and return the top-k results.

Example. Let us consider the example in Fig. 5(b). Suppose $g(q) = (h_1(q), h_2(q)) = 10$, $T = 3$ and the generated probing sequence is 10,00,01. For the first probe 10, we find the signature in the tree, we load the linked page, add p_5 as a candidate and insert 10 into *Probed_Tree*. For the second probe 00, we find its prefix 0. Since 00 is in the *HV* set ($00 \in \{00, 01\}$), we load the linked page, add p_1 and p_4 as candidates and insert 0 into *Probed_Tree*. For the last probe 01, we find its prefix 0 in the tree *Probed_Tree*, meaning that the bucket has already been probed, so we don't need an extra I/O for this probe. In this example, instead of loading 3 pages for the 3 probed buckets, we only loaded 2 pages.

Properties. The DMLSH method proposed in this paper has two important properties compared to the original Multi-Probe LSH. *P1: Under the same parameter setting, the number of I/Os made by DMLSH for a given query q is no more than that made by Multi-probe LSH. P2: Under the same parameter setting, the accuracy of the K-NN search made by DMLSH for a given query q is not lower than that of Multi-probe LSH.* They could be easily proved theoretically, since in DMLSH, 1) several non-full probed buckets may share the same prefix key and they are stored in a single disk page; 2) the candidate set is a superset of that produced by MLSH.

4   Experimental Evaluation

4.1   Methods under Evaluation

DMLSH is an I/O efficient version of Multi-probe LSH, so we will compare these two methods by varying the different parameters: the number of hash functions M, the number of hash tables L and the number of probes T.

Our method could be also combined with basic LSH and its variants mentioned in related work, by organizing each hash table as a DMLSH tree. However, the impact in this case is less important than with Multi-probe LSH, because a single bucket is accessed in each table; also these methods have less practical utility because of the high number of tables. Consequently, we limit our study to the more effective Multi-probe LSH method.

4.2   Dataset

We choose two datasets for our experimental evaluation, widely used in the related work. They are: **Color Data**. The Color dataset contains 68040 vectors of 32 dimensions, which are the color histograms of images in the Corel collection[1]. The dimension values are real numbers with at most 6 decimal digits ranging from 0 to 1. We randomly choose 100 vectors as query examples. **Audio Data**. The audio dataset contains 54387 vectors of 192 dimensions. It is extracted from the LDC SWITCHBOARD-1 collection[2]. The values are real numbers between -1 and 1. *We increase the size of both datasets to be 1 million by inserting noise vectors for the following experiments.* We randomly choose 100 vectors as query examples.

4.3   Evaluation Metrics

We adopt two metrics to measure our method: query efficiency and query accuracy. Since the space consumption of our method is about the same with Multi-probe LSH, we do not consider this metric.

Query Efficiency. Since the vectors are stored in the secondary storage, we evaluate the query efficiency in terms of I/O cost. In the experiments, we set the page size as

---

[1] http://kdd.ics.uci.edu/databases/CorelFeatures/
[2] http://www.cs.princeton.edu/cass/audio.tar.gz

the size of 100 vectors. Note that DMLSH introduces a CPU overhead for distance computation, since the number of candidates it produces is larger than for MLSH. However, the measures in this case indicate only a small difference (5%), not significant compared with the I/O saving.

Query Accuracy. We measure the average recall ratio of the 100 K-NN queries for $K=20$. Given a query object q, let $E(q)$ be the set of exact K-NN objects, and $F(q)$ the set of found K-NN objects. Then the recall ratio is defined as follows:

$$\text{Recall} = \frac{|E(q) \cap F(q)|}{|E(q)|} \tag{1}$$

4.4   Experimental Results

In this section, we compare DMLSH and MLSH by varying the number of hash functions M, the number of hash tables L, respectively the number of probes T.

Impact of the Number of Hash Functions M. We measured the I/O cost and the recall ratio of the first two methods by varying the maximum number of hash functions (M) used for each hash table. For both datasets, the number of hash tables L is set to 3 and the number of probes T is set to 100. The results are shown in Fig. 6 and Fig. 7.

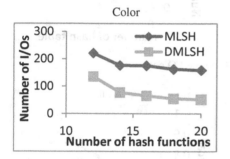
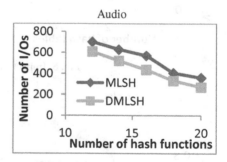

Fig. 6. Impact of M on the I/O cost

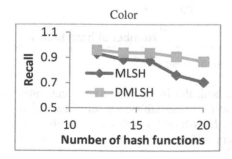
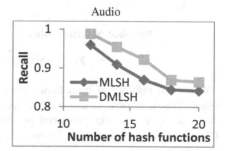

Fig. 7. Impact of M on the recall ratio

For the Color dataset, we set $W = 0.6$. DMLSH reduces the I/O cost by 39% - 67% and increases the recall ratio by 3% - 23%. For the Audio dataset, we set $W = 3.5$. DMLSH reduces the I/O cost by 13% - 25% and increases the recall ratio by 3% - 6%.

We can see that the overall trend is that, when the number of hash functions grows, both the I/O cost and the recall ratio decrease. This is because when we add a new hash function, 1) the average size of each bucket is decreased and 2) more empty buckets are probed.

Impact of the Number of Hash Tables L. Fig. 8 and Fig. 9 show the impact of the number of hash tables L on the I/O cost and on the recall ratio. The number of probes T is set to 100.

For the Color dataset, we set $M = 14$ and $W = 0.6$. DMLSH reduces the I/O cost by 53% - 62% and increases the recall ratio by 3% - 10%. For the Audio dataset, we set $M = 18$ and $W = 3.5$. DMLSH reduces the I/O cost by 16% - 33% and increases the recall ratio by 1% - 9%.

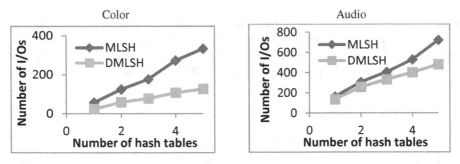

Fig. 8. Impact of L on the I/O cost

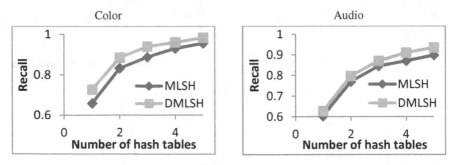

Fig. 9. Impact of L on the recall ratio

When the number of hash tables grows, both the I/O cost and the recall ratio increase. This is normal, because when we use $L+1$ hash tables, the set of candidates is always a superset of that produced by using L hash tables. The choice of the number of hash tables is a trade-off between query efficiency, query accuracy and space consumption.

We observed that, to achieve the same recall ratio, our method DMLSH needs fewer hash tables than MLSH, hence consumes less space.

Impact of the Number of Probes *T*. In Fig. 10 and Fig. 11, we vary the number of probes from 10 to 170. For both datasets, the number of hash tables *L* is set to 3.

For the Color dataset, we set $M = 14$ and $W = 0.6$. DMLSH reduces the I/O cost by 33% - 60%. The bigger the number of probes, the higher the reduction of I/O cost. With the same number of probes, DMLSH increases the recall ratio by 4% - 16%. For the Audio dataset, we set $M = 18$ and $W = 3.5$. DMLSH reduces the I/O cost by 2% - 24% and increases the recall ratio by 3% - 5%. To achieve the same recall ratio, our method DMLSH needs fewer probes than MLSH.

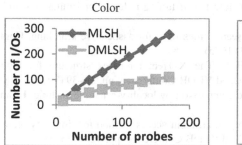
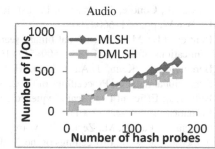

Fig. 10. Impact of T on the I/O cost

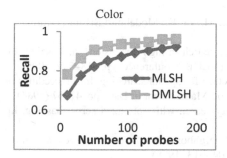
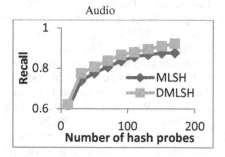

Fig. 11. Impact of T on the recall ratio

5   Conclusion

This paper presents the Dynamic Multi-probe LSH indexing method, which is a more I/O efficient version of the Multi-probe LSH. It dynamically varies the granularity of buckets in order to adapt the number of objects they contain to the size of a disk page. For the construction of the index, it uses initially one hash function and adds a new hash function only when the bucket size exceeds the page size. In the final hash table, the buckets are built by using a different number of hash functions; consequently they have signatures of different length. Bucket signatures are indexed by a slightly modified B+ tree to accelerate the search speed.

For a given query, we first generate the probing sequence and then we access the probed buckets. Since several probed buckets may share the same prefix key and are stored in the same physical page, we need only one single I/O to access these buckets. Thus, the total number of disk accesses is reduced. In addition, since the candidate set is a superset of that produced by Multi-probe LSH, the recall ratio of the approximate K-NN query results is always higher than or equal to that of the Multi-probe LSH.

References

1. Bawa, M., Condie, T., Ganesan, P.: Lsh forest: self-tuning indexes for similarity search. In: WWW, pp. 651–660 (2005)
2. Bentley, J.L.: Multidimensional binary search trees used for associative searching. Communications of the ACM 18(9), 509–517 (1975)
3. Berchtold, S., Keim, D.A., Kriegel, H.P.: The X-Tree: an index structure for high-dimensional data. In: Proceedings of the 22nd VLDB Conference, pp. 28–39 (1996)
4. Buhler, J.: Efficient large scale sequence comparison by locality-sensitive hashing. Bioinformatics 17, 419–428 (2001)
5. Ciaccia, P., Patella, M., Zezula, P.: M-tree an efficient access method for similarity search in metric spaces. In: Proceedings of the 23rd VLDB Conference, pp. 426–435 (1997)
6. Comer, D.: The ubiquitous B-tree. ACM Computing Surveys 11(2), 121–137 (1979)
7. Datar, M., Immorlica, N., Indyk, P., Mirrokni, V.S.: Locality-sensitive hashing scheme based on p-stable distributions. In: Proceedings of the Twentieth Annual Symposium on Computational Geometry, pp. 253–262 (2004)
8. Dong, W., Wang, Z., Josephson, W., Charikar, M., Li, K.: Modeling LSH for performance tuning. In: CIKM 2008, pp. 669–678 (2008)
9. Gionis, A., Indyk, P., Motwani, R.: Similarity search in high dimensions via hashing. In: Proceedings of the 25th Very Large Database (VLDB) Conference, pp. 518–529 (1999)
10. Guttman, A.: R-Trees: A dynamic index structure for spatial searching. In: Proceedings of the ACM SIGMOD International Conference on Management of Data, pp. 47–57 (1984)
11. He, J., Liu, W., Chang, S.: Scalable similarity search with optimized kernel hashing. In: ACM SIGKDD, pp. 1129–1138 (2010)
12. Indyk, P., Motwani, R.: Approximate nearest neighbor: towards removing the curse of dimensionality. In: Proceedings of STOC, pp. 604–613 (1998)
13. Jegou, H., Amsaleg, L., Schmid, C., Gros, P.: Query adaptive locality sensitive hashing. In: ICASSP 2008, pp. 825–828 (2008)
14. Katayama, N., Satoh, S.: The SR-tree: an index structure for high-dimensional nearest neighbor queries. In: SIGMOD Conference, pp. 369–380 (1997)
15. Lv, Q., Josephson, W., Wang, Z., Charikar, M., Li, K.: Multi-probe LSH: efficient indexing for high-dimensional similarity search. In: Proceedings of the 33rd International Conference on Very Large Data Bases (VLDB), Vienna, Austria, pp. 950–961 (2007)
16. Pan, J., Manocha, D.: Bi-level locality sensitive hashing for k-Nearest Neighbor computation. In: ICDE, pp. 378–389 (2012)
17. Raginsky, M., Lazebnik, S.: Locality-sensitive binary codes from shift-invariant kernels. In: Advances in Neural Information Processing Systems, pp. 1509–1517 (2009)
18. Satuluri, V., Parthasarathy, S.: Bayesian locality sensitive hashing for fast similarity search. PVLDB 5(5), 430–441 (2012)
19. Weber, R., Schek, H., Blott, S.: A quantitative analysis and performance study for similarity-search methods in high-dimensional spaces. In: VLDB, pp. 194–205 (1998)

Revisiting the Term Frequency in Concept-Based IR Models

Karam Abdulahhad[1], Jean-Pierre Chevallet[2], and Catherine Berrut[3]

[1] Université de Grenoble
[2] UPMF-Grenoble 2
[3] UJF-Grenoble 1
LIG Laboratory, MRIM Group, Grenoble, France
{karam.abdulahhad,jean-pierre.chevallet,catherine.berrut}@imag.fr

Abstract. Indexing documents and queries using concepts, instead of word-based indexing, is an alternative approach, and it supposes to give a more meaningful indexing. However, this way of indexing needs to revisit some hypotheses of classical Information Retrieval. Therefore, we propose a new concept weighting approach, namely *Relative Weight*, which weights concepts with respect to their corresponding text in the documents or queries. In other words, it assigns to each concept a *relative weight* with respect to the other concepts in the same context. We explore interesting experimental results of our new weighting approach, compared to the classical approaches, through studying the retrieval performance of some classical IR models.

1 Introduction

Classical Information Retrieval (IR) systems are now well developed and experimented. They all use the Luhn conjecture on word importance [11], and establish that each word is weighted according to the importance it has in the corpus, and inside the document. The goal of this paper is to study the side-effects of moving from the word-space to the concept-space on classical IR models. In other words, studying the side-effects of using concepts, instead of words as indexing terms, on classical IR models.

When mapping text to concepts from knowledge bases, the way that the weight is computed must be adapted to concepts for different reasons. The most important one is based on the fact that a phrase is generally mapped to much more concepts than words, due to the specificity of indexing that is produced by mapping text to concepts process using a specific knowledge base. For example, the word '*x-ray*' is linked to six different UMLS' concepts, and Table 1 shows the different concepts linked to '*lobar pneumonia x-ray*' phrase using the Metamap[1] tool on UMLS[2].

---

[1] http://metamap.nlm.nih.gov/

[2] Unified Medical Language System (www.nlm.nih.gov/research/umls/)

H. Decker et al. (Eds.): DEXA 2013, Part I, LNCS 8055, pp. 63–77, 2013.

We claim in this paper, that classical IR models can be used, when using concepts, if at least the concept weighting is revisited, but still based on Luhn conjecture [11].

In this study, a word is the smallest linguistic element that has a semantic and can stand by itself, e.g. *'information'*. A phrase is a sequence of one or more words, e.g. *'information retrieval'*. A concept is an entry ID in a knowledge base and it is associated to a set of phrases or words that describe it. For example, the concept *'C0004238'* in UMLS is associated to two synonymous phrases *'Atrial Fibrillation'* and *'Auricular Fibrillation'*. Sometimes, the word "concept" is used for referring to words and phrases [4].

Table 1. Variants of *'lobar pneumonia x-ray'* generated by MetaMap, their related candidate UMLS' concepts, and the corresponding part of the original phrase

Variants	Candidate Concepts	Original Part
'lobar pneumonia x-ray'	–	*'lobar pneumonia x-ray'*
'lobar pneumonia'	*C0032300, C0155862*	*'lobar pneumonia'*
'lung x-ray'	*C0581647*	*'pneumonia x-ray'*
'lung	*C0024109, C 1278908*	*'pneumonia'*
'pneumonia'	*C0032285*	*'pneumonia'*
'pulmonary'	*C2707265, C2709248*	*'pneumonia'*
'lobar'	*C1522010*	*'lobar'*
'lobe'	*C1428707*	*'lobar'*
'lobus'	*C0796494*	*'lobar'*
'x-ray'	*C0034571, C0043299, C0043309* *C1306645, C1714805, C1962945*	*'x-ray'*

This paper is organized as follow: In section 2, we present the motivations behind using concepts instead of words. The general process of conceptual indexing and some examples of concept-based IR models are depicted in Section 3. Section 3 presents the motivations, hypotheses, and definitions of our proposed weighting approach. We present, in Section 5, some experimental results, and the paper is concluded in Section 6.

2   Why Concepts

Words have been used for a long time in IR, and this type of indexing terms proved its effectiveness in most of IR applications, especially web search engine. Using concepts is motivated by the following reasons:

- the availability of rich and large knowledge bases e.g. UMLS and WordNet[3],
- the multilingualism and multi-modality of content [5] [13]. It is possible to abandon the translation step in a multilingual context, because concepts suppose to be language-independent, e.g. the English word *'lung'* and the French

---

[3] WordNet is a lexical database of English (http://wordnet.princeton.edu/)

word *'poumon'* correspond to the same concept *'C0024109'* in a knowledge base like UMLS,

- some new semantic-based IR applications e.g. Semantic Web [13] and song indexing and retrieval [6],
- some well-known IR problems like the term-mismatch problem [7], where two different words could express the same meaning, e.g. *'atrial'* vs. *'auricular'*. In the ideal case, each concept should be associated to all phrases that have the same meaning in a specific context,

all reasons above lead to the emergence of a new IR field that uses concepts as indexing terms instead of, or besides, words.

3   Conceptual Indexing and Concept-Based IR Models

Conceptual indexing first integrates the process of mapping text to concepts. The main principle of conceptual mapping tools is to extract *phrases* from the text of documents and queries, and then try to map them to one or more *candidate concepts* from a knowledge base. More precisely, the general process for conceptual mapping consists of the following steps [5]:

1. *Morphology and syntax*: extracting noun phrases from text.
2. *Variation*: constructing a list of variants for each noun phrase. Variants could be derivational variants, synonyms, acronyms, etc.
3. *Identification*: for each variant, all concepts that could correspond to it are retrieved from the knowledge base. The retrieved concepts called *candidate concepts*.
4. *Evaluation*: for each candidate concept, a measure is used for evaluating the precision of mapping process, and then the set of candidate concepts is ordered according to this measure. In other words, the measure computes the degree of correctness of mapping a noun-phrase to a concept.
5. *Disambiguation*: choosing the most appropriate concepts, among the candidate concepts, that well correspond to the related noun-phrase. This operation normally depends on the context.
6. *Weighting*: like in word-based indexing, each concept has a weight reflecting its indexing usefulness.

Actually, there are many examples of mapping tools. MetaMap [2], for example, maps medical text to UMLS' concepts. Fast Tagging [8] is a method of tagging medical terms in legal, medical, and news text, and then mapping the tagged terms to UMLS' concepts. Baziz [3] and Maisonnasse [12] built their own mapping tools. In general, concepts are a part of a knowledge base, it is thus mandatory to link mapping tools to some knowledge bases, e.g. UMLS, WordNet, DBpedia[4], etc.

When the mapping is done, the indexing process must continue by first selecting and sometimes weighting the concepts, and then representing documents

---

[4] dbpedia.org

and queries. The system is then able to process the matching of a query and a document.

Baziz [3] proposes a context-based disambiguation approach. This approach maps each phrase to only one concept among the candidate concepts. According to Baziz, concepts are WordNet's *synsets*, and for a phrase he chooses the concept, among the candidate concepts, that most fit the candidate concepts of the other phrases in the same context. He uses a WordNet's relation-based measure to estimate the semantic similarity between concepts or synsets. He also proposes to represent documents and queries as trees of concepts, and then using a sort of fuzzy implication for matching.

Maisonnasse [12] proposes two ways to build a concept-based IR system. The first one is to represent a document as a graph and a query as a graph, where nodes are UMLS' concepts and edges are the semantic relations between the semantic types of UMLS. The matching between theses two graphs is a kind of Conceptual Graph projection. The second one is to represent a document as a set of graphs, one for each sentence, and a query as one graph, and then the matching is a graph-based language model.

Diem [10] and Abdulahhad et al. [1] map documents and queries to UMLS' concepts, and then they position documents and queries on a Bayesian Network. They exploit the relations between concepts to link documents' concepts and queries' concepts. The matching is the actual inference mechanism of Bayesian Networks.

Though interesting, these systems are complex, due to the underlying representation used for documents, queries, and matching. We propose here a simple alternative to concept-based IR, by using classical information retrieval systems for conceptual indexing, and re-thinking the way the weighting of the concepts. The advantage of this proposal is to allow conceptual indexing in well-known IR systems.

4   A New Weighting for Conceptual IR: Concepts Relative Weight

4.1   Introduction

In classical IR models, documents and queries are bags of terms. In addition, they compute the Relevance Status Value $RSV(d, q)$ between a document d and a query q, which depends, one way or another, on the coordination level between d and q. In all classical IR models, the weight of a term t in a document d is a consequence of Luhn conjecture and follows the two following rules:

- **Rule 1**: the weight of t is proportional to the frequency of t in d.
- **Rule 2**: the weight of t is inversely proportional to the frequency of t in the corpus.

Therefore, the weight w_t^d of an indexing term t in a document d consists of two main parts *local* and *global*:

$$w_t^d = f(w_t^l, w_t^g)$$

where, f determines the way of combining w_t^l and w_t^g and it is model-dependent. The local weight w_t^l is the weight of t within a document d. It depends on the frequency of t in d $(tf_{t,d})$ and the length of d $(|d|)$.

When moving from words to concepts, we need a new weighting mechanism, because the general assumption in the word-based IR models is that if a document d contains two words w_1 and w_2 then d should be represented by the meaning of w_1 *and* the meaning of w_2. However, in our point of view, this is not the case in the concept-based IR models, because a word w in a document d is mapped to a set of concepts $\{c_1, \ldots, c_n\}$ where each concept represents a possible meaning of w, then d should be represented by one of these meanings or concepts that best corresponds to the context of d. In other words, d should be represented by c_1 *or* c_2 *or* ... *or* c_n. Therefore, we need to revisit the classical concept weighting mechanism.

4.2 General Description of Our Proposal

For each phrase of document, we generate its variants and map the variants to sets of concepts (step 1 Figure 1). Depending on the output of step 1, we generate a phrase-hierarchy which gives an overview of the concepts of the phrase (step 2 Figure 1). From the hierarchy of the concepts of a phrase, we compute the relative weight of each concept (step 3 Figure 1). Globally, each phrase of each document is processed as described above. The indexed document is computed as the set of all concepts of its initial phrases. In the following sub-sections, we present the definitions of the previous steps, and how to rebuild a classical concept-based IR model depending on concepts relative weight.

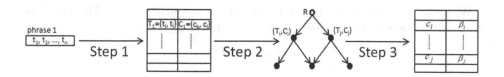

Fig. 1. The general process of the phrase processing

4.3 Concept Weighting Hypotheses

It is possible to keep classical IR models using concepts, and suppose that a document is a bag of concepts. However, when transforming text to concepts, we know that the number of proposed concepts can be high (see the example in Table 1), and still the set of concepts must represent the text of the document.

Our weighting method, namely *Relative Weight*, respects the three following hypotheses:

- **Hypothesis 1:** Concepts that correspond to a longer text should receive larger weight.

- **Hypothesis 2:** The weight of a concept should be inversely proportional to the size of the set of concepts of the text that it belongs to. The bigger the set of concepts is for a text, the less important weight its concepts receive.
- **Hypothesis 3:** As in classical IR, the length of documents correspond to their text length and we propose a model which re-distribute the length of the document on its concepts.

These three hypotheses validate rules 1 and 2.

In the rest of this paper and to illustrate the main idea, we will use the same example of mapping the '*lobar pneumonia x-ray*' to UMLS' concepts using MetaMap. Table 1 shows the variants of the phrase '*lobar pneumonia x-ray*' that are generated by MetaMap, their related candidate UMLS' concepts, and the corresponding part of the original phrase. However, in this study, we regroup all variants that correspond to the same part of the original phrase in only one variant, or equivalently, we regroup rows 4-5-6, and rows 7-8-9.

4.4   Step 1

We start by describing our solution at the level of phrases, and then we will generalize it to the level of documents, where a document d is a sequence of phrases.

We define the set of words W, the set of phrases P, and the set of concepts C. Each phrase $p \in P$ is a sequence of words or equivalently a set of terms. We define the set of terms $T = W \times \mathbb{N}^*$, which is a set of tuples, and each tuple $(w, i) \in T$ links a word $w \in W$ with a number $i \in \mathbb{N}^*$. We also define the set of nodes $N = 2^T \times 2^C$, which links a set of terms with a set of concepts, where 2^T is the power set of T and 2^C is the power set of C.

We define two functions to link terms, phrases, and concepts. The function *trm* returns the set of terms that appear in a phrase, where

$$trm : P \to 2^T$$

For example, assume p is the '*lobar pneumonia x-ray*' phrase then:

$$trm(p) = \{(\text{lobar}, 1), (\text{pneumonia}, 2), (\text{x-ray}, 3)\}$$

where 1, 2, and 3 are the positions of the words in the phrase p, starting by 1. We define $|p|_T = |trm(p)|$ as the length of a phrase $p \in P$ in the word-space. For example, a phrase p like '*lobar pneumonia x-ray*' has a length $|p|_T = 3$.

The function *map* maps a phrase $p \in P$ to its variants and their concepts. This function fits to be a representation of any mapping tool, where

$$map : P \to 2^N$$

$map(p)$ is the set of all variants of p with their candidate concepts. Therefore,

- $\forall p \in P, \forall (T_i, C_i) \in map(p), T_i \subseteq trm(p)$, or in other words, each variant is a sub-phrase of the original phrase.

- $\forall p \in P, \forall (T_i, C_i) \in map(p)$, C_i are the set of concepts that the variant T_i is mapped to.
- $\forall p \in P, \forall (T_i, C_i) \in map(p), C_i \neq \phi$, or in other words, we only consider the variants that have concepts.
- $\bigcap_{(T_i,C_i)\in map(p)} C_i = \phi$, a same concept does not appear more than one time in a phrase.

Table 2. The output of applying the function map to the phrase '*lobar pneumonia x-ray*', where map stand for MetaMap

Terms	Candidate concepts
$T_1 = \{(\text{lobar}, 1), (\text{pneumonia}, 2)\}$	$C_1=\{C0032300,\ C0155862\}$
$T_2 = \{(\text{pneumonia}, 2), (\text{x-ray}, 3)\}$	$C_2=\{C0581647\}$
$T_3 = \{(\text{lobar}, 1)\}$	$C_3=\{C1522010,\ C1428707,\ C0796494\}$
$T_4 = \{(\text{pneumonia}, 2)\}$	$C_4=\{C0024109,\ C1278908,\ C0032285$ $C2707265,\ C2709248\}$
$T_5 = \{(\text{x-ray}, 3)\}$	$C_5=\{C0034571,\ C0043299,\ C0043309$ $C1306645,\ C1714805,\ C1962945\}$

For example, assume p is the '*lobar pneumonia x-ray*' phrase then (Table 2):

$$map(p) = \{(T_1, C_1), (T_2, C_2), (T_3, C_3), (T_4, C_4), (T_5, C_5)\}$$

4.5 Steps 2 and 3: The Relative Weight Function

We define the *Relative Weight* function rw that weights each concept of a phrase.

$$rw : P \rightarrow 2^{C \times \mathbb{R}^{+*}}$$

where $\forall p \in P, rw(p) = \{(c_1, \beta_1), \dots, (c_r, \beta_r)\}$, c_i is a concept, β_i is the relative weight of c_i, and rw must respect the following points:

- $\bigcup_{(c_i,\beta_i)\in rw(p)} \{c_i\} = \bigcup_{(T_i,C_i)\in map(p)} C_i$, every concept of p must appear in $rw(p)$.
- $\forall (c_i, \beta_i) \in rw(p)$ and suppose that $(T_j, C_j) \in map(p)$ is the node that contains the concept $c_i \in C_j$, then:
  - the relative weight β_i of the concept c_i must be proportional to $|T_j|$ (Hypothesis 1).
  - the relative weight β_i of the concept c_i must be inversely proportional to $|C_j|$ (Hypothesis 2).
- $\sum_{(c_i,\beta_i)\in rw(p)} \beta_i = |p|_T$, we maintain the length in both word-space and concept-space. Maintaining the length of phrases in both the word-space and concept-space implicitly leads to maintaining the length of document (Hypothesis 3).

In order to calculate rw, we need a global overview of the phrase, which is a hierarchy, on which we can process the Relative Weight algorithm for rw. The principle of rw is to build a hierarchy of the concepts of the phrase, and then the length of the phrase is distributed on the concepts respecting the three hypotheses, and the position of concepts within the hierarchy.

4.6 Step 2: A Hierarchy on Concepts Extracted from Text

We define a partial order relation $<$ on the set N, as follow:

$$\forall (T_i, C_i), (T_j, C_j) \in N, (T_j, C_j) < (T_i, C_i) \quad \textit{iff} \quad T_j \subset T_i$$

Using the $<$ partial order relation, two functions ch and pr could be defined. However, we first define an abstract root node $R = (T_R, C_R)$, where $|T_R| = 0$, $|C_R| = 0$, and by definition $\forall (T_i, C_i) \in N, (T_i, C_i) < R$. The function ch returns the *direct children* of any node $n \in N \cup \{R\}$.

$$ch : N \cup \{R\} \to 2^N$$

where, $\forall (T_i, C_i), (T_j, C_j) \in N \cup \{R\}$ then $(T_j, C_j) \in ch((T_i, C_i))$ *iff*

- $(T_j, C_j) < (T_i, C_i)$ and
- $\nexists (T_k, C_k) \in N \cup \{R\}$ satisfying that $(T_j, C_j) < (T_k, C_k) < (T_i, C_i)$

Reversely, we define the function pr that returns the *direct parents* of any node $n \in N \cup \{R\}$.

$$pr : N \cup \{R\} \to 2^{N \cup \{R\}}$$

where, $\forall (T_i, C_i), (T_j, C_j) \in N \cup \{R\}$ then $(T_j, C_j) \in pr((T_i, C_i))$ *iff*

- $(T_i, C_i) < (T_j, C_j)$ and
- $\nexists (T_k, C_k) \in N \cup \{R\}$ satisfying that $(T_i, C_i) < (T_k, C_k) < (T_j, C_j)$

It is possible to define a hierarchy on a phrase. The hierarchy of a phrase $p \in P$ is defined by applying the two functions ch and pr to each node in $map(p) \cup \{R\}$. For example, assume p is the '*lobar pneumonia x-ray*' phrase, Figure 2 shows the hierarchy that is defined on the set $map(p) \cup \{R\}$ (Table 2).

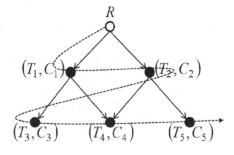

Fig. 2. The hierarchy of the phrase '*lobar pneumonia xray*', and the breadth-first scanning order

4.7   Step 3: The Relative Weight (rw) Algorithm

Assume a phrase $p \in P$ and the node $n = (T_n, C_n) \in map(p) \cup \{R\}$. The node n has $|pr(n)|$ parents and $|ch(n)|$ children. Each node n must distribute a certain amount α_n on its children. If n is the abstract root R then $\alpha_n = |p|_T$ by default.

We should remember that the relative weight of a concept c in a node n must be proportional to $|T_n|$ and inversely proportional to $|C_n|$ (Hypotheses 1 and 2). Therefore, for distributing the amount α_n on the concepts of n and its children, we first compute the portion δ_n of α_n for one single term, where this step takes the number of terms in n and in its direct children $ch(n)$ into account. Each child $n' \in ch(n)$ will receive an amount equals the portion of one term δ_n multiplied by the number of terms in n' ($\delta_n \times |T_{n'}|$). The portion of α_n for the current node n is the portion of one single term δ_n multiplied by the number of terms in n ($\delta_n \times |T_n|$). In a particular node, concepts are equally important, therefore, the relative weight of a concept equals the portion of α_n of the current node divided by the number of its concepts $|C_n|$ or equivalently $\frac{\delta_n \times |T_n|}{|C_n|}$ (see Algorithm 1).

Algorithm 1. RelativeWeight

---

    input : $map(p)$, $|p|_T$
    output: $rw(p) \subseteq C \times \mathbb{R}^{*+}$

1 Initialize $rw(p) = \{\}$;
2 Construct the hierarchy of the set $map(p) \cup \{R\}$;
3 Attach to each node n in $map(p) \cup \{R\}$ a value α_n, where α_n is the total amount that the node n receives from its parents $pr(n)$, and then distributes on its own concepts and its children $ch(n)$;
4 Initializing $\forall n \in map(p), \alpha_n = 0$ and set $\alpha_R = |p|_T$;
5 Scan the hierarchy of $map(p) \cup \{R\}$ in a breadth-first way, starting from the current node $n = R$;
6 Compute the portion δ_n of α_n for one single term: $\delta_n = \frac{\alpha_n}{|T_n| + \sum_{n' \in ch(n)} |T_{n'}|}$;

7 **begin**
8 | **for** *each child $n' \in ch(n)$* **do**
9 | | Compute the amount $\alpha_{n'}$ that must be transferred from n to n': $\alpha_{n'} = \alpha_{n'} + (\delta_n \times |T_{n'}|)$;
10 | **end**
11 | **if** $n \neq R$ **then**
12 | | **for** *each candidate concept $c_i \in C_n$* **do**
13 | | | Compute the relative importance: $\beta_i = \frac{\delta_n \times |T_n|}{|C_n|}$ (*We suppose that in a given node all concepts are equally important*);
14 | | | $rw(p) = rw(p) \cup \{(c_i, \beta_i)\}$;
15 | | **end**
16 | **end**
17 | Change n to the next node according to the breadth-first scan order and go to line 6;
18 **end**

The algorithm starts from the abstract root $n = R$ with $\alpha_n = |p|_T$. The algorithm must achieve a *breadth-first* search on the hierarchy that is defined on $map(p) \cup \{R\}$ (Figure 2).

Finally, the length of a phrase p in the concept-space $|p|_C$ is equal to the sum of the relative weights of all concepts of p:

$$\forall p \in P, |p|_C = \sum_{(c_i, \beta_i) \in rw(p)} \beta_i$$

Knowing that the algorithm always starts at the abstract root and the input amount for the abstract root is the length of the phrase in the word-space. Knowing that at any node, we do not produce any new amount, we just distribute the received amount on concepts and children. Knowing that at any node, we do not lose any amount, or in other words, the whole received amount is distributed on concepts and children, and for leaves the whole amount is distributed on concepts. It is easy to verify that we maintain the length of phrases in both word-space and concept-space. In other words, we know that $\forall p \in P, |p|_C = \sum_{(c_i, \beta_i) \in rw(p)} \beta_i = |trm(p)| = |p|_T$.

Returning to our previous example in Figure 2, the process starts in R with the input amount $\alpha_R = |p|_T = 3$. The scanning order will be $\langle R, n_1, n_2, n_3, n_4, n_5 \rangle$:

- $(n = R)$ and $(\alpha_n = \alpha_R = 3)$: We compute the portion δ_n of α_n for one single term $\delta_n = \frac{\alpha_n}{|T_R| + |T_1| + |T_2|} = \frac{3}{4}$. Now, compute the amount that must be transferred to each child: $\alpha_{n_1} = \delta_n \times |T_{n_1}| = \frac{3}{2}$ and $\alpha_{n_2} = \frac{3}{2}$. The root R does not contain any concept.
- $(n = n_1)$ and $(\alpha_n = \alpha_{n_1} = \frac{3}{2})$: In the same way, $\delta_n = \frac{\alpha_n}{|T_1| + |T_3| + |T_4|} = \frac{3}{8}$, and also, $\alpha_{n_3} = \delta_n \times |T_{n_3}| = \frac{3}{8}$ and $\alpha_{n_4} = \frac{3}{8}$. Each concept $c_i \in C_1$ in n_1 will have the relative weight $\beta_i = \frac{\delta_n \times |T_n|}{|C_n|} = \frac{\frac{3}{8} \times 2}{2} = \frac{3}{8}$. In other words, the relative weight of the concept $C0032300$ is $\frac{3}{8}$.
- By continuing in this way, the final output of our algorithm will be:
  $rw(p) = \{(C0032300, \frac{3}{8}), (C0155862, \frac{3}{8}), (C0581647, \frac{3}{4}),$
  $(C1428707, \frac{1}{4}), (C1522101, \frac{1}{8}), (C0796494, \frac{1}{8}),$
  $(C0024109, \frac{3}{20}), (C1278908, \frac{3}{20}), (C2707265, \frac{3}{20}),$
  $(C2709248, \frac{3}{20}), (C0032285, \frac{3}{20}), (C0034571, \frac{1}{16}),$
  $(C0043299, \frac{1}{16}), (C0043309, \frac{1}{16}), (C1306645, \frac{1}{16}),$
  $(C1714805, \frac{1}{16}), (C1962945, \frac{1}{16})\}$

We can verify that: $\sum_{(c_i, \beta_i) \in rw(p)} \beta_i = 3 = |ph|_T$

4.8 Indexing a Document with Concepts

A document d is a sequence of phrases $P_d = \langle p_1, \ldots, p_{n_d} \rangle$. We represent the indexed document d as a set of concepts with their relative weight, where the relative weight of a concept c in a document d, noted $tf_{c,d}$, is the sum of its relative weights within all phrases of d.

$$d = \{(c, tf_{c,d}) | tf_{c,d} = \sum_{p_i \in P_d, (c, \beta_i) \in rw(p_i)} \beta_i\}$$

Queries are indexed as documents, like in all classical IR models.

In view of our document and query representation, the other components of classical IR models become:

- The term frequency of a concept c in a corpus D is the sum of the relative weights of c in all documents of D, $tf_{c,D} = \sum_{d_i \in D} tf_{c,d_i}$.
- Document length is $|d| = \sum_{(c,tf_{c,d}) \in d} tf_{c,d} = |trm(d)|$.
- Query length is $|q| = \sum_{(c,tf_{c,q}) \in q} tf_{c,q} = |trm(q)|$.
- Corpus length is $|D| = \sum_{d_i \in D} |d_i|$.

5   Experiments

The main goal of our experiments is to show the validity of our new way of weighting (*Relative Weight*). We check this validity through comparing the retrieval performance of some classical IR models using classical concept weighting method, namely *tf.idf*, to their performance using our new way of weighting. In order to achieve this goal we apply four classical IR models to seven corpora of medical contents.

More precisely, we apply four classical IR models belonging to different mathematical frameworks, e.g. probabilistic, vector space, language models. Concerning the weighting approach, we use, besides our proposed approach, the classical approach (*tf.idf*). For moving from words to concepts, there are many tools [3,2,8,12], which essentially map text to concepts of a knowledge base. Each tool proposes some concepts for a certain piece of text, and some of these tools also achieve a supplementary step to filter or disambiguate the proposed concepts for providing more precise list of concepts [3]. In this study, we use MetaMap that provides the basic mapping functionality, and instead of the supplementary filtering or disambiguation step, we keep all concepts proposed by MetaMap, but we give to each concept a relative weight reflecting its importance in the original text.

In order to compare the retrieval performance of IR models, we use the Mean Average Precision (MAP) metric, which is both recall and precision metric, and also the precision at the first ten documents (P@10) metric, which is a pure precision metric. As statistical significance test, we use Fisher's Randomization test at the 0.05 level [17].

5.1   Experimental Setup

We apply four different IR models to seven medical-content corpora form ImageCLEF[5].

Retrieval Models. In order to obtain more valuable and reliable results, we choose models belonging to different mathematical frameworks:

---

[5] www.imageclef.org/2012/medical

- From probabilistic framework: we choose ($bm25$) [14] with the following parameter values $k_1 = 1.2$, $b = 0.75$, and $k_3 = 1000$ [9].
- From language models framework: we choose Dirichlet model (dir) with $\mu = 2000$, and Jelinek-Mercer model (jm) with $\lambda = 0.1$ for short queries or $\lambda = 0.7$ for long queries [18].
- From vector space framework: we choose Pivoted Normalization Method (piv) with $s = 0.2$ [16]. As our counting approach could produce term frequencies less than 1, we thus make a small modification to the piv model through adding 1 to the term frequency component. The retrieval performance of this modified version of piv is equivalent to the original one.

Data. Table 3 shows some statistics about the seven corpora that we use. We have four corpora with *short queries*: *image09*, *image10*, *image11*, and *image12*. We also have three corpora with *long queries*: *case11*, *case12(T+A)*, and *case12*, where *case12(T+A)* is the same corpus as *case12* but we only use the *title* and *abstract* parts of each document.

As we mentioned the indexing terms are concepts. We use MetaMap to map the text of documents and queries to UMLS' concepts. Table 3 clearly shows a difference in the length of documents between the word-space and the concept-space.

Table 3. Corpora statistics. *avdl* and *avql* are the average length of documents and queries. (#d) number of documents and (#q) number of queries

		image09	image10	image11	image12	case11	case12(T+A)	case12
#d		74901	77497	230088	306530	55634	74654	74654
#q		25	16	30	22	10	26	26
avdl	Words	62.16	62.12	44.83	47.16	2594.5	160.51	2731.24
	Concepts	157.48	157.27	101.92	104.26	5752.38	376.14	5971.21
avql	Words	3.36	3.81	4.0	3.55	19.7	24.35	24.35
	Concepts	10.84	12.0	12.73	9.41	57.5	63.73	63.73

5.2   Results and Discussion

We separate our discussion according to the length of queries, in order to study the effects of our concept weighting approach on both long and short queries. In general, we show the validity of our method of weighting, namely *Relative Weight*, through comparing the retrieval performance of IR models using, or without using, the Relative Weight. More precisely, we weight concepts in two ways: the classical way *tf.idf* and the relative way (Algorithm 1).

Short Queries. Table 4 shows the retrieval performance of some IR models using, and without using, our weighting approach. IR models are applied to corpora of short queries. The table shows that using the concepts Relative Weight approach with classical IR models generally improves their performance. In addition, sometimes the improvements are statistically significant.

Table 4. (c) refers to *classical* weight and (r) refers to *relative* weight. The 'Gain' column shows the gain in MAP, and † refers to statistically significant improvement.

		image09			image10			image11			image12		
		MAP	P@10	Gain	MAP	P@10	Gain	MAP	P@10	Gain	MAP	P@10	Gain
piv	c	0.288	0.536	+36%	0.242	0.381	+20%	0.159	0.320	+12%	0.107	0.164	+27%
	r	0.391†	0.616		0.290†	0.394		0.178	0.333		0.136	0.246†	
bm25	c	0.267	0.460	+26%	0.213	0.294	+20%	0.155	0.310	-3%	0.103	0.150	+11%
	r	0.336	0.540		0.256	0.356		0.150	0.290		0.114	0.159	
dir	c	0.268	0.464	+26%	0.246	0.363	+12%	0.123	0.233	+2%	0.086	0.136	+13%
	r	0.338	0.576		0.275	0.438		0.126	0.237		0.097	0.146	
jm	c	0.306	0.528	+31%	0.245	0.375	+14%	0.158	0.280	+15%	0.102	0.160	+29%
	r	0.401†	0.612		0.280	0.450†		0.182†	0.337†		0.132†	0.241†	

Long Queries. Table 5 shows the retrieval performance of some IR models using, and without using, our weighting approach. IR models are applied to corpora of long queries. The table shows that using the concepts Relative Weight approach with classical IR models does not make a big difference, where the performance of most of models is equivalent in both approaches of concept weighting.

Table 5. (c) refers to *classical* weight and (r) refers to *relative* weight. The 'Gain' column shows the gain in MAP, and † refers to statistically significant improvement.

		case11			case12(T+A)			case12		
		MAP	P@10	Gain	MAP	P@10	Gain	MAP	P@10	Gain
piv	c	0.096	0.140	-10%	0.079	0.115	+10%	0.190	0.219	+4%
	r	0.087	0.150		0.087	0.119		0.198	0.204	
bm25	c	0.096	0.140	+11%	0.085	0.119	+6%	0.178	0.189	-6%
	r	0.106	0.190		0.090	0.123		0.167	0.185	
dir	c	0.104	0.150	-3%	0.083	0.112	+8%	0.189	0.181	-1%
	r	0.101	0.170		0.089	0.119		0.188	0.192	
jm	c	0.120	0.160	-2%	0.086	0.115	+5%	0.184	0.181	+6%
	r	0.118	0.180		0.091	0.131		0.194	0.189	

Discussion. In general, the obtained results, concerning short and long queries, is compatible with the main conclusion of applying disambiguation to concept-based indexing, where the gain of disambiguation with short queries is clearer and more considerable [15]. This is logically correct because long queries are less ambiguous than short ones.

We should also clarify that in general, the experimental results of any concept based IR model are highly depended on:

- The quality and the completeness of the knowledge base that contains concepts (UMLS in our case).
- The accuracy of the text-concepts mapping tool (MetaMap in our case).
- The amount of information that is used beside concepts, such as the relations between concepts. In other words, to which limit we profit from the content of the knowledge base.

Accordingly, we should, in future, make in-depth study of MetaMap accuracy, and also try using other tools.

6   Conclusion

In this study, we revisit weight approaches of classical IR models when using concepts. The main motivation of this revisit is that in the word-space when a document d contains two words then d should contain the meaning of both words. Whereas, in the concept-space when a piece of text of d is mapped to a set of concepts, then d should, in our point of view, contain one of these meanings or concepts.

Disambiguation could be one possible solution to this problem, but, in this paper, we propose another solution depending on maintaining the document and query length in both the word-space and concept-space.

The main idea of *Relative Weight*, is to distribute the length of a phrase on its candidate concepts. The Relative Weight approach assigns a relative weight to each concept, where the distribution (or assignation) process must respect three hypotheses.

We integrate our new weighting approach into classical IR models. Then, we compare the performance of these models when using the classical weighing approach (*tf.idf*), with their performance when using the Relative Weight approach. To achieve this goal, we apply four IR models to seven medical-content corpora already annotated by UMLS' concepts using the MetaMap tool.

In general, the experimental results are promising, especially in the case of very short queries. However, in the case of long queries, our approach produces equivalent performance with respect to the classical weighting approach. Anyway, the performance of any concept-based approach highly depends on the accuracy of the mapping tools (MetaMap in our case).

The next step in this study is to test our weighting approach with other mapping tools, in order to study the effects of tools accuracy on the retrieval performance.

Acknowledgments. This work was partly realized as part of the Quaero Program funded by OSEO, French State agency for innovation. This work was supported in part by the french project VideoSense ANR-09-CORD-026 of the ANR.

References

1. Abdulahhad, K., Chevallet, J.-P., Berrut, C.: Solving Concept mismatch through Bayesian Framework by Extending UMLS Meta-Thesaurus. In: CORIA 2011, Avignon, France, pp. 311–326 (March 2011)
2. Aronson, A.R.: Metamap: Mapping text to the umls metathesaurus (2006)
3. Baziz, M.: Indexation conceptuelle guidée par ontologie pour la recherche d'information. Thèse de doctorat, Université Paul Sabatier, Toulouse, France (Décembre 2005)
4. Bendersky, M., Metzler, D., Bruce Croft, W.: Parameterized concept weighting in verbose queries. In: SIGIR 2011, Beijing, China, pp. 605–614 (2011)
5. Chevallet, J.-P., Lim, J.-H., Le., D.T.H.: Domain knowledge conceptual inter-media indexing: application to multilingual multimedia medical reports. In: CIKM 2007, Lisbon, Portugal, pp. 495–504 (2007)
6. Codocedo, V., Lykourentzou, I., Napoli, A.: Semantic Indexing and Retrieval based on Formal Concept Analysis. Technical report (June 2012)
7. Crestani, F.: Exploiting the similarity of non-matching terms at retrieval time. Inf. Retr. 2(1), 27–47 (2000)
8. Dozier, C., Kondadadi, R., Al-Kofahi, K., Chaudhary, M., Guo, X.S.: Fast tagging of medical terms in legal text. In: ICAIL, pp. 253–260 (2007)
9. Fang, H., Tao, T., Zhai, C.: A formal study of information retrieval heuristics. In: SIGIR 2004, Sheffield, United Kingdom, pp. 49–56 (2004)
10. Le., T.H.D.: Utilisation de ressources externes dans un modèle Bayésien de Recherche d'Information. Application à la recherche d'information multilingue avec UMLS. Theses, Université Joseph-Fourier - Grenoble I (May 2009)
11. Luhn, H.P.: The automatic creation of literature abstracts. IBM J. Res. Dev. 2(2), 159–165 (1958)
12. Maisonnasse, L.: Les supports de vocabulaires pour les systèmes de recherche d'information orientés précision: application aux graphes pour la recherche d'information médicale. Theses, Université Joseph-Fourier - Grenoble I (May 2008)
13. Ren, F., Bracewell, D.B.: Advanced information retrieval. Electron. Notes Theor. Comput. Sci. 225, 303–317 (2009)
14. Robertson, S.E., Walker, S.: Some simple effective approximations to the 2-poisson model for probabilistic weighted retrieval. In: SIGIR 1994, Dublin, Ireland, pp. 232–241 (1994)
15. Sanderson, M.: Word sense disambiguation and information retrieval. In: SIGIR 1994, Dublin, Ireland, pp. 142–151 (1994)
16. Singhal, A., Buckley, C., Mitra, M.: Pivoted document length normalization. In: SIGIR 1996, Zurich, Switzerland, pp. 21–29 (1996)
17. Smucker, M.D., Allan, J., Carterette, B.: A comparison of statistical significance tests for information retrieval evaluation. In: CIKM 2007, Lisbon, Portugal, pp. 623–632 (2007)
18. Zhai, C., Lafferty, J.: A study of smoothing methods for language models applied to ad hoc information retrieval. In: SIGIR 2001, New Orleans, Louisiana, United States, pp. 334–342 (2001)

BioDI: A New Approach to Improve Biomedical Documents Indexing

Wiem Chebil[1,2], Lina Fatima Soualmia[1], and Stéfan Jacques Darmoni[1]

[1] Normandie Univ, CISMeF Team, LITIS-TIBS EA 4108,
Rouen University and Hospital, France
[2] Research Unit MARS, Monastir University, Tunisia
wiem.chebil@yahoo.fr,
{Lina.Soualmia,Stefan.Darmoni}@chu-rouen.fr

Abstract. The partial match between biomedical documents and controlled vocabularies allows to find in the documents more terms variants than those existing in the dictionaries. However, it generates irrelevant information. We propose a new approach for indexing biomedical documents with the Medical Subject Headings (MeSH) thesaurus that aims to overcome the limitation of the partial match. In fact, our indexing approach proposes to restrict the stemming process in the step of pretreatment. The step of the descriptors extraction is based essentially on the vector space model and combines semantic and statistic methods to compute a score to estimate the relevance of a descriptor given a document. The knowledge provided by the Unified Medical Language System (UMLS) is used then for filtering. The filtering method aims to keep only relevant descriptors. The experiments of our approach that have been carried out on the OHSUMED collection, showed very encouraging results.

Keywords: Partial match, biomedical documents, stemming, MeSH term, term weight, UMLS.

1 Introduction

The permanent increase of biomedical documents in the internet makes the task of their manual indexing with the biomedical controlled vocabularies become more difficult. To replace the tedious task of the human indexers, several approaches of biomedical documents indexing were proposed. Some of these approaches were based on an exact match [1-3] between the controlled vocabularies and documents which allows to find in the document only terms in the dictionaries. Other approaches were based on a partial (or approximate) match [4-7] which allows to (i) find in the document other terms variants than those existing in the dictionaries by applying the stemming, which reduces words (in the document and in the controlled resource) to their stems (or roots) (e.g. reacts, reacting, reacted, are reduced to react), or lemmatization, which reduces words to their based form (e.g. operation, operated are reduced to operate) (ii) extract multi-word terms that share a subset of their words with the document. The terms extracted in the two cases (i) and (ii) may be relevant which

H. Decker et al. (Eds.): DEXA 2013, Part I, LNCS 8055, pp. 78–87, 2013.

leads to improve the recall. But also they may be irrelevant which leads to decrease the precision. For examples: in the case of (i) a short stem may be confused with an acronym such as *"kid"* which is an acronym of term *"Keratitis, Ichthyosis, and Deafness"* and also a stem of term *"kidding"*. In addition, the existing tools for the lemmatization may don't recognize the exact grammatical classes (verb, noun..) of the biomedical vocabulary. In the case of (ii) the term *"breast cancer"* in a document may yield the Medical Subject Headings (MeSH) [8] terms *"testicular cancer"* and *"stomach cancer"* because the three terms share the word "cancer" [9].

In this paper, we propose a new approach for indexing biomedical documents using MeSH thesaurus denoted Biomedical Document Indexing (BioDI) that aims to overcome the limitation of partial match based approaches. Our first contribution is to restrict the stemming process. In addition, to enhance the relevance estimation of a term[1], we compute a semantic, statistic and structure based score that gives an importance to the position of a word in the document as well as to the occurrence of the terms words in the same phrase. Another main contribution of our approach is to exploit the knowledge provided by the Unified Medical Language System (UMLS) [10] to filter the extracted descriptors. The filtering allows to keep relevant descriptors among those extracted in the case (ii).

The paper is organized as follow: the second section presents the related work. The section 3 details the steps of our indexing approach. In the section 4, we describe the experiments and the generated results that are discussed in the section 5. Finally, in the section 6 we conclude and present our future work.

2   Related Work

Several research approaches for indexing biomedical documents have been proposed. We focus on some of them. Pouliquen *et al.* [1] computed a statistic weight based on TF-IDF for each term automatically extracted from the document using a method based on NLP (Natural Language Processing). These terms are then matched to the terms of the ADM (assistance with the medical diagnosis) dictionary. Jonquet *et al.* [2] applied the Mgrep tool for extracting concepts from 200 biomedical ontologies, and computed a score for each generated annotation according to its origin (preferred term, non-preferred term, synonym term …etc.). Mukherjea *et al.* [3] developed BioAnnotator a new tool for indexing biomedical documents. It uses a parser to identify noun phrases from a document and then matches them to the UMLS concepts using a rule engine. Zhou *et al.* [4] proposed to annotate documents with only the most significant words in the UMLS Meta-thesaurus. Ruch [5] proposed an indexing approach denoted by Eagl that combined two models: the Vector Space Model (VSM) and a regular expression pattern matcher. The indexing technique of Aronson *et al.* [6] is based on three methods: the first uses MetaMap (software tool for English that allows mapping document to the UMLS concepts), the second is the tri-gram method and the last one is the KNN (the k-Nearest Neighbors). Majdoubi *et al.* [7] used the VSM to extract MeSH terms and then computed a statistic and semantic weight for ranking these terms.

---

[1] In this paper we denoted by terms all preferred and no preferred terms in MeSH thesaurus.

3   The Steps of Our Proposed Approach

Our approach BioDI is based on VSM [11] which was initially applied in Information Retrieval (IR) to compute a similarity between user's query and the document. In our approach, as in [5] and [7] the query is replaced by a term. Our method is composed of 4 steps: pretreatment, descriptors extraction, filtering and final ranking.

3.1   Step 1: Pretreatment

The step of pretreatment consists of 4 tasks: (i) dividing the document into phrases (ii) removing punctuation (iii) pruning stop words (iv) stemming. The last three tasks are applied also on MeSH terms. Let *"The binding of acetaldehyde to the active site of ribonuclease: alterations in catalytic activity and effects of phosphate."* a title of a document, after the pretreatment this title become *"bind acetaldehyd activ site ribonucleas alter catalyt activ effect phosphat"*. For the stemming, we chose to use PORTER Algorithm [12]. During the stemming process a short stem can be confused with an acronym. Thus, we propose to restrict the applying of the stemming process only on words that the length of their stems is equal or upper than a threshold Ts which is fixed experimentally (see section 4).

3.2   Step 2: Descriptors Extraction

The step of descriptors extraction begins with extracting all the preferred and no preferred terms. To do, we compute a similarity between each term and the document using cosine similarity. The terms candidates are those having a similarity upper or equal than a tuned Threshold Tcos. Then, we compute a weight for each extracted term. The final score of each selected term is the sum of its similarity with the document and its weight. After, the corresponding descriptors are assigned to the terms. The score of a descriptor is the score of its term. As in [13], if a descriptor corresponds to more than one term among the terms candidates, it will have the highest score. The term that gives its score to the descriptor is denoted the Representative Term (RT).

Similarity between a Term and a Document
Let $\{T_1 \dots T_i \dots T_z\}$ the set of MeSH terms. Each term T_i is composed of a set of words $T_i \{ wdt_1 .. wdt_k ... wdt_t \}$ with t the number of words in a term. T_i is represented by the vector VT ($WWT_1 \dots WWT_k .. . WWT_t$), WWT_k is the weight of word wdt_k in the MeSH. The document DOC is represented by the vector VDOC ($WWDoc_1 \dots WWDoc_k \dots WWDoc_t$), $WWDoc_k$ is the weight of the word wdt_k in the DOC. The cosine similarity is computed then between VT and VDOC (1) and denoted $Sim(T_i, DOC)$. We consider that WWT-WWDoc is the weight combination of wdt_k.

$$Sim(T_i , DOC) = \frac{\sum\limits_{k=1}^{t} WWT_k \times WWDoc_k}{\sqrt{\sum\limits_{k=1}^{t} \left(WWT_k\right)^2 \times \left(WWDoc_k\right)^2}} \tag{1}$$

Weight of Word in the Document (WWDoc)
For computing WWDoc, we use a weight based on the frequency of a word in the document that takes into consideration the position of the word in the document (in the title, in the abstract or in the paragraphs). This weight is denoted Word Average Frequency in the Document (WAFDoc) (2). We consider that the key words are more dissipate in the paragraphs and mixed with non relevant words (comparing to the abstract and title), while key words are more condensed in the title (comparing to abstract and paragraphs). Thus, we assign the following coefficients to each position in document: Position Coefficient (PC) =8 to the title, PC=4 to the abstract, PC=2 to the paragraphs.

$$WWDoc_k = WAFDoc_k = \frac{\sum\limits_{p=1}^{r} FQ\left(wdt_k, P\right) \times PC_p}{\sum\limits_{p=1}^{r} PC_p} \tag{2}$$

— FQ(wdt_k, P): Frequency of wdt_k in the position P
— P=1:Title;   P=2: Abstract;   P=3: Paragraph
— PC_p: The coefficient of the position P.
— r: The number of the positions coefficients

Weight Word in Term (WWT). We consider WF_k-IDF_k^2(Word frequency – Inverse document frequency) [11] is the weight of the word wdt_k in the term.

$$WWT_k = WF_k \times IDF_k \tag{3}$$

$$WF_k = \frac{FWT_k}{\max\limits_{e:1->t}\left(FWT_e\right)} \tag{4}$$

— t: is the number of words in a term
— FWT_k: Frequency of wdt_k in a term

We consider that the normalized frequency of a word wdt_k in the term is equal to its frequency in the descriptor containing this term because this term may be the RT of the descriptor. The frequency of a word in a descriptor is its frequency in all the terms of the descriptor. We consider also that the IDF of wdt_k is equal to the logarithm of the number of the descriptors containing in their terms at least one occurrence of wdt_k divided by the total number of the descriptors in MeSH.

$$IDF_k = -\log\left(\frac{FWDM_k}{ND}\right) \tag{5}$$

---

[2] Instead of using "TF" (term frequency) we used "WF" (word frequency) because we consider that a term can be composed of one word or can be a multi-word term.

— ND: The total Number of Descriptors in MeSH
— FWM_k : The Frequency of the Word wdt_k in MeSH (the number of descriptors having at least one occurrence of wdt_k).

Weight of a Term in the Document (WTDoc)

We propose a new weight of a term T_i in the document denoted TAFDoc. This weight is based on WAFDoc and it is equal to the sum of the weights WAFDoc of all the words of T_i (the t words) divided by t. The results is majored by a coefficient cof>1 if all the t words of T_i are at least one time in the same phrase in the document. In fact, we hypothesize that words in the same phrase are more likely to cover the same meaning. The coefficient cof is experimentally tuned.

$$\text{WTDoc}_i = \text{TAFT}_i = \frac{\sum\limits_{k=1}^{t} \text{WAFDoc}\left(wdt_k\right)}{t} * cof \tag{6}$$

-cof >1 if the term words are in the same phrase at least one time in the document.
-cof= 1 if the term words are not in the same phrase

The Score of a Descriptor

The score of a descriptor is the maximum score of its terms (7). The term having the maximum score is the Representative Term (RT). The score of a term is the sum of its similarity with the document and its weight in the document (8).

$$Score(D) = \max_{j:1->n}(Score(T_j)) \tag{7}$$

$$\text{Score}(T_i) = \text{WTDoc}_i + Sim(T_i, DOC) \tag{8}$$

n: The number of terms of a descriptor D

3.3   Step 3: Filtering

The aim of this step is to keep only the relevant descriptors among those having a multi-word RT that at least one of its words doesn't occur in the document. In fact, we classified the no extraction of these relevant descriptors as a category of indexing errors in [14]. This step consists of dividing the set of MeSH descriptors generated in the previous step into two sets of descriptors: the first set is denoted Principal Index (PI) and the second is denoted Secondary Index (SI). The PI contains the descriptors that their RT terms have all their words in the document. These Descriptors are denoted Principal Descriptors (PD). The SI contains the descriptors that their RT terms have a subset of their words in the document. These descriptors are denoted Secondary Descriptors (SD). We separate the PD and SD because we are based on the assumption that MeSH terms having all their words in the document are more likely to be correct. Then the relevant descriptors in SI are added from the SI to the PI. To do this task, first of all, the PD in PI are ranked using the score (7). Thus we have

PI= {$PD_1, \ldots PD_i \ldots PD_v$}, PD_i is a principal descriptor having the rank i and v is the number of PD in PI. Then, we propose to compute a score S for each SD (9). This score S is based on the co-occurrences of MeSH descriptors in MEDLINE and the semantic relations between MeSH descriptors provided by the semantic work of UMLS [10]. In fact, our assumption is that the SD is more likely to be correct if it is more co-occurrent or/and have more semantic relations with exactly the L first PD in PI that are considered the most relevant. L is the length of a window that contains the L first PD. For example, according to the proposed formula of S (9) if we fix L=1, that means S(SD) is equal to the sum of the number of co-occurrences and relations between the SD and the PD having the rank 1(PD_1). If L=2, that means S(SD) is equal to the sum of the number of the co-occurrences and the semantic relations between the SD and the two PD having the rank 1 and 2(PD_1 and PD_2). If SD doesn't co-occur or doesn't have any semantic relation with one of the L PD, or if the SD has a score S lower than a tuned threshold T, it isn't be added to PI. The threshold T was tuned according to the value of L.

$$S(SD) = \sum_{i=1}^{L} CF(SD, PD_i) + \sum_{i=1}^{L} NR(SD, PD_i) \tag{9}$$

CF: Co-occurrence Frequency;   NR: Number of the semantic Relations

3.4   Step 4: Final Ranking

The SD selected in the previous step will be added to PI, the final index (FI) is thus constructed. The descriptors of FI are re-ranked using the score (7).

4   Experiments and Results

To test our approach we selected randomly 6,000 citations among the OHSUMED collection[3] composed of 4,591,015 MEDLINE citations. Each selected citation is composed of title and an abstract. The content of the title is merged with the content of the abstract when indexing the citations. We don't consider the sub-headings in our approach. To evaluate BioDI, we used the classical measures of Precision (P), Recall (R) and F-score (Fs). The precision is the number of correct descriptors divided by the total number of descriptors automatically generated. The recall is the number of correct descriptors divided by the number of descriptors manually extracted. F-score combines precision and recall with an equal weight [15].

4.1   Evaluation of the Terms Extraction

The different cases experimented in order to fix the adequate value of Ts are: Ts>=3, Ts >=4, Ts>=5 and Ts>=6. We experimented also the stemming without considering the stem length and the case where we didn't stem the words. For each of these cases, we applied the cosine similarity between the MeSH terms and the document and we

---

[3] http://trec.nist.gov/data/t9_filtering.html

tested the performance of the proposed weight combination WFIDF-WAFDoc as well as others combinations: 1-1 (assigning 1 to the weight of word in the document if the word exist in the document, 0 else), IDF-WFIDF, WFIDF-WFIDF. When computing the cosine similarity a big number of terms are extracted, thus, only those having a similarity upper than a tuned threshold Tcos equal to 0.8 were selected as candidates for indexing the document. In order to generate the results of these experiments we affected for each extracted term its correspondent descriptor because the manual indexing has been carried out using descriptors. The table 1 illustrates the obtained results of the experiments described above.

Table 1. Results of terms extraction[4]

	1-1 (or 0)	IDF-WFIDF	WFIDF-WFIDF	WFIDF -WF	WFIDF-WAFDoc
	P-R- Fs	P-R- Fs	P-R- Fs	P-R- Fs	P-R- Fs
A	0.180-0.30- 0.225	0.174-0.310- 0.199	0.175-0.330- 0.228	0.177-0.340- 0.232	0.179-0.360 -0.239
B	0.170-0.32- 0.222	0.161-0.320- 0.214	0.163-0.350- 0.222	0.165-0.400- 0.233	0.168-0.410- 0.238
C	0.159-0.48- 0.238	0.148-0.520- 0.223	0.150-0.521- 0.230	0.155-0.535- 0.240	**0.158-0.570- 0.246**
D	0.121-0.520- 0.196	0.113-0.550- 0.187	0.115-0.560- 0.190	0.117-0.580- 0.194	0.119-0.600- 0.198
E	0.112-0.57- 0.187	0.106-0.605- 0.180	0.107-0.610- 0.182	0.109-0.620- 0.185	0.110-0.630- 0.187
F	0.100-0.60- 0.171	0.090-0.615- 0.157	0.092-0.620- 0.160	0.094-0.630- 0.164	0.099-0.650- 0.172

A: Without stemming, B: Ts>= 6, C: Ts>=5, D: Ts>=4, E: Ts >=3, F: Stemming without considering the length of word stem.

4.2   Experiments and Results of Generating the PI

The aim of these experiments is to compute the precision, recall and f-score of the PI where descriptors are ranked using the score (7) that takes into account the similarity between the MeSH terms and the document and also the weight of the terms in the document. In the first experiment (section 5.2) we evaluated the performance of proposed similarity. In this experiment we tested the performance of the proposed weight TAFDoc through two experiments. First of all, we varied the value of the coefficient cof and we compute the TAFDoc. We carried out this test, in order to find the best value of the coefficient cof. Then, we evaluated the performance of BM25 term weighting model used in [16] to compute the weight of concepts, which is compared

---

[4] We kept three numbers after the point because the results are very close to each other.

Table 2. Results of generating PI with varying cof and comparing TAFDoc to BM25

	BM25	TAFDoc		
		cof=1	cof=1.5	cof=1.6
P-R-Fs(rank1)	0.61-0.17-0.26	0.68-0.19-0.28	**0.71-0.21-0.31**	0.70-0.18-0.28
P-R-Fs(rank10)	0.17-0.43-0.23	0.23-0.43-0.28	**0.29-0.40-0.33**	0.28-0.37-0.23
P-R-Fs(rank15)	0.19-0.47-0.25	0.21-0.45-0.27	**0.25-0.43-0.30**	0.24-0.40-0.29

to the performance of TAFDoc. For each one of the two experiments, a new score (7) was computed with keeping always the proposed similarity, and PI is re-generated.

Table 2 presents the results of these experiments at ranks 1, 10 and 15[5].

4.3 Evaluation of the Filtering Step and Final Ranking

In order to evaluate the step of filtering we generate final results for different values of L. For each value of L a new value of T is experimentally tuned. These results are shown in table 3.

Table 3. Results after filtering and final ranking at rank 1, 10 and 15

	L=1/T=70	L=2/T=50	L=3/T=10	L=4/T=4	L=5/T=5	L=6/T=6
P-R-Fs(rank1)	0.71-0.21-0.31	0.71-0.21-0.31	**0.71-0.21-0.31**	0.71-0.21-0.31	0.71-0.21-0.31	0.71-0.21-0.31
P-R-Fs(rank10)	0.31-0.52-0.38	0.35-0.51-0.40	**0.41-0.50-0.45**	0.37-0.48-0.42	0.35-0.45-0.38	0.30-0.40-0.34
P-R-Fs(rank15)	0.26-0.55-0.34	0.32-0.54-0.39	**0.36-0.52-0.42**	0.34-0.49-0.39	0.30-0.48-0.35	0.27-0.45-0.33

4.4 Evaluation of Some Other Approaches

To highlight the effectiveness of our indexing approach, we compared the performance of BioDI to the performance of some other approaches. In fact, we evaluated MaxMatcher [4], and Eagl [5] which are partial match based approaches and BioAnnotator [3] which is an exact match based approach. The results of this evaluation are detailed in table 4.

Table 4. Evaluation of MaxMatcher, Eagl and BioAnnotator at ranks 1, 10 and 15

	MaxMatcher	Eagl	BioAnnotator	BioDI
P-R-Fs(rank1)	0.69-0.18-0.27	0.62--0.18-0.27	0.70-0.14-0.22	0.71-0.19-0.29
P-R-Fs(rank10)	0.32-0.46-0.37	0.25-0.40-0.30	0.33-0.24-0.26	0.41-0.50-0.45
P-R-Fs(rank15)	0.27-0.50-0.35	0.17-0.54-0.25	0.29-0.27-0.26	0.36-0.52-0.42

---

[5] We didn't test other ranks upper than 15 because the average number of keywords in MEDLINE citations is 15 [5].

5   Discussion

The table 1 shows that, for all the weights combinations, the precision of terms extraction is higher without stemming, and then it decreases when the stemming is applied with considering Ts. The more Ts decrease the more the precision also decreases. In addition, we can observe that the recall is very low without applying stemming and its value is significantly higher when Ts>=6. Moreover, according to the values of f-score we can deduce that the stemming process performs well when Ts>=5. When analyzing table 2, we can see that the performance of the VSM is better (according to the f-score value) when applying the weight combination WFIDF-WAFDoc than the 4 others weights combinations though 1-1(or 0) gives a slightly higher precision. We can deduce also that WAFT when combined with WF-IDF performs well than WF and WF-IDF. The table 2 shows that the best results of generating PI when applying the weight TAFDoc are scored when cof=1.5. We can conclude also according to table 2 that TAFDoc is more effective than BM25. These results show the well interest of: (i) taking into account the word position in the document (ii) giving more importance to terms having their words in the same phrase. According to the table 3 (final results), we can observe that there is no change in the performance of BioDI after PI's expansion when the first descriptor is retrieved. Nonetheless, at rank 10 and 15 an improvement of results can be seen. Obviously, descriptors having a part of the words of their RT doesn't occur in the document don't have the best weight. We can see also, that the expansion method performs better at L=3 than at the other values of L. In addition, when L=6 we have a remarkable decrease of results. Indeed, at L>5 it's less possible to find a SD which is co-occurent or have semantic relations with exactly the L first PD. The evaluation of Maxmatcher, Eagle and BioAnnotator (table 4) confirms the effectiveness of BioDI which out performs the three other approaches in the different ranks and in term of precision, recall and F-score when L is equal to 3, 4 and 5. Thus, we can deduce that the performance of our approach is closely dependent on the parameters L, cof and Tcos that must be well tuned to allow BioDI to outperform the other approaches.

6   Conclusion and Future Work

We presented in this paper our indexing approach that proposes to improve the partial match between biomedical documents and the controlled vocabularies. Our main contributions are: (i) restricting the stemming process to the words that their stem length is equal or upper than 5 (ii) computing a new score to estimate the relevance of a MeSH descriptor given a document. This score takes into account the position of a word in the document and gives more importance to terms having all their words in the same phrase (iii) filtering the index using the semantic and statistic resources of UMLS in the aim of keeping only relevant descriptors among those having a subset of their RT in the document. The several experiments carried out on the OHUMED corpus showed that BioDI allows improving partial match as well as exact match between biomedical documents and biomedical terminologies. We aim after these encouraged results to test the proposed approach with computing the score (9) between SD and all possible combinations of the first PD. In addition, we aim to compare our approach to more others approaches. We are working also on applying our approach on the corpus of the catalog and index of french-language health internet resources (CISMeF)[6].

References

1. Happe, A., Pouliquen, B., Burgun, A., Cuggia, M., Beux, P.L.: Automatic concept extraction from spoken medical reports. I. J. Medical Informatics 70(2-3), 255–263 (2003)
2. Jonquet, C., LePendu, P., Falconer, S.M., Coulet, A., Noy, N.F., Musen, M.A., Shah, N.H.: NCBO Resource Index: Ontology-based search and mining of biomedical resources. J. Web Sem. 9(3), 316–324 (2011)
3. Mukherjea, et al.: Enhancing a biomedical information extraction system with dictionary mining and context Disambiguation. IBM Journal of Research and Development 48(5/6), 693–701 (2004)
4. Zhou, X., Zhang, X., Hu, X.: MaxMatcher: Biological concept extraction using approximate dictionary lookup. In: Yang, Q., Webb, G. (eds.) PRICAI 2006. LNCS (LNAI), vol. 4099, pp. 1145–1149. Springer, Heidelberg (2006)
5. Ruch, P.: Automatic assignment of biomedical categories: toward a generic approach. Bioinform. J. 22(6), 658–664 (2006)
6. Aronson, A.R., Mork, J.G., Gay, C.W., Humphrey, S.M., Rogers, W.J.: The NLM indexing initiative's medical text indexer. Med. Health Info. 11(1), 268–272 (2004)
7. Majdoubi, J., Tmar, M., Gargouri, F.: Using the MeSH thesaurus to index a medical article: combination of content, structure and semantics. In: International Conference on Knowledge-Based and Intelligent Information & Engineering Systems, KES, vol. (1), pp. 277–284 (2009)
8. Nelson, S.J., Johnson, W.D., Humphreys, B.L.: Relationships in Medical Subject Heading. In: Relationships in the Organization of Knowledge, pp. 171–184. Kluwer Academic Publishers (2001)
9. Trieschnigg, D., Pezik, P., Lee, V., et al.: MeSH Up: effective MeSH text classification for improved document retrieval. Bioinformatics 25(11), 1412–1418 (2009)
10. Bodenreider, O.: The Unified Medical Language System (UMLS): integrating biomedical terminology. Nucleic Acids Research 32(4), 267–270 (2004)
11. Singhal, A.: Modern information retrieval: a brief overview. IEEE Data Eng. Bull. 24(4), 35–43 (2001)
12. Porter, M.: An algorithm for suffix stripping. Program 14(3), 130–137 (1981)
13. Couto, F.M., Silva, M.J., Coutinho: Finding genomic ontology terms in text using evidence content. BMC Bioinformatic 6, (S-1) (2005)
14. Chebil, W., Soualmia, L.F., Dahamna, B., Darmoni, S.J.: Automatic indexing of health documents in French: Evaluating and analysing errors. IRBM BioMedical Engineering and Research 33(2), 129–136 (2012)
15. Manning, C.D., Schütze, H.: Fondations of statistical natural language processing, pp. 534–536. MIT Press, Cambridge (1999)
16. Dinh, D., Tamine, L.: Towards a context sensitive approach to searching information based on domain specific knowledge sources. Web Semantics: Science, Services and Agents on the World Wide Web 12-13, 41–52 (2012)

---

[6] http://www.chu-rouen.fr/cismef/

Discovering Semantics from Data-Centric XML

Luochen Li[1], Thuy Ngoc Le[1], Huayu Wu[2],
Tok Wang Ling[1], and Stéphane Bressan[1]

[1] School of Computing, National University of Singapore
{luochen,ltngoc,lingtw,step}@comp.nus.edu.sg
[2] Institute for Infocomm Research, Singapore
huwu@i2r.a-star.edu.sg

Abstract. In database applications, the availability of a conceptual schema and semantics constitute invaluable leverage for improving the effectiveness, and sometimes the efficiency, of many tasks including query processing, keyword search and schema/data integration. The Object-Relationship-Attribute model for Semi-Structured data (ORA-SS) model is a conceptual model intended to capture the semantics of object classes, object identifiers, relationship types, etc., underlying XML schemas and data. We refer to the set of these semantic concepts as the ORA-semantics. In this work, we present a novel approach to automatically discover the ORA-semantics from data-centric XML. We also empirically and comparatively evaluate the effectiveness of the approach.

1 Introduction

To improve the conceptual quality, we needs to discover the intended semantics in the logical XML schemas and data. This requires finding such semantic information as object classes, relationship types, object identifiers (OIDs), etc., as present in conceptual models for semi-structured data such as Object-Relationship-Attribute for Semi-Structured data (ORA-SS) [6]. We refer to this semantics as the ORA-semantics. Once discovered, the ORA-semantics is useful not only for users to understand the data and schemas but also for improving both the effectiveness and efficiency of processing. Let us use the XML document in Fig. 1 to illustrate how the availability of such semantics help applications.

XML Query Processing

To process an XPath query, e.g. *//Student[Matric# ='HT001']/Name*, most approaches match the query pattern to the data to find all occurrences. However, if we have the semantics that *Matric#* is the OID of student, after getting an answer, we can stop searching the rest of data.

XML Keyword Search

The use of semantics in current keyword search approaches [7] is still on object level. For a query {CS5201, CS5208} to find common information of two courses, only by knowing there is a relationship type between object classes *Student* and *Course*, one can infer the meaningful answer should be all students taking these two courses. Otherwise, the root node will be returned by most LCA-based XML keyword search approaches [12].

H. Decker et al. (Eds.): DEXA 2013, Part I, LNCS 8055, pp. 88–102, 2013.
© Springer-Verlag Berlin Heidelberg 2013

Schema/Data Integration

Most existing approaches [1] integrate elements based on their structural and linguistic similarities. *Grade* is an attribute of the relationship type between *Course* and *Student*. Without this semantics, when we integrate this schema with another, in which *Student* has an object attribute *Grade* which means the year of his study in school, we may wrongly integrate these two different attributes with the same attribute name *Grade* and the same parent node *Student*, because of their high structural and linguistic similarities.

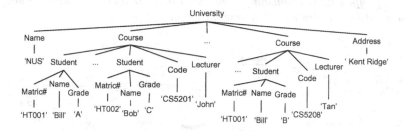

Fig. 1. An XML data tree

However, most practical applications are semantics-less, as most existing XML schema languages, e.g., DTD and XSD, cannot fully represent the semantics such as object class, relationship type, OID, etc. Despite the existence of semantically rich XML models, e.g., ORA-SS, they still requires manual provision of semantics from the initial design or model transformation. We believe only if the automatic semantics discovery technique is developed to a satisfactory level, the achievements in semantics-based query optimization, keyword search, schema/data integration, etc., will be widely adopted by different applications.

In this paper we present a novel approach to automatically discover the ORA-semantics from data-centric XML schemas and data. Different from the existing approaches that only focus on object identification, we consider a comprehensive set of ORA-semantics, including OID, relationship type as well as the distinction between object attribute and relationship attribute.

2   Preliminary

We refer the tree structure derived from XML schemas as *XML schema trees*. For ease of description, all following concepts are defined on XML schema trees.

In XML schema tree, **object class** is an internal node representing a real world entity or concept. An object class has a set of **object attributes** to describe its properties. Each object class has an **object identifier (OID)** to uniquely identify its instance. Several object classes may be connected through a **relationship type** which may or may not explicitly appear in the XML schema tree. We call them **explicit relationship type** and **implicit relationship** *type*. A relationship type may have a set of **relationship attributes**. **Aggregational node** aggregates its child nodes with identical/similar meaning. **Composite attribute** is an object/relationship attribute containing multiple components, each of which can be a single attribute or a composite attribute.

Based on the semantic concepts mentioned above, we define the *ORA-Semantics*, which is the scope of the semantic concepts we consider in this paper.

Concept 1. *ORA-semantics (Object-Relationship-Attribute-semantics)*
In an XML schema tree, the ORA-semantics is the identification of object class, OID, object attribute, aggregational node, composite attribute and explicit/implicit relationship type with relationship attributes. Each particular semantic concept in ORA-semantics is called an ORA-semantic concept.

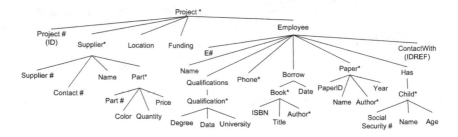

Fig. 2. An XML schema Tree

Example 1. In Fig. 2, we can infer the internal nodes *Project, Supplier, Part, Employee, Book, Paper* and *Child* are object classes, with their OIDs *Project#, Supplier#, Part#, E#, ISBN, PaperID* and *SocialSecurity#*. The internal node *Borrow* is an explicit relationship type between *Employee* and *Book* with a relationship attribute *Date*; the internal node *Has* is an explicit relationship type between *Employee* and *Child* without any relationship attribute. The leaf node *Price* is a relationship attribute of the binary relationship type between *Supplier* and *Part*, and the leaf node *Quantity* is a relationship attribute of the ternary relationship type among *Project, Supplier* and *Part*. The internal node *Qualifications* is an aggregational node, aggregating its child node *Qualification*, which is a composite attribute. All other leaf nodes are object attributes.

3   ORA-Semantics Discovery

We use properties of ORA-semantics, heuristics and data mining techniques to discover the ORA-semantics in data-centric XML schema/XML data. The properties used in our approach conform to the design of the corresponding ORA-SS model or ER model, and the heuristics are summarized based on the characteristics and our observations of different ORA-semantics concepts. In case an XML schema is not available with XML data, XML schema summarization/extraction has been studied in [3]. Fig. 3 shows the road map of our approach.

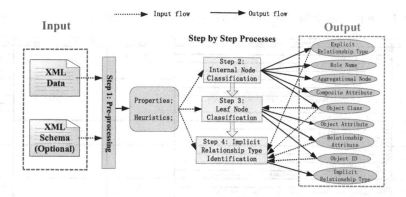

Fig. 3. General process of our automatic semantics discovery approach

3.1   Step 1: Pre-processing

We summarize properties of ORA-semantics concepts. Properties of an ORA-semantics concept are its necessary conditions, which means given an ORA-semantics concept, it must satisfy its properties. E.g, object class has property *'Having more than one child node in its XML schema tree'*, which also conforms to its design in ORA-SS model. We also identify sufficient conditions, by which we can identify a particular ORA-semantics concept, e.g. *'Having an ID attribute in its XML schema as its child node.'* is a sufficient condition for object class. We also proposed heuristics related to ORA-semantics concepts. Some are abstracted from XML schema based on the common way of schema design, and some are discovered from XML data using data mining techniques. We list the properties, sufficient conditions and heuristics for each ORA-semantics concept in Table 1.

In an XML schema tree, a node must be either an internal node or a leaf node. Based on the properties of each ORA-semantics concept in Table 1, internal nodes can be object class, role name, composite attribute, aggregational node and explicit relationship type; while leaf nodes can be OID, object attribute and relationship attribute. We will identify them in 3.2 and 3.3 respectively. There is another ORA-semantics concept, implicit relationship type, which is not explicit shown in the XML schema or XML schema tree. We will identify it in 3.4.

3.2   Step 2: Internal Node Classification

To classify internal nodes, we build a decision tree, Fig. 4, with the properties, sufficient conditions and heuristics in Table 1. We use bottom-up approach so that the category of an internal node can help to identify the category of its parent node. We will explain the decision tree using following rules:

Table 1. Properties, sufficient conditions and heuristics of ORA-semantics concepts

ORA-semantics	Properties (Necessary Conditions)	Sufficient Conditions	Heuristics / Observations	Examples
Object Class	O1) It is an internal node; O2) It has more than one child node; O3) It has at least one FD/MVD among its EDLNs; O4) Not all nodes in the LHS of each of its FDs/MVDs are IDREF attribute;	A) It has ID attribute in its XML schema;(E.g. Project)		Supplier Employee Part etc.
Explicit Relationship Type	E1) It is an internal node; E2) It has at least one object class, IDREF(S) attribute or role name as descendant node. E3) If it has at least one FDs/MVDs among its EDLN(s), then all nodes in the LHS of each of its FDs/MVDs are IDREF attributes; E4) Its EDLN(s) should be relationship attribute;		H1) Its tag name can be a verb form.	Borrow Has RentBy Buy
Aggregational Node	A1) It is an internal node; A2) It has only one child node; A3) Its child node is a repeatable node;		H2) Its tag name is the plural form of the tag name of its only child node;	Qualifica-tions
Composite Attribute	C1) It is an internal node; C2) It has more than one child node; C3) It does not have FD/MVD among its EDLNs; C4) It hasn't any object class, IDREF(s) attribute or role name as its descendant node;			Qualification
OID of object class	OID1) It is a leaf node; OID2) Together with OID(s) of some(zero or more) of its ancestor object class(es), they can functionally multi-valued determine all EDLN(s) of the object class;	B) It is specified as ID attribute in XML schema; (E.g. Project #)		Project# ISBN etc.
Object Attribute	OA1) It is a leaf node; OA2) It can be functionally/multi-valued determined by the OID of its lowest ancestor object class; OA3) Its lowest ancestor object class is the object class it belongs to;			Location Address Author etc.
Relationship Attribute	RA1) It is a leaf node; RA2) It cannot be functionally/multi-valued determined by the OID of its lowest ancestor object class; RA3) It can be functionally/multi-valued determined by OIDs of all object classes involved in the relationship type to which the relationship attribute belongs; RA4) It is an EDLN of an explicit relationship type or EDLN of the lowest object class that involves in an implicit relationship type to which the relationship attribute belongs;			Quantity Price
Role Name	R1) It is an internal node; R2) It has only one child node; R3) Its child node is not a repeatable node; R4) Its child node is an IDREF(S) attribute;		H3) Its tag name shares high linguistic similarity with or being a specialization of the tag name of the object class which the IDREF(S) attribute references;	Landlord Tenant

Rule 1. *[Object Class vs. OID] Given an XML schema tree, if an internal node has an ID attribute[1] specified in its XML schema as its child, then this internal node is an object class, and the ID attribute is the OID of the object class.*

Rule 1 is obvious. However, some OIDs may not or cannot be specified as ID attribute in the corresponding XML schema because of the limitation of XML schema language. In XML data, the value of an ID attribute is required to be unique for the corresponding object in the whole document, which makes it impossible for some object classes to have ID attribute being specified in their XML schemas. E.g, in Fig. 2, *Project#* is specified as OID for object class *Project* by ID attribute, but *Supplier#* and *Part#* cannot. Otherwise, a supplier can only supply one project and a part can only be supplied by one supplier. Because of this, we use following rules to classify the rest of the internal nodes.

Concept 2. *Exclusive Descendant Leaf Node (EDLN) In XML schema tree, an exclusive descendant leaf node of an internal node i is a leaf node, which is also a descendant node of i, but not a descendant node of any other object class which is also a descendant node of i.*

---

[1] ID attribute is specified in DTD. In XSD there is a similar concept, key element, which can also be used to identify object class and its OID. For simpleness, Rule 1 is illustrated using ID attribute, but key element also applies. Detail of key element in XSD is given in our technical report [5].

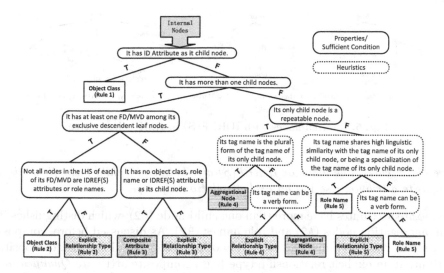

Fig. 4. Decision Tree for Internal Node Classification

The intuitive meaning of EDLN is: given an internal node i, each EDLN of i is a leaf node under i, but there is no other object class between the EDLN and i. E.g, in Fig. 2, the EDLNs of object class *Project* are: *Location* and *Funding*.

Rule 2. *[Object Class vs. Explicit Relationship Type] Given an XML schema tree, let i be an internal node with more than one child nodes and there is at least one Functional/Multi-valued Dependency(FD/MVD) among its EDLNs. If for each FD/MVD, there is a left hand side (LHS) node which is not an IDREF attribute or role name then i is an object class, else i is an explicit relationship type.*

An object class must have more than one child node (**O2** in Table 1), which conflicts with the properties of aggregation node (**A2**) and role name (**R2**), and at least one FD/MVD among its EDLNs (**O3**) that conflict with composite attribute (**C3**). However, explicit relationship type also has these characteristics. To distinguish object class from explicit relationship type, we check whether there is a LHS node in FD/MVD which is not an IDREF attribute or role name (**O4, E3**). This is because the FD/MVD among the EDLNs of an explicit relationship type must involve a relationship attribute, which is functionally/multi-valued determined by the OIDs of all involved object classes, and these OIDs can only be represented as IDREF attributes or role names if it is an EDLN of the explicit relationship type. On the other hand, for FD/MVD among the EDLNs of an object class, its LHS should contain the OID of this object class, which is not an IDREF attribute or role name. FDs/MVDs in XML can be identified in [13].

Rule 3. *[Composite Attribute vs. Explicit Relationship Type] Given an XML schema tree, let i be an internal node with more than one child node and there is no FD/MVD among its exclusive descendant leaf nodes. If i does not*

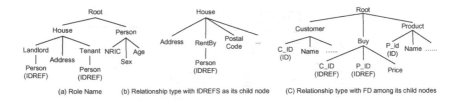

Fig. 5. Internal nodes with IDREF(S) in XML schema tree

have object class, role name or IDREF(S) attribute as its child node, then i *is a composite attribute, else* i *is an explicit relationship type.*

Composite attributes have more than one child node (C2), which distinguishes it from aggregation nodes (A2) and role names (R2). As discussed before, composite attribute has been distinguished from object class. To distinguish composite attribute from explicit relationship type (e.g., composite attribute *Qualification* and explicit relationship type *Borrow* in Fig.2), we check whether it has object class, role name or IDREF(S) attribute as its child node (C4, E2). This is because explicit relationship types should have at least one object class, IDREF(S) attribute or role name as its descendant nodes to represent the involved object class, while composite attributes should not.

Because of the space limit, we leave the rules to identify aggregational node and role name in our technical report [5]. In the following section, we aim to classify them into OID, object attribute and relationship attribute.

3.3   Leaf Node Classification

OID Discovery. As stated in Rule 1, OID can be explicitly specified in the XML schema with ID attribute, which is a sufficient condition to identify OID. Here we only consider the case that the single-attributed OID is not specified in the XML schema (e.g., *ISBN* in Fig. 2), or the OID contains multiple attributes. Before we explain our approach, we first introduce a concept named *Super OID*.

Concept 3. *Super OID The super OID of an object class o is a minimal set of nodes which contains a subset of the exclusive descendant leaf nodes of o and the OIDs of some ancestor object classes of o. Super OID of o can functionally/multi-valued determine all exclusive descendant leaf nodes of o.*

In an XML schema tree, given an object class o, its EDLNs may be object attributes of o, or attributes of a relationship type in which o participates. Based on the definition of super OID, the properties of OID and object/relationship attribute (i.e. OID2, OA2, RA2 and RA3 in Table.1), the super OID of o can functionally/multi-valued determine both object attributes and relationship attribute of o, while the OID of o can only functionally/multi-valued determine the object attributes of o. The rationale of our approach to identify OIDs is that given an object class o, there is a minimal attribute set S formed by the OID of

o and the OIDs of some ancestor object classes of o, which functionally/multi-valued determine all EDLNs of o. If there is no relationship attribute being EDLN of o, no OID of ancestor object class of o will be included in S. Thus, from the super OID of object class o, its corresponding OID can be derived by excluding all OID(s) of the ancestors object class(es) of o.

We proposed a top-down approach (the OID of an ancestor object class may be needed to identify the OIDs of its descendant object classes) shown in Algorithm 1 to identify the OID of each identified object class without ID attribute being specified in its XML schema. Given an object class o, we create a set $SupEDLN_o$, which is a superset of its exclusive descendant leaf nodes, denoted as $EDLN(o)$, and include OIDs of all its ancestor object classes. In $SupEDLN_o$, we identify the super OID of o. There may be more than one super OID for o. For each super OID, we can get an OID candidate for o by excluding all OID(s) of ancestor object class(es) of o. (Details about choosing which OID candidate as the OID will be discussed later.) For the object class without ancestor object class, its only super OID will be the same as its OID.

Algorithm 1. Candidate OID Discovery

Input: Identified object classes $\mathbb{O}$; $EDLN(o)$ for each identified object class $o \in \mathbb{O}$;
Output: Candidate OID id_o for each identified object class $o \in \mathbb{O}$

```
1  foreach identified object class o ∈ 𝕆 do
2      SupEDLN_o = EDLN(o);
3      foreach o_i ∈ 𝕆, which is ancestor object class of o do
4          SupEDLN_o = SupEDLN_o ∪ id_oi;          //id_oi is the OID of object class o_i
5      foreach SID_o ⊂ SupEDLN_o do
6          if ∀e ∈ EDLN(o), such that SID_o → e or SID_o ↠ e then
7              if ∄S ⊂ SID_o, such that ∀e ∈ EDLN(o), such that S → e or S ↠ e then
8                  if ∀o_j ∈ 𝕆 with its OID id_oj ∈ EID_o, such that ∃o_k ∈ 𝕆 with its OID
                       id_ok, AD(o_j,o_k) and AD(o_k,o), then id_ok ∈ SID_o then
9                      foreach e ∈ SID_o do
10                         if e ∈ EDLN(o) then
11                             e ∈ id_o;
12                     return id_o as a candidate OID of o.
```

Example 2. In Fig. 2, considering 3 identified object classes *Project*, *Supplier* and *Part* with their EDLNs, suppose we get the following full FDs from the XML data: $\{Project\#\} \rightarrow \{Location,Funding\}$, $\{Supplier\#\} \rightarrow \{Contact\#,Name\}$, $\{Part\#\} \rightarrow \{Color\}$, $\{Supplier\#,Part\#\} \rightarrow \{Price\}$, $\{Project\#,Supplier\#,Part\#\} \rightarrow \{Quantity\}$. For object class *Project*, we can identify *Project#* as its OID by Rule 1, as it is an ID attribute. For object class *Supplier*, as the attribute *Supplier#* functionally determines all its EDLNs, we identify *Supplier#* as its OID, the same as its super OID. For object class *Part*, we combine its EDLNs and OIDs of its ancestor object classes *Supplier* and *Project*, and use the above given FDs to discover the minimal subsets that functionally determine all its EDLNs to be its super OID, which are {Project#, Supplier#, Part#}, {Supplier#, Part#, Quantity} and {Part#, Quantity, Price}. Then we get {Part#}, {Part#, Quantity} and {Part#, Quantity, Price} as OID candidates of object class *Part*.

We use the following heuristics summarized from our observations to choose the best OID from all OID candidates returned by Algorithm 1.

Observation 1. *[OID] In XML schema tree, given an object class* o, *its OID* id_o *is likely to be designed with some of the following features: (1)* id_o *is a single attribute of* o; *(2) The first child node of* o *is (part of)* id_o; *(3)* id_o *contains substring 'Identifier', 'Number', 'Key' or their abbreviations in its tag name; (4)* id_o *has numeric as (part of) its value, and the numerical part is in sequence.*

Observation 1 is based on structural/linguistic characteristics of OIDs designed in real world. Besides, we have two more observations: (1) the number of object classes without relationship attribute is more than the number of object classes with relationship attributes; (2) the number of relationship attributes of binary relationship type is more than the number of relationship attributes of ternary relationship type, and so on. Based on these observations, we collect 204 object classes with their OIDs being manually specified in XML schemas and extract the statistics mentioned above. Using such statistics, we train a Bayesian Network to rank all OID candidates, and choose the best one as its OID. In Example 2, for the object class *Part*, among all its OID candidates, {Project#, Supplier#,Part#} get the highest ranking using our Bayesian Network ranking model. Thus, {*Part#*} is identified as the OID of object class *Part*. More details of our Bayesian Network ranking model can be found in our technical report [5].

Object Attribute and Relationship Attribute Discovery. For an explicit relationship type, we identify its EDLNs that are not role names, as its relationship attributes based on its property (i.e. E4 and RA4 in Table 1). For implicit relationship type, its relationship attributes should appear as EDLNs of the lowest object class participating in the relationship type (RA4 in Table 1), together with the object attributes of that object class. Based on these, we propose Rule 4 to distinguish object attributes and relationship attributes among the EDLNs of each identified object class with OID identified. We use the properties that object attribute can be functionally/multi-valued determined by OID of the object class it belongs to, while relationship attribute can not, to differentiate them.

Rule 4. *[Object Attribute vs. Relationship Attribute] Given an object class* o *and its OID, if an exclusive descendant leaf node* e *of* o *can be functionally/ multi-valued determined by the OID of* o, *then* e *is an attribute of* o, *otherwise it is an attribute of an implicit relationship type which* o *involves in.*

Example 3. In Figure 2, given the object class *Part* with its OID *Part#*, its child node *Color* is functionally dependent on its OID, while *Quantity* and *Price* are not. Thus, we identify *Color* as an object attribute of *Part*, while *Quantity*, *Price* as relationship attributes of some relationship types that *Part* involves in. The corresponding relationship types will be discovered in the following Step 4.

3.4   Step 4: Implicit Relationship Type Discovery

Recall that explicit relationship type can be identified by Rule 2, 3 in Step 2 in Section 3.2. However, there are some implicit relationship types which are not explicitly represented as any node in its XML schema tree. In this section, we classify implicit relationship type into four categories: (1) Implicit relationship type with at least one relationship attribute; (2) Implicit relationship type with IDREF(S) attribute; (3) Implicit relationship type with no relationship attribute and no IDREF(S) attribute, and (4) Identifier Dependency (IDD) Relationship Type [6]. Because of the space limit, we only discuss the first two categories in this paper. The other two categories are discussed in our technical report [5].

Implicit Relationship Type with at Least One Relationship Attribute. For each relationship attribute discovered in Section 3.3 (except those being EDLNs of explicit relationship type), there must be an implicit relationship type it belongs to. Based on the property that relationship attribute should be functionally/multi-valued determined by the OIDs of all object classes involved in the implicit relationship type, to which the relationship attribute belongs (i.e. RA4 in Table 1), we proposed a bottom-up approach, Algorithm 2, to identify the implicit relationship type with its degree, and all involved object classes.

Example 4. In Fig. 2, given a relationship attribute *Price*, object classes *Project, Supplier, Part*, and their OIDs. By Algorithm 2, we find out {*Supplier#,Part#*} → {*Price*}. Then, there is an implicit binary relationship type between *Supplier* and *Part*, with relationship attribute *Price*. For another relationship attribute *Quantity*, {*Supplier#,Part#*} cannot functionally/multi-valued determine it. Then we add in the OID of object class *Project*, and get {*Project#, Supplier#,Part#*}→ {*Quantity*}. Then there is an implicit ternary relationship type among *Project, Supplier* and *Part*, with relationship attribute *Quantity*.

Algorithm 2. Implicit Relationship Type with Relationship Attribute

Input: Relationship attribute A; Object classes $\mathbb{O}$, with OIDs; XML schema tree; XML data
Output: Relationship type $r(\mathbb{C})$ with its involved object classes $\mathbb{C}$ and degree $|\mathbb{C}|$, for each identified relationship attribute in A

1 **foreach** *identified relationship attribute* $ra \in$ A **do**
2   o_i = the lowest ancestor object class of ra.
3      $\mathbb{C} = \{o_i\}$;
4      $SemID_{ra} = id_{oi}$; //id_{oi} is the OID of object class o_i;
5   **foreach** *identified object class* $o_j \in \mathbb{O}$, *along the path from* o_i *to the root in its XML schema tree in bottom-up order* **do**
6          $SemID_{ra} = SemID_{ra} \cup id_{oj}$; //$id_{oj}$ is the OID of object class o_j;
7          $\mathbb{C} = \mathbb{C} \cup \{o_j\}$;
8   **if** $SemID_{ra} \nrightarrow ra$ or $SemID_{ra} \twoheadrightarrow ra$; **then break**;
9   return implicit relationship type $r(\mathbb{C})$ to which ra belongs, object classes in $\mathbb{C}$ as its involved object classes and $|\mathbb{C}|$ as its degree;

Implicit Relationship Type with IDREF(S) Attribute. In XML schema, some designers may design an implicit relationship type by specifying an IDREF(S) attribute under an object class, which references other object class(es). Thus, if an object class has a child node specified as an IDREF(S) attribute, we identify an implicit relationship type between the object class and the object class(es) the IDREF(S) attribute refers to. For some XML schema language (e.g.,DTD), we do not know to which object class(es) the IDREF(S) attribute refer. Based on the property and the heuristic of IDREF(S) attribute listed in Table 1, there are two ways to identify the object classes involved in implicit relationship type: (1) [H3] Tag name of the IDREF(S) attribute may share high linguistic similarity with the tag name of the object class(es) to which it refers, or the corresponding OID(s). We can identify them by research work [9] comparing linguistic similarity. E.g., given two object classes *Department* and *Staff* with their OIDs *Dept#* and *Staff#* respectively, if there is an IDREF(S) attribute under *Staff* with its tag name as *Dept#*, we identify an implicit relationship type between *Department* and *Staff*; (2) If we cannot find high linguistic similarity between the IDREF(S) attribute and any object class or OID, we can use the XML data to identify which object class(es) the IDREF(S) attribute references. A property of IDREF(S) attribute is that [I1] the value range of the IDREF(S) attribute in its XML data must be a subset of the value range of the OID of the object class(es) which it references. E.g., in Fig. 2, if we know that every value of the IDREFS attribute *ContactWith* is also found as a value of OID of object class *Supplier*, there is a high possibility that there is an implicit relationship type between *Employee* and *Supplier*. Furthermore, as the property of IDREF(S) attribute, I1 is also used to verify H3. Although neither of the two ways can 100% guarantee that the object class we discover is the corresponding object class which the IDREF(S) references, we will show the accuracy in our experiments.

4   Experiment

We evaluate the proposed approach for discovering the ORA-semantics in the given XML schemas. The experimental data includes 15 real world data-centric XML schemas, e.g., mondial[2] and XMark[3], etc. For all XML schemas used in our experiments, the average number of internal node is 11 and the average maximal depth is 5. To evaluate the accuracy of our approach, we measure precision, recall and F-measure[4] against a gold standard provided by 8 evaluators. Divergence in their opinions is accounted for by means of an uncertainty factor weighting the results. Further details are given in [5].

4.1   Accuracy of Internal Node Classification

There are totally 512 internal nodes in our input XML schema trees, with their ORA-semantics being labelled (i.e., object class, role name, explicit relationship type, aggregational node or composite attribute). Table 2 shows that the overall accuracy of our rules achieves almost 95% of precision, recall and F-measure. The low precision and recall for explicit relationship type as well as low precision for role name and aggregational node are because the related heuristics used are not as accurate as the properties used in our rules. In Table 3, we show the

---

[2] http://www.cs.washington.edu/research/xmldatasets/www/repository.html
[3] http://www.xml-benchmark.org/
[4] F-measure = 2 * precision * recall/(precision + recall)

Table 2. Precision, recall and F-measure of internal node classification

	Object Class	Role Name	Explicit Relationship Type	Aggregational Node	Composite Attribute	Overall
Precision	99.4%	85.0%	82.9%	81.0%	96.3%	94.7%
Recall	98.4%	94.4%	69.4%	94.4%	96.3%	94.7%
F-measure	98.9%	89.7%	76.2%	87.7%	96.3%	94.7%

Table 3. Statistic information of the internal node in experiment data

	Object Class	Role Name	Explicit Relationship Type	Aggregational Node	Composite Attribute	Total
# of nodes	311	18	49	54	80	512
Percentage	60.7%	3.5%	9.6%	10.5%	15.6%	100%

number and percentage of each ORA-semantics concept in all our collected data sets. Object class is one of the most important ORA-semantics concepts, and its identification helps many XML applications to increase their efficiency and effectiveness as introduced in Section 1. There are 311 object classes among all 512 internal nodes, which take up around 60% of all internal nodes. Other ORA-semantics concepts only take up a small percentage of the internal nodes, especially for role name, which takes up less than 5% of the internal nodes.

We also used a machine learning approach to classify the internal nodes for comparison purpose. We use the properties listed in Table 1 with all our experimental data to train a classification model to classify the internal nodes. In order to avoid bias because the selection of training data, we use 3 folds cross-verification with all the input internal nodes. In 3 folds cross-validation, the original input data is randomly partitioned into 3 portions. Of the 3 portions, a single portion is retained as the validation data for testing the trained classification model, and the remaining 2 portions are used as training data. The cross-validation process is then repeated 3 times, with each of the 3 portions used exactly once as the validation data. The 3 results of precision, recall and F-measure then can be averaged to get the overall accuracy of the trained classification model. We compare the accuracy of our rules with the trained classification model in Fig. 6. We choose the frequently used decision tree algorithm C4.5 [10] to build the classification models. The results show our rules work better than the trained classification models, especially for discovering the explicit relationship types. This is because the explicit relationship type can be designed with different structures in XML schema tree; the classification of a descendent node also cannot help classifying ancestor node as in our approach using rules. Furthermore, the decision trees trained from different training data sets are quite different from each other and most of their branches are not as meaningful as our rules, which shows that they are heavily dependent on the training data. More details about the internal node classification test can be found in [5].

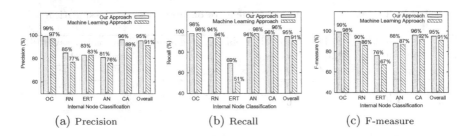

Fig. 6. Comparison of internal node identification (OC: Object Class; RN: Role Name; ERT: Explicit Relationship Type; AN: Aggregational node; CA: Composite Attribute)

4.2   Accuracy of Leaf Node Classification

Recall that our approach may return more than one OID candidates for each object class, thus we build a Bayesian Network to rank all its OID candidates, and choose the highest ranked candidate as its OID. There are 311 object classes with their OIDs in our experimental data. We randomly choose 2/3 of them for training and the rest for testing. From the training data, we collect the statistics of the features mentioned in Observation 1, and build a Bayesian Network, which returns us the probability of an OID candidate being the correct OID based on the statistics. More details of building the Bayesian Network are discussed in [5].

In our step-by-step approach, outputs of the previous step will work as the inputs for a latter step. The accuracy of the latter step is affected by the accuracy of its previous steps. To show the accuracy of each step, we conduct two groups of experiments to evaluate the precision, recall and F-measure of our approach for leaf node classification, one with user verification, which means all object classes have been correctly labelled in XML schema trees, and the other one based on the results of our internal node classification without user verification.

Fig. 7 shows our approach for leaf node classification get above 90% of precision, recall and F-measure. Even without user verification, the precision/recall only drop slightly, as our approach for discovering object class also gets high precision and recall. The low precision of implicit relationship attribute is because

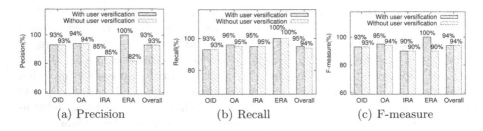

Fig. 7. Precision, Recall and F-measure of Leaf Node Classification (OA: Object Attribute, ERA: Explicit Relationship Attribute, IRA: Implicit Relationship Attribute)

its identification is heavily depended on FDs/MVDs among the corresponding XML data, which may not be large enough to return all the correct FDs/MVDs.

4.3   Accuracy of Implicit Relationship Type Discovery

We also conduct experiments on our approach for implicit relationship type discovery. Similar to the leaf node classification, we conduct two groups of experiments to evaluate its accuracy, one with user verification, and the other one without user verification. Fig. 8 shows that our approach to discover implicit relationship types has high precision, recall and F-measure. Because of the space limit, more detailed breakdown is given in [5].

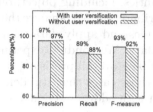

Fig. 8. Precision, Recall, F-measure of Implicit Relationship Type Discovery

5   Related Work

To the best of our knowledge, only a few research works have frontally addressed the problem of automatically discovering the implicit semantics embedded in XML schema and XML data. Most existing works in semantics discovery in XML data only focus on objects. In [2], in the context of view design, all internal nodes in an XML schema tree are considered as object classes. In the context of keyword search, XSeek [7] also infers semantics from XML schemas to identify return nodes. This work infers semantics of objects by using the repeatable node. In [11], the authors build a data graph from an XML document. However, they just focus on objects and properties, still missing lots of meaningful semantics. Compared to our work, the existing works have two major drawbacks. First, they only consider the semantics of object, ignoring many other important ORA-semantics concepts which may play an important role in XML applications, as illustrated. Second, even for object, the inference accuracy of the existing works is quite low. For example, most of them will treat relationship attribute as object attribute when the relationship is implicit. In contrast, our work focuses on a comprehensive set of ORA-semantics concepts, and has high inference accuracy.

Semantics are also captured in other domains. In [8] authors proposed a form-driven approach, which firstly transforms the relational database to a set of form model schemas, each of which is essential a view on the underlying database, and then extracts the corresponding ER schema from them; [4] resolves the reference ambiguation problem, which means an attribute is actually referencing another attribute but cannot be detected due to the inconsistent name issue, by also considering their neighbor attributes. However, as the underlying data is flat in relational database, and these approaches only try to identify relationships between relations through key-foreign key constraints, they still cannot identify ternary or n-nary relationships as well as the relationship attributes.

6   Conclusion and Future Work

ORA-semantics is important for many XML applications. Existing works in semantics discovery only focus on object, ignoring many other important concepts such as relationships. In this paper we present a novel approach to identify a comprehensive set of ORA-semantics, including object, object ID, explicit/implicit relationship, relationship attribute, etc. We analyze the properties of each semantics concepts, and propose rules and apply data mining techniques to discover them in XML schema and data. We conduct experiments to demonstrate our approach can achieve almost 95% overall precision, recall and F-measure.

We are now investigating those cases that still defeat our proposed approach and consider its combination with additional domain knowledge and ontologies.

References

1. Aumueller, D., Do, H.H., Massmann, S., Rahm, E.: Schema and ontology matching with COMA++. In: SIGMOD Conference, pp. 906–908 (2005)
2. Chen, Y.B., Ling, T.W., Lee, M.L.: Designing valid XML views. In: Spaccapietra, S., March, S.T., Kambayashi, Y. (eds.) ER 2002. LNCS, vol. 2503, pp. 463–477. Springer, Heidelberg (2002)
3. Hegewald, J., Naumann, F., Weis, M.: Xstruct: Efficient schema extraction from multiple and large XML documents. In: ICDE Workshops, p. 81 (2006)
4. Kalashnikov, D.V., Mehrotra, S.: Domain-independent data cleaning via analysis of entity-relationship graph. ACM Trans. Database Syst. 31(2), 716–767 (2006)
5. Li, L., Le, T.N., Wu, H., Ling, T.W., Bressan, S.: Discovering semantics from data-centric XML. Technical Report TRA6/13, National University of Singapore
6. Ling, T.W., Lee, M.L., Dobbie, G.: Semistructured database design (2005)
7. Liu, Z., Chen, Y.: Identifying meaningful return information for XML keyword search. In: SIGMOD Conference, pp. 329–340 (2007)
8. Mfourga, N.: Extracting entity-relationship schemas from relational databases: A form-driven approach. In: WCRE, pp. 184–193 (1997)
9. Mizuta, S., Hanya, K.: Specifications of word set in linguistic approach for similarity estimation. In: BICoB, pp. 25–29 (2010)
10. Quinlan, J.R.: C4.5: Programs for Machine Learning. Morgan Kaufmann (1993)
11. Y.S.: A personal perspective on keyword search over data graphs. In: ICDT (2013)
12. Xu, Y., Papakonstantinou, Y.: Efficient lca based keyword search in XML data. In: EDBT, pp. 535–546 (2008)
13. Yu, C., Jagadish, H.V.: XML schema refinement through redundancy detection and normalization. VLDB J. 17(2), 203–223 (2008)

Finding Image Semantics from a Hierarchical Image Database Based on Adaptively Combined Visual Features

Pritee Khanna[1], Shreelekha Pandey[1], and Haruo Yokota[2]

[1] PDPM Indian Institute of Information Technology, Design and Manufacturing Jabalpur,
Dumna Airport Road, Jabalpur 482-005 M.P. India
{pkhanna,shreelekha}@iiitdmj.ac.in
[2] Graduate School of Information Science and Engineering, Tokyo Institute of Technology,
2-12-1 Ookayama, Meguro-kuTokyo, 152-8552 Japan
yokota@cs.titech.ac.jp

Abstract. Correlating image semantics with its low level features is a challenging task. Although, humans are adept in distinguishing object categories, both in visual as well as in semantic space, but to accomplish this computationally is yet to be fully explored. The learning based techniques do minimize the semantic gap, but unlimited possible categorization of objects in real world is a major challenge to these techniques. This work analyzes and utilizes the strength of a semantically categorized image database to assign semantics to query images. Semantics based categorization of images would result in image hierarchy. The algorithms proposed in this work exploit visual image descriptors and similarity measures in the context of a semantically categorized image database. A novel 'Branch Selection Algorithm' is developed for a highly categorized and dense image database, which drastically reduces the search space. The search space so obtained is further reduced by applying any one of the four proposed 'Pruning Algorithms'. Pruning algorithms maintain accuracy while reducing the search space. These algorithms use an adaptive combination of multiple visual features of an image database to find semantics of query images. Branch Selection Algorithm tested on a subset of 'ImageNet' database reduces search space by 75%. The best pruning algorithm further reduces this search space by 26% while maintaining 95% accuracy.

1 Introduction

Cognitive psychology defines categories by grouping "similar objects" and super-categories by grouping "similar categories". Semantic categories form clusters in visual space, and visual similarity is correlated to semantic similarity [1]. Humans can easily correlate these similarities which gives them enormous power to distinguish a large number of objects. Content Based Image Retrieval (CBIR) systems use only visual similarity obtained in terms of low level image features to interpret images [2-4]. The lack of coincidence between the high level semantic and the low-level features of an image is known as semantic gap [5]. In an attempt to reduce semantic gap, proposed work aims to correlate visual similarity and semantics of images in a

H. Decker et al. (Eds.): DEXA 2013, Part I, LNCS 8055, pp. 103–117, 2013.

semantically categorized large image database. Semantic based categorization of an image database would result in categories and subcategories of images. Visual features of images in such a database forms a semantics based hierarchical search space. This tree is searched to assign semantics to query images. For efficient search, it is not advisable to traverse the entire tree or even an entire branch. A novel 'Branch Selection Algorithm' effectively traverses this hierarchical search space and selects a few subtrees to search. Pruning Algorithm further reduces this search space, while maintaining the accuracy. An adaptive combination of multiple visual features and similarity measures are used to design branch selection and pruning algorithms. To ensure the applicability of the proposed algorithms, their performance has been tested on a subset of ImageNet database.

The paper is organized as follows. A review of the related research is given in Section 2. Section 3 emphasizes on correlating visual and semantic similarity. Section 4 gives an insight of related databases. Proposed system is explained in Section 5. Experimental setup is given in Section 6. Section 7 summarizes results and discussion on related issues. Finally, Section 8 concludes the work.

2   Related Work

Computer Vision and Machine Learning approaches use learning based systems to reduce semantic gap [6-7]. It has already been recognized that learning accompanied by object extraction produces good results [8]. In [9], semantic templates are automatically generated during the process of relevance feedback. WordNet is used to construct a network of such semantic templates, which helps in retrieving images based on semantic. The system works on 500 images from categories like human, animal, car, etc. A statistical modeling approach for automatic linguistic indexing of pictures is introduced in [6]. Each of the 600 concepts is represented by a two-dimensional multi-resolution hidden Markov model and is trained using categorized images. A likelihood function measures the extent of the association between an image and the textual description of a concept. The model given in [10] learns visual recognition from semantic segmentation of photographs. For efficient labeling of object classes, a combination of integral image processing and feature sharing is employed. The developed classifier reports 70.5% region-based recognition accuracy on a 21-class database. The work presented in [7] focuses on using a few training images for quick learning. Generative probabilistic models of object categories are learned using a Bayesian incremental algorithm. The system quoted a feasible real-time learning rate for 101 object categories. A region-based image retrieval system with high-level semantic learning is given in [11]. The system uses a decision tree based image semantic learning algorithm but learns natural scenery image semantics only.

Besides in literature, one can find a few more learning based techniques to minimize the semantic gap [5]. Unlimited number of concepts in the real world is a major hindrance for learning based approaches. Most of the works have considered non-hierarchical image database with thousands of images. The proposed work uses a hierarchical image database to correlate visual similarity with semantic similarity. Such a correlation would be an asset to people working in image processing and computer vision. The main aim of this study is to efficiently assign semantics to images through such correlations. Instead of using any of the available learning techniques, this work exploits the inherited features of a hierarchical image database.

3   Correlating Visual Similarity with Semantics of Images

Humans have natural instinct in distinguishing object categories, both in visual as well as in semantic space, but to accomplish this computationally is yet to be fully explored. The semantic based categorization of images would give a hierarchical tree structure having images of different categories at various levels. The focus of this work is to explore whether semantic categories (e.g. dog, flower, mountains etc.) can also be visually segregated.

A semantically categorized database may contain images belonging to a domain or spread over multiple domains. It becomes difficult for a common user to search such database if the nature of its classification or the exact semantics required for the search is unknown. For example, medical terminology is an obvious choice for categorizing medical images but it is very difficult for a common user to understand semantics of these categories and hence finding proper keywords to search the database. In another scenario, a categorized database may have ten categories corresponding to dog based on their breeds, tail, coat etc. Looking at the image of a white dog with black spots, a user may not exactly know its breed name i.e. 'Dalmatian'. The user can derive such knowledge (semantics) by our approach.

Proposed approach utilizes visual features of a categorized image database of any depth and height to determine semantics of images. Huge size of the search space demands algorithms which keep only the desired categories/subcategories in consideration during search. A novel 'Branch Selection Algorithm' has been designed and tested on a large hierarchical image database.

4   Related Databases and Database Used for Experimentation

The nature and scope of image data influences the performance of retrieval algorithms. For decades, in the absence of standard test data, researchers used self-collected images to show their results. Many domain specific and uncategorized databases came into existence lately for example, WANG, UW, IRMA 10000, ZuBuD, and UCID [3]. Some more challenging datasets are Caltech 101/256 [7], Coral Image, Tiny Image, ESP, LabelMe, Lotus Hill, and ImageNet [12].

A publicly available, densely populated, and semantically organized hierarchical image database covering a wide range of domains was required for experimentation. With large number of images for nearly all object classes, ImageNet serves the purpose. Built upon the backbone of the WordNet structure, a subset of ImageNet 2011 Winter Release given in Table 1 is used for experimentation. A category in ImageNet corresponds to a synonym set (synset) in WordNet. Fig. 1 shows some representative images of ImageNet.

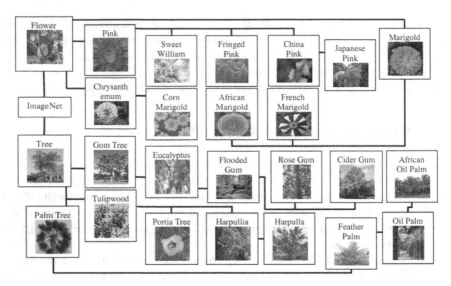

Fig. 1. A snapshot of Flower and Tree subtrees of ImageNet 2011 Winter Release

Table 1. Subset of ImageNet database used for experimentation

Subtree	Width	Depth	# of Synsets	# of Images (K)
Animal	9	9	32	38
Appliance	4	4	29	32
Fabric	2	5	12	11.5
Flower	9	3	24	26
Fruit	6	5	42	30.5
Geological Formation	5	5	50	55
Person	12	4	34	16.5
Sport, Athletic	5	4	23	30.5
Structure	6	6	36	33
Tree	7	6	42	24
Vegetable	6	5	41	35
Total (on an average 910 images per synset)			**365**	**332 K**

5   Methodology

The work visualizes categories/subcategories of a semantically categorized image database as nodes in the image tree. The flow of execution shown in Fig. 2 starts with an offline extraction of visual features of images. Visual features of images belonging to a node form visual signatures of that node. On the basis of the distance between query image and visual signatures of nodes, the Branch Selection Algorithm selects some subtrees to search. This search space is further reduced by pruning algorithms. Retrieval module assigns semantics of the nodes at lower distances to the query image. The proposed system supports both types of searches, i.e. aimed search to get a specific semantic; and category search to find a group of similar semantics.

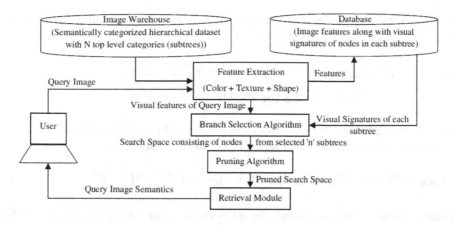

Fig. 2. Work flow of the proposed system

5.1   Feature Extraction Techniques

Conventionally, color, texture, and shape features are used to measure visual similarity of images. The combination of these features gives better results [13].

Color Features. Color is one of the most widely used low-level visual features. It is invariant to size and orientation of image [2]. It shows the strongest similarity to human eye [14]. Color histogram is the most commonly used representation. Various versions of histogram e.g. cumulative histograms, quantized color space histograms have been proposed [3, 15-16].

This work uses a histogram with perceptually smooth color transition in HSV color space [17]. When applied on an image as a whole, the Global Color Histogram (GCH) feature is obtained. Five color histograms corresponding to five regions (central ellipsoidal region and four surrounding regions) are concatenated to form a Local Color Histogram (LCH) feature. In general, GCH and LCH are represented as (1).

$$F_{GCH} = (h_1, h_2, \dots, h_{51}, i_{52}, \dots, i_{68}); h, i \; represent \; hue \; and \; intensity \; resp.$$

$$F_{LCH} = (h_1^1, h_2^1, \dots, h_{51}^1, i_{52}^1, i_{53}^1, \dots, i_{68}^1, \dots h_1^5, h_2^5, \dots, h_{51}^5, i_{52}^5, i_{53}^5, \dots, i_{68}^5).$$

(1)

Another popular color feature is a statistical model of color representation [18-19]. Color distribution of each channel of an image is uniquely characterized by its three central moments i.e. average (E_i), variance (σ_i) and skewness (s_i) as given in (2).

$$E_i = \frac{1}{N}\sum_{j=1}^{N} p_{ij}, \sigma_i = \left(\frac{1}{N}\sum_{j=1}^{N}(p_{ij} - E_i)^2\right) \; and \; s_i = \left(\frac{1}{N}\sum_{j=1}^{N}(p_{ij} - E_i)^{\frac{1}{3}}\right).$$

(2)

p_{ij} = value of ith color channel at j^{th} image pixel and N = number of image pixels. Images are compared by taking a weighted sum of differences of corresponding color moments. Similarity between two images with r color channels and color moments $(E_{i1}, \sigma_{i1}, s_{i1})$ and $(E_{i2}, \sigma_{i2}, s_{i2})$ is given in (3).

$$d_{mom} = \sum_{i=1}^{r} w_{i1}|E_{i1} - E_{i2}| + w_{i2}|\sigma_{i1} - \sigma_{i2}| + w_{i3}|s_{i1} - s_{i2}| .$$

$$\text{where, } w_{kl} \geq 0 \text{ is specified by the user [19].}$$

(3)

Similar to histogram, Global Color Moment (GCM) and Local Color Moment (LCM) features of an image in HSV color space are obtained as shown in (4).

$$F_{GCM} = (E_1, \sigma_1, s_1, E_2, \sigma_2, s_2, E_3, \sigma_3, s_3); \; F_{LCM} = (E_1^1, \dots, s_3^1, \dots, E_1^5, \dots, s_3^5) .$$ (4)

Texture Features. Texture captures the information of patterns lying in an image. An image may contain textures of different degrees of detail. Grey level co-occurrence matrices (GLCM) and Tamura Features are popular single scale texture features. Multi-resolution texture features include Pyramidal Wavelet Transform (PWT), Tree-Structured Wavelet Transform (TSWT), Discrete Cosine Transform (DCT), Gabor filters, and ICA Filters [14]. The most frequently used Gabor filter is given by (5).

$$g(x,y) = \left(\frac{1}{2\pi\sigma_x\sigma_y}\right) exp\left[-\frac{1}{2}\left(\frac{x^2}{\sigma_y^2} + \frac{y^2}{\sigma_y^2}\right)\right] + 2\pi jWx .$$

$$g_{mn}(x,y) = a^{-m}g(x',y'); \; m,n = int, m = 0,1,\dots,S-1,$$

$$x' = a^{-m}(x \cos\theta + y \sin\theta), y' = a^{-m}(-x \sin\theta + y \cos\theta) .$$

(5)

Suitable dilations and rotations of the Gabor function g(x,y) through the generating function g_{mn} give a self-similar filter dictionary. Here $\theta = n\pi/K$, K = total number of orientations, S = number of scale, a = (U_h/U_l)-1/(S-1). U_h and U_l are upper and lower centre frequencies of interest [20]. This work uses Gabor filter with four scales and six orientations. For retrieval purposes the most commonly used measures are mean μ_{mn} and standard deviation σ_{mn} of the magnitude of the wavelet transform coefficients. The resulting Gabor Texture (GT) feature vector is given in (6).

$$F_{GT} = (\mu_i, \sigma_i), \quad i = 1,2,3,\dots,24$$ (6)

Shape Features. Shape features are powerful descriptors in image retrieval. Generic Fourier Descriptors, Zernike and Pseudo Zernike Moments, and Wavelet Descriptors are some popular representations [14]. Recent researches focus on computationally efficient local image descriptors. Scale Invariant Feature Transform (SIFT) extract large number of keypoints from image that leads to robustness in extracting small objects among clutter [8, 21]. This work uses SIFT with 4 octaves and 5 levels. K-means clustering forms 32 clusters per image [3]. For each cluster, count, mean and variance form a SIFT Shape (SS) feature vector given in (7).

$$F_{SS} = (CV_1, CV_2, \ldots, CV_3), (MV_{1,1}, MV_{1,2}, \ldots MV_{32,128}), (VV_{1,1}, VV_{1,2}, \ldots VV_{32,128}) . \quad (7)$$

5.2   Construction of Visual Signature of a Node/Category

To correlate low-level visual features and high level semantics of images belonging to a node, a visual signature is attached to each node. Feature vector of an image is a combination of GCH, LCH, GCM, LCM, GT and SS. Mean feature vectors of all the images in a node, GCHmean, LCHmean, GCMmean, and LCMmean, GTmean, and SSmean form its visual signature. To get the semantics of an image, Branch Selection and Pruning algorithms make use of the similarity measures summarized in Table 2.

Table 2. Visual signatures and similarity measures

Visual Signature	Similarity Measure
GCHmean, LCHmean	Vector Cosine Distance
GCMmean, LCMmean	City Block Distance
GTmean	Euclidean Distance
SSmean	Earth Mover's Distance

5.3   Branch Selection Algorithm

The work proposes a novel Branch Selection Algorithm given in Fig. 3.

Steps to find the subtrees, semantically similar to query image, at each level are as follows:

Step 1:   Calculate feature vector of the query image.

Step 2:   Let there are N nodes at this level. Calculate the distance of query image with N nodes. For each feature, select n subtrees (n ≤ N) having minimum distance from the query image. This results in three lists, one corresponding to each feature, containing n entries. It gives rise to any of the three possibilities:

a. If subtree X is 1st choice in all the three lists, then select only this subtree for search. As root of this subtree X has the closest distance with query image with respect to all the feature vectors considered. Go to Step 3.

b. If subtree X is 1st choice for any two lists, then select this subtree X for search. In addition,

  i.   Select (n-1) more subtrees having maximum frequency of appearance in the two lists where X is 1st choice, and go to Step 3. In case, subtrees have same frequency then go to Step (ii).

  ii. Select one/more subtrees which have minimum sum of distances based on all 3 features.

  iii. Go to Step 3, if (n-1) subtrees are selected by now, otherwise go to Step (i).

c. If 1st choice of subtrees for all 3 lists is different, then select top n subtrees based on the maximum frequency of their appearance in these 3 lists. In case of a tie, select one/more subtrees which have minimum sum of distances based on all 3 features.

Step 3:   Repeat Step 2 for subtrees at every level.

Fig. 3. Branch Selection Algorithm

Branch Selection Algorithm selects a few subtrees (n) out of 'N' available at the first level of the image tree. Only limited nodes that belong to these n subtrees are searched to find semantics of the query image. The algorithm aims to reduce the search space as much as possible, without compromising the accuracy of the system. The sum of distances based on GCH, LCH, GCM, and LCM is color distance. Distance based on GT is texture distance, and sum of distances based on SIFT Mean and Variance is shape distance. The algorithm prepares three lists corresponding to these distances and 'adaptively' selects a branch.

Performance of the system greatly depends on the value of n chosen. Experimental results for n=N/4, allows 75% pruning of the actual search space in terms of subtrees. Initial pruning for more than this results in rejection of the target subtree most of the time and therefore it is not fruitful to generate further results on its output.

An output of this algorithm for n=3 is shown in Fig. 4, where query image "n00450866_898" has been taken from "pony-trekking" synset. At the first level 11 subtrees are used for experimentation. The algorithm selects 3 subtrees (concepts) i.e. Geological Formation, Tree, and Sport, Athletic. At the subsequent levels, synsets of these three high level semantics are chosen to get the complete search space for this query image. In this case, algorithm selects only 51 synsets out of the total 365 synsets in the image tree. Thus search space is reduced by 86% w.r.t. number of synsets to be searched, still keeping the desired subtree in consideration.

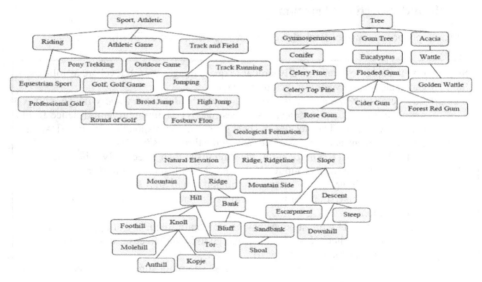

Fig. 4. Output of the Branch Selection Algorithm (n=3) for query image "n00450866_898"

5.4   Pruning Algorithms

Branch Selection Algorithm applied on an image database results in any number of nodes depending on the height and width of the n subtrees chosen in its step 2. Pruning of this search space would further improve the performance. This pruning helps in retaining good nodes while discarding the bad nodes of a selected subtree. A good node is the one that lies on the path leading to the node containing images semantically similar to the query image, while a bad node leads to either a different path in the same subtree or a different subtree. Ideally, bad nodes and their subtrees are to be pruned.

Goodness of a node is tested in terms of distances explained later in this section. In "strict pruning", the whole subtree is pruned if its root fails to prove itself good. While developing pruning approaches, it is observed that often a particular node on the path does not fulfills the criteria of being good one but the query image belongs to some lower level node of that path. Based on this observation, a "soft pruning" is proposed, which removes only the so-called bad node from the path and not the entire subtree following it. The children of this bad node become the children of its parent.

Fig. 5 explains these approaches with the same query image n00450866_898. Strict pruning shown in Fig. 5(a) loses the target synset "pony-trekking" because its parent "riding" fails to prove itself good. A less restricted soft pruning approach shown in Fig. 5(b) preserves the target synset even if its parent is being neglected. This less restrictive approach for pruning is followed in this work.

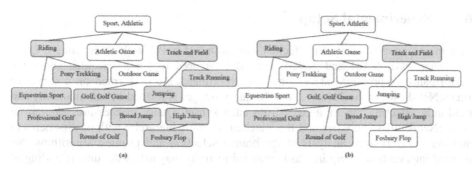

Fig. 5. (a) Strict Pruning. (b) Soft Pruning Approaches (gray nodes are the pruned ones).

Fig. 6 shows proposed pruning algorithms working on the distances corresponding to the dominant visual feature. Dominant feature of a subtree is the feature which gives top rank to this subtree. Dqs_i is the distance (already calculated) between query image and i^{th} node of the subtree (having N_s nodes) w.r.t. dominant feature. Dmean and Dmed are mean and median of Dqs. If a subtree having 10 nodes is given top ranking by texture feature, then mean and median of the GT based distances between query image and each of these 10 nodes are calculated. Additionally, extended mean distance (Dmeanx) and extended median distance (Dmedx) are calculated as shown in (8). Dmeanx is the sum of Average Absolute Deviation (AAD) in Dmean. Dmedx is the sum of Median Absolute Deviation (MAD) in Dmed. AAD and MAD are less affected by extreme observations than are variance and standard deviation [22].

$$Dmeanx = Dmean + \sum_{i=1}^{N_s}(|Dqs_i - Dmean|/N_s) .$$

$$Dmedx = Dmed + median(|Dqs_i - Dmed|)$$

(8)

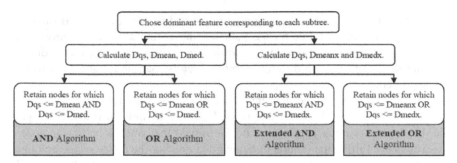

Fig. 6. Pruning Algorithms

In the quest of good pruning algorithms, the four possible combinations shown in Fig. 6 are exhaustively tested. The performance of pruning algorithms is judged on two parameters: The number of nodes retained to be searched in retrieval module and the ability to preserve the nodes appearing in the path ending at the target synset. The nodes retained by pruning algorithm are used to assign semantics to the query image.

6   Experimental Setup

The most common computing facility consisting of a PC with Intel Core 2 Quad processor, 8GB RAM and 500GB hard disk is used to get a fair idea about the performance of proposed algorithms. All experiments are performed on a subset of ImageNet shown in the Table 1. A set of query images is formed by automatically and randomly selecting 5% of images from each synset. Query images are taken from ImageNet database only because the attached semantic hierarchy with images helps to automate performance analysis of the branch selection and pruning algorithms. No manual intervention is required as human subjectivity may affect the understanding of correlation between visual and semantic similarity.

Feature vector of images and visual signatures of nodes in the database are generated through an offline procedure. During experimentation new images are not inserted in the database. In real life scenario, database may be kept in the updated mode. This insertion intensiveness can be easily handled in the online version. Insertion of an image requires its feature extraction and re-computation of the visual signature of the node to which this image is added.

7   Results and Discussion

The following discussion establishes a correspondence between visual similarities and semantic similarity in a semantically categorized hierarchical image database.

7.1   Performance of Branch Selection Algorithm

Branch Selection Algorithm prunes the search space but the precise selection of target subtree based on the query image is to be ensured. This selection is expressed in terms of 'Precision' that denotes the selection of target subtree in terms of percentage. The graph in Fig. 7(a) shows the performance of Branch Selection Algorithm on 11 subtrees of ImageNet (Table 1) for n=3. This prunes the search space by 75%. Out of 11 hierarchies tested, 9 give more than 50% precision, while precision of 70% or more is achieved for 6 hierarchies. The algorithm out-performs if a query image is from 'Appliance' (94%), but opposite is the case if it is 'Fabric' or 'Sports, Athletic' (30%). This happens due to the nature of the images that constitutes these categories. Fig. 8 gives a glimpse of some of the images at the top level categories for these synsets. Appliance synset contains visually as well as semantically closer images, while 'Fabric' or 'Sports, Athletic' consist of images poorly related on the semantics. Visual signatures of nodes having dissimilar images do not represent these nodes well. This greatly affects the performance of the algorithm. Average precision @ 3 over all 11 branches is 65.36%, but on removing the two outliers i.e. 'Fabric' and 'Sports, Athletic', it becomes 73.22%, which is fairly acceptable. The target subtree is selected for approximately 75%, while pruning the search space by the same amount.

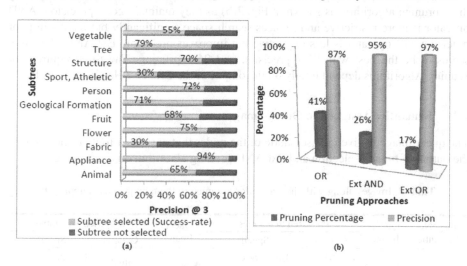

Fig. 7. Performance of (a) Branch Selection Algorithm. (b) Pruning Algorithm.

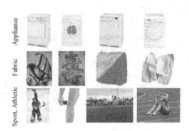

Fig. 8. Some representative images of three categories

The execution time of the algorithm is significantly affected by the width and depth of the subtrees selected at level 1. If the width of subtrees is more, then it would lead to selection of more subtrees and searching of; while higher depth means more iteration. On an average, execution time of the algorithm is 70 sec. For an online search this time is high but considering the computational facility used and the absence of an appropriate indexing of image features with this size of database, results are encouraging. In the real time environment algorithm would be executed at the server end with indexed feature vectors, which will significantly reduce the time.

7.2   Performance of Pruning Algorithms

It is desirable to have pruning algorithms with high pruning percentage and high precision. It is difficult to achieve high precision with high pruning percentage as they are inversely proportional to each other. The best algorithm would be the one that gives the highest pruning percentage with the desired precision. Performance of only three pruning algorithms is shown in Fig. 7(b) as they maintain good precision. AND operator is more restrictive and reduces search space significantly but results in poor precision. OR operator is less restrictive but improves precision. Ext AND algorithm seems to be the best with 95% precision and 26% pruning. The time required for Pruning Algorithms depend totally on the output of the Branch Selection Algorithm.

7.3   Semantics Assigned to Query Images

The query image is given the semantic of the nodes that are closer to it. Table 3 shows the output of branch selection and Ext AND pruning algorithm on query images.

Table 3. Images along with the semantics assigned to them by proposed approach

ImageNet Semantic	Query Image	Proposed Semantics		Query Image	Proposed Semantics	
		General	Specific		General	Specific
Animal		Animal Vegetable Geological Formation	Live stock, Ravine, Insectivore, Draw		Animal Tree Sports	Ungulate, Pachyderm, Animal, Ming tree
Appliance		Appliance Structure Person	Clothes dryer, Refrigerator, Coffee maker, Electric range		Appliance	Deep freeze, Clothes dryer, Oven, Wringer
Fabric		Fabric Fruit Structure	Hand towel, Viscos rayon, Towel, Honeydew		Fabric Appliance Sports	Rayon, Fabric, Towel, Pony-trekking
Person		Person Appliance Vegetable	Optimist, Personification, Neutral, Refrigerator		Person Fruit Sports	Neutral, Master of ceremonies, Entertainer, Person

The output shows the top four semantics assigned to the query image. A query image from any category; say Person, retrieves not only a general semantic 'Person' but also a number of specific semantics like 'Optimist, Personification, Neutral', etc. Presence of misclassified images in the database adversely affects the performance of proposed algorithms.

Table 4 lists some conflicting images in ImageNet. For example, the first image in the Table 4 belongs to 'animal' category while visually it seems to be a 'structure'. Proposed approach keeps it closer to the 'structure' semantics. The proposed approach also helps to identify such cases and reclassification of these images will further improve the performance.

Table 4. Some misclassified images and their correct classification by the proposed approach

Image	ImageNet Semantic	Proposed Semantics	
		General	Specific
	Animal	Structure, Animal, Vegetable	Parapet, Otter shrew, Support, Elephant
	Fruit	Tree, Sports, Flower	Gum tree, Gymnospermous, Conifer, Eucalyptus
	Fruit	Tree	Gymnospermous, Gum tree, Rose gum, Tree
	Flower	Tree, Geological Formation, Vegetable	Ravine, Forest red gum, Rose gum, Eucalyptus

7.4   Other Issues

Size of the Visual Signature of a Node. The size of the visual signature of a node although large for an online application, but the algorithms assign efficient semantics to the images. In future, efforts would be made to obtain compact visual signatures.

Lack of Comparative Evaluations. As most of the available image databases are flat in nature, the performance of proposed algorithms cannot be compared. Due to lack of hierarchy, subtrees selected by the Branch Selection Algorithm contain only a single node, which serves as both the root and the leaf. Pruning algorithms are also insignificant for flat structures. Further, most of the work done in this field is based on the personal databases and thus, it is not possible to get the results of the proposed algorithms on those databases.

In the present work, for the purpose of comparison, WANG database is categorized at the top level. Table 5 shows the performance of the proposed Branch Selection Algorithm on WANG and compares it with other related work. It gives an overall precision of 94.2% with 75% reduction in the search space. As a result the retrieval process is much faster in comparison to other approaches.

Table 5. A comparison on WANG database using average precision values

Category	Proposed Approach	F. Malik et al. [23]	R. Gali et al. [24]	P. Kinnaree et al. [25]
Reduction in search space	75%	0%	0%	0%
Africa	0.93	1	0.76	1
Beach	0.9	0.58	0.587	1
Bus	0.96	0.61	0.963	1
Dinosaur	1	0.71	1	1
Elephant	0.96	0.49	0.741	1
Flower	0.97	0.58	0.945	1
Food	0.9	0.48	0.733	1
Horse	0.95	0.72	0.941	1
Monument	0.9	0.57	0.714	1
Mountain	0.95	0.47	0.457	1
Average	0.942	0.621	0.7841	1

8   Conclusion and Future Scope

The paper discusses an open ended problem of semantic gap and proposes some algorithms to correlate visual and semantic similarity. The algorithms are developed for semantically categorized image database. The experiments show that visual features based on the adaptive combination of multiple low level features of image may serve well for a semantically categorized large image database. It shows that if categorized properly, low level features of the images can be combined with their semantics. The selection of good nodes by proposed algorithms ensures better performance of the system. Derived semantics can be used for effective image retrieval as a future research. Proper indexing of visual signatures can significantly reduce the time required for Branch Selection Algorithm. Inclusion of user feedback will also enhance the performance of retrieval system.

Acknowledgements. The authors acknowledge the support provided by JSPS to carry out this work under JSPS Invitation Fellowship for Research in Japan (Long-Term).

References

1. Sternberg, R.J.: Cognitive Psychology, 5th edn. Wadsworth Cencage Learning, Belmont (2008)
2. Datta, R., Joshi, D., Li, J., Wang, J.Z.: Image Retrieval: Ideas, Influences, and Trends of the New Age. ACM Comput. Surv. 40(2), 5–60 (2008)
3. Deselaers, T., Keysers, D., Ney, H.: Features for Image Retrieval: An Experimental Comparison. Inf. Retr. 11(2), 77–107 (2008)
4. Liu, Y., Zhang, D., Lu, G., Ma, W.Y.: A survey of content-based image retrieval with high-level semantics. Pattern Recogn. 40(1), 262–282 (2007)
5. Wang, H.H., Mohamad, D., Ismail, N.A.: Semantic Gap in CBIR: Automatic Objects Spatial Relationships Semantic Extraction and Representation. IJIP International Journal of Image Processing 4(3), 192–204 (2010)

6. Li, J., Wang, J.Z.: Automatic Linguistic Indexing of Pictures by a Statistical Modeling Approach. IEEE Trans. Pattern Anal. Mach. Intell. 25(9), 1075–1088 (2003)
7. Fei-Fei, L., Fergus, R., Perona, P.: Learning generative visual models from few training examples: An incremental Bayesian approach tested on 101 object categories. Comput. Vis. Image Underst. 106(1), 59–70 (2007)
8. Lowe, D.G.: Object recognition from local scale-invariant features. In: International Conference on Computer Vision, ICCV 1999, vol. 2, pp. 1150–1157 (1999)
9. Zhuang, Y., Liu, X., Pan, Y.: Apply Semantic Template to Support Content-based Image Retrieval. In: Proceeding of IS&T and SPIE Storage and Retrieval for Media Databases, San Jose, California, USA, January 23-28, pp. 442–449 (2000)
10. Shotton, J., Winn, J., Rother, C., Criminisi, A.: *TextonBoost*: Joint appearance, shape and context modeling for multi-class object recognition and segmentation. In: Leonardis, A., Bischof, H., Pinz, A. (eds.) ECCV 2006, Part I. LNCS, vol. 3951, pp. 1–15. Springer, Heidelberg (2006)
11. Liu, Y., Zhang, D., Lu, G.: Region-based image retrieval with high-level semantics using decision tree learning. Pattern Recogn. 41(8), 2554–2570 (2008)
12. Deng, J., Dong, W., Socher, R., Li, L.J., Li, K., Fei-Fei, L.: ImageNet: A Large-Scale Hierarchical Image Database. In: IEEE Conference on Computer Vision and Pattern Recognition, pp. 248–255 (June 2009)
13. Wang, X.Y., Yu, Y.J., Yang, H.Y.: An effective image retrieval scheme using color, texture and shape features. Computer Standards & Interfaces 33(1), 59–68 (2011)
14. Vassilieva, N.S.: Content-based Image Retrieval Methods. Program. Comput. Softw. 35(3), 158–180 (2009)
15. Liu, G.H., Yang, J.Y.: Content-based image retrieval using color difference histogram. Pattern Recogn. 46(1), 188–198 (2013)
16. Pandey, K.K., Mishra, N., Sharma, H.K.: Enhanced of color matching algorithm for image retrieval. International Journal of Computer Science Issues 8(3), 529–532 (2011)
17. Sural, S., Qian, G., Pramanik, S.: A Histogram with Perceptually Smooth Color Transition for Image Retrieval. In: 4th International Conference on Computer Vision, Pattern Recognition and Image Processing, Durham, North Carolina, pp. 664–667 (2002)
18. Shih, J.L., Chen, L.H.: Colour image retrieval based on primitives of colour moments. IEEE Proceedings on Vision, Image and Signal Processing 149(6), 370–376 (2002)
19. Stricker, M., Orengo, M.: Similarity of Color Images. In: SPIE Conference on Storage and Retrieval for Image and Video Databases III, San Jose, CA, USA, vol. 2420, pp. 381–392 (1995)
20. Manjunath, B.S., Ma, W.Y.: Texture Features for Browsing and Retrieval of Image Data. IEEE Trans. Pattern Anal. Mach. Intell. 18(8), 837–842 (1996)
21. Lowe, D.G.: Distinctive image features from scale-invariant keypoints. Int. J. Comput. Vision 60(2), 91–110 (2004)
22. NIST/SEMATECH: e-handbook of statistical methods (2012), http://www.itl.nist.gov/div898/handbook/eda/section3/eda356.htm
23. Malik, F., Baharudin, B.: Quantized histogram color features analysis for image retrieval based on median and Laplacian filters in DCT domain. In: International Conference on Innovation Management and Technology Research (ICIMTR), Malacca, Malaysia, May 21-22, pp. 624–629 (2012)
24. Gali, R., Dewal, M.L., Anand, R.S.: Genetic Algorithm for Content Based Image Retrieval. In: International Conference on Computational Intelligence, Communication Systems and Networks (CICSyN), Phuket, Thailand, July 24-26, pp. 243–247 (2012)
25. Kinnaree, P., Pattanasethanon, S., Thanaputtiwirot, S., Boontho, S.: RGB Color Correlation Index for Image Retrieval. Procedia Engineering (8), 36–41 (2011)

Formalization and Discovery of Approximate Conditional Functional Dependencies

Hiroki Nakayama[1], Ayako Hoshino[2], Chihiro Ito[2], and Kyota Kanno[2]

[1] NEC Informatec Systems, Ltd., 2-6-1 Kitamikata, Takatsu-ku, Kawasaki-shi, Japan
[2] NEC Knowledge Discovery Research Labs., 1753 Shimonumabe, Nakahara-ku,
Kawasaki-shi, Japan
{h-nakayama@cj,a-hoshino@cj,c-ito@az,k-kanno@ah}.jp.nec.com

Abstract. We propose efficient and precise discoveries of approximate Conditional Functional Dependencies (CFDs), by providing a precise formalization of approximate CFDs and presenting three discovery algorithms approxCFDMiner, approxCTANE and approxFastCFD as extensions of existing algorithms with renewed techniques. First, approxCFDMiner introduces a global FP-tree traversal for finding Right-hand Side items. Second, approxCTANE uses a modified pruning strategy. Third, approxFastCFD adopts a minimal coverset that is used to exclude non-minimal approximate CFDs. For these algorithms, we theoretically proved the correctness and experimentally evaluated the performances.

Keywords: Conditional Functional Dependency, Approximate CFD, Discovery Algorithms.

1 Introduction

Several rules with different degrees of specificity have been studied to express regularity among attributes in a database. Functional Dependency (FD) $X \rightarrow A$, stating that the values of attributes X uniquely determine the value of an attribute A, ranges over all tuples in the dataset. Association Rule (AR) [1] $(X, x) \Rightarrow (A, a)$, stating that for every tuple satisfying $X = x$, the value of A should be a, is relevant only with part of the dataset where $X = x$. Moreover, Conditional FD (CFD) [6] $([X^c, X^v] \rightarrow A, t_p)$, stating that for the tuples satisfying $X^c = t_p[X^c]$, an FD $X^v \rightarrow A$ holds, can express types of regularity that include not only FDs and ARs but also their intermediates [11].

Although such rules are useful in themselves for data profiling and cleansing [2], allowing some exceptions against data makes it possible to apply more intensive investigations to the data [3,4,6,8]. Such approximated rules in terms of data profiling can highlight implicit essences in databases, and for data cleansing, exceptions to rules may inform us of errors in data with their revised value candidates [6]. An approximate FD (AFD) [9] is for such a purpose.

The discoveries of such rules are not trivial and several algorithms have been proposed. In the 1990's, Agrawal et al. proposed Apriori [1] for discovery of ARs. Subsequently, TANE [9] and FastFD [12], which are level-wise and depth-first approaches, respectively, were proposed for discovering FDs. In 2007, Fan

H. Decker et al. (Eds.): DEXA 2013, Part I, LNCS 8055, pp. 118–128, 2013.

Table 1. Instance r of purchase log

	FLG	FN	LN	PAYMENT	COMMODITY	DATE	SHOP
t_1	1	Mike	Scott	card	bread	20120428	XXX
t_2	1	Mike	Scott	cash	egg	20120429	YYY
t_3	1	Mike	Scott	cash	bread	20120430	ZZZ
t_4	1	Mike	Scot	cash	milk	20120501	ZZZ
t_5	1	Emmy	Smith	cash	bread	20120410	YYY
t_6	0	Lisa	Davis	card	cheese	20120410	XXX

et al. [7] proposed the three algorithms, CFDMiner, CTANE and FastCFD, of which the latter two extend TANE and FastFD. Constant CFD inference [5] also works for finding constant CFDs. However, the discovery of approximate CFDs has lately gained attention [3,8] and there are no established algorithms with a sufficient theoretical foundation.

Example 1. Let us consider three rules, an FD $\varphi_1 = (\text{SHOP} \rightarrow \text{PAYMENT})$; an AR $\varphi_2 = ((\text{FN}, \text{Mike}) \Rightarrow (\text{LN}, \text{Scott}))$; and a CFD $\varphi_3 = ((\text{FN}, \text{PAYMENT}) \rightarrow \text{COMMODITY}, (\_, \text{card} \parallel \_))$, for instance r in Table 1.

Rule φ_1 means that *the value of* SHOP *determines that of* PAYMENT. As every tuple is under the effect of φ_1, no violations occur. Rule φ_2 means that *if* FN=Mike *then* LN=Scott. Tuples from t_1 to t_4 are the target, where t_4 violates against φ_2. This implies that Scot should be modified to Scott. The φ_3 states that *for tuples such that* PAYMENT=card, *the value of* FN *determines that of* COMMODITY. No violations occur for its target tuples t_1 and t_6.

1.1 Our Contributions and Organization of This Paper

Our contributions consist of (1) the first precise formalization of approximate CFDs, (2-4) the three algorithms for discovering approximate CFDs, and (5) experimental studies. We have assigned one section to each contribution.

(1, Section 2) Formalization of approximate CFDs and their discovery problem: We consider *sufficiently confident* CFDs called *approximate CFDs* that hold with high confidence. With the definition of the minimality, the problems are stated as enumerations of *minimal, frequent* and *confident* CFDs.

(2, Section 3) approxCFDMiner: This is for constant approximate CFD discovery. It finds Right-hand Side (RHS) of each Left-hand Side (LHS) using FP-trees.

(3, Section 4) approxCTANE: This explores a level-wise approach to both constant and variable approximate CFD discovery. We adopt a relaxed strategy that allows non-exact patterns.

(4, Section 5) approxFastCFD: It uses a depth-first approach for finding both constant and variable approximate CFDs. We adopt a *minimal coverset* to exclude non-minimal CFDs.

(5, Section 6) Experiments: We describe the performance of each algorithm using a synthetic dataset and real-life datasets. The synthetic one is used to show the time scalability of algorithms. The real-life ones are used to inspect the computation time and number of CFDs when the threshold confidence varies.

Finally, we conclude the paper in Section 7.

2   Statement of Discovery Problem

Let $\mathbf{attr}(R)$ be a set of all attributes in a relation schema R and $\mathbf{dom}(A)$ be the domain of an attribute $A \in \mathbf{attr}(R)$. We introduce *variable* '$\_$' as a symbol matching every value in an instance r of R, and we call every non-variable symbol a *constant*. For attribute A and attribute set $X \subseteq \mathbf{attr}(R)$, $t_p[A] \in \mathbf{dom}(A) \cup \{\_\}$ and its set $t_p[X]$ are called *pattern tuples* over A and X, respectively.

Definition 1. *(Items and Itemsets) For attribute A and value $t_p[A]$, we call the pair $(A, t_p[A])$ an item. An itemset $(X, t_p[X])$, or simply (X, t_p), is defined as a set of items where attributes in X are different. We can call an item with $t_p[A] \in \mathbf{dom}(A)$ a constant item and one with $t_p[A] = $ '$\_$' a variable item.*

Definition 2. *(Order between itemsets [7]) A partial order $\leq$ on one attribute A is defined as follows: if (1) $t'_p[A]$ is a variable, or (2) $t_p[A] = t'_p[A]$, we have $t_p[A] \leq t'_p[A]$. We write $t_p[A] \ll t'_p[A]$ if $t_p[A] \leq t'_p[A]$ but $t_p[A] \neq t'_p[A]$. Its extension on multiple attributes X is trivial.*

Definition 3. *(Support of itemsets) For instance r and an itemset (X, t_p), the support $\mathrm{supp}(X, t_p, r)$, or simply $\mathrm{supp}(X, t_p)$, is defined as the number of tuples in r that match t_p on X, i.e., $|\{t \in r \mid t[X] \leq t_p[X]\}|$.*

2.1   Approximate CFDs

To provide a notion of *approximate CFDs*, we introduce the *confidence* of CFDs which was referred to in [7].

Definition 4. *(CFDs) A CFD φ is given as a rule $(X \rightarrow A, (t_p[X] \parallel t_p[A]))$ where the LHS is an itemset $(X, t_p[X])$ and RHS is an item $(A, t_p[A])$. If no misleading occurs, we abbreviate it to $(X \rightarrow A, (t_p \parallel a))$ or simply $(X \rightarrow A, t_p)$.*

$\varphi = (X \rightarrow A, t_p)$ is called a *constant CFD* if every value in t_p is constant, else is called a *variable CFD*. If we need to distinguish between constant and variable items, we also write it as $([X^c, X^v] \rightarrow A, t_p)$. An instance r *satisfies* φ, denoted as $r \models \varphi$, iff "for each pair of tuples t_1, t_2 in r, if $t_1[X] = t_2[X] \leq t_p[X]$ then $t_1[A] = t_2[A] \leq t_p[A]$".

Based on notions of *support* and *confidence* for instance r and CFD φ, we introduce approximate CFDs. Support represents how many tuples are relevant to the CFD, and confidence indicates the rate of tuples satisfying the CFD.

Definition 5. *(Support of CFDs) The support $\mathrm{supp}(\varphi, r)$ is the number of tuples in r matching both $t_p[X]$ and $t_p[A]$, which is equal to $\mathrm{supp}(X \cup A, t_p, r)$.*[1]

---

[1] $X \cup \{A\}$ is abbreviate to $X \cup A$, likewise $X \setminus \{A\}$ to $X \setminus A$.

Definition 6. *(Confidence) The confidence* $\mathrm{conf}(\varphi, r)$ *is the ratio* $\max_{r'} |r'|/|r|$ *such that* $r' \models \varphi$ *for some* $r' \subseteq r$. *We can say that* $\mathrm{conf}(\varphi, r) = 1$ *is equivalent to* $r \models \varphi$.

Definition 7. *(Approximate CFDs) We say that an approximate CFD* φ *holds in a relation* r *w.r.t. a threshold value* p *iff* $\mathrm{conf}(\varphi, r) \geq p$. *This is said that* r *approximately satisfies* φ *and denoted as* $r \models_{approx} \varphi$.

2.2   Discovery Problem for Approximate CFDs

To avoid a large amount of unnecessary rules, we focus on *minimal* CFDs, i.e., those containing no redundant items and covering most tuples. Minimal exact CFDs have been defined by Fan et al. [7], and we extend it to approximate ones.

Definition 8. *(Minimal approximate CFDs) An approximate CFD* $(X \rightarrow A, (t_p[X] \parallel t_p[A]))$ *is minimal if the following three conditions are satisfied; (1, Non-triviality)* $A \notin X$, *(2, LHS minimality) for any* $Y \subset X$, $r \not\models_{approx}$ *(*$Y \rightarrow A, (t_p[Y] \parallel t_p[A])$*), and (3, Most generality) for any* t'_p *with* $t_p \ll t'_p$, $r \not\models_{approx} (X \rightarrow A, (t'_p[X] \parallel \_))$.

We introduce two parameters k and p, as inputs of discovery algorithms, where k means *minsup*, which is the threshold of support, and p means *minconf*, which is that of confidence. For a given instance r, a CFD $\varphi = (X \rightarrow A, t_p)$, k and p, φ is called k-*frequent* if $\mathrm{supp}(\varphi, r) \geq k$ and called p-*confident* if $\mathrm{conf}(\varphi, r) \geq p$. Moreover, we call φ *valid* if φ is both k-frequent and p-confident.

We now can give a statement on the discovery problem of approximate CFDs. Afterward, the word "approximate" is omitted as long as no misleading occurs.

Problem 1. (Approximate CFD Discovery Problem) *For a given instance* r *of* R, *minsup* k, *and minconf* p, *enumerate all minimal and valid CFDs.*

3   ApproxCFDMiner

approxCFDMiner discovers all constant CFDs. In contrast to the existing CFD-Miner [7], the mapping approach between a closed itemset and free itemsets is not available. To overcome this difficulty, FP-tree traversal approach is used.

3.1   Free Itemset and FP-Tree

First we introduce *free itemsets*, referred as generators in [10]. Since supports are anti-monotonic, any removal of items keeps or increases its support. If the support is unchanged, called *not-free*, such items can be regarded as redundant.

Definition 9. *(Free itemsets [10]) An itemset* (X, t_p) *is called free if the support increases by removing any item from* (X, t_p). *That is, there exists no sub-itemset* $(Y, t_p[Y]) \subset (X, t_p)$ *for which* $\mathrm{supp}(X, t_p) = \mathrm{supp}(Y, t_p[Y])$.

GrGrowth [10] is an exhaustive enumeration algorithm of free itemsets for minsup k. It uses FP-trees [10] for compactly storing comprehensive information on k-frequent itemsets, which have supports being equal to or more than k.

Algorithm 1. approxCFDMiner

Input: Instance r, minsup k, and minconf p
Output: All k-frequent and p-confident minimal constant CFDs for r
 1: Enumerate all k-frequent free itemsets (X, t_p) of r and store them in list L in
     ascending order of the size $|X|$
 2: **for** each free itemset $(X, t_p) \in L$ **do**
 3:    Find any item (A, a) such that $A \notin X$, $\text{supp}(X \cup A, (t_p \cup a)) \geq k$,
       and $\text{supp}(X \cup A, (t_p \cup a)) \geq p \cdot \text{supp}(X, t_p)$
 4:    Output $\varphi = (X \rightarrow A, (t_p \parallel a))$
 5:    For every free super-itemset such that $(Y, s_p) \supset (X, t_p)$ in L, preclude (A, a)
       from RHS candidates for (Y, s_p)

3.2 approxCFDMiner Algorithm

We provide Proposition 1, which shows the correctness of the approxCFDMiner, as a modification to that by Fan et al. [7].

Proposition 1. *For instance r and any valid (i.e., k-frequent and p-confident) minimal constant CFD $\varphi = (X \rightarrow A, (t_p \parallel a))$, $r \models_{approx} \varphi$ iff (1) the itemset (X, t_p) is free and does not contain the item (A, a), (2) the itemset $(X \cup A, (t_p \cup a))$ is k-frequent, (3) $\text{supp}(X \cup A, (t_p \cup a)) \geq p \cdot \text{supp}(X, t_p)$, and (4) (X, t_p) does not contain a smaller free set $(Y, t_p[Y])$ with this property.*

Proposition 1 ensures that every minimal constant CFD has a free itemset as its LHS. Then, by traversing the global FP-tree, we sum up the appearance of items in each path that contains every LHS item. Consequently, RHS candidates are obtained as the co-occurrence of (X, t_p) with each item (A, a).

    Now we provide approxCFDMiner in Algorithm 1. The validity of the CFD generated at Line 3 is ensured by Proposition 1. Furthermore, the minimality of CFDs output at Line 4 is also assured by RHS candidate removals at Line 5.

4 ApproxCTANE

Based on the existing CTANE, we present approxCTANE, which is a level-wise algorithm for discovering CFDs. Along with the modifications to the pruning strategy for FDs indicated in TANE, we provide an explicit strategy for CFDs.

    First, we introduce a *generality* relation to itemsets in the same way as [7]. We say that an itemset (Y, s_p) is *more general* than another itemset (X, t_p), denoted as $(X, t_p) \preceq (Y, s_p)$, if $Y \subseteq X$ and $t_p[Y] \leq s_p$.

    Similar to CTANE, approxCTANE generates each itemset (X, s_p) accompanied with an RHS candidate set $C^+(X, s_p)$, which satisfies the following conditions:

(C1): If $A \in X$, then $c_A = s_p[A]$.
(C2.1): For all $B \in X \setminus A$, $r \not\models (X \setminus \{A, B\} \rightarrow B, (s_p[X \setminus \{A, B\}] \parallel s_p[B]))$.
(C2.2): For all $C \in X \setminus A$, $r \not\models_{approx} (X \setminus \{A, C\} \rightarrow A, (s_p[X \setminus \{A, C\}] \parallel s_p[A]))$.

Algorithm 2. approxCTANE

Input: Instance r, minsup k, and minconf p
Output: All k-frequent and p-confident minimal CFDs for r

1: Initially let $L_1 = \{(A, \_) \mid A \in \mathbf{attr}(R)\} \cup \{(A, a) \mid \text{supp}(A, a) \geq k, A \in \mathbf{attr}(R)\}$
2: Let $C^+(\emptyset) = L_1$ and $\ell = 1$
3: **while** $L_\ell \neq \emptyset$ **do**
4:     Sort itemsets in L_ℓ in descending order $\succ$ of generality
5:     **for each** $(X, s_p) \in L_\ell$ **do** $C^+(X, s_p) = \bigcap_{B \in X} C^+(X \setminus B, s_p[X \setminus B])$
6:     **for each** $(A, c_A) \in C^+(X, s_p)$ of $(X, s_p) \in L_\ell$ with $A \in X$ and $s_p[A] = c_A$ **do**
7:         Generate CFD $\varphi = (X \setminus A \to A, (s_p[X \setminus A] \parallel c_A))$
8:         Let u_p be any tuple pattern such that $u_p[A] = c_A$ or $\_$, and $u_p[X \setminus A] \leq s_p[X \setminus A]$
9:         **if** $r \models_{approx} \varphi$ **then**
10:            Output φ, and remove $(A, *)$ from $C^+(X, u_p)$ for every $(X, u_p) \in L_\ell$
                // '*' consists of all values for A, including the variable
11:           **if** $r \models \varphi$ **then**
12:               Remove $(B, *)$ from $C^+(X, u_p)$ for every $B \in \mathbf{attr}(R) \setminus X$, $(X, u_p) \in L_\ell$
                // '*' consists of all values for B, including the variable
13:     **for each** $(X, s_p) \in L_\ell$ **do**
14:         **if** $C^+(X, s_p) = \emptyset$ **then** Remove (X, s_p) from L_ℓ
15:     Let $L_{\ell+1} = \emptyset$
16:     **for each** pair of $(X, s_p), (Y, t_p) \in L_\ell$ that agrees on just $\ell - 1$ elements **do**
17:         Let $(Z, u_p) = (X \cup Y, (s_p, t_p[Y \setminus X]))$
18:         **if** (Z, u_p) is k-frequent and for all $A \in Z$, $(Z \setminus A, u_p[Z \setminus A]) \in L_\ell$ **then**
19:             Add (Z, u_p) to $L_{\ell+1}$
20:     $\ell = \ell + 1$

(C3): For all $B \in X \setminus A$, $r \not\models_{approx} (X \setminus A \to A, (s_p^B[X \setminus A] \parallel s_p[A]))$, where $s_p^B[C] = s_p[C]$ for all $C \neq B$ and $s_p^B[B] = \_$.

(C1) ensures the consistency; (C2.1) and (C2.2) ensure the minimality; and (C3) ensures the most generality. Compared to the corresponding conditions for CTANE given in [7], the second condition is separated into two cases, 2.1 (exactly satisfied) and 2.2 (approximately satisfied).

We introduce the Lemma 1, as modifications of Lemma 2 by Fan et al. [7].

Lemma 1. *Let $X \subseteq \mathbf{attr}(R)$, s_p be a pattern over X, $A \in X$ and assume that $r \models_{approx} \varphi = (X \setminus A \to A, (s_p[X \setminus A] \parallel s_p[A]))$. Then φ is minimal iff for all $B \in X$ we have $(A, s_p[A]) \in C^+(X \setminus B, s_p[X \setminus B])$.*

Lemma 1 implies that $C^+(X, s_p)$ is anti-monotonic: if $(X, s_p) \supseteq (Y, t_p)$, $C^+(X, s_p) \subseteq C^+(Y, t_p)$ holds. In particular, if $C^+(Y, t_p) = \emptyset$, every (X, s_p) that contains (Y, t_p) is pruned because no minimal CFDs are generated from (X, s_p).

We now provide approxCTANE in Algorithm 2. At Line 5, $C^+(X, s_p)$ for each frequent itemset (X, s_p) are initialized. From Line 6 to 12, generated CFDs are tested for their validity. If valid, the CFDs are output, and RHS candidate updates are conducted. In Lines 13 and 14, itemsets with no RHS candidates are removed. From Line 16 to 19, we generate larger itemsets. The major difference from CTANE is Lines 11 and 12: only if $r \models \varphi$, the removal of (B, c_B) is executed.

The Lemma 2, given as a modification of Lemma 3 in Fan et al. [7], shows that Algorithm 2 accurately maintains $\mathcal{C}^+(X, s_p)$. From Lemmas 1 and 2, Proposition 2 is derived.

Lemma 2. *Suppose that $\mathcal{C}^+(Y, t_p)$ is accurate for all $(Y, t_p) \in L_\ell$. Then Lines 5, 10, and 12 in* approxCTANE *correctly compute $\mathcal{C}^+(X, s_p)$ for all $(X, s_p) \in L_{\ell+1}$.*

Proposition 2. approxCTANE *enumerates all valid minimal CFDs.*

5   ApproxFastCFD

Similar to the existing FastCFD, approxFastCFD finds CFDs $([X^c, X^v] \to A, t_p)$ by two steps: (1) determine X^c and A, and (2) search X^v with depth-first search.

However, several properties used in FastCFD no longer available. First, *difference sets*, whose minimal covers correspond to the LHSs of minimal CFDs, is not reasonably extendable. Second, since confidence is not anti-monotonic, validity of immediate subsets of CFDs is not sufficient for minimality proof.

5.1   Minimal Coverset

To overcome above problems, we introduce a *minimal coverset*. Intuitively, it works as a set cover of attributes X^v inducing non-minimality of CFDs, and prevents the non-minimal CFD search.

Definition 10. *(Minimal Coverset) For every pair of a free itemset $(X^c, t_p[X^c])$ and an item $(A, t_p[A])$, minimal coverset L stores a set of attributes, and is updated each time a CFD is discovered as stated below.*

1. *If constant CFD $(X^c \to A, (t_p[X^c] \parallel a))$ is discovered, then for every free itemset $(Y^c, s_p[Y^c]) \prec (X^c, t_p[X^c])$, update $L((Y^c, s_p[Y^c]), (A, a)) = L((Y^c, s_p[Y^c]), (A, \_)) = \{\emptyset\}$.*
2. *If variable CFD $([X^c, X^v] \to A, (t_p[X^c], \_, \ldots, \_ \parallel \_))$ is discovered, then for every free itemset $(Y^c, s_p[Y^c]) \prec (X^c, t_p[X^c])$, add a set of attributes $Y = X^v \setminus (Y^c \setminus X^c)$ to $L((Y^c, s_p[Y^c]), (A, \_))$.*

As we verify the minimality of CFD $\varphi' = ([Y^c, Y^v] \to A, (s_p[Y^c], \_, \ldots, \_ \parallel s_p[A]))$, φ' is found to be non-minimal iff $Y \in L((Y^c, s_p[Y^c]), (A, s_p[A]))$ such that $Y \subseteq Y^v$, Then, no further searches from φ' are done.

Proposition 3. *When CFD $\varphi = ([X^c, X^v] \to A, t_p)$ is found, the above update of the minimal coverset avoids having to search every non-minimal CFD $\varphi' = ([Y^c, Y^v] \to A, s_p)$ against φ in which $X^c \subset Y^c$.*

Algorithm 3. approxFastCFD

Input: Instance r, minsup k, and minconf p
Output: All k-frequent and p-confident minimal CFDs for r
1: Enumerate all k-frequent free itemsets of r and store them in list L in descending order of generality $\succ$
2: **for** each attribute $A \in \mathbf{attr}(R)$ **do**
3:   **for** each free itemset$(X^c, t_p[X^c])$ where $A \notin X^c$ **do** RECURSIVE$((X^c, t_p[X^c]), A, \emptyset)$

4: **procedure** RECURSIVE(k-frequent free itemset $(X^c, t_p[X^c])$, attribute $A \notin X^c$, and attributes $X^v \subseteq \mathbf{attr}(R) \setminus (X^c \cup A))$
5:   $\Phi = \emptyset$ // Candidates of CFDs
6:   **if** $X^v = \emptyset$ **then**
7:     Generate all CFDs candidates $\varphi_c = (X^c \to A, (t_p[X^c] \parallel a \in \mathbf{dom}(A)))$, and if φ_c is k-frequent and p-confident, add each φ_c to Φ
8:   **else**
9:     Generate the only CFD candidate $\varphi_v = ([X^c, X^v] \to A, (t_p[X^c], \_, \ldots, \_ \parallel \_))$, and if φ_v is (obviously k-frequent) p-confident, add φ_v to Φ
10:   **if** $\Phi \neq \emptyset$ **then**
11:     **if** $X^v = \emptyset$ **then**
12:       **for** each candidate $\varphi_c = (X^c \to A, (t_p[X^c] \parallel a))$ in Φ **do**
13:         **if** $L((X^c, t_p[X^c]), (A, a)) \neq \{\emptyset\}$ **then**
14:           Output φ_c, and update the minimal coverset
15:     **else** // $\varphi_v = ([X^c, X^v] \to A, (t_p[X^c], \_, \ldots, \_ \parallel \_))$ is the only CFD candidate
16:       **if** $X \in L((X^c, t_p[X^c]), (A, \_))$ such that $X \subseteq X^v$ **then return**   ▷ non-minimal
17:       **else**
18:         **for** each attribute $B \in X^v$ **do**
19:           Let $\varphi'_v = ([X^c, X^v \setminus B] \to A, (t_p[X^c], \_, \ldots, \_ \parallel \_))$
20:           **if** φ'_v is p-confident **then return**   ▷ non-minimal
21:         Output φ_v, and update the minimal coverset
22:   **else**
23:     **for** every $x_v \in \mathbf{attr}(R) \setminus (X^c \cup X^v \cup A)$ **do** RECURSIVE$((X^c, t_p[X^c]), A, X^v \cup x_v)$

5.2 approxFastCFD Algorithm

Algorithm 3 describes approxFastCFD, where a sub-routine *Recursive* is called for each pair of a free itemset $(X^c, t_p[X^c])$ and attribute A.

Proposition 4. *approxFastCFD finds all valid minimal CFDs.*

6 Experiments

We discuss the performance of our three algorithms approxCFDMiner, approx-CTANE, and approxFastCFD described in Sections 3, 4 and 5. Our experiments were conducted from two points of view: (1) scalability on data size (i.e. number of tuples) and arity (i.e. number of attributes), (2) response time and number of CFDs for varied minconf p. Note that this paper is the first to discuss the performance of a CFD discovery for varied p.

Table 2. Experimental Datasets

Dataset	Arity	No. of tuples
synthetic datasets	6 to 20	1,000 to 1,000,000
Wisconsin Breast Cancer (WBC)	11	699
Chess	7	28,056

6.1   Settings

The experiments were conducted on both the synthetic and real-life datasets summarized in Table 2. The synthetic datasets were generated by randomly assigning a value that ranges in ten distinct values in each field. We adopted the Wisconsin Breast Cancer (WBC) and Chess datasets from the UCI machine learning repository (http://archive.ics.uci.edu/ml/) as the real-life data.

The CFD discovery algorithms and GrGrowth were implemented in Java and run using a Quad Core Xeon E5420 (2.5 GHz) with 8 GB of memory. We called GrGrowth to find the free itemsets used in approxCFDMiner and approxFastCFD.

6.2   Scalability on Synthetic Data

We evaluated the performance of our algorithms by varying the arity and number of tuples of the synthetic data for fixed $k = 0.05 \times$ (No. of tuples) and $p = 0.8$.

Figures 1 and 2 indicate that the performance of approxCTANE greatly depends on the arity. approxCFDMiner is much faster than the other two algorithms. In comparing approxCTANE and approxFastCFD, the former causes an explosion of computation time as the arity increases, while the latter is relatively durable against the increase in the number of tuples, unlike the existing FastCFD.

6.3   Real Data Experiments

We used two datasets WBC and Chess, which are dissimilar with respect to the arity and the number of tuples to inspect our algorithms more closely. We fixed k and varied p from 0.5 to 1.

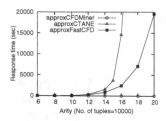
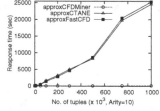
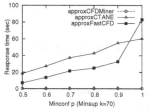

Fig. 1. Scalability w.r.t. Arity

Fig. 2. Scalability w.r.t. #Tuples

Fig. 3. Time for WBC w.r.t. p

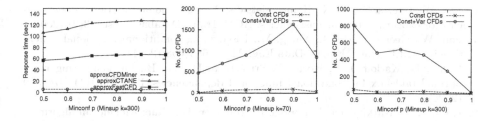

Fig. 4. Time for Chess w.r.t. p

Fig. 5. #CFDs for WBC w.r.t p

Fig. 6. #CFDs for Chess w.r.t. p

The response time is plotted in Figure 3 for WBC and in Figure 4 for Chess. As the value of p increases, the resulting CFDs tend to have more items; thus, the response time becomes longer. With $p = 1$ in particular, approxFastCFD takes longer than approxCTANE since exact CFDs in WBC have many variable items and the minimal coverset does not work well.

The number of CFDs may decrease or increase as p varies for the following reason: while high confidence reduces the number of valid CFDs, turning a CFD invalid may cause other larger CFDs to be newly minimal. The behaviors in response to different p values are in Figure 5 for WBC and in Figure 6 for Chess.

7   Conclusions

We have provided the first formalization of approximate CFDs and presented three algorithms for their discovery: approxCFDMiner was quite useful if only constant CFDs were needed. To obtain also variable CFDs, approxFastCFD outperformed approxCTANE where the arity was large.

We expect to obtain useful approximate CFDs for efficient data profiling and cleansing by using these three algorithms. Both time and space complexity analyses of our algorithms should be conducted.

References

1. Agrawal, R., Srikant, R.: Fast algorithms for mining association rules in large databases. In: VLDB, pp. 487–499 (1994)
2. Bohannon, P., Flaster, M., Fan, W., Rastogi, R.: A cost-based model and effective heuristic for repairing constraints by value modification. In: SIGMOD, pp. 143–154 (2005)
3. Chiang, F., Miller, R.J.: Discovering data quality rules. PVLDB 1(1), 1166–1177 (2008)
4. Cong, G., Fan, W., Geerts, F., Jia, X., Ma, S.: Improving data quality: Consistency and accuracy. In: VLDB, pp. 315–326 (2007)
5. Diallo, T., Novelli, N., Petit, J.M.: Discovering (frequent) constant conditional functional dependencies. IJDMMM 4(3), 205–223 (2012)

6. Fan, W., Geerts, F., Jia, X., Kementsietsidis, A.: Conditional functional dependencies for capturing data inconsistencies. ACM Trans. Database Syst. 33(2) (2008)
7. Fan, W., Geerts, F., Li, J., Xiong, M.: Discovering conditional functional dependencies. IEEE Trans. Knowl. Data Eng. 23(5), 683–698 (2011)
8. Golab, L., Karloff, H.J., Korn, F., Srivastava, D., Yu, B.: On generating near-optimal tableaux for conditional functional dependencies. PVLDB 1(1), 376–390 (2008)
9. Huhtala, Y., Kärkkäinen, J., Porkka, P., Toivonen, H.: Tane: An efficient algorithm for discovering functional and approximate dependencies. Comput. J. 42(2), 100–111 (1999)
10. Liu, G., Li, J., Wong, L.: A new concise representation of frequent itemsets using generators and a positive border. Knowl. Inf. Syst. 17(1), 35–56 (2008)
11. Medina, R., Nourine, L.: A unified hierarchy for functional dependencies, conditional functional dependencies and association rules. In: Ferré, S., Rudolph, S. (eds.) ICFCA 2009. LNCS, vol. 5548, pp. 98–113. Springer, Heidelberg (2009)
12. Wyss, C., Giannella, C., Robertson, E.: FastFDs: A heuristic-driven, depth-first algorithm for mining functional dependencies from relation instances - extended abstract. In: Kambayashi, Y., Winiwarter, W., Arikawa, M. (eds.) DaWaK 2001. LNCS, vol. 2114, pp. 101–110. Springer, Heidelberg (2001)

Parallel Partitioning and Mining Gene Expression Data with Butterfly Network

Tao Jiang, Zhanhuai Li, Qun Chen, Zhong Wang, Wei Pan, and Zhuo Wang

School of Computer Science and Technology,
Northwestern Polytechnical University, 710072, Xi'an, China
{jiangtao,zhongwang}@mail.nwpu.edu.cn,
{lizhh,chenbenben}@nwpu.edu.cn

Abstract. In the area of massive gene expression analysis, Order-Preserving Sub-Matrices have been employed to find biological associations between genes and experimental conditions from a large number of gene expression datasets. While many techniques have been developed, few of them are parallel, and they lack the capability to incorporate the large-scale datasets or are very time-consuming. To help fill this critical void, we propose a Butterfly Network based parallel partitioning and mining method (BNPP), which formalizes the communication and data transfer among nodes. In the paper, we firstly give the details of OPSM and the implementations of OPSM on MapReduce and Hama BSP and their shortcomings. Then, we extend the Hama BSP framework using Butterfly Network to reduce the communication time, workload of bandwidth and duplicate results percent, and call the new framework as BNHB. Finally, we implement a state-of-the-art OPSM mining method (OPSM) and our BNPP method on top of the framework of naïve Hama BSP and our BNHB, and the experimental results show that the computational speed of our methods are nearly one order faster than that of the implementation on a single machine and the proposed framework has better effectiveness and scalability.

Keywords: Gene Expression Data, Data Partitioning, Butterfly Network, BSP model, MapReduce, Parallel Processing, OPSM, Hadoop, Hama.

1 Introduction

The rapid advances in high-throughput technologies, such as microarrays, enable simultaneous measurement of the expression levels of all genes in a given organism, which accumulates massive gene data [1-6]. These data can be viewed as an $n \times m$ matrix with n gene (rows) and m experimental conditions (columns), in which each entry denotes the expression level of a given gene under a given experimental condition. Recently, Order-Preserving Sub-Matrices (OPSMs) [1-6], which plays an important role in inferring gene regulatory networks, has been accepted as a significant tool for gene expression data analysis. The objective is to discover a subset of rows and columns in a data matrix where all the rows exhibit a similar pattern of rises and falls in the values of entries. For example, Fig. 1 shows the expression levels of the two sets of genes under four experimental conditions [6],

H. Decker et al. (Eds.): DEXA 2013, Part I, LNCS 8055, pp. 129–144, 2013.

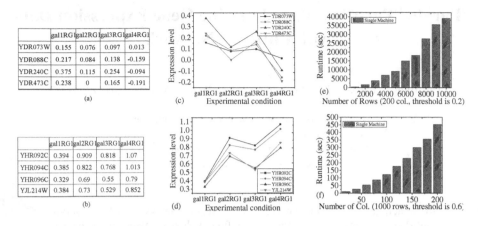

Fig. 1. Example of OPSMs and Performance of OPSM on a single machine

where Fig. 1(c) and 1(d) are the graphical representations of two datasets plotted in Fig. 1(a) and 1(b), respectively.

Recently, some approaches have been proposed to discover significant OPSMs. [1] observes that small group genes are tightly co-regulated under many conditions, thus, they propose the *KiWi* framework which substantially reduces the search space and problem scale. To cope with the noise in datasets, [3] utilizes the measurements collected from repeated experiments; [4] proposes a noise-tolerant model AOPC; [5] employs bucket to relax OPSM model by considering linearity relaxation. However, time-consuming of the methods on large-scale datasets is very long and cannot be tolerated.

As the high-rate increasing of the numbers and sizes of gene expression datasets, there is an increasing need for the fast mining techniques to handle the massive gene datasets. However, rapidly mining OPSMs is a challenging task due to several reasons below [1-6]. First, there are a large number of genes and physiological conditions (experimental conditions) in species, and the computational complexity of OPSMs mining with respect to the number of genes and experimental conditions is $O(m^2n^2)$, where m is the number of experimental conditions and n is the number of genes. For example, there are thousands of genes in any complex organism, the organism number of Homo sapiens is more than 90, thus the gene number of Homo sapiens is tens of thousands (the accurate number is 27,000). As shown in Fig. 1(e), the running time on one third of the number of Homo sapiens genes (10,000 genes) is more than 10 hours. As shown in Fig. 1(f), for 1,000 genes, the running time of 20 and 200 experimental conditions are 11 and 453 seconds, respectively. Second, to reduce the impact of the inherent noise of microarray measurements, OPSM-RM method [3] is proposed, time complexity of which is $O(km^2n^2)$, where k is the number of replicates. However, the memory of a single machine is not enough to incorporate large number of replicates.

Although it is a hard problem to address, distributed and parallel processing techniques [7] can be employed in the rapid analysis of gene expression data. For example, Hadoop, an open source implementation of MapRduce (MR), is a framework that allows for the distributed processing of large datasets across computer clusters using a simple programming model. However, there are no communication mechanisms among both Mappers and Reducers in MR [8], thus, it needs a lot of iterations of MR to make the final results complete, which is very time-consuming. Hama [11] is a pure Bulk Synchronous Parallel (BSP) [9, 10] computing framework on top of Hadoop Distributed File System (HDFS) for massive scientific computations. Although Hama is superior to MR, it has some disadvantages for gene expression data mining. First one is each node exchanges data with all the rest nodes in a super-step, which makes the nodes on Hama produce more duplicate results. Second one is that although we can reduce the duplicate results by starting only one node for reduce, the other nodes in the cluster are not employed sufficiently, and it also needs one super-step which consumes a long time. Third one is that the node for reduce receives all the data from other nodes, which may make the node has not enough space to save these data and consumes more time and bandwidth. Thus, it is necessary to transfer data moderately. Fortunately, Hama provides a flexible, simple, and easy-to-use small APIs, we can use which to address these issues. To fill this void, we extend the naïve Hama BSP framework using Butterfly Network to reduce the communication time, workload of bandwidth and percent of duplicate results, which makes nodes exchange data moderately in $log_2 N$ steps and produce smaller percent of duplicate results.

We conduct extensive experiments of our methods with java on a single machine and Hama; the experimental results show that our methods are nearly one order faster than the implementation on a single machine, and also show that our framework has better effectiveness and scalability. The main contributions of the paper are as follows:

1) We give the preliminaries, details of OPSM method, and the implementations of OPSM on MapReduce and Hama BSP and their shortcomings. (Sec. 2).

2) To reduce communication time and workload of bandwidth, we extend Hama BSP with Butterfly Network (BNHB) (Sec. 3.1). To reduce the percent of duplicate results, we propose a deduplication method based on distributed hash tables (Sec. 3.3).

3) Based on BNHB framework, we propose a BNPP method (Sec. 3.3).

4) We implement one state-of-the-art OPSM mining method (OPSM) and our method BNPP on top of the naïve Hama BSP and new framework BNHB (Sec. 4).

2   Preliminary and Analysis

2.1   Preliminary

In this subsection, we introduce some notations and definitions used throughout the paper, which are illustrated in Table 1.

Definition 1. OPSM(order-preserving sub-matrix): Given a dataset ($n \times m$ matrix) $D(G, T)$, an OPSM is a pair (g, t), where g is a subset of the n rows and t is a permutation of a subset of the m columns which satisfies the condition: for each row in g, the data values x are monotonically increasing/decreasing with respect to the permutation of the indexes of columns, i.e., $x_{i1} < x_{i2} < \ldots < x_{ij} < \ldots < x_{ik}$ ($x_{i1} > x_{i2} > \ldots > x_{ij} > \ldots > x_{ik}$), where $(i1, \ldots, ij, \ldots, ik)$ is the permutation of the indexes of columns $(1, \ldots, j, \ldots, k)$.

Definition 2. Core of OPSM: Given an OPSM $M_i(g, t)$, the core of OPSM, which is denoted as $Core(M_i)$, is defined as the longest common subsequence $LCS(g, t)$.

Definition 3. Similarity: Consider a gene g_i and an OPSM $M_j(g, t)$, we define the similarity of them as the ratio between the length of the intersection of $LCS(g_i \cup g, t)$ with $Core(M_j)$ and the number of columns of dataset (m), which is denoted by $S(g_i, M_j) = |LCS(g_i \cup g, t) \cap Core(M_i)| / m$. Similarly, similarity of OPSMs $M_j(g', t')$ and $M_k(g'', t'')$ is defined by $S(M_j, M_k) = |Core(M_j) \cap Core(M_k)| / m$. Throughout the paper, if there are no specific notifications, we use terms *similarity* and *threshold* interchangeably.

Table 1. Notations Used in This Paper

Notation	Description	Notation	Description
G	Set of genes	x	Set of gene expression values
g	Subset of G	x_{ii}	An entry of gene expression value
g_i	A gene of G or g	$D(G, T)$	A given gene expression dataset
T	Set of experi. conditions	$LCS(g, t)$	Longest common subsequence on g, t
t	Subset of T	$Core(M_i)$	Core of OPSM of $M_i(g, t)$
t_i	An experi. condition	$S(g_i, M_i)$	Similarity of g_i and an OPSM $M_i(g, t)$
$M_i(g, t)$	An OPSM	S_{min}	Threshold of Similarity

2.2   OPSM Mining on a Single Machine, MR and Hama BSP

Example 1. (OPSM mining on a single machine) There is a gene expression data which has 16 rows and 4 columns illustrated in Fig. 2(a). And the threshold is 0.6. The procedure and results on a single machine are illustrated in Fig. 2(b), and 2(c).

Example 1 shows the OPSM mining on a single machine, and Algorithm 1 gives the details of OPSM. The machine firstly sorts the expression values of each row, the permutation of which is represented by the indexes of related column (line 1), Then, it finds $LCSs$ between each pair of genes (lines 2-5). If the length of LCS is less than $m \times \rho$, where ρ is the threshold, we prune it. Due to $LCS()$ (finding $LCSs$) is well-known, we do not present it. Finally, it reduces $LCSs$ (line 6). As we know, regardless of how many nodes a gene expression data is partitioned onto, the mining results should be the same as that produced by a single machine. Thus, we can use this criterion to evaluate the performance of our methods.

	0	1	2	3		0	1	2	3
g0	21	33	42	54	g8	5	7	2	4
g1	11	23	37	46	g9	44	76	23	31
g2	2	7	10	18	g10	22	35	6	17
g3	3	10	15	27	g11	14	24	4	11
g4	13	2	5	9	g12	10	24	31	5
g5	19	3	8	15	g13	9	15	23	4
g6	25	8	13	17	g14	24	38	47	9
g7	37	12	18	26	g15	22	37	43	7

(a)

Indexes of columns					Indexes of columns				
g0	0	1	2	3	g8	2	3	0	1
g1	0	1	2	3	g9	2	3	0	1
g2	0	1	2	3	g10	2	3	0	1
g3	0	1	2	3	g11	2	3	0	1
g4	1	2	3	0	g12	3	0	1	2
g5	1	2	3	0	g13	3	0	1	2
g6	1	2	3	0	g14	3	0	1	2
g7	1	2	3	0	g15	3	0	1	2

(b)

gene	OPSMs			
g0-g3	0	1	2	3
g4-g7	1	2	3	0
g8-g11	2	3	0	1
g12-g15	3	0	1	2
g0-g3,g12-g15	0	1	2	
g0-g7	1	2	3	
g4-g11	2	3	0	
g8-g15	3	0	1	

(c)

Fig. 2. Gene expression data and OPSM results on a single machine

---

Algorithm 1. OPSM mining on a single machine

Input: $n \times m$ data matrix $D(G, T)$, threshold ρ; Output: OPSMs $M_i(g, t)$

1. sort expression values in each row and denote with column No.
2. *for* $i=0$; $i<n-1$; $i++$
3.   *for* $j=i+1$; $j<n$; $j++$
4. $LCS(g_i, g_j, g_i.\text{length}, g_j.\text{length}, b[][], c[][])$; $lcs = PrintLCS(g_i, g_j)$;
5.   *if* $lcs.\text{length}<m \times \rho$ *then* prune lcs; *else* $LCSs.\text{add}(lcs)$;
6. summarize the $LCSs$;

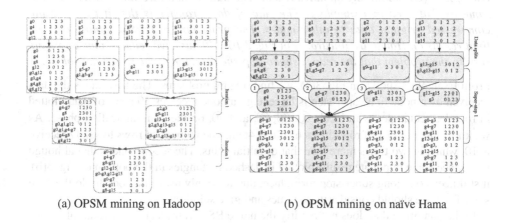

(a) OPSM mining on Hadoop   (b) OPSM mining on naïve Hama

Fig. 3. OPSM mining on Hadoop and naïve Hama and Example

Example 2. (OPSM mining on Hadoop) The dataset and threshold are the same as that used in Example 1. The procedure of OPSM mining and the final results on Hadoop are illustrated in Fig. 3(a), and the true final results are presented in Fig. 2(c).

Fig. 3(a) shows the OPSM mining of Example 2 processing on Hadoop. In the 1st iteration, it starts 4 nodes. And we only describe the procedure of the 2nd node; the others are the same as that of the 2nd one. The 2nd node reads 4 rows of data, the original data of g1, g5, g6, g7, which is illustrated in Fig. 2(a) and omitted in Fig. 3(a). Then, it generates the permutation of indexes of columns, which are "g1: 0, 1, 2, 3", "g5-g7: 1, 2, 3, 0". Further, it finds the *LCSs* among the 2 rows, the results of which is shown in the 2nd rectangle in 2nd row, due to "$g5$-$g7$: 2, 3, 0" is contained by "$g5$-$g7$: 1, 2, 3, 0", we omit it. In the 2nd iteration, it starts 2 nodes, and we only describe the procedure of the 2nd node. The 2nd node reads the files, i.e., the data illustrated in the 3rd and 4th rectangles in the 2nd row. After pairwise comparison, it outputs the results, which is illustrated in the 2nd rectangle in the 3rd row. In the 3rd iteration, it only starts one node. And it reads the data in the two rectangles in the 3rd row, and outputs the final results, which is illustrated in the bottom rectangle in Fig. 3(a).

In Example 2, we omit the results that can be derived from other results simply, and retain the results that need many pairwise comparisons. For example, we omit "$g13$-$g15$: 3, 0, 1", but retain "$g3$, $g13$-$g15$: 0, 1, 2", in the 4th rectangles in the 2nd row in Fig. 3(a). Due to the former one can be easily derived from other results, but the latter one needs pairwise comparison which is time-consuming in computational resource constrained situation. If in IO constrained situation, we omit the two kinds of results.

Due to that there are no communication mechanisms in MR and using only one iteration of MR cannot guarantee final results is complete, we run a flow of customized map() and reduce(). However, it spends a long time on the restart of MR. Thus, it is urgent to extend the MapReduce framework or propose a novel one.

Example 3. (OPSM mining on naïve Hama) The dataset and threshold utilized are the same as that used in Example 1. The rule of data transfer is that it transfers the compressed original data to other nodes. The procedure of OPSM mining and the final results on Hama are illustrated in Fig. 3(b). Although we reduce some duplicate results, the problem that the reduce super-step spends a long time is still not solved.

From Example 3 shown in Fig. 3(b), we find several problems. The first one is the data received by a node in a super-step is large, e.g., a node receives 10 rows of data plotted in rounded rectangles with labels ①, ②, ③, and ④, on the communication phase. As we know, the number of rows in each split is 4, and the number of rows to send / receive is 10, which is two times more than that of initial splits. The second one is that although we reduce some duplicate results plotted in dashed rectangles in the bottom of Fig. 3(b), it still needs one long super-step for reduce, due to it only uses one node to do the final step. Thus, we should consider these issues and give solutions.

In the condition that does not modify the naïve BSP framework, i.e., Example 3, we only can reduce duplicate results described above, but the other nodes in the cluster are not employed sufficiently, and it also needs a long time super-step.

From the above analysis, we get the idea that we should solve the problems based on sufficiently use the nodes on the cluster, rather than use one node and stop other nodes. Thus, the problem is changed to reduce the amount of data to transfer and percent of duplicate results. And we extend the naïve Hama BSP in Section 3.

3   Parallel Partitioning Methods

In this section, we firstly introduce the BNHB framework. Then, we propose a BNPP method to formalize the data transfer among nodes and a distributed hash tables (*DHT*) based deduplication method. Finally, we give some theorems.

3.1   Butterfly Network Based Hama BSP Framework and Example

In order to guarantee the nodes on Hama have enough memory to save the transfer data, have less workload of bandwidth, and produce less duplicate results, we propose a BNHB framework illustrated in Fig. 4(a).

From Example 3 shown in Fig. 3(b), we find that if each node communicates and transfers data with all of the rest nodes in the super-step for reduce, the results is duplicate. If we use pairwise communication which guarantees the percent of duplicate is smaller, some duplicate results can be avoided. Further, it sufficiently employs all the nodes on the cluster. Thus, the new framework achieves the goal that we mentioned earlier. Now what we are worried about is the completeness of final results after pairwise communication, and we will give the proof about the completeness in Theorem 1.

Fig. 4(a) illustrates the BNHB framework, which inherits the basic framework of BSP model and utilizes HDFS to store the original data and final results. In BNHB, the nodes play the same roles as the naïve Hama. In each super-step, each node firstly receives the split from Master node, then does OPSM mining, further communicates with the rest nodes, and goes into barrier synchronization in the final phase of a super-step. Finally, each node outputs mining results to HDFS. Certainly, the new framework has some differences with naive Hama. First one is that our new framework needs no more than log_2N super-steps instead of one or uncertain super-steps, where N is the number of nodes. Second one is that each node only needs to communicate with one of the rest nodes instead of all nodes in a super-step. Third one is that all the communication nodes are not same for a node in no more than log_2N super-steps. Fourth one is that each node has no more than one communication node in a super-step.

In the following, we firstly present an example to describe how OPSM mining is processed on the new framework and then summarize the data transfer rules.

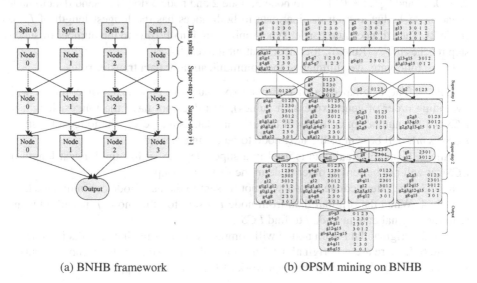

(a) BNHB framework   (b) OPSM mining on BNHB

Fig. 4. Butterfly Network based Hama BSP framework

Example 4. (OPSM mining on BNHB) The dataset and threshold utilized are the same as that used in Example 1. The rules of data transfer are given later. The procedure of OPSM mining and the final results on BNHB are illustrated in Fig. 4(b). Obviously, the mining results are the same as that in Fig. 2(c), which is complete.

Fig. 4(b) illustrates the procedures of Example 4. Firstly, each node reads 1 split as its input, and then enters no more than log_2N super-steps. In super-step 1, the four nodes do pairwise comparison with local data illustrated in 1st row of rectangles, and then generate intermediate results, which are presented in the 2nd row of rectangles. Further the four nodes on the cluster are divided into 2 ((log_24) $/2^{1-1}$=2) groups, and the number in each group is 2, i.e., 4/2=2. Each group is divided into two sub-groups, i.e., in 1st group, node 0 and 1 are divided into node 0 and node 1 two sub-groups, and the division of 2nd group is the same as the 1st one, the step length to communicate between two sub-groups in each group is half group size 1, i.e., 2 / 2=1 (The division method will be given in Algorithm 2). In the communication phase, node 0 sends "$g0$: 0, 1, 2, 3", "$g4$: 1, 2, 3, 0", "$g8$: 2, 3, 0, 1" and "$g12$: 3, 0, 1, 2" to node 1, node 1 sends "$g1$: 0, 1, 2" to node 0, node 2 sends "$g2$: 0, 1, 2, 3" to node 3, and node 3 sends "$g3$: 0, 1, 2, 3" to node 2 (Rule 1, 2, 3, which is discussed later). To wait the nodes to finish data transfer, it goes into barrier synchronization. In super-step 2, each node firstly does pairwise comparison between received data and original data (Rule 4), then does pairwise comparison between the received data and intermediate results (Rule 5). After local computation, four nodes are divided into 1 ($log_24/2^{2-1}$=1) group, and the group is divided into 2 sub-groups, i.e., the 1st sub-group includes node 0 and node 1, the 2nd sub-group contains node 2 and node 3. And the step length to communicate between two sub-groups is half group size 2, i.e., 4 / 2 = 2. In the communication phase, node 0 sends "$g4$: 1, 2, 3, 0", "$g8$: 2, 3, 0, 1" and "$g12$: 3, 0, 1, 2" to node 2, node 1 sends "$g8$: 2, 3, 0, 1" and "$g12$: 3, 0, 1, 2" to node 3, node 2 and node 3 does not send data to node 0 and node 1, due to each original data in both nodes has the longest length of *LCSs* (Rule 2). Then, it goes into barrier synchronization. Finally, due to the number of super-step reaches log_2N (Rule 6), four nodes output the results.

In the following, we summarize the communication or data transfer rules.

Rule 1. The original data of one gene (permutation of indexes of columns with ascending order of gene expression values), that does not have the longest length of *LCSs* (if the number of columns of the original data is m, we say the longest length of *LCSs* is m), will be sent to the next node to be communicated.

Rule 2. If one original data is used in a super-step, i.e., it has the longest length of *LCSs*, it will not be sent to other node in the rest super-steps.

Rule 3. The intermediate results will not be sent to the next node communicated.

Rule 4. When the original data from node i is sent to other node j, it will compare with the original data in node j to find *LCSs*.

Rule 5. Original data from node i will compare with intermediate results in node j.

Rule 6. If there are no original data in each node to be sent or the number of super-step reaches log_2N, the computational work of Hama can be stopped.

3.2   Distributed Hash Tables Based Deduplication Method and Algorithm

Before presenting the example of OPSM mining, we give the details about how to partition data, reduce the amount of data transfer, and communicate among nodes.

BNHB uses the default hash partitioning function to partition data.

An example of summarizing *LCSs* and generating the count of each *LCS* by *DHT* is given in Fig. 5(a). The data is the 4th split in Fig. 4(b). When it finds one $LCS(a_i a_{i-1}...a_1 a_0)$, it utilizes hash function $hash(LCS, No.)$, where *No.* denotes the distinct number of *LCSs*, to compute the hash address in *Array[hash]* which saves the address of *LCSs* that has *m* elements in a List or *Array1[hash]* which saves the address of *LCSs* that has less than *m* elements in a List, where *m* is the number of column in original data. *DHT* tests whether each *LCS* already exists in the List. If yes, it increases the count of this *LCS* by one with *ArrayNo[hash]*. For example, it reads the 1st row "g3: 0 1 2 3" from the split in Fig. 5(a), and computes its hash address, which is 3 $((0*10^3 + 1*10^2 + 2*10^1 + 3) \bmod 4 = 3)$, then it saves "0 1 2 3" as the 1st element in List of *LCSs* that has 4 elements, and records address of "0 1 2 3" in the List into *Array[3]*, i.e., it changes *Array[3]* = -1 to *Array[3]* =0. And it increases the count of "0 1 2 3" by one in *ArrayNo[3]*, i.e., *ArrayNo[3]*=1. The procedure of other rows is the same as the 1st row. Similarly, mining of *LCSs* that have less than 4 elements is the same as that of *LCSs* that have 4 elements, illustrated in the bottom of Fig. 5(a).

An example to reduce the amount of data transfer is presented in Fig. 5(b). In the framework of naïve Hama BSP, each node sends all the intermediate results to other nodes, which is not applicable to gene expression datasets, this is because the amount of intermediate results in this application is large. In Fig. 5(b), *ArrayNo[hash]* records the counts of all of the largest length of *LCSs* (*m*), which is the implementation of Rule 1. The usages of other five rules are presented in Fig. 4(b). If the number saved in *ArrayNo[hash]* is 1, it records the row number in a row set *rowSend*. When it finishes the local computation, it sends the original data saved in *rowSend* to one node. For example, in Fig. 5(b), the row numbers recorded in the row set on both

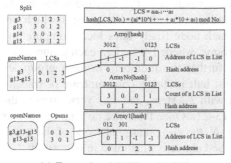

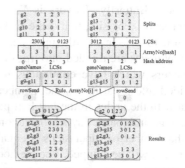

(a) Example of DHT on BNHB   (b) Example of Sending Data on BNHB

Fig. 5. Example of DHT and Sending Data on BNHB

Algorithm 2. Butterfly Network based Parallel Partitioning (BNPP)
1. *geneName* := Φ; *LCSs* := Φ; *Array*[] := -1; *ArrayNo*[] := 0;
2. *opsmName* := Φ; *OPSMs* := Φ; *Array1*[] := -1; *LCS.getData()*;
3. compare between received data and original data; /*Local computation*/
4. *if Array*[*hash*(*LCS*)] == -1 *do* /*Rule 4*/
5.*geneName.add*(*LCS.name*); *LCSs.add*(*LCS*); *Array*[*hash*] := *LCSs.size()*-1; *ArrayNo*[*hash*]++;
6. *else geneName.get*(*hash*).*add*(*LCS.name*); *ArrayNo*[*hash*]++;
7. compare between received data and intermediate results;
8. *if Array1*[*hash*(*LCS*)] == -1 *do* /*Rule 5*/
9.   *opsmName.add*(*LCS.name*); *OPSMs.add*(*LCS*); *Array1*[*hash*] := *LCSs.size()*-1;
10. *else opsmName.get*(*hash*).*add*(*LCS.name*);
11. *if ArrayNo*[*i*] == 1 *do rowSend.add*(*Array*[*i*]); /*Rule 1, 2, 3*/
12. *if rowSend.getData()*!=*null then flag* = **true**; *else flag*=**false**; /*Rule 6*/
13. *step*=1; **while** *step*≤*log$_2$N*&&*flag* == **true** /*Process communication*/
14.   *grpSz* := 2^{step}, *hfGrpSz* := 2^{step-1};   //size of whole & half group
15.   **for** *i*=0; *i*<*N*; *i*=*i*+*grpSz* //divided into groups
16.   *grpLt* := *i*, *grpMid* := *i*+*hfGrpSz*, *grpRt* := *i*+*grpSz*-1;
17.   **for** *j* = *grpLt*; *j* < *grpMid*; *j*++   //left half group
18.   node *j* sends data in *rowSend* to node *j*+*hfGrpSz*; and vice versa;
19.   **for** *k* = *grpRt*; *k* ≥ *grpMid*; *k*--   //right half group
20.   node *k* sends data in *rowSend* to node *k*-*hfGrpSz*; and vice versa;
21.   *step*++;
22. node.sync(); /*Barrier synchronization*/

nodes are the row 0, due to this is the 1st super-step, i.e., the step length is 1 (super-step number, step length and node groups will be given later), both nodes send its local original data of row 0 to each other. The data to be sent is one 4th of the local data, which is sharply smaller than the intermediate mining results.

From the example, we know procedure of BNPP is as follows. The number of nodes of Butterfly Network is N, which is 2^n, where n is the maximum number of super-steps. For simplicity, the names of nodes are denoted by integers which are from 0 to 2^{n-1}. In the ith super-step ($i≥1$), each node firstly does local computation, then N nodes are divided into $(log_2N)/2^{i-1}$ groups, where $1≤i≤n$, i.e., each group has 2^i members which have continuous integers, further the members in each group are divided into 2 partitions, the members in first half partition communicate or transfer data with the nodes in the last half partition with 2^{i-1} steps, and vice versa, finally the nodes go into barrier synchronization (line 13-22). Once there are no data to transfer or the number of super-steps is equal to log_2N, the computational work of the nodes on Hama will be stopped. The method described above is illustrated in Algorithm 2.

3.3   Theorem

Theorem 1. Final results after pairwise communication or data transfer are complete.

Proof: From table 1, we know that one row of gene expression values of gene g_i can be represented by $D(g_i, T)$, and $D(g_i, T) = (x_{i0}, x_{i1}, ..., x_{im})$. Further, we give permutation of indexes of columns with ascending orders of gene expression values of gene g_i, which is denoted by g_i and $g_i = (e_{i0}, e_{i1}, ..., e_{im})$, where e_{ij} is an integer, the range of which is from 0 to $m-1$. We assume v is a subset of g_i, and $v = (e_{io}, ..., e_{ip}, ..., e_{iq})$, where $0 \le o \le p \le q \le m-1$. The number of subsets of length k ($1 < k \le m$) is c_m^k, thus, $g_i = \bigcup_{i=0}^{c_m^i} v_i$. We assume one dataset with n genes is divided into 2^τ splits, where $\tau = log_2 N$ and N is the number of nodes. The split in node i is "$g_{i0}, g_{i1}, ..., g_{ij}$".

Due to the number of super-steps is no more than $log_2 N$, we firstly give the proof of $log_2 N$ super-steps, then give the proof of less than $log_2 N$ super-steps.

(I) $log_2 N$ super-steps: The maximum number of steps is $\tau + 3$, which includes τ super-steps, 1 step for original data saving, preprocessing and summary, respectively. The results of step i in node j is denoted by R_{ij}, and final results is represented by $R_{\tau+2}$.

The completeness of $LCSs$ of length m can be guaranteed by Rule 1 and Rule 2. To prove the completeness of $LCSs$ of length k ($1 < k < m$), which can be guaranteed by Rule 1 to Rule 6, we should consider N^2 situations. Now we give the N situations of node 0, the other situations of node i are similar with node 0.

(1) If v_{nk} in g_{ij} is same with $v_{n'k'}$ in $g_{i'j'}$, and $g_{ij}, g_{i'j'}$ are in the same node (node 0), then we get $v_{nk} \in R_{00}$, $v_{n'k'} \in R_{00} \Rightarrow v_{nk}$, $v_{n'k'} \in R_{10} \Rightarrow v_{nk}$, $v_{n'k'} \in R_{\tau+2}$.

(2) If v_{nk} in g_{ij} is same with $v_{n'k'}$ in $g_{i'j'}$ and $g_{ij}, g_{i'j'}$ are in node 0, 1, then we get that $v_{nk} \in R_{00}$, $v_{n'k'} \in R_{01} \Rightarrow v_{nk} \in R_{10}$, $v_{n'k'} \in R_{00}$ or $R_{10} \Rightarrow v_{nk}$, $v_{n'k'} \in R_{20} \Rightarrow v_{nk}$, $v_{n'k'} \in R_{\tau+2}$.

(3) If v_{nk} in g_{ij} is same with $v_{n'k'}$ in $g_{i'j'}$, and $g_{ij}, g_{i'j'}$ are in node 0, 2, then we get that $v_{nk} \in R_{00}$, $v_{n'k'} \in R_{02} \Rightarrow v_{nk} \in R_{10}$, $v_{n'k'} \in R_{02} \Rightarrow v_{nk} \in R_{20}$, $v_{n'k'} \in R_{20} \Rightarrow v_{nk} \in R_{20}$, $v_{n'k'} \in R_{00}$ or R_{10} or $R_{20} \Rightarrow v_{nk}$, $v_{n'k'} \in R_{30} \Rightarrow v_{nk}$, $v_{n'k'} \in R_{\tau+2}$.

(4) If v_{nk} in g_{ij} is same with $v_{n'k'}$ in $g_{i'j'}$, and $g_{ij}, g_{i'j'}$ are in node 0 and i, then <1> if $i \in [2^\xi, 2^\xi+1]$ ($1 \le \xi \le \tau$) and i is even, we get $v_{nk} \in R_{00}$, $v_{n'k'} \in R_{0i} \Rightarrow v_{nk} \in R_{10}$, $v_{n'k'} \in R_{0i}$ or $R_{1i} \Rightarrow ... \Rightarrow v_{nk} \in R_{\xi-1\,0}$, $v_{n'k'} \in R_{00}$ or R_{10} or ... or $R_{\xi-1\,0} \Rightarrow v_{nk}$, $v_{n'k'} \in R_{\xi 0} \Rightarrow v_{nk}$, $v_{n'k'} \in R_{\tau+2}$; <2> if $i \in [2^\xi, 2^\xi+1]$ and i is odd, we get $v_{nk} \in R_{00}$, $v_{n'k'} \in R_{0i} \Rightarrow v_{nk} \in R_{10}$, $v_{n'k'} \in R_{0i-1}$ or $R_{1i-1} \Rightarrow ... \Rightarrow v_{nk} \in R_{\xi-1\,0}$, $v_{n'k'} \in R_{00}$ or R_{10} or ... or $R_{\xi-1\,0} \Rightarrow v_{nk}$, $v_{n'k'} \in R_{\xi 0} \Rightarrow v_{nk}$, $v_{n'k'} \in R_{\tau+2}$.

(5) If v_{nk} in g_{ij} is same with $v_{n'k'}$ in $g_{i'j'}$ and $g_{ij}, g_{i'j'}$ are in node 0, $N-1$, then we get $v_{nk} \in R_{00}$, $v_{n'k'} \in R_{0N-1} \Rightarrow v_{nk} \in R_{10}$, $v_{n'k'} \in R_{0N-2}$ or $R_{1N-2} \Rightarrow v_{nk} \in R_{20}$, $v_{n'k'} \in R_{0n-2}^2$ or R_{1n-2}^2 or $R_{2n-2}^2 \Rightarrow ... \Rightarrow v_{nk} \in R_{\tau 0}$, $v_{n'k'} \in R_{00}$ or R_{10} or ... or $R_{\tau 0} \Rightarrow v_{nk}$, $v_{n'k'} \in R_{\tau+1\,0} \Rightarrow v_{nk}$, $v_{n'k'} \in R_{\tau+2}$.

(II) less than $log_2 N$ super-steps: The maximum number of steps is $\zeta+3$ ($\zeta < \tau$), which includes ζ super-steps, 1 step for original data saving, preprocessing and summary. The results after step i in node j is denoted by R_{ij}, final results is represented by $R_{\zeta+2}$.

(1) If $\zeta=0$, it only has the data split phase, and there are no data to be transferred. If the v_{nk} in g_{ij} is same with $v_{n'k'}$ in $g_{i'j'}$, and $g_{ij}, g_{i'j'}$ are in the different node (node p, q), we have that v_{nk} in g_{ij} is same with v_{mk} in $g_{ij'}$ in node p, the $v_{n'k'}$ in $g_{i'j'}$ is same with $v_{m'k'}$ in $g_{i'j''}$ in node q, and then we get $v_{nk} \in R_{0p}$, $v_{n'k'} \in R_{0q} \Rightarrow v_{nk}$, $v_{n'k'} \in R_{\zeta+2}$.

(2) If $\zeta=1$, it has 1 data split phase, super-step and data transfer, the group size and communication step of which are 2 and 1. If v_{nk} in g_{ij} is same with $v_{n'k'}$ in $g_{i'j'}$, and $g_{ij}, g_{i'j'}$ are in different groups (p, q), due to there are no transfers between group p and q, we have the $v_{nk}(v_{n'k'})$ in $g_{ij}(g_{i'j'})$ is same with $v_{mk}(v_{m'k'})$ in g_{mj} ($g_{m'j''}$) in group $p(q)$, and then we get $v_{nk} \in R_{0i}$, $v_{n'k'} \in R_{0j} \Rightarrow v_{nk} \in R_{12}^{p-1}$, $v_{n'k'} \in R_{12}^{q-1} \Rightarrow v_{nk}$, $v_{n'k'} \in R_{\zeta+2}$.

(3) If $\zeta=i$, it has 1 data split phase and i super-steps, the group size and communication step of which are 2^i and 2^{i-1}. If the v_{nk} in g_{ij} is same with $v_{n'k'}$ in $g_{i'j'}$, and g_{ij}, $g_{i'j'}$ are in the different groups (group p, q), due to there are no data to be transferred between group p and q, we have the $v_{nk}(v_{n'k'})$ in $g_{ij}(g_{i'j'})$ is same with $v_{mk}(v_{m'k'})$ in $g_{mj'}$ ($g_{m'j''}$) in group $p(q)$, and then we get $v_{nk} \in R_{0i}$, $v_{n'k'} \in R_{0j} \Rightarrow v_{nk} \in R_{i2}^{i*(p-1)}$, $v_{n'k'} \in R_{i2}^{i*(q-1)} \Rightarrow v_{nk}$, $v_{n'k'} \in R_{\zeta+2}$. Thus, we prove the theorem.

Theorem 2. When each node has no data to transfer, i.e., each of the local data has the longest length of *LCSs*, the computational work of Hama can be stopped.

Proof: The theorem can be transformed to Theorem 1, we do not give the proof.

Theorem 3. The nodes that have communicated with node i in earlier super-steps do not need to communicate with node i in the later super-steps.

Proof: (proof based on data locality). When node j has communicated with node i, node j has owned the data of node i. Even if the data of node i has changed, due to it only saves mining result which is similar with node i, thus we prove it.

Theorem 4. The maximum number of super-steps of BNHB framework is $log_2 N$.

Proof: We assume the maximum number of super-steps is n. Based on the property of Butterfly Network, step length in ith super-step is 2^{i-1}. Due to the number of nodes is 2 time of the nth step length, node number in Hama platform is 2^n, i.e., $N = 2^n$. And due to $n = log_2 2^n$, thus, the number of super-steps of BNHB framework is $log_2 N$.

4   Experimental Evaluation

All algorithms are implemented using Java and compiled with Eclipse 3.6 on Ubuntu 11.04, and experiments are conducted on n ($1 \le n \le 8$) nodes with 1.8GHz CPU, 16G memory and 120G disk. The nodes are connected by a gigabit Ethernet. Besides, the versions of Hadoop [7] and Hama [11] are 0.20.2 and 0.4.0, respectively. Note the programs are run on Hama system, input and output data are saved in HDFS [10].

We evaluate the methods with a real microarray dataset that used in some previous studies [12]. It is from the study of clustering Lung Cancer genes, which has 1000 genes and 197 experimental conditions, and we insert / delete some rows / columns.

4.1   Comparison of OPSM and BNPP Methods

In this subsection, we evaluate the efficiency of OPSM and BNPP running on a single machine and BNHB (4 node). First, we test runtime of methods on gene expression data with 1000 rows and columns varying from 20 to 200, shown in Fig. 6(a). Then we test the runtime of methods on gene expression data with 200 columns and rows varying from 1000 to 10000, shown in Fig. 6(b). *threshold* denotes ratio of columns.

Fig. 6(a) shows the runtime of OPSM and BNPP with respect to number of columns, running on a single machine and BNHB (4 nodes), respectively. Row number is 1000, threshold is 0.6, and number of columns varies from 20 to 200. When the number of columns is smaller, differentials of runtime of two methods are relatively smaller. With the increasing of columns, differentials of runtime become larger (about 10). From the test, we can see that when the column number is larger, the performance of BNPP on BNHB is relatively obvious. Fig. 6(b) shows runtime of OPSM and BNPP with respect to number of rows, running on a single machine and BNHB (4 nodes), respectively. The number of columns is 200, threshold is 0.2, and number of rows varies from 1000 to 10000. As same as runtime shown in Fig. 6(a), when the number of rows is smaller, differentials of runtime are relatively smaller. However, with the increasing of rows, differentials of runtime become larger. OPSM on a single machine cannot finish in 3 hours, but BNPP running on BNHB can finish in 1 hour even when row number reaches to 10000. From the test, we can see that BNHB is much superior.

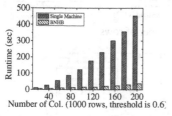 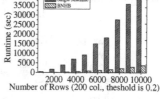

(a) Runtime w.r.t. number of columns   (b) Runtime w.r.t. number of rows

Fig. 6. Comparison of a single machine and BNHB

4.2   Comparison of BSP and BNHB

In this subsection, we will evaluate the scalability of the naïve Hama BSP framework and BNHB framework running on a cluster with 4 nodes. We test the runtime of OSPM and BNPP methods with respect to the number of columns, the number of rows and the number of nodes respectively, which are illustrated in Fig. 7.

As the results shown in Fig. 7(a), when varying the number of columns from 20 to 200 and keeping the number of rows to be 1000, the running time of BNHB is smaller than that of BSP when the number of columns is relatively larger, and nearly the same as that of BSP when the number of columns is relatively smaller. This is because that as the increasing of number of columns, the number of *LCSs* above the threshold increases dramatically and there is larger number of rows to be sent on BSP than that on BNHB. When varying the number of columns from 20 to 200 and keeping the number of rows to be 5000, although the performance of both frameworks is nearly the same, BNHB scales better than BSP. When varying the number of columns from 20 to 200 and keeping the number of rows to be 10000, the improved performance on BNHB is obviously better than that on BSP, due to that there are a lot of original data to be sent on BSP than that on BNHB.

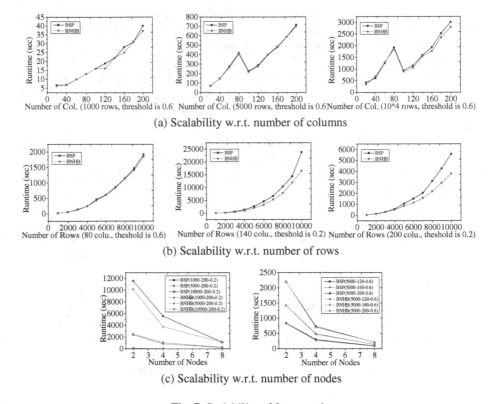

(a) Scalability w.r.t. number of columns

(b) Scalability w.r.t. number of rows

(c) Scalability w.r.t. number of nodes

Fig. 7. Scalability of frameworks

Fig. 7(b) shows the execution times on BSP and BNHB over various numbers of rows, which are ranging from 1000 to 10000. Three figures plotted in Fig. 7(b) indicate that BNHB constantly performs well for various numbers of rows and columns. As pointed out earlier, this very feature of BNHB is made feasible in practice due the early reduction on the size of *rowSend*, and turns out to be very powerful especially when there are large number of rows and columns. When varying the number of rows from 1000 to 10000, and keeping the number of columns to be 80 and threshold to be 0.6, the runtime on BNHB is much smaller than that on BSP. When varying the number of rows from 1000 to 10000, and keeping the number of columns to be 140 or 200 and threshold to be 0.2, BNHB scales much better than BSP obviously.

Fig. 7(c) gives the scalability of frameworks of BNHB and BSP. As the number of nodes on the cluster grows up from 2 to 8, both BSP and BNHB show well scalability. When keeping the number of rows to be 5000 and threshold to be 0.6, varying the column number from 120 to 200, BNHB scales much better than BSP on 4 nodes, and both frameworks show nearly the same performance on 2 and 8 nodes. When keeping the column number to be 200 and threshold to be 0.2, varying the row number from 1000 to 10000, BNHB scales much better than BSP on 4 nodes, and both framework show nearly the same performance on 2 and 8 nodes. This is because that the number of rows to be sent on both frameworks is same or nearly same when the number of nodes is 2 or 8. The experiments imply that if both frameworks scale better with the growth of nodes, but when the number of nodes is too small or too large, the performance of BNHB is not obvious than that of BSP.

5   Related Work

OPSM Mining or Clustering Genes: [1] introduces the *KiWi* mining framework for massive datasets, which substantially reducing the search space and problem scale. [2] devises a method called Affinity Propagation. In which, Real-valued messages are exchanged between data points until a high-quality set of exemplars and clusters emerges. To cope with data noise, [3] proposes a more robust version of OPSM (OPSM-RM). [4] presents a noise-tolerant model called AOPC, [5] employs bucket to relax OPSM model. However, these methods are developed for a single machine.

Parallel Processing Techniques: In [13], Kang et al. observe many graph mining operations are essentially a repeated matrix-vector multiplication. Thus, they describe an important primitive PEGASUS, which is a Peta Graph Mining library implemented on the Hadoop platform [7]. In [14], Zhou et al. describe how reasoning about data partitioning is incorporated into SCOPE optimizer. Pregel [15] is a distributed system for processing large graph datasets. [16] introduces a lightweight extension of Hadoop (CoHadoop) that allows applications to control where data are stored.

6   Conclusion

To address the problem mining gene expression data lacks the capability to incorporate the largest data sets and consumes a long time, the paper proposes a Butterfly Network based parallel partitioning method. Further, we prove our method can guarantee the mining results are complete in theory. Finally, we implement one state-of-the-art OPSMs mining methods (OPSM) and our parallel partitioning method (BNPP) on top of the framework of naïve Hama BSP and our BNHB, and experimental results demonstrate the scalability and effectiveness of our methods.

Acknowledgments. This work is partly supported by the National Basic Research Program (973) of China (No. 2012CB316203), the Natural Science Foundation of China (No. 61033007, 61272121), the National High Technology Research and Development Program (863) of China (No. 2012AA011004), graduate starting seed fund of Northwestern Polytechnical University (No. Z2012128, Z2013125, Z2013126).

References

1. Gao, B.J., et al.: Discovering Significant OPSM Subspace Clusters in Massive Gene Expression Data. In: Proceedings of KDD, pp. 922–928. ACM Press, New York (2006)

2. Frey, B.J., Dueck, D.: Clustering by Passing Messages between Data Points. Science 315(5814), 972–976 (2007)
3. Chui, C.K., Kao, B., et al.: Mining Order-Preserving Submatrices from Data with Repeated Measurements. In: Proceedings of ICDM, pp. 133–142. IEEE Press, Cancun (2008)
4. Zhang, M., Wang, W., Liu, J.: Mining Approximate Order Preserving Clusters in the Presence of Noise. In: Proceedings of ICDE, pp. 160–168. IEEE Press, Cancun (2008)
5. Fang, Q., Ng, W., Feng, J., Li, Y.: Mining Bucket Order-Preserving SubMatrices in Gene Expression Data. IEEE Trans. on Know. and Data Engin. 24(12), 2218–2231 (2012)
6. Gene Data, http://genomebiology.com/content/supplementary/gb-2003-4-5-r34-s8.txt
7. Dean, J., et al.: MapReduce: Simplified Data Processing on Large Clusters. In: Proceedings of OSDI, pp. 137–150. USENIX Press, California (2004)
8. Ding, L., Xin, J., Wang, G., Huang, S.: ComMapReduce: An Improvement of MapReduce with Lightweight Communication Mechanisms. In: Lee, S.-G., Peng, Z., Zhou, X., Moon, Y.-S., Unland, R., Yoo, J. (eds.) DASFAA 2012, Part II. LNCS, vol. 7239, pp. 150–168. Springer, Heidelberg (2012)
9. Feldmann, R., Unger, W.: The Cube-Connected Cycles Network is a Subgraph of the Butterfly Network. Parallel Processing Letters 2(1), 13–19 (1992)
10. Bulk Synchronous Parallel, http://en.wikipedia.org/wiki/Bulk_synchronous_parallel
11. Apache Hama, http://hama.apache.org
12. Cancer Program Data Sets, http://www.broadinstitute.org/cgi-bin/cancer/datasets.cgi
13. Kang, U., et al.: PEGASUS: A Peta-Scale Graph Mining System-Implementation and Observations. In: Proceedings of ICDM, pp. 229–238. IEEE Press, Florida (2009)
14. Zhou, J., Larson, P.A., et al.: Incorporating Partitioning and Parallel Plans into the SCOPE Optimizer. In: Proceedings of ICDE, pp. 1060–1071. IEEE Press, California (2010)
15. Malewicz, G., et al.: Pregel: A System for Large-scale Graph Processing. In: Proceedings of SIGMOD, pp. 135–146. ACM Press, Indiana (2010)
16. Eltabakh, M.Y., Tian, Y., et al.: CoHadoop: Flexible Data Placement and its Exploitation in Hadoop. In: Proceedings of VLDB, pp. 575–585. ACM Press, Washington (2011)

Parallel and Distributed Mining of Probabilistic Frequent Itemsets Using Multiple GPUs

Yusuke Kozawa[1], Toshiyuki Amagasa[2], and Hiroyuki Kitagawa[2]

[1] Graduate School of Systems and Information Engineering, University of Tsukuba
kyusuke@kde.cs.tsukuba.ac.jp
[2] Faculty of Engineering, Information and Systems, University of Tsukuba
{amagasa,kitagawa}@cs.tsukuba.ac.jp

Abstract. Probabilistic frequent itemset mining, which discovers frequent itemsets from uncertain data, has attracted much attention due to inherent uncertainty in the real world. Many algorithms have been proposed to tackle this problem, but their performance is not satisfactory because handling uncertainty incurs high processing cost. To accelerate such computation, we utilize GPUs (Graphics Processing Units). Our previous work accelerated an existing algorithm with a single GPU. In this paper, we extend the work to employ multiple GPUs. Proposed methods minimize the amount of data that need to be communicated among GPUs, and achieve load balancing as well. Based on the methods, we also present algorithms on a GPU cluster. Experiments show that the single-node methods realize near-linear speedups.

1 Introduction

Uncertain data management is attracting considerable interest due to inherent uncertainty in real-world applications such as sensor-monitoring systems. In the area of uncertain data management, frequent itemset mining [1] from uncertain databases is one of the important research issues. Since the uncertainty is represented by probability, this problem is called *probabilistic frequent itemset mining*. Many algorithms have been proposed to tackle probabilistic frequent itemset mining [4,10,11]. However, existing algorithms suffer from performance problems because the computation of probability is highly time-consuming. It is thus necessary to accelerate this computation in order to handle large uncertain databases.

To this end, GPGPU (General-Purpose computation on Graphics Processing Unit) is an attractive solution. GPGPU refers to performing computation on GPUs (Graphics Processing Units), which are originally designed for processing 3D graphics. GPGPU has received much attention from not only the field of high performance computing but also many other fields such as data mining [3,5,9]. This is because GPUs have more than hundred processing units and can process many data elements with high parallelism. It is also known that GPUs are energy-efficient and have higher performance-to-price ratios than CPUs.

H. Decker et al. (Eds.): DEXA 2013, Part I, LNCS 8055, pp. 145–152, 2013.

By leveraging such an emerging processor, our previous work [6] accelerated an algorithm of probabilistic frequent itemset mining. Meanwhile, GPU clusters, which are computer clusters where each node has one or multiple GPUs, have emerged as a powerful computing platform, such as TSUBAME2.0[1] and Titan.[2] Thus, it is increasingly important to harness the power of multiple GPUs and GPU clusters. The utilization of multiple GPUs gives us further parallelism and larger memory spaces. However, employing multiple GPUs has the problem that each GPU has a separate memory space. If one GPU requires data that reside in another GPU, the data need to be communicated via PCI-Express bus. Since the PCI-Express latency is much higher than the GPU-memory latency, it is probable that the communication becomes a bottleneck. It is therefore desirable to reduce data dependencies among data fragments on different GPUs.

This paper proposes multi-GPU methods that take into account the above concerns. First, we develop methods on a single node with multiple GPUs, and then we extend the methods to use a GPU cluster. The proposed methods reduce data dependencies by distributing candidates of probabilistic frequent itemsets among GPUs. In addition, the methods consider load balancing, which is also an important issue to achieve scalability. Experiments show that the single-node methods realize near-linear speedups.

The rest of this paper is organized as follows. Section 2 explains preliminary knowledge of proposed methods. Then Section 3 describes our proposed methods that utilize multiple GPUs. The methods are empirically evaluated in Section 4. Section 5 reviews related work, and Section 6 concludes this paper.

2   Preliminaries

We describes necessary knowledge to understand proposed methods. Section 2.1 defines probabilistic frequent itemsets. A baseline algorithm [10] is explained in Section 2.2.

2.1   Probabilistic Frequent Itemsets

Let I be a set of all items. A set of items $X \subseteq I$ is called an *itemset*, and a k-itemset means an itemset that contains k items. An *uncertain transaction database* is a set of *transactions*, each of which is a triplet of an ID, an itemset, and an *existential probability*. An existential probability stands for the probability that a transaction really exists in the database.

Since transactions have existential probabilities, the support of an itemset (the number of transactions that include the itemset) becomes a random variable. The probability mass function of the support of an itemset X is called a *Support Probability Mass Function (SPMF)* f_X. The function f_X takes an integer $k \in \{0, 1, ..., |\mathcal{U}|\}$, and $f_X(k)$ represents the probability that the support of X equals k. An itemset X is called a *Probabilistic Frequent Itemset (PFI)* if

---

[1] http://www.gsic.titech.ac.jp/en/tsubame2
[2] http://www.olcf.ornl.gov/titan

$$P(\sup(X) \geq \text{minsup}) = \sum_{k=\text{minsup}}^{|\mathcal{U}|} f_X(k) \geq \text{minprob}, \tag{1}$$

where minsup and minprob $\in (0, 1]$ are user-specified thresholds.

2.2   pApriori Algorithm

Sun et al. [10] proposed a pApriori algorithm, which adapts the classical Apriori algorithm [2] to uncertain databases. The pApriori algorithm comprises two main procedures:

1. Generating a set of *candidate* k-itemsets $\mathcal{C}_k$ from a set of $(k-1)$-PFIs $\mathcal{L}_{k-1}$
2. Extracting a set of k-PFIs $\mathcal{L}_k$ from $\mathcal{C}_k$

The pApriori algorithm continues these procedures alternately with incrementing k by one until no additional PFIs are detected. Note that, in the beginning of the algorithm, k's value is one and each candidate 1-itemset is an itemset whose element is merely one item in an input database $\mathcal{U}$. Eventually the pApriori algorithm returns all the PFIs extracted from $\mathcal{U}$.

In order to determine whether or not an itemset X is a PFI, the SPMF of X needs to be computed and assigned to Equation 1. For computing SPMFs, Sun et al. [10] proposed two algorithms: dynamic-programming and divide-and-conquer approaches. We adopted the latter algorithm for GPU implementation because it is more suitable for parallel processing.

Owing to the high time complexity of computing SPMFs, it is desirable to prune infrequent itemsets without computing SPMFs. Let $\text{cnt}(X)$ be the number of transactions that include an itemset X in an uncertain transaction database $\mathcal{U}$ regardless of existential probabilities. Besides, let $\text{esup}(X)$ be the expected value of $\sup(X)$. With the two values, two lemmas are proved [10]. The lemmas enable us to prune candidates in $O(|\mathcal{U}|)$ time, while the computation of SPMFs requires $O\left(\text{cnt}(X) \log^2(\text{cnt}(X))\right)$ time.

3   Multi-GPU Parallelization

In a multi-GPU system, each GPU has a separate memory space. If data dependencies exist among GPUs, GPUs need to communicate with each other via PCI-Express bus. Thus it is important how to distribute data to be processed on GPUs, so that data dependencies are minimized.

Multi-GPU systems can be considered as a kind of distributed-memory systems. Meanwhile, there exists much work on frequent itemset mining for distributed-memory systems. Zaki [12] classified data-distribution schemes that existing algorithms employ into three: count distribution, data distribution, and candidate distribution. In this paper, we propose methods employing the candidate-distribution scheme, because this scheme enable GPUs to compute SPMFs independently, unlike the other two schemes.

Section 3.1 describes methods on a single node equipped with multiple GPUs. Then, Section 3.2 extends the methods to use a GPU cluster.

3.1   Single-Node Methods

Prefix Distribution. We here describe an algorithm based on a naïve candidate-distribution scheme. This algorithm is called *Prefix Distribution (PD)*, because the algorithm distributes candidates by exploiting the property that two candidates with different prefixes are not joinable. PD is divided into three phases: initialization, distribution, and loop phases.

In the initialization phase, PD extracts 1-PFIs. PD firstly generates candidate 1-itemsets from an input uncertain transaction database, and evenly distributes the candidates among GPUs. Then GPUs extract 1-PFIs by using the single-GPU method [6]. These PFIs are transferred to the CPU for generating next candidates.

In the distribution phase, PD generates candidate 2-itemsets from the 1-PFIs on the CPU, and distributes the candidates into GPUs according to their prefixes. For example, consider four GPUs and eight 1-PFIs {a}, {b}, {c}, {d}, {e}, {f}, {g}, and {h}. Then, candidate 2-itemsets are assigned into the GPUs in a zigzag fashion, in order to evenly partition the candidates among GPUs. More specifically, candidates with prefixes a, b, c, and d are assigned to GPUs 1, 2, 3, and 4, respectively. Next candidates with prefix e are assigned to GPU 4, candidates with prefix f are assigned to GPU 3, and so on.

Subsequently, each GPU extracts k-PFIs from the assigned candidates by using the single-GPU method. The CPU gathers the k-PFIs from each GPU. The collected k-PFIs on the CPU are broadcasted to all the GPUs, because the k-PFIs are necessary in candidate generation on GPUs. Then k is incremented and GPUs generate candidate k-itemsets from the $(k-1)$-PFIs found on each GPU. PD continues these operations until no candidates are found.

The proposed method PD reduces data dependencies among GPUs by distributing candidates according to prefixes. However, the distribution by prefix may result in load imbalance. This is because the number of PFIs differs depending on a prefix. As a result, some GPUs may need to compute many SPMFs and other GPUs may compute a few SPMFs. Thus PD is not desirable from the viewpoint of load balancing.

Round-Robin Distribution. *Round-Robin Distribution (RRD)* distributes candidates in a round-robin fashion. While RRD has an almost identical algorithm to PD, major differences lie in the distribution and loop phases. In the distribution phase, candidates are distributed among GPUs in round robin, instead of using prefixes. In the loop phase, RRD generates next candidates on the CPU and distributes these candidates into GPUs in round robin, thereby balancing loads of GPUs at each iteration.

Count-Based Distribution. The algorithm of *Count-Based Distribution (CBD)* is identical to RRD except for its candidate-distribution scheme. CBD assigns candidates to GPUs by taking into account candidates' cnt values. The rationale is that the cnt values determine the computing time of SPMFs [10], and the computing time of SPMFs dominates the processing time of candidates

Table 1. Characteristics of datasets

Dataset	Type	Number of items	Avg. size of transactions	Number of transactions	Density
Accidents	real	468	33.8	340,183	7.2%
T25I10D500K	synthetic	7558	25	499,960	0.33%
Kosarak	real	41270	8.1	990,002	0.020%

on GPUs. Therefore we can achieve load balancing if candidates are distributed by considering cnt values.

3.2 A Method on a GPU Cluster

A method on a GPU cluster is an extension of single-node methods. For the sake of simplicity, we here assume that all the nodes hold an input database. As in the single-node methods, candidates are distributed and are processed in parallel. More precisely, candidates are distributed among nodes, and candidates within a node are distributed among GPUs. For both of the candidate distribution, the schemes of single-node methods can be used. Note that PFIs extracted on nodes need to be broadcasted to other nodes in order to generate next candidates.

4 Experiments

4.1 Experimental Setup

We implemented the proposed methods using CUDA [13], OpenMP, and MPI. Experiments were conducted on a GPU cluster of eight nodes, each of which is equipped with two GPUs. The nodes are connected via InfiniBand QDR (Quad Data Rate). This paper presents only the results on a single node due to the limitation of space.

Table 1 summarizes the three datasets used in the experiments. The density of a dataset is computed as the average length of transactions divided by the number of items. Accidents and Kosarak are real datasets that are accessible on Frequent Itemset Mining Implementations (FIMI) Repository.[3] While Accidents is the densest dataset, Kosarak is the sparsest dataset. T25I10D500K is a synthetic dataset, generated by a data generator.[4] The default values of minsup on Accidents, T25I10D500K, and Kosarak are 33%, 0.65% and 0.2%, respectively. Existential probabilities for the datasets are randomly drawn from a normal distribution with mean 0.5 and variance 0.02. The value of minprob is fixed to 0.5 for all the experiments.

4.2 Results on a Single Node

This section evaluates the three methods on a single node (PD, RRD, and CBD). Figures 1(a)–1(c) show the speedups of the three methods compared to the

---

[3] http://fimi.cs.helsinki.fi/

[4] http://miles.cnuce.cnr.it/~palmeri/datam/DCI/datasets.php

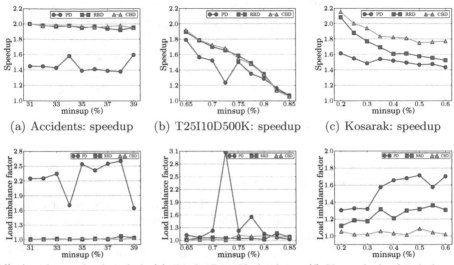

(a) Accidents: speedup (b) T25I10D500K: speedup (c) Kosarak: speedup

(d) Accidents: load imbal- (e) T25I10D500K: load im- (f) Kosarak: load imbalance
ance between two GPUs   balance between two GPUs between two GPUs

Fig. 1. Results on a single node with two GPUs

single-GPU method [6], with varying `minsup` values. From the charts, we can make the following three observations:

1. The multi-GPU methods become faster than the single-GPU method, as the `minsup` value decreases.
2. RRD and CBD are generally faster than PD.
3. CBD outperforms RRD on Kosarak, while CBD and RRD exhibit similar performance on Accidents and T25I10D500K.

The first observation is due to the fact that small `minsup` values make the number of candidates large, and thus GPUs have much more workload to be done in parallel. Although the number of candidates is small on Accidents, Figure 1(a) shows near-linear speedups. This is because the cnt values become very large (nearly 250,000) on Accidents, and hence the computation of SPMFs on Accidents is far more computationally demanding than the computation on T25I10D500K and Kosarak. As a result, the overall processing time on GPUs is dominated by the time of computing SPMFs, which can be performed in parallel.

The second observation results from the load imbalance in PD. To verify this observation, we use a *load imbalance factor* defined as the ratio of the longest processing time of GPUs to the shortest processing time of GPUs. The factor takes the value of 1 if the load is completely balanced, and the factor increases as the load is more imbalanced. Figures 1(d)–1(f) show the load imbalance factors between two GPUs as a function of `minsup` value. These charts reveal that RRD and CBD achieve better load balancing than PD on all the datasets.

The third observation emerges from a characteristic of datasets and the better capability of CBD for load balancing. The characteristic is related to the dispersion of cnt values. For instance, candidates in Kosarak have a wide range of cnt values (2,000–600,000). Since CBD distributes candidates by taking into account cnt values, CBD accommodates to such variability, unlike RRD, which statically assigns candidates to GPUs. As a result, CBD achieves better load balancing than RRD, and realizes higher speedup ratios, as shown in Figures 1(c) and 1(f). On the other hand, CBD and RRD on Accidents and T25I10D500K exhibit similar performance as shown in Figures 1(a) and 1(b). This is because the cnt values (i.e., processing times) of candidates in Accidents and T25I10D500K do not vary much. Thus it is sufficient to distribute candidates in round robin in order to achieve load balancing.

5 Related Work

Frequent itemset mining. The problem of frequent itemset mining was firstly introduced by Agrawal et al. [1]. Since the introduction, many algorithms have been proposed to accelerate the mining. Parallelization of these algorithms has been also widely studied. In the late 1990s, distribute-memory systems were mainly used as the underlying architecture. Zaki summarized such algorithms [12]. More recently, Özkural et al. [8] introduced a data distribution scheme based on graph theory that divides the frequent itemset mining task in a top-down manner. Frequent itemset mining on GPUs has been also studied [5,3,9]. Fang et al. [5] proposed two approaches, GPU-based and CPU-GPU hybrid methods. Amossen et al. [3] presented a novel data layout *BatMap* to represent bitstrings. Then they make use of BatMap to accelerate frequent itemset mining. Silvestri et al. [9] proposed a GPU version of a state-of-the-art algorithm *DCI* [7].

Probabilistic frequent itemset mining. While the above-mentioned parallel algorithms work well for the conventional *certain* transaction databases, they cannot effectively process frequent itemset mining from uncertain databases, which gains increasing importance in order to handle data uncertainty. To mine frequent itemsets with taking into account the uncertainty, a number of algorithms have been proposed. Bernecker et al. [4] proposed an algorithm to find probabilistic frequent itemsets under the attribute-uncertainty model, where existential probabilities are associated with items. On the other hand, Sun et al. [10] considered the tuple-uncertainty model, where existential probabilities are associated with transactions.

Several attempts to accelerate these algorithms also exist. Wang et al. [11] developed an algorithm to approximate the probability that determines whether itemsets are PFIs or not. Our previous work [6] presented an algorithm using a single GPU based on the work by Sun et al. [10]. We have extended this method to use multiple GPUs in this paper.

6   Conclusions

This paper has proposed methods of probabilistic frequent itemset mining using multiple GPUs. The methods run in parallel by distributing candidates among GPUs. The proposed method PD assigns candidates to GPUs according to their prefixes. Then we have described RRD and CBD, which take into consideration load balancing. RRD distributes candidates to GPUs in a round-robin fashion, while CBD uses the cnt values of candidates to achieve better load balancing. We have also presented methods on a GPU cluster by extending RRD and CBD. Experiments on a single node showed that CBD achieves the best load balancing and results in the fastest algorithm. Future work involves developing a CPU-GPU hybrid method that utilizes CPUs and GPUs simultaneously. In addition, more sophisticated multi-node methods should be considered.

Acknowledgments. This work was supported by JSPS KAKENHI Grant Number 24240015 and HA-PACS Project for advanced interdisciplinary computational sciences by exa-scale computing technology.

References

1. Agrawal, R., Imieliński, T., Swami, A.: Mining Association Rules between Sets of Items in Large Databases. In: SIGMOD, pp. 207–216 (1993)
2. Agrawal, R., Srikant, R.: Fast Algorithms for Mining Association Rules. In: SIGMOD, pp. 207–216 (1993)
3. Amossen, R.R., Pagh, R.: A New Data Layout for Set Intersection on GPUs. In: IPDPS, pp. 698–708 (2011)
4. Bernecker, T., Kriegel, H.-P., Renz, M., Verhein, F., Zuefle, A.: Probabilistic Frequent Itemset Mining in Uncertain Databases. In: KDD, pp. 119–128 (2009)
5. Fang, W., Lu, M., Xiao, X., He, B., Luo, Q.: Frequent Itemset Mining on Graphics Processors. In: DaMoN, pp. 34–42 (2009)
6. Kozawa, Y., Amagasa, T., Kitagawa, H.: GPU Acceleration of Probabilistic Frequent Itemset Mining from Uncertain Databases. In: CIKM, pp. 892–901 (2012)
7. Orlando, S., Palmerini, P., Perego, R., Silvestri, F.: Adaptive and Resource-Aware Mining of Frequent Sets. In: ICDM, pp. 338–345 (2002)
8. Ozkural, E., Ucar, B., Aykanat, C.: Parallel Frequent Item Set Mining with Selective Item Replication. IEEE TPDS 22(10), 1632–1640 (2011)
9. Silvestri, C., Orlando, S.: gpuDCI: Exploiting GPUs in Frequent Itemset Mining. In: PDP, pp. 416–425 (2012)
10. Sun, L., Cheng, R., Cheung, D.W., Cheng, J.: Mining Uncertain Data with Probabilistic Guarantees. In: KDD, pp. 273–282 (2010)
11. Wang, L., Cheung, D.W., Cheng, R., Lee, S.D., Yang, X.S.: Efficient Mining of Frequent Item Sets on Large Uncertain Databases. IEEE TKDE 24(12), 2170–2183 (2012)
12. Zaki, M.J.: Parallel and Distributed Association Mining: A Survey. IEEE Concurrency 7(4), 14–25 (1999)
13. NVIDIA, CUDA C Programming Guide (October 2012)

Taming Elephants,
or How to Embed Parallelism into PostgreSQL*

Constantin S. Pan and Mikhail L. Zymbler

South Ural State University, Chelyabinsk, Russia

Abstract. The paper describes the design and the implementation of PargreSQL parallel database management system (DBMS) for cluster systems. PargreSQL is based on PostgreSQL open-source DBMS and exploits partitioned parallelism. Presented experimental results show that this scheme is worthy of further development.

1 Introduction

Currently open-source PostgreSQL DBMS is one of reliable alternatives for commercial DBMSes [9]. There are many practical database applications based upon PostgreSQL and research projects devoted to extension and improvement of PostgreSQL.

One of the research goals is to adapt PostgreSQL for parallel query processing. In this paper we describe the architecture and design of PargreSQL [8] parallel DBMS for analytical data processing on cluster systems. PargreSQL represents PostgreSQL with embedded partitioned parallelism.

The paper is organized as follows. Section 2 gives a description of the PargreSQL architecture in comparison with PostgreSQL's one. Section 3 introduces the implementationt principles of PargreSQL DBMS. The results of experiments on the current implementation are shown in section 4. Section 5 briefly discusses related work. Section 6 contains concluding remarks and directions for future work.

2 PargreSQL Design

PargreSQL utilizes the idea of partitioned parallelism [2] in cluster systems (see fig. 1). This form of parallelism supposes partitioning of relations and their distribution among the disks of the cluster.

The way the partitioning is done is defined by a *fragmentation function*, which for each tuple of the relation calculates the number of the processor node where this tuple should be placed. A query is executed in parallel on all processor nodes as a set of parallel *agents*. Each agent processes its own fragment and generates a partial query result. The partial results are merged into the result relation.

---

* The reported study was partially supported by the Russian Foundation for Basic Research, research projects No. 12-07-31217 and No. 12-07-00443.

H. Decker et al. (Eds.): DEXA 2013, Part I, LNCS 8055, pp. 153–164, 2013.

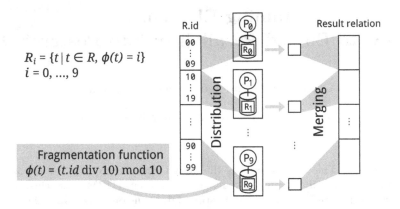

$$R_i = \{t \mid t \in R,\ \phi(t) = i\}$$
$$i = 0, ..., 9$$

Fragmentation function
$\phi(t) = (t.id\ \mathrm{div}\ 10)\ \mathrm{mod}\ 10$

Fig. 1. Query processing with partitioned parallelism

2.1   Client-Server Model

PostgreSQL is based on the client-server model. A session involves three processes into interaction: a frontend, a backend and a daemon (see fig. 2a; here k is a number of clients).

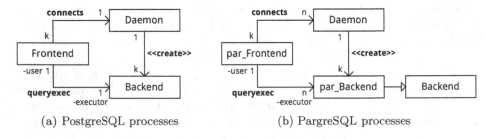

(a) PostgreSQL processes   (b) PargreSQL processes

Fig. 2. DBMS processes

The daemon handles incoming connections from frontends and creates a backend for each one. Each backend executes queries received from the related frontend. The architecture of PargreSQL, in contrast with PostgreSQL, assumes that a client connects to two or more servers (see fig. 2b; here n is a number of cluster nodes).

The interaction sequence is shown in fig. 3. As opposed to PostgreSQL there are many daemons running in PargreSQL. The frontend connects to each of them, sends the same query to many backends, and receives the result relation.

2.2   Deployment Scheme

The application library *libpq* implements the interaction protocol between the client and the server and consists of two parts: the frontend (*libpq-fe*) and the

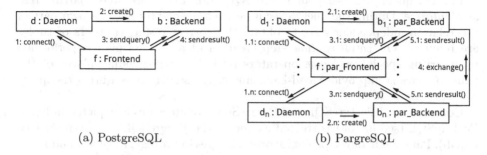

(a) PostgreSQL   (b) PargreSQL

Fig. 3. Interaction of clients and servers

backend (*libpq-be*). The former is deployed on the client side and serves as an API for the end-user application. The latter is deployed on the server side and serves as an API for *libpq-fe*, as shown in fig. 4a.

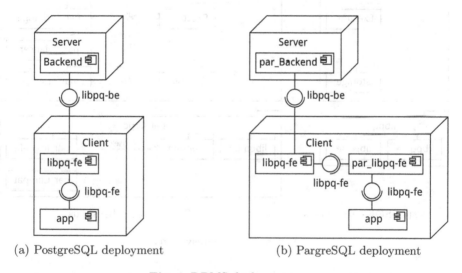

(a) PostgreSQL deployment   (b) PargreSQL deployment

Fig. 4. DBMS deployment

PargreSQL deployment scheme is depicted in fig. 4b. The only difference of deployment schemes (see fig. 4b) is that in case of PargreSQL there is one more component on the client side — the *libpq-fe* wrapper.

2.3   PargreSQL Subsystems

There are following steps of query processing in PostgreSQL: *parse*, *rewrite*, *plan/optimize*, and *execute*.

Parallel query processing in PargreSQL adds two more steps: **parallelize** and **balance**. During the query execution each agent processes its own part of the relation independently so, to obtain the correct result, transfers of tuples are required. On *parallelization* step creation of a parallel plan is performed by inserting special *exchange* operators into the corresponding places of the plan. *Balance* step provides load-balancing of the server nodes during the query execution process.

Comparison of PostgreSQL and PargreSQL architectures is depicted in fig. 5. PostgreSQL (see fig. 5a) is treated as one of the PargreSQL's subsystems (see fig. 5b). PargreSQL development involves changes in Storage, Executor and Planner subsystems of PostgreSQL.

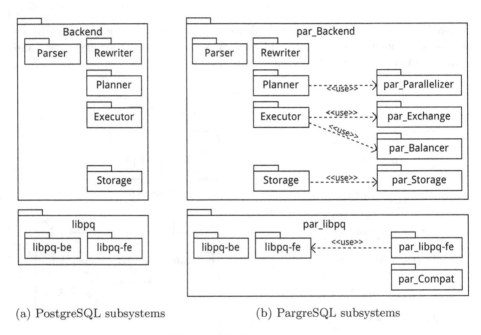

(a) PostgreSQL subsystems   (b) PargreSQL subsystems

Fig. 5. DBMS subsystems

Parser checks the syntax of the query string and builds a parse tree. *Rewriter* processes the tree according to the rules specified by the user (e.g. view definitions). *Planner* creates an optimal execution plan for this query tree. *Executor* takes the execution plan and processes it recursively from the root. *Storage* provides functions to store and retrieve tuples and metadata.

The changes in the PostgeSQL's source code are needed to integrate it with the new subsystems. *par_Storage* is responsible for storing partitioning metadata of relations. *par_Exchange* encapsulates the implementation of the *exchange* operator. *Exchange* operator is meant to compute the exchange function ψ for each tuple of the relation, send "alien" tuples to the other nodes, and receive "native" tuples in response. In section 3.2 we will describe *exchange* operator in detail.

There are new subsystems which do not require any modifications to the PostgreSQL's source code: *par_libpq-fe* and *par_Compat*. *par_libpq-fe* is a wrapper around *libpq-fe*, it is needed to propagate queries from an application to many servers. *par_Compat* makes this propagation transparent to the application. Section 3.1 further describes implementation details of *par_libpq* subsystem.

3   PargreSQL Implementation

In this section we describe the implementation principles of some of the PargreSQL subsystems depicted in fig. 5b.

3.1   par_libpq

Since the frontend in PargreSQL has to initiate a connection to every of the database daemons, some modifications were introduced into the *libpq* application library. The modified version is called *par_libpq*. The purpose of this library is to serve as a replacement for the original *libpq* and to allow the applications to use PargreSQL without much effort.

par_libpq consists of *par_libpq-fe* library and a set of macros (*par_Compat*). *par_libpq-fe* is a library to be linked with frontend applications instead of original PostgreSQL's *libpq-fe*, around which it is a wrapper. Its implementation is illustrated with a class diagram in fig. 6. The idea is to use the original *libpq-fe* for connecting to many servers simultaneously.

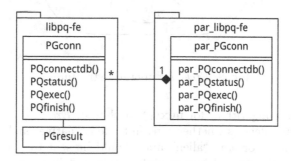

Fig. 6. PargreSQL libpq-fe wrapper

par_Compat is a set of C preprocessor definitions for transparent usage of *par_libpq-fe*. An example of these macros is given in fig. 7.

These macros change the original API calls into the new API calls, so by including them an application programmer can switch from PostgreSQL to PargreSQL without global changes in the application code.

```
#define PGconn par_PGconn
#define PQconnectdb(X) par_PQconnectdb()
#define PQfinish(X) par_PQfinish(X)
#define PQstatus(X) par_PQstatus(X)
#define PQexec(X,Y) par_PQexec(X,Y)
```

Fig. 7. PargreSQL compatibility macros

3.2   Exchange Operator

In order to compute the correct results the DBMS instances running in parallel
have to send tuples to each other, because a tuple stored on one node could
be processed on another node, e.g. in case of an aggregation with group-by on
attribute A while the fragmentation attribute is B. To resolve such situations
we should implement an operator that would move tuples from one point in the
query plan to the same point on another node's plan.

Exchange operator [3,10] serves as an exchange point for transfering tuples
between parallel agents. It is inserted into the query plans by *par_Parallelizer*
subsystem (which will be discussed further). The operator's structure is presented in fig. 8.

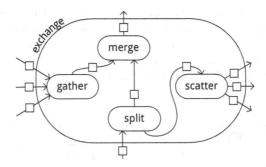

Fig. 8. *Exchange* operator structure

Fig. 9 shows the algorithms for **next()** method of *Exchange* suboperators.

Split (see fig. 9a) decides whether a given tuple is "native" and should be kept
on the current node, or it is "alien" and should be sent to appropriate node.
"Native" tuples are returned immediately whereas "alien" tuples and NULLs
(meaning that scanning of tuples is over) are put into *Scatter*'s buffer for sending
to appropriate nodes.

Gather (see fig. 9b) provides receiving of tuples from other processor nodes.
Having received a tuple *Gather* starts a receive operation again. NULL value
received means that the corresponding node has finished its work. As soon as a
NULL is received from every node *Gather* finishes its work.

Scatter (see fig. 9c) sends tuples coming from *Split* to other processor nodes.
Non-NULL tuple should be sent to a node with a number calculated by means

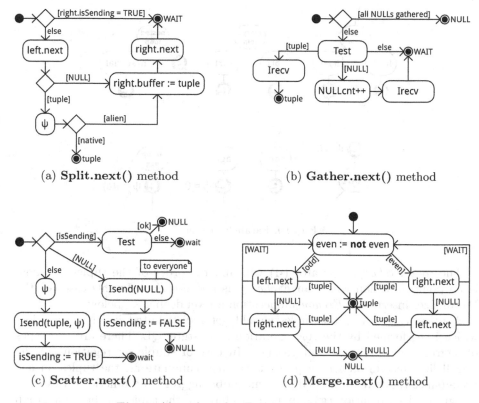

(a) **Split.next()** method   (b) **Gather.next()** method

(c) **Scatter.next()** method   (d) **Merge.next()** method

Fig. 9. Algorithms for *Exchange* suboperators

of fragmentation function. In case of NULL value *Scatter* sends NULLs to all the other nodes.

Merge (see fig. 9d) merges tuples from *Gather* and *Split* in an even-odd manner.

The asynchronous MPI methods Isend, Irecv are used by *Exchange* to transmit tuples and the Test method to check whether the appropriate transmission finished.

3.3   Parallelizer

par_Parallelizer subsystem prepares the query plan for parallel execution. The cases in which the *par_Parallelizer* inserts *Exchange* operators into the plan are shown in fig. 10.

The *Join* operation executed independently on multiple nodes will miss some tuples, unless we move the tuples matching the join qualifier to the same node. That is performed by the *Exchange* operators, which are inserted under the *Join* in cases **(a)**, **(b)**, and **(c)**.

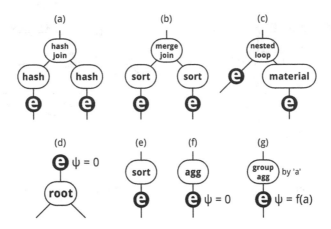

Fig. 10. Parallelizer cases

The root *Exchange* operator (**d**) will mix the order of the tuples, so there is no point in a plan where an *Exchange* is above a *Sort*. In this case (**e**) the Parallelizer inserts the *Exchange* operation a level deeper — below the Sort.

Another case where the tuples should get redistributed is an aggregation, which is performed by the *Agg* operation in PostgreSQL. There are two types of aggregation — simple and grouped. In case of simple aggregation (**f**) the Parallelizer inserts an *Exchange* that would accumulate all the tuples on one node (since they are all needed for some global aggregating function). However, the grouped aggregation (**g**) only needs to have all the tuples of the same group located together.

3.4   Data Manipulation Operations

The algorithms for the exchange subnodes shown above will only work for SE-LECT statements. In order to support data manipulation queries the execution process needs to become a bit trickier.

When PostgreSQL executes an UPDATE or DELETE query, the resulting tuples coming from the root of the plan have a special hidden attribute — the CTID. It is the address of this tuple inside the storage of PostgreSQL, with this CTID PostgreSQL tells which tuples are to be deleted or updated. The other attributes contain the updated values or, in case of a DELETE, there are no other attributes.

No changes are needed in order for DELETE to work in PargreSQL. But INSERT and UPDATE should have additional logic — since a tuple only needs to be inserted on one node and can move from one node to another during an UPDATE.

There are two places in the PargreSQL code that were changed in order to implement that behaviour for UPDATE queries — the *Split* operator, and the executor.

When *Split* meets an alien tuple, it creates a copy of the tuple and passes one instance to the *Merge* (with the "delete me" bit set inside the CTID) and the other — to the *Scatter* (with the "insert me" bit set inside the CTID). The schematic for tuple flow in *Exchange* is shown in fig. 11.

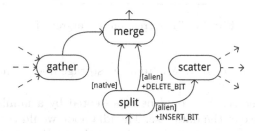

Fig. 11. Tuple flow during an exchange

The executor in its turn checks for these bits and reacts accordingly. So, if the "delete me" bit is set, it performs the delete routines, and if the "insert me" bit is set — the insert routines. If neither bit is set the tuple is considered native to the node, so the executor behaves as in PostgreSQL and updates the tuple in the local storage.

For plain INSERT queries the parallelizer appends an additional condition to the plan, that is equivalent to a `WHERE fragattr % nodes == this_node` clause. With this condition a tuple only gets inserted if it is native to the node. Complex INSERT queries like `INSERT INTO dest SELECT columns FROM src` do not need this additional condition.

3.5   Data Definition Operations

In order to provide data partitioning in PargreSQL we establish an additional storage parameter for PostgreSQL tables, named *fragattr* (fragmentation attribute). An application programmer is to specify an int-valued attribute of a table as *fragattr* on the table's creation. It is equivalent to defining the table's partitioning with $\psi(t) = t.fragattr \bmod n$ fragmentation function, where n is a number of nodes and **mod** denotes the *modulo* operation. The parameter is specified in the WITH clause of the CREATE TABLE query (see fig. 12).

3.6   Load Balancing (Future Work)

We are planning to implement a load balancing scheme proposed in [6]. The scheme is based on partial data replication. The last portion of tuples from each fragment are copied to several other nodes in case the native node would get delayed processing its fragment. When a node manages with its own fragment, and some other node has not, the idle node can start processing the corresponding copy, thus freeing the other node from some work. The "last" portion here

```
create table Person (
    id int,
    name varchar(30),
    gender char(1),
    birth date
) with (fragattr = id);
```

Fig. 12. Table creation in PargreSQL

means the tuples that get read last from the storage due to their physical order or an index.

In PargreSQL the scheme could be represented by a number of tables in a dedicated namespace of the database. The idle node would communicate to the busy one and ask where to start processing of the partial copy from. After that the two nodes would know where to start and to stop and would have rouhgly the same amount of tuples to process.

4   Experimental Evaluation

To evaluate our approach we performed a series of experiments on SKIF-Aurora SUSU supercomputer[1] based on Intel® Xeon 5680 processors and liquid cooling. We executed a query carrying out a natural join of two synthetical tables comprising of 60 mln. and 1.5 mln. records respectively and distributed uniformly among the computer nodes. To form values of tables' fragmentation attributes, a probabilistic model proposed in [6] was used.

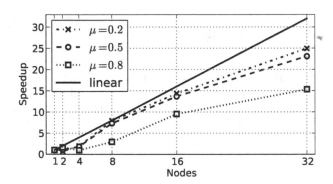

Fig. 13. PargreSQL speedup

Fig. 13 depicts the experimental results. Here μ is the portion of "alien" tuples at every fragment of the relations, e.g. $\mu = 0.75$ means that during the query

---

[1] http://www.hpcwire.com/hpcwire/2011-12-08/skif_aurora_susu_supercomputer_is_most_energy-efficient_hpc_system_in_russia.html

execution every agent was obliged to send 75% of the tuples stored at the agent's node to other nodes. As we can see PargreSQL demonstrates a quite acceptable speedup.

5 Related Work

The research on adaptation of PostgreSQL for parallel and distributed query processing includes the following.

In [5] authors introduce their work on extending PostgreSQL to support distributed query processing. Several limitations in PostgreSQL's query engine and corresponding query execution techniques to improve performance of distributed query processing are presented.

ParGRES [7] is an open-source database cluster middleware for high performance OLAP query processing. ParGRES exploits intra-query parallelism on PC clusters and uses adaptive virtual partitioning of the database.

GParGRES [4] exploits database replication and inter- and intra-query parallelism to support OLAP queries in a grid. The approach has two levels of query splitting: grid-level splitting, implemented by GParGRES, and node-level splitting, implemented by ParGRES.

In [1] building a hybrid between MapReduce and parallel database is explored. The authors created a prototype named HadoopDB on the basis of Hadoop (communication layer) and PostgreSQL (database layer), that is as efficient as parallel DBMS, but as scalable as MapReduce systems.

Our contribution is embedding partitioned parallelism into PostgreSQL on the basis of the methods for parallel query processing, proposed in [2,3,6,10]. We believe that our approach could be applied to other serial relational open-source DBMSes (e.g. MySQL) to implement their parallel versions.

6 Conclusion

In this paper we have described the design and implementation of PargreSQL parallel DBMS for cluster systems. PargreSQL is based upon PostgreSQL open-source DBMS and exploits partitioned parallelism. This approach is applicable to other open-source relational DBMSes. The results of preliminary experiments show that this scheme is worthy of further development.

As future work we plan to implement load-balancing based upon partial data replication, parallel execution of subqueries and stored procedures, and conduct advanced experiments to analyze PargreSQL performance on complex queries.

References

1. Abouzeid, A., Bajda-Pawlikowski, K., Abadi, D.J., Rasin, A., Silberschatz, A.: HadoopDB: An Architectural Hybrid of MapReduce and DBMS Technologies for Analytical Workloads. PVLDB 2(1), 922–933 (2009)

2. DeWitt, D.J., Gray, J.: Parallel Database Systems: The Future of High Performance Database Systems. Commun. ACM 35(6), 85–98 (1992)
3. Graefe, G.: Encapsulation of parallelism in the volcano query processing system. In: Garcia-Molina, H., Jagadish, H.V. (eds.) SIGMOD Conference, pp. 102–111. ACM Press (1990)
4. Kotowski, N., Lima, A.A.B., Pacitti, E., Valduriez, P., Mattoso, M.: Parallel query processing for OLAP in grids. Concurrency and Computation: Practice and Experience 20(17), 2039–2048 (2008)
5. Lee, R., Zhou, M.: Extending PostgreSQL to Support Distributed/Heterogeneous Query Processing. In: Kotagiri, R., Radha Krishna, P., Mohania, M., Nantajeewarawat, E. (eds.) DASFAA 2007. LNCS, vol. 4443, pp. 1086–1097. Springer, Heidelberg (2007)
6. Lepikhov, A.V., Sokolinsky, L.B.: Query processing in a DBMS for cluster systems. Programming and Computer Software 36(4), 205–215 (2010)
7. Paes, M., Lima, A.A.B., Valduriez, P., Mattoso, M.: High-Performance Query Processing of a Real-World OLAP Database with ParGRES. In: Palma, J.M.L.M., Amestoy, P.R., Daydé, M., Mattoso, M., Lopes, J.C. (eds.) VECPAR 2008. LNCS, vol. 5336, pp. 188–200. Springer, Heidelberg (2008)
8. Pan, C.: Development of a parallel dbms on the basis of postgresql. In: Turdakov, D., Simanovsky, A. (eds.) SYRCoDIS. CEUR Workshop Proceedings, vol. 735, pp. 57–61. CEUR-WS.org (2011)
9. Paulson, L.D.: Open source databases move into the marketplace. IEEE Computer, 13–15 (2004)
10. Sokolinsky, L.B.: Organization of Parallel Query Processing in Multiprocessor Database Machines with Hierarchical Architecture. Programming and Computer Software 27(6), 297–308 (2001)

Effectively Delivering XML Information in Periodic Broadcast Environments

Yongrui Qin[1], Quan Z. Sheng[1], Muntazir Mehdi[2], Hua Wang[3], and Dong Xie[4]

[1] School of Computer Science,
The University of Adelaide, Adelaide, SA 5005, Australia
{yongrui,qsheng}@cs.adelaide.edu.au
[2] Department of Computer Science,
TU Kaiserslautern, Gottlieb-Daimler-Strasse, Kaiserslautern 67663, Germany
muntazir.75@gmail.com
[3] Department of Mathematics & Computing,
University of Southern Queensland, QLD 4350, Australia
hua.wang@usq.edu.au
[4] Department of Computer Science and Technology,
Hunan University of Humanities, Science and Technology, Loudi 417000, China
dong.xie@hotmail.com

Abstract. Existing data placement algorithms for wireless data broadcast generally make assumptions that the clients' queries are already known and the distribution of access frequencies of their queries can be obtained a priori. Unfortunately, these assumptions are not realistic in most real life applications because new mobile clients may join in anytime and clients may be reluctant to disclose their queries (due to privacy concerns). In this paper, we study the data placement problem of periodic XML data broadcast in mobile wireless environments. This is an important issue, particularly when XML becomes prevalent in today's ubiquitous Web and mobile computing devices. Taking advantage of the structured characteristics of XML data, we are able to generate effective broadcast programs based purely on XML data on the server without any knowledge of the clients' access patterns. This not only makes our work distinguished from previous studies, but also enables it to have broader applicability. We discuss structural sharing in XML data which forms the basis of our novel data placement algorithm. The proposed placement algorithm is validated through a set of experiments and the results show that our algorithm can effectively place XML data on air and significantly improve the overall access efficiency.

1 Introduction

Wireless technology has become deeply embedded in everyday life. At the end of 2011, there were 6 billion mobile subscriptions, estimated by the International Telecommunication Union (2011). That is equivalent to 87 percent of the world population, and is a huge increase from 5.4 billion in 2010 and 4.7 billion mobile subscriptions in 2009.

H. Decker et al. (Eds.): DEXA 2013, Part I, LNCS 8055, pp. 165–179, 2013.
© Springer-Verlag Berlin Heidelberg 2013

Broadcast is one of the basic ways of information access via wireless technologies. In a wireless data broadcast system, the server broadcasts public information to all mobile devices within its transmission range via a downlink broadcast channel. Mobile clients "listen" to the downlink channel and access information of their interest directly when related information arrives. Broadcast is bandwidth efficient because all mobile clients can share the same downlink channel and retrieve data from it simultaneously. Broadcast is also energy efficient at the client ends because downloading data costs much less energy than sending data [27].

Wireless data broadcast services have been available as commercial products for many years, e.g. StarBand and Hughes Network. Recently, there has been a push for such systems from the industry and various standard bodies. For example, born out of the ITU "IMT-2000" initiative, the Third Generation Partnership Project 2 is developing Broadcast and Multicast Service in CDMA2000 Wireless IP network. Systems for Digital Audio Broadcast (DAB) and Digital Video Broadcast (DVB) are capable of delivering wireless data services. Recent news also reported that XM Satellite Radio (www.xmradio.com) and Raytheon have jointly built a communication system, known as the Mobile Enhanced Situational Awareness Network (MESA), that would use XM satellites to relay information to soldiers and emergency responders during a homeland security crisis.

On the other hand, information expressed in semi-structured formats is widespread over the past years. XML has rapidly gained popularity as a de facto standard to represent semi-structured information. Most Internet browsers provide support for XML in their newer versions and nearly all the major IT companies (e.g., Microsoft, Oracle, and IBM) have integrated XML into the software products. Delivering information in XML format is also popular in Web services and in different kinds of Publish/Subscribe systems. Consequently, XML has attracted attentions from database community recently and there has been a large body of research work focusing on XML, such as XML filtering, querying and indexing [17,26].

Combining both trends of the proliferation of mobile computing technologies and XML data, broadcasting information in XML format in a wireless environment would be a preferable way of information delivering and sharing. Consequently, the research of XML data broadcast is of great importance and in fact it has been attracting more and more research interests [20,7,24,19,18]. To further demonstrate practicability of XML data broadcast, we will present a potential application of it by detailing a real life scenario in Section 2.

There are two typical data broadcast modes: (i) *Periodic Broadcast Mode* and (ii) *On-Demand Broadcast Mode* [27]. In the periodic broadcast mode, data are periodically broadcasted on a downlink channel via which the server sends data to clients. Clients only "listen" to that channel and download data they are interested in. In the on-demand broadcast mode, clients send their queries to the server via an uplink channel and then the server considers all submitted requests and decides the contents of next broadcast cycle. In this work, we focus on the periodic broadcast mode since it has many benefits such as saving uplink bandwidth and power at the client ends by avoiding uplink transmissions and effectively delivering information to an unlimited number of clients simultaneously.

Data placement algorithms determine what data items to be broadcasted by the server and the order of data items on wireless channels, aiming to reduce average waiting time for mobile clients. To a large extent, the data placement problem of XML data is similar to that in multi-item contexts [25,4] where mobile clients may request multiple items each time. However, there are drawbacks of existing data placement approaches in traditional data broadcast.

Firstly, previous work on multi-item placement problems generally makes assumptions that the clients' queries are already known and the distribution of access frequencies of these queries can be obtained in advance [1,2,25,4]. For example, it is proposed to allow the clients to provide a profile of their interests to the servers [1,2], but this can lead to privacy concerns. These assumptions significantly limit the practicability of the proposed placement algorithms in real situations because: (i) new mobile clients may join in the network at anytime; and (ii) mobile users may be reluctant to disclose their queries to the server via uplink channel due to expensive communication cost and privacy concerns.

Secondly, in traditional data broadcast systems, appropriate placement can hardly be generated based only on information of data items themselves on the server. Hence, the above assumptions are inevitable for the design of data placement algorithms. Alternatively, some work applies data mining techniques to discover association rules from the history access patterns of a set of data [3]. This avoids to obtain access patterns of mobile clients on-the-fly. However, the availability of such history access patterns of mobile clients is a necessity.

By contrast, in XML data broadcast, data items (or XML documents) usually share parts of their structure. Taking structural sharing between XML documents into consideration, we are able to analyze and estimate clients' access patterns via the analysis of this structural sharing. Then we can effectively place XML data on wireless channels based purely on XML data on the server, which is important for practical usage. In summary, the main contributions of this paper can be described as follows:

- By taking advantage of the structural characteristics of XML data, we are able to generate appropriate data placement results based only on XML data on the server.
- A novel data placement algorithm which organizes XML data on air is presented.
- Extensive experiments are conducted to show the effectiveness of our proposed data placement algorithm.

The remainder of this paper is organized as follows. Section 2 describes background knowledge of this work, including an application scenario, the system model and XML similarity background. Section 3 discusses the structural sharing property of XML data and proposes a novel data placement algorithm. Section 4 presents our experimental study for evaluating the performance of the proposed data placement algorithm. Finally, Section 5 discusses related work and Section 6 gives some concluding remarks.

2   Application Scenario, System Model and XML Similarity

In this section, we first describe an application scenario. Then we show the system model of this work and introduce background knowledge of XML similarity.

2.1   Application Scenario

We use the following scenario to show potential applications of XML data broadcast in real life.

Consider a live basketball game. Information about the game and the players on the court is usually the interest of a large number of audience. In this context, data broadcast is a preferable way of delivering latest information to the audience. Meanwhile, some audience could be outside of the stadium, such as basketball fans who are watching live text information about the game via the Internet at their homes. Therefore, the game information could also be delivered via the Internet to online audience and other Web service providers who have subscribed this basketball game. Using XML format to represent game information can satisfy all these needs and realize simplicity, generality, and usability of game information at the same time.

2.2   Periodic XML Data Broadcast System Model

Fig. 1 shows the model of our wireless XML data broadcast system. The system includes an XML Data Center (the broadcast server), a broadcast program scheduler, broadcast listeners (mobile clients) and a downlink channel (the server sends information to mobile clients via it). The downlink channel can be shared by all mobile clients. But mobile clients can not send their individual queries to the server in this model as no uplink channel is available.

From the figure, we can see that the XML Data Center could be connected to the Internet and deliver information to online users, Web service providers and Publish/Subscribe systems, etc. With the use of XML format data, these different applications can be integrated seamlessly with our wireless XML data broadcast system for the purpose of sharing and delivering same information to different kinds of users.

2.3   XML Similarity

Our goal is to place XML documents on the broadcast channel based only on the information at the server side. We propose to explore relatedness between different XML documents and place documents according to the relatedness results.

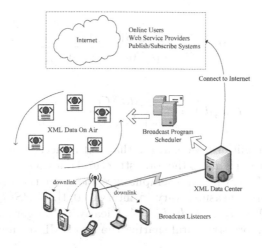

Fig. 1. A wireless XML data broadcast system

For XML documents, structural similarity is well studied and can be applied in our broadcast system as a way to calculate relatedness between documents.

Some existing work on measuring structural similarity between XML documents can be found in [23,11]. The main idea of their work is based on the concept of *path sets*. Here, a path set of an XML document contains all full paths (paths that are from root element to leaves) and their subpaths. A simple example is presented in Fig. 2. The path set of this example is: {/player/name, /player/position, /player/nationality, /player/college, /player, /name, /position, /nationality, /college}. We denote a path set of an XML document d as $PS(d)$.

Fig. 2. An XML structure tree

Different types of measure can be adopted, such as Jaccard measure [16,10], Dice's coefficient [9] and Lian's measure [15], to measure the similarity between two XML documents d_i and d_j. The exact forms of these measures based on PS are as follows (Jaccard measure denoted as $J(d_i, d_j)$, Dice's coefficient denoted as $D(d_i, d_j)$ and Lian's measure denoted as $L(d_i, d_j)$):

$$J(d_i, d_j) = \frac{|PS(d_i) \bigcap PS(d_j)|}{|PS(d_i) \bigcup PS(d_j)|} \tag{1}$$

$$D(d_i, d_j) = \frac{2 \cdot |PS(d_i) \bigcap PS(d_j)|}{|PS(d_i)| + |PS(d_j)|} \tag{2}$$

$$L(d_i, d_j) = \frac{|PS(d_i) \bigcap PS(d_j)|}{max\{|PS(d_i)|, |PS(d_j)|\}} \tag{3}$$

From the above definitions, we can see that both Jaccard measure and Dice's coefficient give more weights on the total structural information of two comparing documents while Lian's measure emphasizes more on the difference of these documents. All three measures vary in interval $[0, 1]$. If $PS(d_i) = PS(d_j)$, we have $J(d_i, d_j) = D(d_i, d_j) = L(d_i, d_j) = 1$. Clearly, the larger the values of these measures are, the more structural sharing the two XML documents have.

3 Data Placement Algorithm

In this section, we introduce our data placement algorithm for periodic XML data broadcast. We first discuss the structural sharing property of XML data which we use to estimate the potential access patterns of mobile clients, i.e., the probability of accessing a small set of similar XML documents simultaneously. Then we put forward a novel data placement algorithm based on it.

3.1 Structural Sharing in XML Data

In the literature, two critical metrics, namely *access time* and *tuning time*, are used to measure the system's performance [12]. Data placement mainly affects access time because tuning time depends on the total content downloaded by mobile clients but not on the order of data. Hence, we use access time as our metric in this analysis. In periodic broadcast, queries are used to describe the interests of mobile clients and help mobile clients to skip irrelevant data on air, but they are not actually submitted to the broadcast server.

Intuitively, for any two given XML documents, we can utilize one of the three similarity measures described in Section 2.3 to calculate the similarity between them and the similarity results can be used to approximate the probability that a specific query is matched with both documents at the same time. For example, if two XML elements are under structurally similar paths, then it is more likely that either both elements or none satisfy a given query [23]. In fact, query issuers hardly have thorough knowledge about the broadcasted content and XPath queries usually contain * and // which would match similar structure. Therefore, if two XML documents are with larger structural similarity, i.e. d_1 and d_2, then they would have a higher probability to be required simultaneously. However, there are still three other cases to be considered, such as requiring d_1 but not d_2, requiring d_2 but not d_1 and requiring neither of d_1 and d_2. Therefore, the

Table 1. Matching Cases for Document d_1 and d_2 in a document set $\mathcal{D}$

Case	Probability	Effect on AT_{exp}
Matched both d_1, d_2	$Pr(d_1 \bigcap d_2)$	Positive
Matched none of d_1, d_2	$1 - Pr(d_1 \bigcup d_2)$	Positive
Matched d_1, but not d_2	$Pr(d_1 - d_2)$	Negative
Matched d_2, but not d_1	$Pr(d_2 - d_1)$	Negative

above similarity measures consider only successful match probabilities of both XML documents but do not consider unsuccessful match probabilities of them.

Nonetheless, unsuccessful match cases have effects on the expected access time as well (but the query may still be satisfied by other documents). In order to have better access efficiency, the distance between any two documents required by the same query should be as less uniform as possible on air. Based on this, we can infer that in the above example, cases of required d_1 but not d_2 and required d_2 but not d_1 are likely to generate more uniform distances while other two cases (required both documents or neither) are likely to have less uniform distances. Observing this, we define a new similarity measure called *Cohesion* to give a more accurate estimation of access patterns of mobile clients in the following.

Note that, for any query q requiring at least one of the documents in $\mathcal{D}$, q must match some paths in $PS(\mathcal{D})$ and it has a probability of $\frac{|PS(d)|}{|PS(\mathcal{D})|}$ to match d. If a query q fails to match any document in $\mathcal{D}$, the issuer of q will not be waiting the result to be broadcasted. Hence, we only consider satisfied queries (this means at least one document is matched) in this work.

Now suppose we have a set of n XML documents $\mathcal{D} = \{d_1, d_2, \ldots, d_n\}$ on the server, we can approximate access probability of any document d for queries which successfully match at least one document in set $\mathcal{D}$ as follows:

$$Pr(d) = \frac{|PS(d)|}{|PS(\mathcal{D})|} \qquad (4)$$

and for any i, j $(1 \leq i, j \leq n)$

$$Pr(d_i - d_j) = \frac{|PS(d_i) - PS(d_j)|}{|PS(\mathcal{D})|} \qquad (5)$$

Here, $PS(\mathcal{D}) = \bigcup_{i=1}^{n} PS(d_i)$.

There would be many different matching cases for a given set $\mathcal{D}$. Take two XML documents d_1 and d_2 in $\mathcal{D}$ as an example. As mentioned previously, there would be four cases of matching of them and the probability of each case is shown in Table 1. In this table, we also include positive and negative effects on the expected access time (AT_{exp}) for each case.

Based on Table 1, we define Cohesion $C(d_i, d_j)$ of XML documents d_i and d_j as follows:

$$C(d_i, d_j) = \frac{Pr(d_i \cap d_j) \cdot (1 - Pr(d_i \cup d_j))}{max\{Pr(d_i - d_j), Pr(d_j - d_i)\}} \tag{6}$$

Here d_i and d_j are both in set $\mathcal{D}$. It is easy to see that $C(d_i, d_j) = C(d_j, d_i)$. According to Equation (4), Equation (5) and Equation (6), we can calculate $C(d_i, d_j)$ after finding path sets of d_i, d_j and $\mathcal{D}$. Cohesion values can vary in a wide range which exceeds interval $[0, 1]$. Strictly speaking, Cohesion values only vary in interval $[0, \frac{|PS(\mathcal{D})|}{4}]$ given that $C(d_i, d_j) = \frac{|PS(\mathcal{D})|}{4}$ when $PS(d_i) = PS(d_j)$. The lower bound 0 is trivial. In order to obtain the upper bound, we only consider cases that have $PS(d_i) \neq PS(d_j)$, from which we can infer that $max\{|PS(d_i - d_j)|, |PS(d_j - d_i)|\} \geq 1$. Without loss of generality, let $|PS(d_i)| \geq |PS(d_j)|$, according to Equation (4) and Equation (5), we can rewrite Equation (6) as follows:

$$C(d_i, d_j) \leq \frac{\frac{|PS(d_i \cap d_j)|}{|PS(\mathcal{D})|} \cdot (1 - \frac{|PS(d_i \cup d_j)|}{|PS(\mathcal{D})|})}{\frac{1}{|PS(\mathcal{D})|}}$$

$$< \frac{-(|PS(d_i)| - \frac{|PS(\mathcal{D})|}{2})^2 + \frac{|PS(\mathcal{D})|^2}{4}}{|PS(\mathcal{D})|}$$

$$\leq \frac{|PS(\mathcal{D})|}{4}$$

Then the above result gives the upper bound of Cohesion $C(d_i, d_j)$. Now we can normalize Cohesion values to interval $[0, 1]$ in the following:

$$C'(d_i, d_j) = \begin{cases} \frac{4 \cdot C(d_i, d_j)}{|PS(\mathcal{D})|} & PS(d_i) \neq PS(d_j) \\ 1 & PS(d_i) = PS(d_j) \end{cases} \tag{7}$$

We can also infer that $C'(d_i, d_j) = 1$ if and only if $PS(d_i) = PS(d_j)$. Similar to other three similarity measures, the larger the value of Cohesion is, the more structural sharing the two comparing XML documents have.

3.2   The Data Placement Algorithm

Based on the discussion of structural sharing in XML data, we can generate a broadcast program for periodic data broadcast in a greedy way. From previous discussions, we can see that the more the structural sharing of two XML documents is, the larger probability of matching both XML documents simultaneously. As a result, our Greedy Data Placement Algorithm (GDPA) places XML documents with most structural sharing together first as an initial broadcast program. Then it progressively appends other XML documents to the broadcast program in a descendant order of structural sharing. Detailed steps of GDPA are shown in Algorithm 1 and Algorithm 2.

Algorithm 1. Initialize structural sharing matrix $S[n][n]$

Input: A set of XML documents $\mathcal{D} : \{d_1, d_2, ..., d_n\}$
Output: Structural sharing matrix $S[n][n]$
1. create matrix $S[n][n]$
2. **for** each document d in $\mathcal{D}$ **do**
3.   compute $PS(d)$
4. **end for**
5. **for** each pair of documents $< d_i, d_j >$ in $\mathcal{D}$ $(i < j)$ **do**
6. $S[i][j] \Leftarrow$ structural sharing between d_i and d_j
7. $S[j][i] \Leftarrow S[i][j]$
8. **end for**

Algorithm 2. GDPA

Input: Structural sharing matrix $S[n][n]$
Output: A broadcast program σ for $\mathcal{D}$
1. $\sigma \Leftarrow$ empty sequence
2. select a pair of documents $< d_i, d_j >$ with maximum value $S[i][j]$ in matrix $S[n][n]$

3. **if** Length of $d_i <=$ Length of d_j **then**
4.   add $< d_i, d_j >$ into σ
5. **else**
6.   add $< d_j, d_i >$ into σ
7. **end if**
8. $\mathcal{D}' \Leftarrow \mathcal{D} - d_i - d_j$
9. **while** $\mathcal{D}'$ is not empty **do**
10. $d_{head} \Leftarrow$ the first document in σ
11.   select a pair of documents $< d_{i_{max}}, d_{head} >$ with maximum value $S[i_{max}][head]$
       $(d_{i_{max}} \in \mathcal{D}')$
12. $d_{rear} \Leftarrow$ the last document in σ
13.   select a pair of documents $< d_{j_{max}}, d_{rear} >$ with maximum value $S[j_{max}][rear]$
       $(d_{j_{max}} \in \mathcal{D}')$
14.   **if** $S[i_{max}][head] >= S[j_{max}][rear]$ **then**
15.   append $d_{i_{max}}$ into σ from head
16. $\mathcal{D}' \Leftarrow \mathcal{D}' - d_{i_{max}}$
17.   **else**
18.   append $d_{i_{max}}$ into σ from rear
19. $\mathcal{D}' \Leftarrow \mathcal{D}' - d_{j_{max}}$
20.   **end if**
21. **end while**

Algorithm 1 initializes a structural sharing matrix $S[n][n]$ for n XML documents on the broadcast server. Note that, all four similarity measures defined in subsection 2.3 and 3.1 can be used in Algorithm 1 to compute structural sharing between two documents (Line 6). All of them are symmetric which means for any one of these measures, we must have $S[j][i] = S[i][j]$. Also we have $J(d_i, d_j) = D(d_i, d_j) = L(d_i, d_j) = C'(d_i, d_j) = 1$ if $i = j$. Therefore, we only need to calculate matrix S for entries $S[i][j]$ where $i < j$.

Based on matrix S, Algorithm 2 finds the pair of XML documents with maximum structural sharing and adds them into the initial empty broadcast program σ (Line 2). The sequence of the first pair of XML documents are placed according to the ascendant order of document lengths (Line 3 to 7). Then Algorithm 2 appends the XML document with maximum structural sharing to the head document d_{head} or the rear document d_{rear} of σ. If the maximum structural sharing is derived between document d and document d_{head}, d will be appended into σ from head; otherwise, d will be appended into σ from rear. This process will be repeated until all XML documents are placed into σ (Line 9 to 21).

Regarding the computing complexity of Algorithm 2, the main task of the scheduling is performed from Line 9 to 21. The whole 'while' block has at most n loops. Within this block, Line 11 takes $O(n)$ time. It is similar at Line 13, which also takes $O(n)$ time. Hence the time complexity of the whole 'while' block is $O(n^2 + n^2)$. Meanwhile, the complexity of Line 2 is $O(n^2)$. As a result, the complexity of Algorithm 2 is $O(3n^2)$.

4   Experiments

Since this is the first work that determines broadcast schedules based only on XML data on the server without any knowledge of the clients' access patterns, we compare our algorithm with a common random data placement algorithm (RDPA) and show its efficiency in terms of access time, which is a common measure of performance in data broadcasts. We have not compared tuning time as the comparing data placement algorithms would not affect it.

4.1   Experimental Setup

The experiments are run on three data sets each with 250 XML documents defined by News Industry Text Format (NITF) DTD [13], which is published for news copy production, press releases, and Web-based news organizations. The average depth of the three document sets is between 6 and 8 while the maximum depth is 20.

There are three data sets in the experiments, which are $DS1$, $DS2$ and $DS3$. Data in $DS1$ can be well clustered into 6 clusters. Moreover, for any two documents d_i, d_j in two different clusters of $DS1$, the minimum similarity values, the maximum similarity values and the average similarity values of all four measures (normalized Cohesion is adopted here) are shown in Table 2. We can see that all clusters are quite different from each other and share very little structural information. Data in $DS2$ are miscellaneous. Documents in $DS2$ cannot be classified into fine clusters. Data in $DS3$ are a mix of well-clustered data and miscellaneous data, which include 125 XML documents from $DS1$ and 125 XML documents from $DS2$.

In the experiments, XPath queries are generated using the generator developed by [8]. Queries are allowed to repeat. The generator provides several parameters to generate different types of XPath queries, such as query depth,

Table 2. Similarity between clusters in $DS1$

Measure	Similarity		
	Minimum	Maximum	Average
$Jaccard$	0.0097	0.1102	0.0435
$Dice$	0.0049	0.0583	0.0225
$Lian$	0.0057	0.1039	0.0345
$Cohesion$	0.0229	0.4620	0.1457

Table 3. Workload Parameters for the Experiments

Parameter	Range	Default	Description
$PROB$	5% to 30%	10%	probability(* and //)
QIR	0.1 to 5	1	query incoming rate
MQD	5 to 8	7	maximum query depth

probability of * and // and the maximum depth of generated XPath queries. The probability of * and // appearing in each query's step is between 5% and 30% (denoted $PROB$, and the default value is 10%). Note that, Query Incoming Rate (denoted QIR) means the number of newly issued queries from mobile clients in a unit of time (these queries are only locally issued for data retrieval purpose and are not sent to the broadcast server). We measure this unit of time by the time that mobile wireless system takes to broadcast a block of 1024-byte XML data. The maximum depth of generated XPath queries (denoted MQD) is between 5 and 8. Table 3 shows details of the parameters in the experiments.

The random data placement algorithm (RDPA) is compared with GDPA (using all four similarity measures defined in Equations (1), (2), (3) and (7)). In RDPA, the server broadcasts XML documents in a random order.

We implement both RDPA and GDPA on Java Platform Standard Edition 6 running on Windows 7 Enterprise, 64-bit Operating System. All our experiments are obtained by running 30 consecutive broadcast cycles. When we vary $PROB$, we set QIR and MQD to their default values. When we vary QIR, we set $PROB$ and MQD to their default values. Similarly, when we vary MQD, we set $PROB$ and QIR to their default values.

Regarding air indexing and index distribution strategy, in our experiments, we adopt Compact Index (CI) [24] as our index structure and $(1, m)$ index scheme [12] as our index distribution strategy. This is because CI is the state-of-the-art indexing technique for XML data broadcast and $(1, m)$ index scheme is the most popular index distribution strategy for traditional periodic data broadcast. More details can be found in [24] and [12].

4.2   Performance of GDPA

Our experimental results are shown in Fig. 3, Fig. 4 and Fig. 5. Average access time (AAT) is our performance metric. Also we only consider AAT for

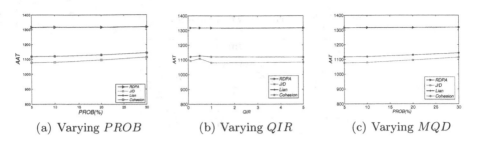

(a) Varying $PROB$ (b) Varying QIR (c) Varying MQD

Fig. 3. Evaluating AAT Performance on $DS1$: well-clustered data set

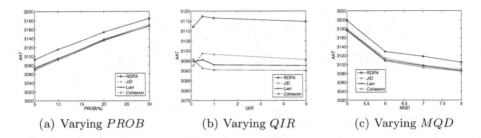

(a) Varying $PROB$ (b) Varying QIR (c) Varying MQD

Fig. 4. Evaluating AAT Performance on $DS2$: miscellaneous data set

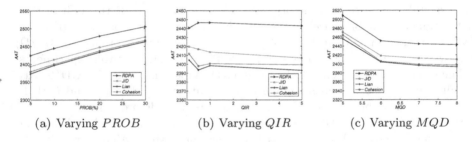

(a) Varying $PROB$ (b) Varying QIR (c) Varying MQD

Fig. 5. Evaluating AAT Performance on $DS3$: a mixed set of well-clustered data and miscellaneous data

all successful matched queries and abandon unsuccessful matched queries. The main reason for this is that, AAT of unsuccessful queries is determined by index distribution but not by data placement results (more details about this can be found in [12]). Note that, GDPA can be implemented with four different similarity measures defined in Section 3, which are Jaccard measure, Dice's coefficient, Lian's measure and our proposed Cohesion. Through our experiments, Jaccard measure and Dice's coefficient always yield the same results. Therefore, we denote GDPA implemented with them as J/D method in all figures. Meanwhile, we denote GDPA implemented with Lian's measure as *Lian* method and denote GDPA implemented with Cohesion as *Cohesion* method.

Fig. 3 shows the results on $DS1$. From the figure we can see that all GDPA methods outperform RDPA significantly. Specifically, J/D method achieves the best results while Lian method and Cohesion method provides similar results. This indicates that J/D method better fits well-clustered data. In Fig. 3(a), GDPA methods become slightly worse when $PROB$ increases. Since $DS1$ is well-clustered, most queries only require documents in the same clusters. Thus $PROB$ has less effect on AAT. In Fig. 3(b), when QIR increases, J/D method becomes slightly better. This indicates that J/D method can achieve better scalability than other methods when accessing well-clustered data. Fig. 3(c) shows that all GDPA methods remain stable as MQD increases. It is interesting to note that for RDPA, AAT always remains stable.

Fig. 4 shows the results on $DS2$. From the figure we can see that all GDPA methods achieve better performance when compared with RDPA. Specifically, Cohesion method achieves the best results while J/D method achieves the worst results among GDPA methods. This indicates that Cohesion method better fits miscellaneous data. In Fig. 4(a), both GDPA methods and RDPA become worse when $PROB$ increases. It is clear that $PROB$ has more effect on AAT for miscellaneous data. In Fig. 4(b), when QIR increases from 0.1 to 0.5, GDPA methods J/D and Lian together with RDPA become worse while Cohesion method still becomes better. After that, when QIR increases, all methods become slightly better. This shows that Cohesion method can achieve best scalability when accessing miscellaneous data.

Fig. 5 shows the results on $DS3$. Similarly, all GDPA methods achieve better performance when compared with RDPA. Specifically, Lian method achieves the best results while J/D method provides the worst results among GDPA methods. This shows that Lian method better fits hybrid data. However, Cohesion method achieves very similar performance of Lian method. In Fig. 5(a), both GDPA methods and RDPA become worse when $PROB$ increases. $PROB$ has more effect on AAT for hybrid data. In Fig. 5(b), when QIR increases, all GDPA methods become slightly better and still Lian method provides the best results.

To sum up, GDPA methods always achieve better AAT when compared with RDPA. When accessing well-clustered data, J/D method achieves the best performance. When accessing miscellaneous data, Cohesion method provides the best performance and finally when accessing hybrid data, Lian method shows the best performance.

5   Related Work

Multi-item data placement problem is related to the data placement problem of XML data which is the focus of our work. It is proved to be a NP-Complete problem [6].

Existing data placement methods for processing multi-item queries in periodic broadcast[5,14,3] generally makes assumptions that the clients' queries are already known and the distribution of access frequencies of these queries can be obtained in advance. However, these assumptions are not true for most applications in real life because the demand is either not known or it may be costly to collect the demand information.

Multi-item data placement problem in on-demand broadcast mode has also attracted lots of interests [25,22]. These approaches are in pure on-demand broadcast mode and strictly require that mobile clients submit their queries to the server for desired data. Otherwise, the server will not broadcast related data on air. This is because the server filters and schedules data solely based on submitted queries. However, frequent use of uplink channel leads to high communication cost via uplink channel, which can shorten battery life of mobile clients dramatically.

The most related work is proposed in [21] where the broadcast schedules are generated based on clustering results of XML data on the server. However, when finding the optimal clustering result, the clustering process requires manually specifying the number of clusters and has to compare different clustering results based on clients' query distribution, which differs from our work in this paper.

6 Conclusion

In this paper, we have studied the data placement problem of periodic XML data broadcast in mobile wireless environments. Taking advantage of the structured characteristics of XML data, we are able to generate effective broadcast programs based only on XML data on the server without any knowledge of the clients' access patterns. This not only makes our work distinguished from previous studies, but also enables it to have broader applicability. Our experiments demonstrated that the proposed algorithm could improve access efficiency and achieve better scalability.

In the future, we plan to further improve system's performance by investigating the insights of structural sharing among XML documents. For example, we may consider details on how to measure structural sharing distribution in an XML document set, how the distribution affects the expected access time of queries and how to choose a similarity measure based on structural sharing distribution in a set of XML documents.

References

1. Acharya, S., Alonso, R., Franklin, M.J., Zdonik, S.B.: Broadcast Disks: Data Management for Asymmetric Communications Environments. In: SIGMOD, pp. 199–210 (1995)
2. Acharya, S., Franklin, M.J., Zdonik, S.B.: Balancing Push and Pull for Data Broadcast. In: SIGMOD Conference, pp. 183–194 (1997)
3. Chang, Y.I., Hsieh, W.H.: An Efficient Scheduling Method for Query-Set-Based Broadcasting in Mobile Environments. In: ICDCS Workshops, pp. 478–483 (2004)
4. Chen, J., Lee, V.C.S., Liu, K.: On the Performance of Real-time Multi-item Request Scheduling in Data Broadcast Environments. Journal of Systems and Software 83(8), 1337–1345 (2010)
5. Chung, Y.D., Kim, M.H.: QEM: A Scheduling Method for Wireless Broadcast Data. In: DASFAA, pp. 135–142 (1999)

6. Chung, Y.D., Kim, M.H.: Effective Data Placement for Wireless Broadcast. Distributed and Parallel Databases 9(2), 133–150 (2001)

7. Chung, Y.D., Lee, J.Y.: An Indexing Method for Wireless Broadcast XML Data. Inf. Sci. 177(9), 1931–1953 (2007)

8. Diao, Y., Altinel, M., Franklin, M.J., Zhang, H., Fischer, P.M.: Path Sharing and Predicate Evaluation for High-Performance XML Filtering. ACM Trans. Database Syst. 28(4), 467–516 (2003)

9. Dice, L.R.: Measures of the Amount of Ecologic Association Between Species. Ecology 26(3), 297–302 (1945)

10. Ganesan, P., Garcia-Molina, H., Widom, J.: Exploiting Hierarchical Domain Structure to Compute Similarity. ACM Trans. Inf. Syst. 21(1), 64–93 (2003)

11. Helmer, S.: Measuring the Structural Similarity of Semistructured Documents Using Entropy. In: VLDB, pp. 1022–1032 (2007)

12. Imielinski, T., Viswanathan, S., Badrinath, B.R.: Data on Air: Organization and Access. IEEE Trans. Knowl. Data Eng. 9(3), 353–372 (1997)

13. IPTC: International Press Telecommunications Council, News Industry Text Format (NITF), http://www.nitf.org

14. Lee, G., Yeh, M.S., Lo, S.C., Chen, A.L.P.: A Strategy for Efficient Access of Multiple Data Items in Mobile Environments. In: MDM, pp. 71–78 (2002)

15. Lian, W., Cheung, D.W.L., Mamoulis, N., Yiu, S.M.: An Efficient and Scalable Algorithm for Clustering XML Documents by Structure. IEEE Trans. Knowl. Data Eng. 16(1), 82–96 (2004)

16. Lin, D.: An Information-Theoretic Definition of Similarity. In: ICML, pp. 296–304 (1998)

17. Miliaraki, I., Koubarakis, M.: FoXtrot: Distributed structural and value XML filtering. TWEB 6(3), 12 (2012)

18. Park, C.S., Park, J.P., Chung, Y.D.: PrefixSummary: A Directory Structure for Selective Probing on Wireless Stream of Heterogeneous XML Data. IEICE Transactions 95-D(5), 1427–1435 (2012)

19. Park, J.P., Park, C.S., Chung, Y.D.: Energy and Latency Efficient Access of Wireless XML Stream. J. Database Manag. 21(1), 58–79 (2010)

20. Park, S.-H., Choi, J.-H., Lee, S.: An Effective, Efficient XML Data Broadcasting Method in a Mobile Wireless Network. In: Bressan, S., Küng, J., Wagner, R. (eds.) DEXA 2006. LNCS, vol. 4080, pp. 358–367. Springer, Heidelberg (2006)

21. Qin, Y., Wang, H., Sun, L.: Cluster-Based Scheduling Algorithm for Periodic XML Data Broadcast in Wireless Environments. In: AINA Workshops, pp. 855–860 (2011)

22. Qin, Y., Wang, H., Xiao, J.: Effective Scheduling Algorithm for On-Demand XML Data Broadcasts in Wireless Environments. In: ADC, pp. 95–102 (2011)

23. Rafiei, D., Moise, D.L., Sun, D.: Finding Syntactic Similarities Between XML Documents. In: DEXA Workshops, pp. 512–516 (2006)

24. Sun, W., Yu, P., Qin, Y., Zhang, Z., Zheng, B.: Two-Tier Air Indexing for On-Demand XML Data Broadcast. In: ICDCS, pp. 199–206 (2009)

25. Sun, W., Zhang, Z., Yu, P., Qin, Y.: Efficient Data Scheduling for Multi-item Queries in On-Demand Broadcast. In: EUC (1), pp. 499–505 (2008)

26. Vagena, Z., Moro, M.M., Tsotras, V.J.: RoXSum: Leveraging Data Aggregation and Batch Processing for XML Routing. In: ICDE, pp. 1466–1470 (2007)

27. Xu, J., Lee, D.L., Hu, Q., Lee, W.C.: Handbook of Wireless Networks and Mobile Computing, pp. 243–265. John Wiley & Sons, Inc. (2002)

GUN: An Efficient Execution Strategy for Querying the Web of Data

Gabriela Montoya[1], Luis-Daniel Ibáñez[1], Hala Skaf-Molli[1],
Pascal Molli[1], and Maria-Esther Vidal[2]

[1] LINA– Nantes University, France
{gabriela.montoya,luis.ibanez,hala.skaf,
pascal.molli}@univ-nantes.fr
[2] Universidad Simón Bolívar, Venezuela
mvidal@ldc.usb.ve

Abstract. Local-As-View (LAV) mediators provide a uniform interface to a federation of heterogeneous data sources to attempt the execution of queries against the federation. LAV mediators rely on query rewriters to translate mediator queries into equivalent queries on the federated data sources. The query rewriting problem in LAV mediators has shown to be NP-complete, and there may be an exponential number of rewritings, making unfeasible the execution or even generation of all the rewritings for some queries. The complexity of this problem can be particularly impacted when queries and data sources are described using SPARQL conjunctive queries, for which millions of rewritings could be generated. We aim at providing an efficient solution to the problem of executing LAV SPARQL query rewritings while the gathered answer is as complete as possible. We formulate the Result-Maximal k-Execution problem (Re-MakE) as the problem of maximizing the query results obtained from the execution of only k rewritings. Additionally, a novel query execution strategy called GUN is proposed to solve the ReMakE problem. Our experimental evaluation demonstrates that GUN outperforms traditional techniques in terms of answer completeness and execution time.

1 Introduction

Querying the Web of Data raises the issue of semantic heterogeneity between a large number of data sources. Local-as-view (LAV) mediation [1] is a well-known and flexible approach to perform data integration over heterogeneous data sources. A LAV mediator relies on views to define semantic mappings between a uniform interface defined at the mediator level, and local schemas or views that describe the integrated data sources. A LAV mediator relies on a query rewriter to translate a mediator query into the union of queries against the local views. Additionally, new data sources can be included into LAV mediators without affecting the definition of the existing ones; thus, LAV mediators are well suitable to integrate sources from the Web of Data [2]. Nevertheless, the query rewriting problem has shown to be NP-complete, and the number of rewritings can be exponential even if mediated queries and local views are conjunctive queries [3,4]. For example, a LAV mediator with 140 conjunctive views

H. Decker et al. (Eds.): DEXA 2013, Part I, LNCS 8055, pp. 180–194, 2013.
© Springer-Verlag Berlin Heidelberg 2013

can generate 10,000 rewritings for a conjunctive query with 8 goals [5]. This query rewriting problem complexity can be exacerbated by the usage of mediator queries and local views defined as SPARQL conjunctive queries. SPARQL queries are commonly comprised of a large number of triple patterns and many of them are defined on general predicates that can be answered by the majority of the data sources, i.e., rdf:type or rdfs:seeAlso. Additionally, these triple patterns can be grouped into chained connected star-shaped sub-queries [6]. Finally, a large number of variables can be projected out. Thus, the conjunction of all these properties impacts on the complexity of the query rewriting problem and conduces to the explosion of the number of query rewritings. For example, a query with 12 triple patterns that comprised three chained star-shaped sub-queries can be rewritten using 300 views in billions of rewritings, if general predicates are used in the triple patterns. This problem is even more challenging considering that statistics cannot be always collected from the sources, and there are not clear criteria to rank or prune the generated rewritings [7].

Therefore, it is not realistic to generate or execute such a huge number of rewritings, and we aim at providing an efficient solution to the problem by considering only k LAV SPARQL query rewritings, where k corresponds to the first k rewritings produced by a LAV rewriter. We devised the Result-Maximal k-Execution Problem (ReMakE) as an extension of the Query-Rewriting-Problem (QRP) as follows: given a subset R_k of size k of a solution R of a QRP for a query Q, the ReMakE problem is to evaluate a set of rewritings R' containing R_k and contained in Q such that R' is result-maximal. Furthermore, we propose the Graph-Union execution strategy (GUN) as a solution to the ReMakE problem. Unlike traditional techniques, GUN relies on wrappers that populate an RDF graph that is locally managed by the execution engine. This approach takes advantage of the relatively low cost of the RDF-Graph union operation to construct an aggregation of the data retrieved from the views. This approach attempts at executing the original mediator query directly on the graph union and consequently, it may find results hidden to the k first rewritings. For a given set of rewritings, GUN always gathers at least all the answers collected by a traditional engine by executing the rewritings independently. If all relevant views identified by the rewriter are in R_k, GUN guarantees to return the complete answer without further processing of rewritings. Thus, the execution time of GUN depends on the number of the relevant views that comprise the rewritings in R_k, which is usually considerably lower than the total number of rewritings.

We compare GUN against traditional strategies in an experiment on synthetic data generated by the Berlin SPARQL benchmark tool [8] and views proposed by Castillo-Espinola [9]. We measure execution time and answer completeness for a benchmark of queries. In the experiments, we can observe that GUN retrieves much more results in less time than existing engines. The amount of main memory required to maintain a GUN graph is in general higher than the one required to execute traditional approaches; however, improvements in execution time and results are substantial enough to consider it a good trade-off.

The paper is organized as follows: Section 2 states preliminaries, while Section 3 formalizes the ReMakE problem. Section 4 presents the GUN query execution strategy as a solution for the ReMakE problem. Section 5 reports our experimental study. Section 6 summarizes related work; and finally, conclusions and future work are outlined in Section 7.

2   Preliminaries

We assume that a federation of data sources is integrated using the mediator-wrapper architecture proposed by Wiederhold [10]. Mediators provide a uniform interface to autonomous and heterogeneous data sources; they also implement the tasks of rewriting an input query into queries against the data sources, and merging data collected from the selected sources. Wrappers are software components that solve interoperability between sources and mediators by translating data collected from the sources into the schema and format understood by the mediators. Particularly, GUN-based mediators rely on wrappers able to solve resource identification and perform the corresponding RDF transformations to conform source data into the mediator RDF schema.

Formally, a conjunctive query Q over a database or mediator schema D has the form $Q(\bar{X})$:- $P_1(\bar{X}_1), \ldots, P_n(\bar{X}_n)$ where Q, P_1, ..., P_n are predicates name of some finite arity, and $\bar{X}$, $\bar{X}_1$, ..., $\bar{X}_n$ are tuples of variables. These predicates constitute the global schema. We define the body of the query as $body(Q) = \{P_1(\bar{X}_1), \ldots, P_n(\bar{X}_n)\}$. Any non-empty subset of $body(Q)$ is called a subgoal of Q, singleton subgoals are called atomic subgoals. Predicates in the body stand for relations of D, while the head Q represents the answer relation of the query over D. We consider queries that are *safe*, i.e., $\bar{X} \subseteq \bigcup_{i=1}^{n} \bar{X}_i$, and call $Q(D)$ the result of executing Q over D.

In the spirit of [5], we define a view v as a safe query over D, we establish the difference between the *extension* of v, denoted $ext(v)$, and its evaluation over D, $v(D)$, and assume the relation $ext(v) \subseteq v(D)$ to state two important hypothesis: there may be data belonging to the database that is not available to the extensions, and the extensions never hold data that is not in the database.

A rewriting of a query Q over a database D with a set of views V is a conjunctive query $r(\bar{x})$:- $v_1(\bar{x}_1), \ldots, v_m(\bar{x}_m)$ where, $v_i \in V$. A query rewriting is *contained* in Q, if for all database D and set of views V over D, the result of executing r in V is contained in the result of executing Q on D, i.e., $r(V) \subseteq Q(D)$.

Maximally Contained Query Rewriting Problem (QRP). Given a conjunctive query Q and a set of views $V = \{ v_1, \ldots, v_n \}$ over a database D, QRP is to find a set of rewritings R, called the solution of the QRP, such that:

- For all extensions of the views in the bodies of all rewritings in R, the union of the results of executing each query rewriting in the views V is contained in the result of executing Q in D, i.e., $\bigcup_{r \in R} r(ext(v_1), \ldots, ext(v_n)) \subseteq Q(D)$
- R is maximal, i.e., there is no other set R', such that:

$$\bigcup_{r \in R} r(ext(v_1), \ldots, ext(v_n)) \subset \bigcup_{r' \in R'} r'(ext(v_1), \ldots, ext(v_n)) \subseteq Q(D)$$

For a set R of rewritings, we define the set of relevant views $\Lambda(R) = \{v \mid v \in body(r) \land r \in R\}$ as the set of views in the rewritings in R, and its execution $R(D) = \bigcup_{r \in R} r(D)$. We also call $ext(\Lambda(R))$, the extension of the elements in $\Lambda(R)$.

The main drawback of existing query rewriting problem solutions for LAV [1,5,4,11] is that the size of the set R can be exponential in the number of query subgoals [3,11]. Therefore, more than executing or generating an enormous number of rewritings, it is more realist to solve the problem of gathering data considering only k rewritings while obtaining an answer as complete as possible.

3   Result-Maximal k-Execution Problem (ReMakE)

In this section, we formalize the problem of obtaining the maximal set of results from a given subset of the rewritings of a query over a set of views.

Result-Maximal k-Execution Problem (ReMakE). Given a subset R_k of size k of a solution R of a QRP of a query Q and a set of views V over a database D, ReMakE is to find a set of rewritings R' over the set of views in the bodies of the rewritings of R_k, such that:

$$\bigcup_{r_k \in R_k} r_k(ext(\Lambda(R_k))) \subseteq \bigcup_{r' \in R'} r'(ext(\Lambda(R_k))) \subseteq Q(D)$$

and that is result-maximal, i.e., that there is no another set R'' such that:

$$\bigcup_{r' \in R'} r'(ext(\Lambda(R_k))) \subset \bigcup_{r'' \in R''} r''(ext(\Lambda(R_k))) \subseteq Q(D)$$

We define this problem over the extensions of the views, as they are the real datasets where the query will be evaluated. It is important to note that the ReMakE problem only uses the query rewritings as an input and they could be obtained using any query rewriter, therefore, it is independent of the approach used to solve QRP. We also highlight that ReMakE is independent of the format of the data inside the extensions of the views, the wrappers would transform any format to the mediator schema.

To illustrate the problem, consider the generic set of rewritings in Figure 1, and suppose we can only execute the first five query rewritings. It could exist a rewriting comprised of some combination of views that were gathered for evaluating these five rewritings but that is not in the first five rewritings. ReMakE aims to consider all the rewritings that could be obtained from the already materialized views, hence in Figure 1 answers for rewriting r_n would be also obtained.

4   GUN: A Solution to the ReMakE Problem

In this section, we explain how to solve the ReMakE problem by taking advantage of the relatively low cost of the RDF-Graph union. We use definitions of SPARQL semantics of [12]:

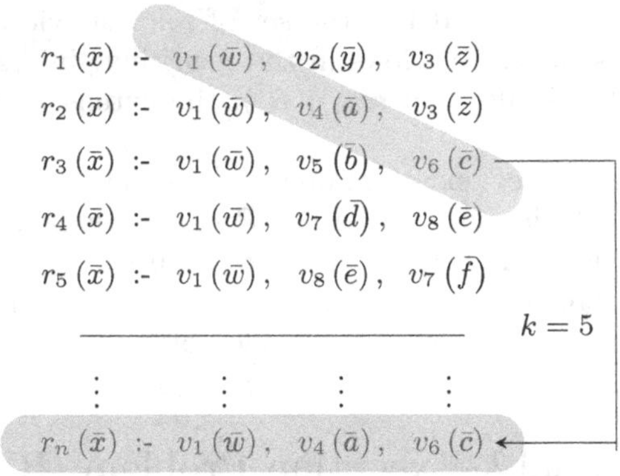

$$r_1\,(\bar{x}) \;:\text{-}\; v_1\,(\bar{w}),\quad v_2\,(\bar{y}),\quad v_3\,(\bar{z})$$
$$r_2\,(\bar{x}) \;:\text{-}\; v_1\,(\bar{w}),\quad v_4\,(\bar{a}),\quad v_3\,(\bar{z})$$
$$r_3\,(\bar{x}) \;:\text{-}\; v_1\,(\bar{w}),\quad v_5\,(\bar{b}),\quad v_6\,(\bar{c})$$
$$r_4\,(\bar{x}) \;:\text{-}\; v_1\,(\bar{w}),\quad v_7\,(\bar{d}),\quad v_8\,(\bar{e})$$
$$r_5\,(\bar{x}) \;:\text{-}\; v_1\,(\bar{w}),\quad v_8\,(\bar{e}),\quad v_7\,(\bar{f})$$

$$k = 5$$

$$\vdots \qquad \vdots \qquad \vdots \qquad \vdots$$

$$r_n\,(\bar{x}) \;:\text{-}\; v_1\,(\bar{w}),\quad v_4\,(\bar{a}),\quad v_6\,(\bar{c})$$

Fig. 1. Illustration of the Result-Maximal k-Execution problem. Some combinations of views materialized during the execution of a subset of rewritings (R_k) could correspond to valid rewritings that does not belong to R_k.

Definition 1. *The Sets I (IRI Identifiers), B (Blank Nodes), L (Literals) and Υ (Variables) are four infinite and pairwise disjoint sets. We also define $T = I \cup B \cup L$. An RDF-Triple is 3-tuple $(s, p, o) \in (I \cup B) \times I \times T$. An RDF-Graph is a set of RDF-Triples.*

Definition 2. *A mapping μ from Υ to T is a partial function $\mu : \Upsilon \to T$. The domain of μ, $dom(\mu)$, is the subset of Υ where μ is defined.*

Definition 3. *A triple pattern is a tuple $t \in (I \cup \Upsilon \cup L) \times (I \cup \Upsilon) \times (I \cup \Upsilon \cup L)$. A Basic Graph Pattern is a finite set of triple patterns. Given a triple pattern t, $var(t)$ is the set of variables occurring in t, analogously, given a basic graph pattern B, $var(B) = \cup_{t \in B} var(t)$. Given two basic graph patterns B_1 and B_2, the expression $B_1\ AND\ B_2$ is a graph pattern.*

Definition 4. *Given a triple pattern t and a mapping μ such that, $var(t) \subseteq dom(\mu)$, $\mu(t)$ is the triple obtained by replacing the variables in t according to μ. Given a basic graph pattern B and a mapping μ such that $var(B) \subseteq dom(\mu)$, then $\mu(B) = \cup_{t \in B} \mu(t)$.*

Definition 5. *Two mappings μ_1, μ_2 are compatible (we denote $\mu_1 \shortparallel \mu_2$) iff for all $?X \in (dom(\mu_1) \cap dom(\mu_2))$, then $\mu_1(?X) = \mu_2(?X)$. This is equivalent to say that $\mu_1 \cup \mu_2$ is also a mapping.*

Definition 6. *Let Ω_1, Ω_2 two sets of mappings. The join between Ω_1 and Ω_2 is defined as: $\Omega_1 \bowtie \Omega_2 = \{\mu_1 \cup \mu_2 \mid \mu_1 \in \Omega_1 \wedge \mu_2 \in \Omega_2 \wedge \mu_1 \shortparallel \mu_2\}$*

Definition 7. *Given an RDF-Graph G, the evaluation of a triple pattern t over G corresponds to: $[[t]]_G = \{\mu \mid dom(\mu) = var(t) \wedge \mu(t) \in G\}$. The evaluation of a basic graph pattern B over G is defined as: $[[B]]_G = \bowtie_{t \in B} [[t]]_G$. The evaluation of a Graph Pattern B' of the form $(B_1\ AND\ B_2)$ over G is as follows: $[[B']]_G = [[B_1]]_G \bowtie [[B_2]]_G$*

We consider that our database is an RDF-Graph G. A conjunctive query over a general database is analogous to the following query over an RDF-Graph:

$$Q(x) = \text{SELECT } x \text{ WHERE } F(p_1(\bar{x}_1)) \text{ AND } \dots \text{ AND } F(p_n(\bar{x}_n))$$

where F is a translation function from predicates to triple patterns as defined in [13] or a customized one. The definitions of variables, head and body are the same. As the definitions of views and rewritings are based on the definition of query, they remain equivalent, together with the definitions of QRP and ReMakE. We define the evaluation of a rewriting $[[r(x)]]_G$ as:

$$[[r(x)]]_G = [[v_1(\bar{x}_1), \dots, v_m(\bar{x}_m)]]_G = ([[p_a(\bar{x}_a)]]_{ext(v_1)} \bowtie \cdots \bowtie [[p_z(\bar{x}_z)]]_{ext(v_1)})$$
$$\bowtie \cdots \bowtie ([[p_\alpha(\bar{x}_\alpha)]]_{ext(v_m)} \bowtie \cdots \bowtie [[p_\beta(\bar{x}_\beta)]]_{ext(v_m)})$$

where $p_a \dots p_z \in body(v_1)$ and $p_\alpha \dots p_\beta \in body(v_m)$. Note that this definition captures the practical implementation of the execution engine, where we materialize each call to a view (or more precisely, to its extension) and then, perform the joins between the sub-results. Traditionally, plans like Left Linear, Right Linear or Bushy Trees [14] are used to evaluate the rewritings over the extension of the views present in each rewriting; but to solve the ReMakE problem, we should ensure that any relevant combinations of obtained views are not missed, even if these combinations are not part of the rewritings in R_k.

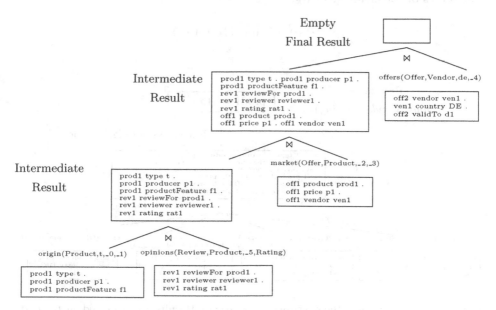

Fig. 2. Left Linear execution of the rewriting r of query Q. Views *origin, opinions, market* and *offers* are loaded, but it is not possible to produce any results since the join for *Offer* is empty. Prefixes are omitted to improve legibility.

Consider a query Q over a dataset generated with the Berlin Benchmark [8], which offers information about products, their offers and users' reviews. Q is defined as: "Products of type t that are sold by vendors from Germany, and their rating evaluation".

$Q(Product, Vendor, Rating)$:- $type(Product, t)$, $product(Offer, Product)$, $vendor(Offer, Vendor)$, $country(Vendor, de)$, $reviewfor(Review, Product)$, $rating1(Review, Rating)$. Considering the following four views: i) *origin*: "Products' type, producer, and features", ii) *market*: "Products' offers, price and vendors", iii) *offers*: "Offers' vendor, countries and validity", iv) *opinions*: "Products' reviews, ratings and reviewers". A possible rewriting of Q is:

$r(Product, Vendor, Rating)$:- $origin(Product, t, \_0, \_1)$, $opinions(Review, Product, \_5, Rating)$, $\qquad market(Offer, Product, \_2, \_3)$, $offers(Offer, Vendor, de, \_4)$.

Figure 2 shows the execution of the rewriting of the query Q following a left linear execution plan. In this execution, the RDF Graphs retrieved from the sources through the views are used to execute this rewriting. Notice that while executing the plan some intermediate results are produced, like those corresponding to the evaluation of the join between *origin* and *opinions*; however, these intermediate results are dismissed without being used to produce answers. A pertinent solution of the ReMakE problem must take advantage of these retrieved data. We now define our solution to the ReMakE problem as follows:

Graph Union (GUN). Given R_k a subset of a set of rewritings R of a query Q over a set of views V, apply Q to the union of the extensions of the views in the bodies of the elements of R_k:

$$GUN(R_k) = [[Q]]_{\bigcup ext(\Lambda(R_k))}$$

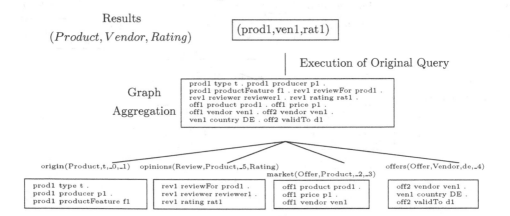

Fig. 3. GUN execution of the rewriting r of query Q. Results are produced, at the cost of building and querying over an aggregated graph. Prefixes are omitted to improve legibility.

Figure 3 shows the execution with GUN associated with the rewriting of query Q. For each view in the rewriting, we retrieve all its answers and store them into an aggregate RDF-Graph. As we are using the Local-As-View approach, views are expressed in terms of the global schema, and we can run the original query and execute joins that are not considered in the rewritings, like the one between *market* and *offers* through *Vendor*. Therefore, GUN takes advantage of all retrieved data to produce an answer for the user. GUN is affordable in the context of the Semantic Web thanks to the simplicity of the RDF data model. Implementing the same idea in relational databases would require to create the universal relation, which may have a prohibitive cost.

Proposition 1. *Graph Union is a solution to the ReMakE problem, i.e.,*

$$\bigcup_{r \in R_k} [[r]]_{ext(\Lambda(R_k))} \subseteq [[Q]]_{\bigcup ext(\Lambda(R_k))} \tag{1}$$

$$[[Q]]_{\bigcup ext(\Lambda(R_k))} \subseteq [[Q]]_G \tag{2}$$

And it is result-maximal.

Proof. As by construction of the views and their extensions $\bigcup_{v \in V}[[v]]_G \subseteq G$, then $\bigcup ext(\Lambda(R_k)) \subseteq G$, making straightforward to see that (2) holds. For (1) note that the set $\Lambda(R_k)$ can be considered as a set of views over the graph $\bigcup \Lambda(R_k)$, then by the containment property, each member of R_k is contained in Q. As we are applying the original query Q, it is clear that there is no rewriting that can return more results than Q, meaning that GUN is maximal.

4.1 GUN's Properties

In this section, we state some properties of GUN. Given a query Q and a set of views V on a database D, let V_R the set of relevant views, R the set of rewritings of Q over V, and R_k a subset of R.

- *Answer Completeness:* If GUN is performed over R_k with $\Lambda(R_k) = \Lambda(R)$, then GUN will produce the complete query answer i.e., $\Lambda(R_k) = \Lambda(R) \Rightarrow GUN(R_k) = Q(D)$. By definition of QRP, only the relevant views contribute to the answer, therefore, if GUN's aggregation graph contains all the relevant views, then we can ensure that GUN will produce the complete answer.
- *Effectiveness:* We define the *Effectiveness* of GUN for a given R_k as:

$$GUNEffect(R_k) = \frac{|GUN(R_k)| - |\bigcup_{r_k \in R_k} r_k(ext(\Lambda(R_k)))|}{|Q(D)| - |\bigcup_{r_k \in R_k} r_k(ext(\Lambda(R_k)))|}$$

intuitively, GUN has more effectiveness if it finds answers that are not found by the execution of $\bigcup_{r_k \in R_k} r_k(ext(\Lambda(R_k)))$. We say that GUN is *effective* for a given R_k if $GUNEffect(R_k) > 0$

If $|Q(D)| - |\bigcup_{r_k \in R_k} r_k(ext(\Lambda(R_k)))| = 0$, then, the effectiveness is defined to be 0. Note that effectiveness and answer completeness are related by the following relation:

Table 1. Queries and their answer size, number of rewritings, number of relevant views (RV) and views size

(a) Query information   (b) Views size

Query	Answer Size	# rewritings	# of RV		Views	Size
Q1	3.33E+07	1.61E+09	260		V1-V20	147,327
Q2	2.99E+05	6.37E+21	260		V21-V40	133,992
Q3	2.03E+05	3.52E+24	280		V41-V60	41,463
Q4	1.42E+02	6.02E+03	240		V61-V80	22,410
Q5	2.82E+05	1.30E+07	240		V81-V100	4,515
Q6	9.84E+04	1.22E+05	100		V101-V120	53,131
Q7	1.12E+05	1.15E+12	180		V121-V140	32,511
Q8	2.82E+05	4.08E+04	100		V141-V160	90,873
Q9	1.41E+04	2.00E+01	20		V161-V180	21,138
Q10	1.49E+06	9.76E+05	260		V181-V200	9,836
Q11	1.49E+06	3.24E+03	80		V201-V220	4,515
Q12	2.99E+05	2.37E+08	260		V221-V240	4,515
Q13	2.99E+05	2.41E+04	260		V241-V260	67,364
Q14	2.82E+05	8.08E+05	180		V261-V280	81,313
Q15	1.41E+05	4.64E+09	280		V281-V300	840,470
Q16	1.41E+05	8.36E+05	100			
Q17	9.84E+04	2.02E+03	100			
Q18	2.82E+05	3.12E+08	240			

$$\bigcup_{r_k \in R_k} r_k(ext(\Lambda(R_k))) \subset Q(D) \Rightarrow (GUNEffect(R_k)=1 \equiv GUN(R_k)=Q(D))$$

When the execution of R_k does not produce the complete answer, then, GUN's effectiveness for R_k equals to one iff GUN produces complete answers.

- *Execution Time Independency of k:* the execution time of GUN does not depend on the number of rewritings executed (k). It depends on the number of relevant views present in R_k. Execution time is the elapsed time between the generation of R_k rewritings and their execution time. This includes the time required to obtain the data from the wrappers, the time required to add the obtained data to the graph and the time required to execute the query plan on the graph.

- *Non-blocking:* GUN solves the ReMakE problem under the assumption that $\bigcup \Lambda(R_k)$ fits in main memory. If not, GUN can only approximate it, for example, by splitting R_k into disjoint R_{k_1} and R_{k_2} such that $\bigcup \Lambda(R_{k_1})$ and $\bigcup \Lambda(R_{k_2})$ fit in memory. Then, execute $GUN(R_{k_1})$, clear the memory, and execute $GUN(R_{k_2})$. Therefore, GUN is a non-blocking execution strategy i.e., running out of memory will not prevent GUN to execute at the expense of non-maximality of the answer, as the combined effectiveness of $GUN(R_{k_1})$ and $GUN(R_{k_2})$ is in general less than this of $GUN(R_k)$.

5   Experimental Evaluation

To setup the experimental evaluation, we used the Berlin SPARQL Benchmark tool (BSBM) [8] to generate a dataset of 5,000,251 triples, using a scale factor of 14,091 products. We used the 18 queries and the 10 views proposed in [9]. These queries are very challenging for a query rewriter since their triple patterns can be grouped into chained connected star-shaped sub-queries, that have between 1 and 13 subgoals, with only distinguished variables.

Table 2. Values of k for obtaining the Complete Answers (CA) for queries Q4-Q6, Q8-Q18; using GUN and Jena. GUN's Effectiveness for different values of k. Effectiveness for Q9 is not reported here since it only has 20 rewritings.

		Q4	Q5	Q6	Q8	Q9	Q10	Q11	Q12	Q13	Q14	Q15	Q16	Q17	Q18
CA	GUN	281	45	>500	381	20	21	29	36	21	>500	20	21	21	56
	Jena	281	>500	>500	383	20	141	119	>500	320	>500	>500	>500	40	>500
GUN's	k=80	0	1	0.0016	0		1	1	1	1	0.0476	1	1	0	1
Effectiveness	k=160	0	1	0.0002	0		0	0	1	1	0.0451	1	1	0	1
	k=320	0	1	0.0018	0		0	0	1	0	0.0406	1	1	0	1
	k=500	0	1	0.0024	0		0	0	1	0	0.0382	1	1	0	1

We defined 5 additional views to cover all the predicates in the queries. From these 15 views, we produced 300 views by horizontally partitioning each original view into 20 parts, such that each part produces 1/20 of the answers given by the original view. Queries and views information is shown in Tables 1a and 1b. The size of the complete answer was computed by loading all the views into a persistent RDF-Store (Jena-TDB) and executing the queries over it. The number of rewritings was obtained using the models counting feature of the SSDSAT [15] rewriter.

As we can see in Table 1a, the number of rewritings may be very large, making unfeasible their full execution. Furthermore, the time to generate the rewritings is not negligible, and in some cases (Q2, Q3 and Q7) SSDSAT could not generate them after 72 hours. We chose to compute 500 rewritings, as this was the best compromise we could find between number of rewritings and generation time, i.e., 500 is the larger number of rewritings (multiple of 50) that could be produced for all queries (but Q2, Q3 and Q7) in less than 15 hours. Additionally, we do not have any statistics about the sources to select the best rewritings or to shrink the set of relevant views. Q1 execution reached a timeout of 48 hours.

Some general predicates like *rdfs:label* are present in the most of the views; therefore, the queries that have a triple pattern with these predicates will have a large number of relevant views, but not all of these views will contribute to the answer. The size of a view corresponds to the number of triples that can be accessed through that view. Detailed information about the definition of the queries and views can be found in the project website[1].

We implemented wrappers as simple file readers. For executing rewritings, we used one named graph per subgoal as done in [16]. The Jena 2.7.4[2] library with main memory setup was used to store and query the graphs. We used the Left Linear Plans implemented by Jena as a representative of traditional query execution techniques.

5.1 Experimental Results

The analysis of our results focus on four aspects: answer completeness, effectiveness, execution time and non-blocking as defined in section 4.1.

To study the answer completeness of GUN, we executed the GUN and Jena strategies over R_k rewritings, we counted the number of rewritings to have the

---

[1] https://sites.google.com/site/graphunion/
[2] http://jena.apache.org/

complete answer. Table 2 shows that GUN is able to achieve the complete answer for 12 queries whereas Jena is able to do so only for 6 queries. For queries Q9, Q11, Q13 and Q17, GUN produced complete answers because at the reported k, $\Lambda(R_k) = \Lambda(R)$. For the rest of the queries, the non aggregated relevant views did not contribute to produce more results. Detailed information about the ratio of relevant views for each R_k can be found in the project's website.

To demonstrate the effectiveness of GUN, we executed the GUN and Jena strategies over R_k with $k \in \{80, 160, 320, 500\}$, counted the number of answers and computed the effectiveness. Table 2 shows that GUN has effectiveness 1 for $k = 80$ for half of the queries, moreover, in 5 of these 7 queries, the maximum effectiveness remains even after Jena executes 500 rewritings. In 4 cases, GUN is not effective because Jena already found the complete answer for this value of k. Finally, in Q6 and Q14, GUN found more results than Jena. Effectiveness values are not monotonic, since they can increase when considering a rewriting that contains a view that contributes to produce results in GUN and not in Jena. However, they can decrease after executing a rewriting that does not add new views to GUN, but produces results for Jena.

Regarding the execution time, we want to: *1)* demonstrate that GUN execution time does not depend on k, but on $|\Lambda(R_k)|$, and *2)* compare GUN's execution time with Jena's. Table 3a shows total execution time, the detailed values of the execution time are available in the project website. For all queries GUN has better execution time, and for all but Q6 with $k = 80$, is more than twice faster. When $k = 500$, the difference is dramatic, varying from almost 4 times faster (Q13) to 680 times faster (Q4). Table 3b shows total execution time and number of loaded views for GUN. Execution time grows linearly in $|\Lambda(R_k)|$, this is particularly visible in Q4 and Q13.

If we compare the times detailed in section 4.1, we notice that the dominating time is the wrapper time. GUN loads views into the aggregated graph only once, whereas Jena reloads them for each executed rewriting. Note that if we try to cache the views in Jena to avoid reloading, it would consume more memory and could consume even more memory than GUN if the views have overlapped information, as it is the case in our setup.

Finally, we analyzed GUN's and Jena's memory consumption to demonstrate that in spite of complex queries and many relevant views: *1)* GUN is not blocking, and *2)* to compare memory used by GUN with respect to Jena. For GUN, we count the number of triples of the aggregated graph. For Jena, we report an upper bound, that is, the maximum number of triples loaded for executing a rewriting in R_k. Table 4 summarizes the results. Neither GUN nor Jena consumes all the available memory (8GB). GUN needs to load more triples than Jena, varying from less than twice to 12 times more, in all cases except for Q10 with $k \geq 320$. GUN's aggregation is in general larger than the sum of the named graphs of the most memory-consuming rewriting in R_k.

In summary, GUN is effective with better execution time at the cost of higher memory consumption. However, in our experimentation GUN never exhausts

Table 3. Execution Time (ET) for GUN and Jena. Impact of Number of Relevant Views (RV) over Execution Time in GUN.

(a) Execution Time for GUN and Jena

Query		Execution Time			
		K=80	K=160	K=320	K=500
Q4	GUN	39	39	63	73
	Jena	167	293	48,943	49,721
Q5	GUN	377	400	400	400
	Jena	1,155	2,302	3,848	5,935
Q6	GUN	336	337	338	339
	Jena	398	798	1,610	2,516
Q8	GUN	41	47	58	64
	Jena	190	377	751	1,278
Q10	GUN	132	132	132	132
	Jena	2,214	5,941	119,137	251,641
Q11	GUN	121	121	121	121
	Jena	1,906	3,707	9,985	16,939
Q12	GUN	28	28	28	28
	Jena	79	146	288	475
Q13	GUN	71	203	478	522
	Jena	146	352	734	2,034
Q14	GUN	328	395	395	395
	Jena	439	842	1,657	2,485
Q15	GUN	358	358	358	358
	Jena	1,207	3,000	5,812	9,160
Q16	GUN	35	35	35	35
	Jena	119	283	596	972
Q17	GUN	69	345	345	345
	Jena	168	965	2,450	4,029
Q18	GUN	324	414	415	415
	Jena	1,149	2,413	4,355	6,808

(b) Execution Time and Number of Relevant Views for GUN

Query		ET and # of RV			
		k=80	k=160	k=320	k=500
Q4	GUN	39	39	63	73
	# RV	23	25	31	38
Q5	GUN	377	400	400	400
	# RV	80	100	100	100
Q6	GUN	336	337	338	339
	# RV	62	63	66	69
Q8	GUN	41	47	58	64
	# RV	24	28	36	40
Q10	GUN	132	132	132	132
	# RV	79	80	80	80
Q11	GUN	121	121	121	121
	# RV	79	80	80	80
Q12	GUN	28	28	28	28
	# RV	80	80	80	80
Q13	GUN	71	203	478	522
	# RV	61	123	240	260
Q14	GUN	328	395	395	395
	# RV	81	101	101	101
Q15	GUN	358	358	358	358
	# RV	60	60	60	60
Q16	GUN	35	35	35	35
	# RV	40	40	40	40
Q17	GUN	69	345	345	345
	# RV	41	100	100	100
Q18	GUN	324	414	415	415
	# RV	80	100	100	100

the available memory in spite of the challenging setup. This makes it a very appealing solution for the ReMakE problem.

6 Related Work

In recent years, several approaches have been proposed for querying the Web of Data [17,18,19,20,21]. Some tools address the problem of choosing the sources that can be used to execute a query [20,21]; others have developed techniques to adapt query processing to source availability [17,20]. Finally, frameworks to retrieve and manage Linked Data have been defined [18,20], as well as strategies for decomposing SPARQL queries against federations of endpoints [7]. All these approaches assume that queries are expressed in terms of RDF vocabularies used to describe the data in the RDF sources; their main challenge is to effectively select the sources, and efficiently execute the queries on the data retrieved from the selected sources. In contrast, our approach attempts to semantically integrate data sources, and relies on a global vocabulary to describe data sources and provide a unified interface to the users. Thus, in addition to efficiently gather and process the data transferred from the selected sources, it decides which of the rewritings of the original query need to execute to efficiently and effectively produce the query answer.

Two main paradigms have been proposed to define the data sources in integration systems. The LAV approach is commonly used because it permits the scalability of the system as new data sources become available [22]. Under LAV, the

Table 4. Maximum number of triples loaded by a rewriting in R_k in Jena. The number of triples of the aggregated graph of GUN.

Query	Maximal Graph Size k=80		Maximal Graph Size k=160		Maximal Graph Size k=320		Maximal Graph Size k=500	
	GUN	Jena	GUN	Jena	GUN	Jena	GUN	Jena
Q4	1,201,671	148,739	1,208,714	148,739	1,753,969	907,775	1,878,666	907,775
Q5	1,993,617	907,905	2,275,437	907,923	2,275,437	907,923	2,275,437	907,923
Q6	1,578,294	850,376	1,583,212	850,376	1,597,964	850,376	1,612,716	850,376
Q8	1,479,686	148,725	1,536,050	148,745	1,648,778	148,745	1,705,142	230,045
Q10	422,269	294,678	422,269	294,678	422,269	442,052	422,269	442,052
Q11	422,268	294,701	422,269	294,701	422,269	294,748	422,269	294,748
Q12	439,946	83,260	439,946	83,260	439,946	83,260	439,946	83,276
Q13	1,713,056	862,917	2,277,638	862,962	2,923,233	862,962	2,923,233	862,962
Q14	2,095,418	912,422	2,279,248	926,356	2,279,248	935,825	2,279,248	935,825
Q15	1,568,458	905,450	1,568,458	905,529	1,568,458	905,529	1,568,458	905,529
Q16	584,792	53,678	584,792	63,411	584,792	63,411	584,792	74,802
Q17	1,496,262	850,331	1,807,718	850,376	1,807,718	850,376	1,807,718	850,376
Q18	2,175,448	907,916	2,275,437	921,840	2,275,437	921,840	2,275,437	921,859

appearance of a new source only causes the addition of a new mapping describing the source in terms of the concepts in the RDF global vocabulary. Under GAV, on the other hand, entities in the RDF global vocabulary are semantically described using views in terms of the data sources. Thus, the extension or modification of the global vocabulary is an easy task in GAV as it only involves the addition or local modification of few descriptions [22]. Therefore, the LAV approach is best suited for applications with a stable RDF global vocabulary but with changing data sources whereas the GAV approach is best suited for applications with stable data sources and a changing vocabulary. Given the nature of the Semantic Web, we rely on the LAV approach to describe the data sources in terms of a global and unified RDF vocabulary, and assume that the global vocabulary of concepts is stable while data sources may constantly pop up or disappear from the Web.

The problem of rewriting a global query into queries on the data sources is one relevant problem in integration systems [23], and several approaches have been defined to efficiently enumerate the query rewritings and to scale when a large number of views exists (e.g., MCDSAT [4], GQR [5], Bucket Algorithm [23], MiniCon [11]). Recently, Le et al, [16] propose a solution to identify and combine GAV SPARQL views that rewrite SPARQL queries against a global vocabulary, and Izquierdo et al [15] extends the MCDSAT with preferences to identify the combination of semantic services that rewrite a user request. A great effort has been made to provide solutions able to produce query writings in the least time possible, however, to the best of our knowledge, the problem of executing the query rewritings against the selected sources still remains open.

We address this problem and propose GUN, a query processing technique for RDF store architectures that provide a uniform interface to data sources that have been defined using the LAV paradigm [1]. GUN assumes that the query rewriting problem has been solved using an off-the-shell query rewriter (e.g., [15,5]), which may produce a large number of query rewritings. GUN implements a query processing strategy able to execute a reduced number of query rewritings of and generate a *more complete* answer than the rest of the engines in less time, as it was observed in our experimental results.

Because GUN is able to answer mediator queries even in presence of a very large space of query rewritings, it constitutes a relevant contribution to the implementation of integration systems, and provides the basis for feasible semantic integration architectures in the Web of Data.

7   Conclusion and Future Work

Performing complex queries on different data sources raises the severe issue of semantic heterogeneity. Local-as-View mediators is one of the main approaches to solve it. However, the high number of rewritings needed to be executed represents a severe bottleneck. We proposed the ReMakE problem, that consists in maximizing the number of results obtained by considering only k rewritings (R_k). We also proposed GUN, a solution to this problem, it uses the RDF data model and takes advantage of the low cost of graph union operation.

Compared to state-of-the-art approaches, GUN provides an alternative way to improve performance at the execution engine level rather than at the rewriter level. This makes GUN usable with any LAV rewriter guaranteeing to achieve greater or equal answer completeness for the same R_k. Our experiments demonstrate that GUN gain is real, i.e., its effectiveness is equal to one for 57% of the queries for the values of k until 80. It remains equal to one for 38% of the queries for the values of k until 500.

Furthermore, GUN consumes considerably less execution time than Jena in all the cases; the difference in execution time is tremendous, up to 681 times. However, this improvement in effectiveness and execution time is at the cost of an additional memory consumption of up to 12 times.

This work opens new perspectives to improve LAV approach for the Semantic Web. First, we would like to measure the effectiveness degradation when executing on low-memory setups, and include some heuristics to minimize it. As GUN creates materialized views for processing rewritings, we plan to evaluate the impact on the effectiveness and the execution time when performing inference tasks on the graph union. Finally, because GUN is mostly dependent on the ratio of views in rewritings divided by the number of relevant views, an interesting perspective is to modify rewriters to optimize the number of views in the first k rewritings.

References

1. Levy, A.Y., Mendelzon, A.O., Sagiv, Y., Srivastava, D.: Answering queries using views. In: Fourteenth ACM SIGACT-SIGMOD-SIGART Symposium on Principles of Database Systems, PODS 1995, pp. 95–104 (1995)
2. Abiteboul, S., Manolescu, I., Rigaux, P., Rousset, M., Senellart, P.: Web data management. Cambridge University Press (2011)
3. Abiteboul, S., Duschka, O.M.: Complexity of answering queries using materialized views. In: Seventeenth ACM SIGACT-SIGMOD-SIGART Symposium on Principles of Database Systems, PODS 1998, pp. 254–263 (1998)
4. Arvelo, Y., Bonet, B., Vidal, M.E.: Compilation of query-rewriting problems into tractable fragments of propositional logic. In: AAAI, pp. 225–230 (2006)

5. Konstantinidis, G., Ambite, J.L.: Scalable query rewriting: a graph-based approach. In: SIGMOD Conference, pp. 97–108 (2011)
6. Vidal, M.-E., Ruckhaus, E., Lampo, T., Martínez, A., Sierra, J., Polleres, A.: Efficiently joining group patterns in SPARQL queries. In: Aroyo, L., Antoniou, G., Hyvönen, E., ten Teije, A., Stuckenschmidt, H., Cabral, L., Tudorache, T. (eds.) ESWC 2010, Part I. LNCS, vol. 6088, pp. 228–242. Springer, Heidelberg (2010)
7. Schwarte, A., Haase, P., Hose, K., Schenkel, R., Schmidt, M.: FedX: Optimization techniques for federated query processing on linked data. In: Aroyo, L., Welty, C., Alani, H., Taylor, J., Bernstein, A., Kagal, L., Noy, N., Blomqvist, E. (eds.) ISWC 2011, Part I. LNCS, vol. 7031, pp. 601–616. Springer, Heidelberg (2011)
8. Bizer, C., Shultz, A.: The berlin sparql benchmark. International Journal on Semantic Web and Information Systems 5, 1–24 (2009)
9. Castillo-Espinola, R.: Indexing RDF data using materialized SPARQL queries. PhD thesis, Humboldt-Universität zu Berlin (2012)
10. Wiederhold, G.: Mediators in the architecture of future information systems. IEEE Computer 25, 38–49 (1992)
11. Halevy, A.Y.: Answering queries using views: A survey. The VLDB Journal 10, 270–294 (2001)
12. Pérez, J., Arenas, M., Gutiérrez, C.: Semantics and complexity of sparql. ACM Transactions on Database Systems (TODS) 34 (2009)
13. Baget, J.-F., Croitoru, M., Gutierrez, A., Leclère, M., Mugnier, M.-L.: Translations between RDF(S) and conceptual graphs. In: Croitoru, M., Ferré, S., Lukose, D. (eds.) ICCS 2010. LNCS (LNAI), vol. 6208, pp. 28–41. Springer, Heidelberg (2010)
14. Chaudhuri, S.: An overview of query optimization in relational systems. In: Seventeenth ACM SIGACT-SIGMOD-SIGART Symposium on Principles of Database Systems, PODS 1998, pp. 34–43 (1998)
15. Izquierdo, D., Vidal, M.-E., Bonet, B.: An expressive and efficient solution to the service selection problem. In: Patel-Schneider, P.F., Pan, Y., Hitzler, P., Mika, P., Zhang, L., Pan, J.Z., Horrocks, I., Glimm, B. (eds.) ISWC 2010, Part I. LNCS, vol. 6496, pp. 386–401. Springer, Heidelberg (2010)
16. Le, W., Duan, S., Kementsietsidis, A., Li, F., Wang, M.: Rewriting queries on sparql views. In: WWW, pp. 655–664. ACM (2011)
17. Acosta, M., Vidal, M.-E., Lampo, T., Castillo, J., Ruckhaus, E.: ANAPSID: An adaptive query processing engine for SPARQL endpoints. In: Aroyo, L., Welty, C., Alani, H., Taylor, J., Bernstein, A., Kagal, L., Noy, N., Blomqvist, E. (eds.) ISWC 2011, Part I. LNCS, vol. 7031, pp. 18–34. Springer, Heidelberg (2011)
18. Basca, C., Bernstein, A.: Avalanche: Putting the Spirit of the Web back into Semantic Web Querying. In: SSWS, pp. 64–79 (2010)
19. Harth, A., Hose, K., Karnstedt, M., Polleres, A., Sattler, K.U., Umbrich, J.: Data summaries for on-demand queries over linked data. In: WWW, pp. 411–420 (2010)
20. Hartig, O.: Zero-knowledge query planning for an iterator implementation of link traversal based query execution. In: Antoniou, G., Grobelnik, M., Simperl, E., Parsia, B., Plexousakis, D., De Leenheer, P., Pan, J. (eds.) ESWC 2011, Part I. LNCS, vol. 6643, pp. 154–169. Springer, Heidelberg (2011)
21. Ladwig, G., Tran, T.: SIHJoin: Querying remote and local linked data. In: Antoniou, G., Grobelnik, M., Simperl, E., Parsia, B., Plexousakis, D., De Leenheer, P., Pan, J. (eds.) ESWC 2011, Part I. LNCS, vol. 6643, pp. 139–153. Springer, Heidelberg (2011)
22. Ullman, J.D.: Information integration using logical views. Theoretical Computer Science 239, 189–210 (2000)
23. Levy, A., Rajaraman, A., Ordille, J.: Querying heterogeneous information sources using source descriptions. In: VLDB, pp. 251–262 (1996)

Complex Matching of RDF Datatype Properties

Bernardo Pereira Nunes[1,2], Alexander Mera[1], Marco Antônio Casanova[1],
Besnik Fetahu[2], Luiz André P. Paes Leme[3], and Stefan Dietze[2]

[1] Department of Informatics - PUC-Rio - Rio de Janeiro, RJ, Brazil
{bnunes,acaraballo,casanova}@inf.puc-rio.br
[2] L3S Research Center - Leibniz University Hannover - Hannover, Germany
{nunes,fetahu,dietze}@l3s.de
[3] Computer Science Institute, Fluminense Federal University,
Niterói, RJ, Brazil
lapaesleme@ic.uff.br

Abstract. Property mapping is a fundamental component of ontology matching, and yet there is little support that goes beyond the identification of single property matches. Real data often requires some degree of composition, trivially exemplified by the mapping of "first name" and "last name" to "full name" on one end, to complex matchings, such as parsing and pairing symbol/digit strings to SSN numbers, at the other end of the spectrum. In this paper, we propose a two-phase instance-based technique for complex datatype property matching. Phase 1 computes the Estimate Mutual Information matrix of the property values to (1) find simple, 1:1 matches, and (2) compute a list of possible complex matches. Phase 2 applies Genetic Programming to the much reduced search space of candidate matches to find complex matches. We conclude with experimental results that illustrate how the technique works. Furthermore, we show that the proposed technique greatly improves results over those obtained if the Estimate Mutual Information matrix or the Genetic Programming techniques were to be used independently.

Keywords: Ontology Matching, Genetic Programming, Mutual Information, Schema Matching.

1 Introduction

Ontology matching is a fundamental problem in many applications areas [10]. Using OWL concepts, *by datatype property matching* we mean the special case of matching datatype properties from two classes.

Concisely, an *instance* of a datatype property p is a triple of the form (s, p, l), where s is a resource identifier and l is a literal. A *datatype property matching* from a *source class* S to a *target class* T is a partial relation μ between sets of datatype properties from S and T, respectively. We say that a match $(A,B) \in \mu$ is $m{:}n$ iff A and B contain m and n properties, respectively. A match $(A,B) \in \mu$ should be accompanied by one or more *datatype property mappings* that indicate how to construct instances of the properties in B from instances of the properties in A. A match $(A,B) \in \mu$ is *simple* iff it is 1:1 and the mapping is a identity function; otherwise, it is complex.

H. Decker et al. (Eds.): DEXA 2013, Part I, LNCS 8055, pp. 195–208, 2013.

In this paper, we introduce a two-phase, instance-based datatype property matching technique that is able to find complex $n{:}1$ datatype property matches and to construct the corresponding property mappings. The technique extends the ontology matching process described in [19] to include complex matches between sets of datatype properties and is classified as instance-based since it depends on sets of instances.

Briefly, given two sets, s and t, that contain instances of the datatype properties of the source class S and the target class T, respectively, the first phase of the technique constructs the Estimated Mutual Information matrix (EMI) [18,19] of the datatype property instances in s and t, which intuitively measures the amount of related information from the observed property instances. The scope of this phase is to identify simple datatype property matches. For example, it may detect that the "e-mail" datatype property of one class matches the "electronic address" datatype property of another class. Additionally, the first phase suggests, for the second phase, sets of candidate datatype properties that can be matched only under more complex relationships, thereby reducing the search space.

The second phase uses a genetic programming approach (GP) to find complex $n{:}1$ datatype property matches. For example, it discovers that the "first name" and "last name" datatype properties of the source class match the "full name" datatype property of the target class, and returns a property mapping function that concatenates the values of "first name" and "last name" (of the same class instance) to generate the "full name" value. The reason for adopting genetic programming is two-fold: it reduces the cost of traversing the search space; and it can be used to automatically generate complex mappings between datatype property sets.

The difficulty of the problem of finding complex matches between sets of datatype properties should not be underestimated since the search space is typically quite large. Therefore, our contribution towards a more accurate and efficient solution lies in proposing a two-phase technique, which deals with the problem of finding complex matches by: (a) using the Estimated Mutual Information matrix (in Phase 1) as a pre-processing stage, limiting the candidate sets of properties for complex matches; (b) adopting a genetic programming strategy to automatically generate complex property mappings. We also give empirical evidence that the combination of both approaches, EMI and GP, yields better results than using either technique in separate.

2   Background

2.1   Vocabulary Matching and Concept Mapping

We decompose the problem of OWL ontology matching into the problem of vocabulary matching and that of concept mapping. In this section, we briefly review these concepts and extend them to account for complex property matching. In what follows, let S and T be two OWL ontologies, and V_S and V_T be their vocabularies, respectively. Let C_S and C_T be the sets of classes and P_S and P_T be the sets of properties in V_S and V_T, respectively.

An *instance* of a class c is a triple of the form $(s, rdf{:}type, c)$, an instance of an object property p is a triple of the form (s, p, o) and an instance of a datatype property d is a triple of the form (s, d, l), where s and o are resource identifiers and l is a literal.

A *vocabulary matching* between S and T is a finite set $\mu \subseteq V_S \times V_T$. Given $(v_1, v_2) \in \mu$, we say that (v_1, v_2) is a *match* in μ and that μ *matches* v_1 with v_2; a *property* (or *class*) *matching* is a matching defined only for properties (or classes).

A *concept mapping* from S to T is a set of transformation rules that map instances of the concepts of S into instances of the concepts of T.

In this paper, we extend *vocabulary matchings* to also include pairs of the form *(A,B)* where A and B are sets of datatype properties in P_S and P_T, respectively. We say that *(A,B)* is an *m:n* match *iff* A and B contain m and n properties, respectively. In this case, a match *(A,B)* must be accompanied by *datatype property mappings*, denoted $\mu[A, B_i]$, which are transformation rules that map instances of the properties in A into instances of the property B_i, for $i = 1, \ldots, n$, where $B = \{B_1, \ldots, B_n\}$. Using "//" to denote string concatenation, the following transformation rule $(s, fullName, v) \leftarrow (s, firstName, n), (s, lastName, f), v = n//f$ indicates that the value of the "fullName" property is obtained by concatenating the values of properties "firstName" and "lastName". We will use the following abbreviated form for mapping rules with the above syntax:

$$\mu[\{firstName, lastName\}, fullName] =$$
$$"fullName \leftarrow firstName//lastName"$$

As an abuse of notation, when A is a singleton $\{A_1\}$, we simply write $\mu[A_1, B_i]$, rather than $\mu[\{A_1\}, B_i]$. Finally, a match *(A,B)* is *simple iff* it is 1:1, that is, of the form $(\{A_1\}, B_1)$, and the mapping $\mu[A_1, B_1]$ is the *identity transformation rule*, defined as "$(s, B_1, l) \leftarrow (s, A_1, l)$"; otherwise, the match is *complex*.

2.2 An Instance-Based Process for Vocabulary Matching

In this section, we very briefly summarize the instance-based process to create *vocabulary matchings* introduced in [19]. The outline of the process is as follows:

S1. Generate a preliminary property matching using similarity functions.
S2. Generate a class matching using the property matching obtained in S1.
S3. Generate an instance matching using the output from S1.
S4. Refine the property matching using the class matching generated in S2 and the instance matching from S3.

The final vocabulary matching is the result of the union of the class matching obtained in S2 and the refined property matching obtained in S4.

The intuition used in all steps of the process is that *"two schema elements match iff they have many values in common and few values not in common"*, i.e. *iff* they are similar above a given similarity threshold.

We obtain the following output from each individual step. S1 generates preliminary 1:1 property matchings based on the intuition that two properties match *iff* their instances share similar sets of values. In the case of string properties, their values are replaced by the tokens extracted from their values. S1 provides evidences on class and instance matchings, explored in the next two steps.

S2 generates class matchings based on the intuition that two classes match *iff* their sets of properties are similar. This step uses the property matchings generated in S1.

S3 generates instance matchings based on the intuition that two instances match *iff* the values of their properties are similar. However, equivalent instances from different classes may be described by very different sets of properties. Therefore, extracting values from all of their properties may lead to the wrong conclusion that the instances are not equivalent. Therefore, Leme et al. [19] propose to extract values only from the matching properties of the instances.

3   Two-Phase Property Matching Technique

In this section, we introduce a technique to partly implement and extend the ontology matching process of Section 2.2 to compute complex n:1 datatype property matches (note that the technique does not cover n:m matches). The technique comprises two phases: Phase 1 uses Estimated Mutual Information matrices, defined in Section 3.1, to compute 1:1 simple matches, while Phase 2 uses genetic programming to compute complex n:1 matches, based on the information returned by Phase 1.

3.1   Phase 1: Computing Simple Datatype Property Matches with Estimated Mutual Information

Let $p=(p_1,\ldots,p_u)$ and $q=(q_1,\ldots,q_v)$ be two lists of sets. The *co-occurrence matrix* of p and q is defined as the matrix $[m_{ij}]$ such that $m_{ij} = |p_i \cap q_j|$, for $i \in [1,u]$ and $j \in [1,v]$. The *Estimated Mutual Information matrix (EMI)* of p and q is defined as the matrix $[EMI_{pq}]$ such that:

$$EMI_{pq} = \frac{m_{pq}}{M} \cdot log \left(M \cdot \frac{m_{pq}}{\sum\limits_{j=1}^{v} m_{pj} \cdot \sum\limits_{i=1}^{u} m_{iq}} \right) \tag{1}$$

where $M = \sum\limits_{i=1}^{u} \sum\limits_{j=1}^{v} m_{ij}$.

We now adapt these concepts to define Phase 1 of the datatype property matching process. Let S and T be two classes with sets of datatype properties $A=\{A_1,\ldots,A_u\}$ and $B=\{B_1,\ldots,B_v\}$, respectively. Let s and t be sets of instances of the properties in A and B, respectively (s and t therefore are sets of RDF triples).

Rather than simply using the cardinality of set intersections to define the co-occurrence matrix $[m_{ij}]$, Phase 1 computes $[m_{ij}]$ using *set comparison functions* that take two sets and return a non-negative integer. Such functions play the role of *flexibilization points* of Phase 1, as illustrated in Section 4.1.

The set comparison functions depend on the types of the values of the datatype properties as well as on whether the functions take advantage of instance matches. For example, given a pair of datatype properties, A_i and B_j, m_{ij} may be defined as the number of pairs of triples (a, A_i, b) in s and (c, B_j, d) in t such that instances a and c match (or are identical) and the literals b and d are equal (or are considered equal, under a *literal comparison function* defined for the specific datatype of b and d).

For instance, Leme et al. [19] adopt the cosine similarity function to compare strings. Thus, m_{ij} is computed as the number of (string) values of triples for property A_i in s whose cosine distance to values of instances for property B_j in t is above a given threshold ($\alpha = 0.8$ in [19]).

To compute simple matches (1:1), the cosine similarity function proved to be appropriate, especially if the strings to be compared have approximately the same number of tokens. However, the cosine similarity function turned out not to be appropriate when using the co-occurrence matrix to suggest complex matches to Phase 2 of the technique. We therefore adopted the Jaccard similarity coefficient to compute the co-occurrence matrix, defined as

$$Jaccard(b, d) = \frac{|b \cap d|}{|b \cup d|} \tag{2}$$

which counts the number of tokens that strings b and d have in common.

Thus, given two properties A_i and B_j, m_{ij} is computed as the sum of $Jaccard(A_i, B_j)$, for all pairs of strings d and b such that there are triples of the form (a, A_i, b) in s and (c, B_j, d) in t.

Phase 1 proceeds by computing the EMI matrix based on the co-occurrence matrix, as in Eq. 1. Next, it computes a 1:1 matching, μ_{EMI}, between the properties in $A = \{A_1, \ldots, A_u\}$ and those in $B = \{B_1, \ldots, B_v\}$ such that, for any pair of properties A_p and B_q, $(A_p, B_q) \in \mu_{EMI}$ iff $EMI_{pq} > 0$ and $EMI_{pj} \leq 0$, for all $j \in [1, v]$, with $j \neq q$, and $EMI_{iq} \leq 0$, for all $i \in [1, u]$, with $i \neq p$. Furthermore, Phase 1 assumes that the property mappings, $\mu_{EMI}[A_r, B_s]$, are always the identity function.

Finally, Phase 1 also outputs a list of datatype properties to be considered for complex matching in Phase 2. For the kth column of the EMI matrix, it outputs the pair (A^k, B_k) as a candidate n:1 complex match, where B_k is the property of T that corresponds to the k_{th} column and A^k is the set of properties A_i of S such that $EMI_{ik} > 0$. Indeed, if $EMI_{ik} \leq 0$, then A_i and B_k have no information in common. However, note that this heuristics does not indicate what is a candidate property mapping $\mu[A^k, B_k]$. This problem is faced in Phase 2.

3.2   Phase 2: Computing Complex Property Matches with Genetic Programming

The second phase of the technique uses genetic programming to create mappings between the properties that have some degree of correlation, as identified in the first phase. Briefly, the process goes as follows.

Recall that *genetic programming* refers to an automated method to create and evolve programs to solve a problem [16]. A *program*, also called an *individual* or a *solution*, is represented by a tree, whose nodes are labeled with functions (concatenate, split, sum, etc) or with values (strings, numbers, etc). New individuals are generated by applying genetic operations to the current population of individuals. Note that genetic programming does not enumerate all possible individuals, but it selects individuals that should be bred by an evolutionary process. The *fitness function* assigns a *fitness value* to each individual, which represents how close an individual is to the solution and determines the chance of the individual to remain in the genetic process.

The process requires two configuration steps, carried out just once. First, certain parameters of the process must be properly calibrated to prevent overfitting problems, to avoid unnecessary runtime overhead, and to help finding good solutions (see Section 4). Once the parameters are calibrated, the second configuration step is to determine the stop criterion. We opted to stop after a predetermined maximum number of generations and return the best-so-far individual to limit the cost of searching for individuals.

We now show how to use genetic programming to compute complex datatype property matches. Let S and T be two classes with sets of datatype properties $A=\{A_1, \ldots, A_u\}$ and $B=\{B_1, \ldots, B_v\}$, respectively. Let s and t be lists of sets of instances of the properties in A and B, respectively.

The genetic programming phase receives as input the candidate matches that Phase 1 outputs and the sets s and t. For each input candidate match, it outputs a property mapping $\mu[A^k, B_k]$, if one exists; otherwise it discards the candidate match.

Let (A^k, B_k) be a candidate match output by the first phase, where A^k is a set of properties in A and B_k is a property in B. The genetic programming phase first generates a random initial population of candidate property mappings. In each iteration step, it creates new candidate property mappings using genetic operations. It keeps the best-so-far individual, and returns it when the stop criterion is reached.

The process depends on the following specifications (see [24] for a concrete example), which should be regarded as flexibilization points.

A candidate property mapping $\mu[A^k, B_k]$ (the individual in this case) is represented as a tree whose leaves are labeled with the properties in A^k and whose internal nodes are labeled with primitive mapping functions.

The maximum population size, $\sigma_{population}$, is a parameter of the process. The initial population consists of $\sigma_{population}$ randomly generated trees. Each tree has a maximum height, defined by the parameter σ_{height}, each leaf is labeled with a property from A^k and each internal node is labeled with a primitive mapping function.

The reproduction operation simply preserves a percentage of the property mappings from one generation to the next, defined by the parameter $\sigma_{reproduction}$.

The crossover operation exchanges subtrees of two candidate property mappings to create new candidate mappings. For example, suppose that $A^k=\{firstName, middleName, lastName\}$ and $B_k=fullName$ and consider the following two candidate property mappings (which use the concatenation operation, "$//$", and are represented using the notation adopted in Section 2.1):

$$\mu_1[A^k, B_k] = \text{``}fullName \leftarrow (lastName//(\textbf{firstName // middleName}))$$

$$\mu_2[A^k, B_k] = \text{``}fullName \leftarrow ((\textbf{middleName // firstName})//lastName)$$

The crossover operation might generate the following two new candidate property mappings (by swapping the sub-expressions in boldface):

$$\mu_3[A^k, B_k] = \text{``}fullName \leftarrow (lastName//(\textbf{middleName // firstName}))$$

$$\mu_4[A^k, B_k] = \text{``}fullName \leftarrow ((\textbf{firstName // middleName})//lastName)$$

The mutation operation randomly alters a node (labeled with a property or with a primitive mapping function) of a candidate property mapping. For example, the node labeled with "middleName" of $\mu_4[A^k, B_k]$ can be mutated to "firstName", resulting in a new candidate property mapping (which is acceptable, but not quite reasonable, since it repeats firstName):

$$\mu_5[A^k, B_k] = \text{``}fullName \leftarrow ((firstName//firstName)//lastName)$$

Finally, recall that s and t are lists of sets of instances of the properties in A and B, respectively. The fitness value of $\mu[A^k, B_k]$ is computed by applying $\mu[A^k, B_k]$ to the instances of the properties in A^k occurring in s, creating a new set of instances for B_k, which is then compared with the set of instances of B_k occurring in t. As in Section 3.1, the exact nature of the fitness function depends on the types of the values of the datatype properties as well as on whether the function takes advantage of instance matches or not (which is possible when implementing S4). For instance, we used the Levenshtein similarity function for string values and KL-divergence measure [2] for numeric values.

The Levenshtein similarity function is normalized to fall into the interval $[0, 1]$, where 1 indicates that a string is exactly equal to the other and 0 that the two strings have nothing in common, while the KL-divergence measure is used to compute the similarity between two value distributions.

Recall that we are given two samples, p and q, of instances of properties of classes P and Q, respectively. Construct the set X of strings that occur as literals of instances of B_k obtained by applying $\mu[A^k, B_k]$ to p, and the set Y of strings that occur as literals of instances of B_k in q. The fitness score for a candidate property mapping is:

$$Fitness_{string}(\mu[A^k, B_k]) = \frac{1}{n} \sum_{\substack{x \in X \\ y \in Y}} Levenshtein(x, y) \tag{3}$$

where n is the number of pairs in $X \times Y$.

In the case of numeric values, construct the set X of numeric values that occur as literals of instances of B_k, obtained by applying $\mu[A^k, B_k]$ to p, and the set Y of numeric values that occur as literals of instances of B_k in q. The fitness score for a candidate property mapping is:

$$Fitness_{numeric}(F, G) = \frac{1}{n} \sum_{\substack{x \in X \\ y \in Y}} ln\left(\frac{F(x)}{G(y)}\right) F(x) \tag{4}$$

where n is the number of pairs in $X \times Y$, $F(x)$ represents the target distribution of instances in X and $G(y)$ is the the set of materialized mapping μ in Y from the source distribution of instances.

4   An Example Implementation

With the help of an example, we illustrate how to implement the two-phase technique. We assume that the implementation is in the context of S1 of the process described in Section 2.2, that is, we will not use instance matches. We start with Phase 1, described in Section 3.1.

Table 1. Example schemas

#	P	#	Q
A_1	FirstName	B_1	FullName (FirstName // LastName)
A_2	LastName		
A_3	E-Mail	B_2	E-Mail
A_4	Address	B_3	FullAddress (Address // Number // Complement // Neighborhood)
A_5	Number		
A_6	Complement		
A_7	Neighborhood		

The example is based on personal information classes, modeled by class P, with 7 properties and class Q with 3 properties. Table 1 shows the properties from the two classes P and Q, and also indicates which properties or sets of properties match. For example, $\{A_1, A_2\}$ matches B_1.

4.1 Phase 1: Computing Simple Property Matches with Estimated Mutual Information

Recall from Section 3.1 that an implementation of Phase 1 requires defining set comparison functions used to compute the co-occurrence matrix $[m_{ij}]$. We discuss this point in what follows, with the help of the running example.

We assume that all property values are string literals and that we are given two samples, p and q, of instances of properties of classes P and Q, respectively (each with 500 instances). As mentioned in Section 3, Leme et al. [19] use the cosine similarity function to compute the co-occurrence matrix, which is able to indicate only simple 1:1 matches. By contrast, we used the Jaccard similarity coefficient that measures the similarity between sets, which is able to find simple 1:1 matches and *suggest* complex matches.

Figure 1 (a) shows the co-occurrence matrix computed using the cosine similarity measure. Note that $m_{43} = 164k$, which is high because the values of A_4 and B_3 come from a controlled vocabulary with a small number of terms (not indicated in Table 1). By contrast, $m_{32} = 500$, which is low because A_3 and B_2 are keys (also not indicated in Table 1).

Figure 1 (b) shows the co-occurrence matrix computed using the Jaccard similarity (see Eq. 2), which measures the similarity and diversity between sets. Thus, the co-occurrence indices are more sparse between the attributes that have values in common.

To clarify, consider A_7 (Neighborhood) and B_3 (FullAddress) and suppose that "Cambridge" is an observed value of A_7 and "* Oxford Street Cambridge MA, United States" of B_3. The cosine similarity of these two strings is 0.37, which is lower than the threshold set by [19] (again, $\alpha = 0.8$). Hence, these two strings are considered not to be similar. However, also observe that "Cambridge" is fully contained in "* Oxford Street Cambridge MA, United States", which might indicate that A_7, perhaps concatenated with the values of other datatype properties, might match B_3. Continuing with this argument, lowering the threshold also proved not to be efficient to account for these situations, since this increases noise in the matching process.

$$
\begin{array}{c|ccc}
 & B_1 & B_2 & B_3 \\
\hline
A_1 & 4 & 1 & 0 \\
A_2 & 0 & 0 & 0 \\
A_3 & 0 & 500 & 0 \\
A_4 & 0 & 0 & 164k \\
A_5 & 0 & 0 & 0 \\
A_6 & 0 & 0 & 0 \\
A_7 & 0 & 0 & 0
\end{array}
\qquad
\begin{array}{c|ccc}
 & B_1 & B_2 & B_3 \\
\hline
A_1 & 4,8k & 0 & 1,6k \\
A_2 & 12,3k & 0 & 5,1k \\
A_3 & 0 & 500 & 0 \\
A_4 & 5,5k & 0 & 55k \\
A_5 & 0 & 0 & 726 \\
A_6 & 797 & 0 & 8,5k \\
A_7 & 750 & 0 & 9,5
\end{array}
$$

(a)   (b)

Fig. 1. Co-occurrence matrices using (a) cosine similarity and (b) Jaccard similarity coefficient

$$
\begin{array}{c|ccc}
 & B_1 & B_2 & B_3 \\
\hline
A_1 & 0,0550 & 0,0 & 0,0040 \\
A_2 & 0,0138 & 0,0 & 0,0020 \\
A_3 & 0,0 & 0,0020 & 0,0 \\
A_4 & 0,0 & 0,0 & 0,0677 \\
A_5 & 0,0 & 0,0 & 0,0090 \\
A_6 & 0,0024 & 0,0 & 0,0094 \\
A_7 & 0,0002 & 0,0 & 0,0114
\end{array}
$$

Fig. 2. EMI matrix: dark gray cells represent simple matches and light gray cells represent possible complex matches for the property in the column

Thus, given two properties A_i and B_j, m_{ij} is computed as the sum of $Jaccard(x, y)$, for all pairs of strings x and y such that there are triples of the form (a, A_i, x) in p and (b, B_j, y) in q (see Figure 1). Once the co-occurrence matrix $[m_{ij}]$ is obtained, we compute the EMI matrix $[EMI_{ij}]$, as described in Section 3.1 (see Figure 2).

The result of Phase 1 therefore is the matching μ_{EMI} between the sets of properties $\{A_1, \ldots, A_u\}$ and $\{B_1, \ldots, B_v\}$, computed as in Section 3.1 (which we recall is $1 : 1$), assuming that, for each $(A_i, B_j) \in \mu_{EMI}$, the property mappings $\mu[A_i, B_j]$ is always the identity function (see Figure 2).

4.2 Phase 2: Computing Complex Property Matches with Genetic Programming

The second phase of the technique was implemented using a genetic programming toolkit [21], (the discussion on calibration is omitted for brevity, see [24] for more details).

The first phase of the technique outputs, for instance, a candidate match between properties A_1, A_2, A_4, A_5, A_6 and A_7 (FirstName, LastName, Address, Number, Complement and Neighborhood, respectively) and property B_3 (FullAddress), see Figure 2. Note that quite frequently streets are named after famous people, which justifies why EMI outputs A_1 and A_2 as candidates properties. Following the example, having 6 properties as input, the genetic process begins the search for the solution.

As the property values are strings, the fitness function selected to find the best individual is the Levenshtein (see Eq. 3). Thus, after randomly generate an initial set of individuals, the fitness function assigns to each individual a score. For each new generation, a new set of individuals is created from those individuals chosen according to a probability based on their fitness value. After a predetermined number of generations, the process stops with an expression that represents a property mapping that maps the concatenation of the properties A_4, A_5, A_6 and A_7, that is, the expression:

$$((Address//Number)//(Complement//Neighborhood))$$

into property B_3 (that is, FullAddress).

5   Evaluation and Results

The first result in this paper is the comparison of the two approaches, Estimated Mutual Information and Genetic Programming, when separately evaluated.

For this evaluation, we used three datasets[1] from three different domains. Table 2 lists and describes the datasets used and their schema information. The "Personal Information" dataset lists information about people, the "Real Estate" dataset lists information about houses for sale, while the "Inventory" dataset describes product inventories.

Column "EMI" of Table 3 indicates that, using only the Estimated Mutual Information approach, we obtained a precision of 1.0 for all datasets, which indicates that none of the matches were mistakenly found; the rate of recall was low, between 0.21 and 0.38, indicating a high rate of missed property matches; and the F-Measure varied from 0.34 to 0.54, hinting that this approach is insufficient to find simple and complex matches. Indeed, out of the 12 simple matches expected for the "Personal Information" dataset, this approach correctly obtained 6 matches only. Likewise, the EMI found 3 out of 4 and 4 out of 6, for the datasets "Inventory" and "Real Estate", respectively.

However, according to the discussion at the end of Section 3.1, as well as by observing the column "EMI" marked with "*" in Table 2, there are several candidate complex matches that were suggested to the GP phase in each approach. Note that amongst those are the exact remaining matches not found by the EMI technique. This is an indication that, although not sufficient in itself, the EMI approach is an effective pre-processing stage to the GP approach, by reducing the complexity of the search space while providing a high quality list of candidate complex matches.

Column GP of Table 3 indicates that, using genetic programming alone, the F-Measure obtained was higher, and that all simple mappings were found. However, precision was 0.8 for the "Personal Information" dataset and 0.96 for the "Inventory" dataset, which indicates that some matches were mistakenly suggested.

Table 3 shows that our two-phase technique resulted in a considerable improvement over the independent use of the EMI and GP approaches when used independently. This improvement is related to the fact that the first phase, using the EMI matrix, correctly found all simple matches and suggested correct complex matches to the second phase.

---

[1] With exception of the "Personal Information" dataset due to privacy reasons, other datasets are available at http://pages.cs.wisc.edu/~anhai/
wisc-si-archive/domains/

Table 2. Mapping results for three datasets in different domains

Datasets	Type		EMI	GP	**EMI+GP**	#Match
Personal Information	String	1:1	6	12	**12**	12
		1:n	11*	1	**4**	5
	Numeric	1:1	0	0	**0**	0
		1:n	0	0	**0**	0
Inventory	String	1:1	3	4	**4**	4
		1:n	18*	2	**4**	4
	Numeric	1:1	6	25	**25**	25
		1:n	18*	1	**3**	4
Real Estate	String	1:1	4	4	**6**	6
		1:n	7*	2	**5**	5
	Numeric	1:1	1	1	**1**	1
		1:n	7*	0	**0**	3

(∗) Complex matches suggested by EMI.

Table 3. P/R/F1 results for three datasets in different domains

	EMI			GP			EMI+GP		
Dataset	P	R	F1	P	R	F1	P	R	F1
Personal Information	1	0.38	0.54	0.8	0.75	0.77	**1**	**0.94**	**0.96**
Inventory	1	0.24	0.39	0.96	0.87	0.91	**0.97**	**0.97**	**0.97**
Real Estate	1	0.33	0.5	1	0.47	0.64	**1**	**0.8**	**0.89**

The fact that the EMI matrix suggests correlated properties helps reduce the solution space considered by the genetic programming algorithm, thus improving its overall performance. In our tests, the run time of the combined approach showed an improvement of approximately 36% when compared with the run time of the genetic programming approach alone.

Furthermore, we also compared our method against state of the art methods. As a baseline we used the iMap system [5], which similar to our approach addresses the problem of 1:1 and n:1 (complex) matchings. From previously reported results in terms of accuracy, iMap obtains 0.84 and 0.55 for 1:1 and 1:n mappings respectively, while we obtain 1 and 0.955 for the "Inventory" dataset. For the "Real Estate" dataset, iMap achieves 0.58 and 0.32, whereas we achieve 1 and 0.72, respectively. We also compared our method against LSD [7], which is able to find only simple 1:1 matchings and achieves an accuracy of 0.67.

6   Related Work

Ontology alignment frameworks implement a set of similarity measures to find the correct mappings. For instance, Duan et al. [9] utilize user feedback to determine the importance of each similarity measure in the final mapping result. Similarly, Ritze et al. [27] introduce ECOMatch that uses alignment examples to define parameters to set the correct mapping strategy. Dhamankar et al. [6] describe iMap that predefines modules of functions to semi-automatically find simple and complex matches by leveraging external knowledge. Likewise, Albagli et al. [1] search for mappings using Markov Networks, which combines different sources of evidence (e.g. human experts, existing mappings, etc). Finally, Spohr et al. [30] use a translation mechanism to discover mappings in cross-lingual ontologies.

A drawback in most approaches is scalability. Duan et al. [8] address the scalability problem using a local sensitivity hashing to match instances inside a cluster. Jiménez-Ruiz and Grau [15] propose an "on the fly" iterative method called LogMap that, based on a set of anchors (exact mappings), creates, extends and verifies mappings using a logical reasoner. Complementary, Wang et al. [31] suggest a method for reducing the number of anchors needed to match ontologies. Recent advances, such as RiMOM [20], offer an automated environment to select an appropriate matching strategy through risk minimization of Bayesian decision, while ASMOV ([14]) uses semantic validation to verify mappings. Falcon [13] applies a divide-and-conquer approach to ontology matching. Several other systems, such as DSSim [22], S-Match [11], Anchor-Flood [12], Agreement-Maker [3], ATOM [26] and SAMBO [17] tackle the alignment for ontologies and schemas relying on lexical, structural and semantical similarity measures. In a recent survey, [29] analyze in more details well-established frameworks and outline future directions and challenges in this field. Additional surveys are provided by [28,25].

Contrasting with the approaches just outlined, we provide an automatic technique that finds simple and complex mappings between RDF datatype properties without prior knowledge that can evolve to adapt to schema and ontology changes, previously described in [23]. Similar to our approach, Carvalho et al.[4] propose a genetic programming approach for deduplication problem. However, as the results show, our two-phase approach achieves better results than those using only the genetic programming approach. Moreover, we extend his work to match simple and complex numeric datatype properties.

7   Conclusion

In this paper, we described an instance-based, property matching technique that follows a two-phase strategy. The first phase constructs the Estimated Mutual Information matrix of the property values to identify simple property matches and to suggest complex matches, while the second phase uses a genetic programming approach to detect complex property matches and to generate their property mappings. This combined strategy proved promising to beat combinatorial explosion. In fact, our experiments prove that the technique is a promising approach to construct complex property matches, a problem rarely addressed in the literature.

Acknowledgement. This work has been partially supported by the European Union Seventh Framework Programme (FP7/2007-2013) under grant agreement No 317620 (LinkedUp).

References

1. Albagli, S., Ben-Eliyahu-Zohary, R., Shimony, S.E.: Markov network based ontology matching. Journal of Computer and System Sciences 78(1), 105–118 (2012)

2. Cover, T.M., Thomas, J.A.: Elements of information theory. Wiley, New York (1991)
3. Cruz, I.F., Antonelli, F.P., Stroe, C.: Agreementmaker: Efficient matching for large real-world schemas and ontologies. PVLDB 2(2), 1586–1589 (2009)
4. de Carvalho, M.G., Laender, A.H.F., Gonçalves, M.A., da Silva, A.S.: A genetic programming approach to record deduplication. IEEE Trans. Knowl. Data Eng. 24(3), 399–412 (2012)
5. Dhamankar, R., Lee, Y., Doan, A., Halevy, A., Domingos, P.: imap: discovering complex semantic matches between database schemas. In: Proceedings of the 2004 ACM SIGMOD International Conference on Management of Data, SIGMOD 2004, pp. 383–394. ACM, New York (2004)
6. Dhamankar, R., Lee, Y., Doan, A., Halevy, A.Y., Domingos, P.: imap: Discovering complex mappings between database schemas. In: SIGMOD Conference, pp. 383–394 (2004)
7. Doan, A., Domingos, P., Levy, A.: Learning Source Descriptions for Data Integration. In: Proceedings of the Third International Workshop on the Web and Databases, Dallas, TX, pp. 81–86. ACM SIGMOD (2000)
8. Duan, S., Fokoue, A., Hassanzadeh, O., Kementsietsidis, A., Srinivas, K., Ward, M.J.: Instance-based matching of large ontologies using locality-sensitive hashing. In: Cudré-Mauroux, P., et al. (eds.) ISWC 2012, Part I. LNCS, vol. 7649, pp. 49–64. Springer, Heidelberg (2012)
9. Duan, S., Fokoue, A., Srinivas, K.: One size does not fit all: Customizing ontology alignment using user feedback. In: Patel-Schneider, P.F., Pan, Y., Hitzler, P., Mika, P., Zhang, L., Pan, J.Z., Horrocks, I., Glimm, B. (eds.) ISWC 2010, Part I. LNCS, vol. 6496, pp. 177–192. Springer, Heidelberg (2010)
10. Euzenat, J., Shvaiko, P.: Ontology matching. Springer (2007)
11. Giunchiglia, F., Autayeu, A., Pane, J.: S-match: An open source framework for matching lightweight ontologies. Semantic Web Journal 3(3), 307–317 (2012)
12. Hanif, M.S., Aono, M.: An efficient and scalable algorithm for segmented alignment of ontologies of arbitrary size. Journal of Web Semantics 7(4), 344–356 (2009)
13. Hu, W., Qu, Y., Cheng, G.: Matching large ontologies: A divide-and-conquer approach. IEEE Trans. Knowl. Data Eng. 67(1), 140–160 (2008)
14. Jean-Mary, Y.R., Shironoshita, E.P., Kabuka, M.R.: Ontology matching with semantic verification. Journal of Web Semantics 7(3), 235–251 (2009)
15. Jiménez-Ruiz, E., Cuenca Grau, B.: LogMap: Logic-based and scalable ontology matching. In: Aroyo, L., Welty, C., Alani, H., Taylor, J., Bernstein, A., Kagal, L., Noy, N., Blomqvist, E. (eds.) ISWC 2011, Part I. LNCS, vol. 7031, pp. 273–288. Springer, Heidelberg (2011)
16. Koza, J.R.: Genetic programming: on the programming of computers by means of natural selection. MIT Press, Cambridge (1992)
17. Lambrix, P., Tan, H.: Sambo - a system for aligning and merging biomedical ontologies. Journal of Web Semantics 4(3), 196–206 (2006)
18. Leme, L.A.P.P., Brauner, D.F., Breitman, K.K., Casanova, M.A., Gazola, A.: Matching object catalogues. ISSE 4(4), 315–328 (2008)
19. Leme, L.A.P.P., Casanova, M.A., Breitman, K.K., Furtado, A.L.: Instance-based OWL schema matching. In: Filipe, J., Cordeiro, J. (eds.) ICEIS 2009. LNBIP, vol. 24, pp. 14–26. Springer, Heidelberg (2009)
20. Li, J., Tang, J., Li, Y., Luo, Q.: Rimom: A dynamic multistrategy ontology alignment framework. IEEE Transactions on Knowledge and Data Engineering 21(8), 1218–1232 (2009)
21. Meffert, K.: Jgap - java genetic algorithms and genetic programming package (2013), http://jgap.sf.net/ (Online; accessed January 31, 2013)
22. Nagy, M., Vargas-Vera, M., Stolarski, P.: Dssim results for oaei 2009. In: Ontology Matching (2009)

23. Nunes, B.P., Caraballo, A.A.M., Casanova, M.A., Breitman, K., Leme, L.A.P.P.: Complex matching of rdf datatype properties. In: Ontology Matching (2011)
24. Nunes, B.P., Mera, A., Casanova, M.A., Breitman, K., Leme, L.A.P.P.: Complex matching of rdf datatype properties. Technical Report MCC-11/12 (September 2011)
25. Rahm, E., Bernstein, P.A.: A survey of approaches to automatic schema matching. VLDB Journal 10(4), 334–350 (2001)
26. Raunich, S., Rahm, E.: Atom: Automatic target-driven ontology merging. In: ICDE Conference, pp. 1276–1279 (2011)
27. Ritze, D., Paulheim, H.: Towards an automatic parameterization of ontology matching tools based on example mappings. In: Ontology Matching (2011)
28. Shvaiko, P., Euzenat, J.: A survey of schema-based matching approaches, pp. 146–171 (2005)
29. Shvaiko, P., Euzenat, J.: Ontology matching: State of the art and future challenges. IEEE Trans. Knowl. Data Eng. 25(1), 158–176 (2013)
30. Spohr, D., Hollink, L., Cimiano, P.: A machine learning approach to multilingual and cross-lingual ontology matching. In: Aroyo, L., Welty, C., Alani, H., Taylor, J., Bernstein, A., Kagal, L., Noy, N., Blomqvist, E. (eds.) ISWC 2011, Part I. LNCS, vol. 7031, pp. 665–680. Springer, Heidelberg (2011)
31. Wang, P., Zhou, Y., Xu, B.: Matching large ontologies based on reduction anchors. In: IJCAI, pp. 2343–2348 (2011)

Coordination Issues in Artifact-Centric Business Process Models

Giorgio Bruno

Politecnico di Torino, Torino, Italy
giorgio.bruno@polito.it

Abstract. In recent years, research in the field of business processes has shown a shift of interest from the activity-centric perspective to the artifact-centric one. The benefits, such as improved communication among the stakeholders and higher potential for flexibility, come from the focus on the key business entities (called artifacts) and on the distribution of the control flow in the life cycles of the artifacts. However, this perspective also entails a number of challenges, such as the coordination between the life cycles of the artifacts. This paper proposes an approach based on correlated transitions, i.e., transitions that belong to different life cycles and must be performed jointly. A new notation called Acta is illustrated with the help of two motivating examples.

Keywords: business processes, artifacts, life cycles, correlations, tasks.

1   Introduction

While it is commonly accepted that a business process is a standard way of organizing work in a business context, the consent on what the starting point in the investigation of the intended business should be is no longer unanimous.

According to the traditional definitions, a business process consists of a number of tasks designed to produce a product or service and is meant to cross functional boundaries in that it may involve members of different departments [5]. This point of view has spurred a fruitful line of research, now labeled as activity-centric, and a standard notation, BPMN [2].

In recent years, the notion of PAIS (Process-Aware Information System) [6], which advocates a tighter integration between the areas of information systems and business processes, has brought about a shift of interest from the activity-centric perspective to the artifact-centric one. The latter emphasizes the identification of the key business entities (called artifacts) and of their life cycles, which show how the artifacts evolve over time through the execution of business operations (a.k.a. tasks).

The analysis (reported in [4]) of the operations of a global financing division resulted in a high-level model consisting of 3 major artifact types, whose life cycles include 18 states and approximately 65 tasks. The major benefit is the right level of granularity, which facilitates communication among the stakeholders and helps them focus on the primary purposes of the business.

H. Decker et al. (Eds.): DEXA 2013, Part I, LNCS 8055, pp. 209–223, 2013.

The major criticism raised against the activity-centric approach is the emphasis placed on the tasks and the control-flow elements, while the business entities are not considered as first-class citizens. As a matter of fact, the data flow in the activity-centric approach is based on process variables and there is no automatic mapping between business entities and process variables. Moreover, this perspective seems to be more suitable for automated processes than for human-centric ones, as it lacks an adequate representation of the situations in which different courses of action are possible and the choices depend on human decisions. The artifact-centric approach has the potential for coping with such issues owing to the emphasis placed on the business entities on which human decisions are grounded.

A critical aspect in the artifact-centric approach is the handling of tasks (called spanning tasks) operating on two or more artifacts in that some form of coordination of their life cycles is needed. This paper grounds the coordination of the life cycles on the notion of correlated transitions, i.e., transitions that belong to different life cycles and must be performed jointly. Three kinds of correlation are examined: they are referred to as generative, selective and direct correlation.

A new notation called Acta (Artifacts, Correlations and TAsks) is also presented. Acta models are made up of two components, the structural model and the dynamic one. The structural model basically shows the artifacts in terms of their properties and the dynamic model is the collection of their life cycles.

Another important point is the impact that the artifact-centric approach has on the structure of work lists. This paper presents a solution in which the work lists are organized on the basis of the artifacts their owners are in charge of.

This paper is organized as follows. Section 2 is an overview of Acta and of coordination issues; section 3 describes two motivating examples concerning a purchase requisition process and a negotiation one. Section 4 illustrates structural models and section 5 analyzes the correlated transitions needed in the dynamic models of the examples. Section 6 presents the structure of work lists, section 7 discusses the related work and section 8 provides the conclusion of the paper.

2   Acta Models and Coordination Issues

Acta models are made up of two components, the structural model and the dynamic one. The structural model shows the entities involved in a business process, in terms of their types, associations and attributes. Emphasis is placed on mandatory relationships and associative attributes, which play an important part in the specification of the tasks, as will be illustrated in the next sections.

The dynamic model of a business process is the collection of the life cycles of the artifact types needed. A life cycle defines the path to be followed over time by any artifact of the type under consideration from the initial state to a final one, and it is represented as a state-transition diagram. The states represent the stages in the path and the transitions cause the artifact to move from the current state to the next one. In this paper, transitions denote human tasks, i.e., actions to be carried out by persons who participate in the process by playing specific roles. Transitions may also be associated with automatic tasks.

A major issue in the artifact-centric perspective is the coordination of the life cycles of the business entities involved. Coordination in Acta is achieved through correlated transitions, i.e., transitions that belong to different life cycles and must be performed jointly.

A group of correlated transitions is the result of a unit of work, called spanning task, which affects artifacts of different types. Three kinds of correlation are examined in this paper: generative, selective and direct correlation. A short account is given in this section, while a more detailed description is provided in the next sections with the help of the motivating examples.

In a group of correlated transitions, one is the master transition and the others are the subordinate transitions; the spanning task is associated with the master transition. The artifact acted on by the master transition is called master artifact and the artifacts affected by the subordinate transitions are called subordinate artifacts.

When the effect of a spanning task is the generation of new artifacts, generative correlation takes place. The correlated transitions are the master transition, i.e., the one affecting the master artifact, and the initial transitions in the life cycles of the new artifacts. Generation acts are usually remembered through associations between the master artifact and the subordinate ones. For example, an account manager who is not able to directly fulfill a purchase requisition coming from a customer generates a number of requests for quote for different suppliers: the spanning task enterRequestsQ operates on a purchase requisition (the master artifact) so as to produce new requests for quote (the subordinate artifacts). The purchase requisition changes state and the requests for quote enter the initial state; in addition, the requests for quote are associated with the purchase requisition.

Selective correlation means that correlation stems from a selection of artifacts. For example, the account manager selects one request for quote from among those that have been fulfilled by the suppliers so as to fulfill the purchase requisition. Both the request for quote and the purchase requisition change state; the selection is remembered by means of a new association established between the artifacts. Selective correlation may be additive or constitutive. In the first case, a new association is established between artifacts, such as the purchase requisition and the request for quote, already connected. In the second case, the artifacts become connected for the first time; an example is a spanning task that enables a broker to match a request and an offer, and the result is a connection between two artifacts that before were unrelated.

Direct correlation means that a spanning task can also bring about a change of state for the artifacts that are not input artifacts but are associated with the input ones. For example, a confirmed purchase requisition makes the account manager produce a confirmed request for quote: the request for quote is the one associated with the input purchase requisition.

Acta models are meant to be conceptual models: tasks specify their intended effects by means of post-conditions, while constraints are expressed by pre-conditions and selections by selection rules.

3   Description of the Examples

This section presents a short description of two motivating examples, whose models will be illustrated in the next sections. The description focuses on the most important aspects and disregards the details, such as the attributes of the entities.

3.1   Purchase Requisition Process

The process enables a selling organization, referred to as the seller, to carry out commercial transactions with its partners (customers and suppliers). Three roles are involved, account manager, customer and supplier. Account manager is an internal role of the seller while the other roles designate external users acting on behalf of the partners. The process runs on a platform that enables all the users to operate on the same information system.

Customers enter purchase requisitions to get the prices for the goods they need. The purchase requisitions are handled by the account managers. The process assumes that each customer is served by one specific account manager, who may serve several customers. While processing a purchase requisition, an account manager has three options: they may fulfill it, reject it or involve a number of suppliers.

In the third case, they enter a number of requests for quote, each one directed to a different supplier; this is a case of generative correlation. Suppliers may fulfill a request for quote or may reject it. The account manager may select one request for quote from among those fulfilled (additive selective correlation) so as to fulfill the corresponding purchase requisition, or they may give up and reject the purchase requisition. In addition, they must withdraw all the requests for quote fulfilled except for the one selected, if any.

The customer may confirm a fulfilled purchase requisition or may withdraw it. If the purchase requisition is based on a request for quote, the account manager is in charge of confirming or withdrawing it, respectively; this is a case of direct correlation.

The details regarding the attributes of the entities are ignored; however, the general meaning of the above-mentioned actions can be inferred from the verbs used. Therefore, fulfilling a request implies providing the information required, e.g., the prices of the goods needed by a customer; confirming (or withdrawing) a request means that the requester is satisfied (or dissatisfied) with the information provided by the recipient of the request.

3.2   Negotiation Process

The process enables brokers to manage sales campaigns. After a broker has started a campaign, sellers may enter offers of products and buyers may enter requests for products. The broker may match one or more requests and one or more offers to produce a transaction; the matching rule is not considered. This is a case of constitutive selective correlation: the selection is made from among peer entities and there are no direct connections between them.

The broker may reject requests and offers at their discretion. After some time, the campaign is closed by the broker.

4   Structural Models

Structural models show the entities involved in business processes, in terms of their types, associations and attributes. They extend UML class models with the purpose of emphasizing mandatory relationships and associative attributes, which play an important part in the specification of the tasks, as will be illustrated in the next section.

Relationships are connections between pairs of types; they represent the associations that may exist between the instances of the types involved. Cardinalities place constraints on the number of connections between instances. The standard cardinality is many-to-many and indicates that any entity of one type may be connected to a number of entities of the other type; this number is not predetermined and may be zero as well.

Associations may be mandatory on one side and optional on the other; in this case, the relationship is shown as an oriented link whose origin is the type for which the association is mandatory.

Mandatory relationships are important in that they determine which associations have to be set when a new entity is generated. Depending on the multiplicity of the relationship, the newly generated entity will be connected to one or more entities of the destination type. On the destination side of a mandatory relationship, the default multiplicity is one; on the source side it is 0 to many.

The structural model of the purchase requisition process is shown in Fig.1a. Types Customer, Supplier and AccountMgr represent the participants in the process in terms of the roles they play; such types are referred to as role types. Types PurchaseR and RequestQ represent purchase requisitions and requests for quote, respectively.

Relationships imply attributes called associative attributes in the entities involved. Such attributes refer to single entities or collections of entities depending on the cardinalities of the corresponding relationships. The names of the associative attributes may be omitted and in such cases they take the name of the type they refer to, in the singular form or in the plural one (with the initial in lower case) depending on the cardinality of the relationship. The name of a relationship is obtained from the names of the types involved (if there are no other relationships between the same pair of types) or from the names of the corresponding associative attributes (either explicit or implicit) in alphabetic order. For example, the name of the relationship between AccountMgr and Customer is accountMgr-customers or AccountMgr-Customer.

The meaning of the mandatory relationships appearing in Fig.1a is as follows. Purchase requisitions are generated by customers, and then a relationship (Customer-PurchaseR) is needed to identify the generator of each purchase requisition. Relationship AccountMgr-Customer connects a customer with the account manager who will take care of their purchase requisitions.

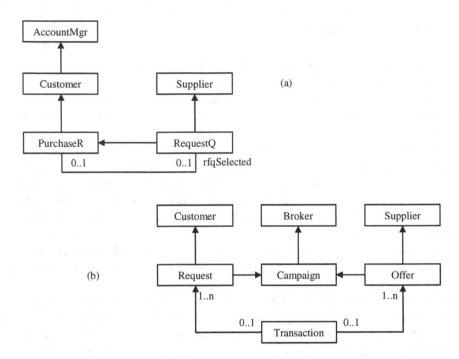

Fig. 1. The structural models of the processes: purchase requisition (a) and negotiation (b)

Requests for quote are complements of purchase requisitions; then when a new request for quote is generated, it is associated with the purchase requisition it is a complement of, and it is also connected to the supplier entity representing the recipient of the request for quote. The relationships involved are purchaseR-requestsQ and supplier-requestsQ, respectively; term requestsQ is the implicit associative attribute on the RequestQ side. Relationship purchaseR-rfqSelected associates a purchase requisition with the request for quote that has been selected from among those fulfilled (if any).

Associative attributes are essential for navigational purposes as will be illustrated in the next section. For example, if pr denotes a certain purchase requisition, navigational expression "pr.customer.accountMgr" returns the AccountMgr entity related to the Customer entity associated with the entity denoted by pr. The syntax of navigational expressions in Acta is based on a simplified version of OCL [16].

In the negotiation process, whose structural model is shown in Fig.1b, types Broker, Customer and Supplier represent the participants. Campaigns are managed by brokers and such associations are represented by relationship Campaign-Broker. Transactions are made up of a number of requests and a number of offers; however, since requests and offers may be rejected, the cardinality of the relationships Request-Transaction and Offer-Transaction is 0 or 1 on the transaction side.

5   Dynamic Models

This section presents the dynamic models of the motivating examples with the purpose of illustrating the three kinds of correlation introduced in section 2. The dynamic model of a business process is the collection of the life cycles of the artifact types needed; for convenience they are shown in the same figure. The dynamic models of the purchase requisition process and of the negotiation process are shown in Fig.2 and in Fig.3, respectively. The following subsections provide the general features of the Acta notation, illustrate initial tasks and post-conditions, and present the examples of the correlations introduced in section 2.

5.1   General Features of the Acta Notation

Tasks are associated with transitions. Human tasks show the role required, or its acronym, before the task name, e.g. "AM: enterRequestsQ". Automatic tasks have no role indications. Spanning tasks appear in two or more life cycles and group a number of correlated transitions. The names of such transitions are identical; however, the master transition can be distinguished from the subordinate ones in that its name appears in bold while the other names are shown in italics. Moreover, role names only appear on master transitions.

The process model specifies the effects and the constraints of the tasks by means of post-conditions and pre-conditions, respectively; they are based on the structure of the entities provided by the information model and are expressed with a simplified form of OCL [16].

A human task is assigned to one role, but several participants may play the same role; it is then necessary to determine whether any participant playing the role required is entitled to perform the task or a specific participant is needed. The performer is generic in the first case, and specific in the second one. The performer of task enterPurchaseR in Fig.2 is generic, as indicated by the qualifier (any) following the role acronym. The performers of all the other human tasks are specific. The specificity is determined by a connection, either direct or indirect, between the role type and the artifact type. The account manager in charge of a given purchase requisition is not a generic one, but the one associated with the customer who issued the purchase requisition. In general, a rule is needed to specify the desired connection; however, in simple situations like those addressed in this paper, there is no need to express these rules explicitly. The simplification is due to the assumption that the connections between roles and artifact types are based on chains of mandatory relationships; what is more, such chains are assumed to be unique. As a matter of fact, on the basis of the information model shown in Fig.1a, there is only one such path from type PurchaseR to type AccountMgr and it consists of types PurchaseR, Customer and AccountMgr.

If a final state, i.e., a state having no output transitions, implies a notification for a specific participant, the role name (or acronym) appears after the state name. For example, when a purchase requisition enters state rejected (shown in Fig.2), a notification has to be sent to the appropriate customer; the same rules introduced for the identification of the task performers apply in this case as well.

5.2   Initial Tasks and Post-Conditions

The purchase requisition process consists of the purchase requisition (PurchaseR) life cycle and the request for quote (RequestQ) one.

Purchase requisitions are entered by customers when they want to. An initial task is then needed: it is named enterPurchaseR and is associated with the initial transition, i.e., the one entering the initial state and having no source icon. Its post-condition "new PurchaseR" appears in the task description section below the life cycle. The "new" operator asserts that a number of new entities of the type specified will exist after the execution of the task. If the multiplicity is 1, it is omitted; multiplicity n (cf. task enterRequestsQ) means that the number is decided by the performer of the task.

Due to mandatory relationship Customer-PurchaseR, the newly generated artifact needs to be connected with a customer entity. The partners of mandatory relationships are automatically searched for among the entities forming the context of the task. In general, the context of a task includes the current artifact, the entity representing the performer of the task (referred to as the performer entity) and the entities selected by the performer. The context of task enterPurchaseR consists of the performer entity only, which, however, fits the requirements of the relationship in that it is a customer entity; therefore, the newly generated purchase requisition will be connected to this entity.

5.3   Generative Correlation

A purchase requisition in the initial state is handled by an account manager who may reject or fulfill it, or enter a number of requests for quote directed to suitable suppliers.

Task enterRequestsQ performs a generative correlation. This effect is implied by post-condition "new n RequestQ". The mandatory relationships related to type RequestQ require a new request for quote to be connected to two entities, i.e., a purchase requisition and a supplier entity. The former entity is matched by the current artifact but for the latter entity the context of the task provides no match; in such cases, it is up to the performer to choose a suitable entity from among those available. The selection may be subjected to the constraints indicated in the requirements, if any.

At the end of the task, the purchase requisition is moved in the pending state.

5.4   Additive Selective Correlation

A newly generated request for quote may be fulfilled or rejected by the supplier. When the purchase requisition is in the pending state, the account manager may reject it with task reject2 or they may fulfill it with task fulfill2 provided that a request for quote is chosen from among those fulfilled by the suppliers.

Task fulfill2 carries out an additive selective correlation in that the quote to be chosen is already connected with the purchase requisition. The master transition is in the PurchaseR life cycle.

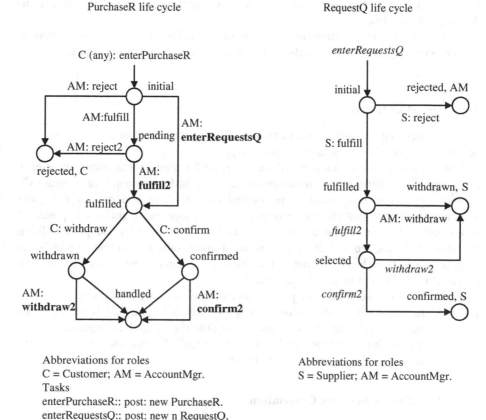

PurchaseR life cycle

RequestQ life cycle

Fig. 2. The dynamic model of the purchase requisition process

The description of task fulfill2 includes two parts, i.e., the selection rule and the post-condition. The selection rule "with requestQ in purchaseR.requestsQ", introduced by keyword with, indicates that one entity of type RequestQ is needed and it is to be selected from among those associated with the input purchase requisition. As a general rule, the entities selected by the performer become part of the context of the task. The state of the requests for quote is not explicitly indicated as it can be found in their life cycle; in fact, it is the input state (fulfilled) of the subordinate transition fulfill2. The selection basis, which is the collection of the entities from among which the choice has to take place, may be empty and in this case the spanning task is not enabled.

The post-condition "purchaseR.rfqSelected == requestQ" indicates that the choice is recorded by means of a new association based on relationship purchaseR-rfqSelected. Since requestQ is the term used in the selection rule to denote the request for quote chosen by the performer, the associative attribute rfqSelected will refer to that request for quote.

A request for quote in state fulfilled may be withdrawn by the account manager; the reason is to prevent the requests for quote not selected from remaining blocked in this state.

5.5   Direct Correlation

The customer may confirm or withdraw a fulfilled purchase requisition, whose state becomes confirmed or withdrawn, respectively.

If the purchase requisition is based on a request for quote, the account manager is in charge of confirming or withdrawing it with tasks confirm2 or withdraw2, respectively. These tasks must be performed only if there is a request for quote associated with the input purchase requisition, i.e., if associative attribute rfqSelected is not null. The conditional nature of the tasks is indicated by the pre-condition "purchaseR.rfqSelected != null". If the pre-condition is false the alternative transitions without labels (which are automatic transitions) are carried out so as to bring the purchase requisition to final state "handled".

Tasks confirm2 and withdraw2 provide an example of direct correlation: if they are enabled, they also act on the request for quote which is obtained from the input purchase requisition through the associative attribute rfqSelected, as expressed by the selection rule "with requestQ as purchaseR.rfqSelected". The request for quote is then moved to state confirmed or to state withdrawn.

5.6   Constitutive Selective Correlation

The negotiation process whose dynamic model and information one are shown in Fig.3 and in Fig.1b, respectively, presents an example of constitutive selective correlation.

The process consists of the Campaign, Offer and Request life cycles. When a campaign is open, suppliers may enter offers and customers may enter requests. They do so with the initial tasks enterOffer and enterRequest, respectively.

Generating a new offer or a new request implies submitting it to an open campaign; therefore the generation of these artifacts takes place within a context, which is determined by the choice of an open campaign, this choice being made by the performer of the task. For this reason, the descriptions of the tasks include the selection rule "with campaign (lifecycleState == open), which means that their performers have to select an open campaign as a contextual entity for the newly generated artifact. The term "lifecycleState" denotes a system attribute that provides the name of the current state in the life cycle of the artifact under consideration.

The broker can reject offers and requests or combine them into transactions. Task genTransaction carries out a constitutive selective correlation in that no previous connections exist between the artifacts combined in the newly generated transaction. The requirements specify no matching rule for the selection of offers and requests; if, instead, one is given, it will be expressed as a pre-condition of the task.

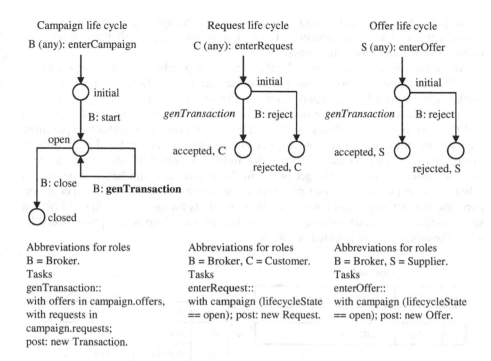

Fig. 3. The dynamic model of the negotiation process

6   Artifact-Centric Work Lists

Business processes are means to organize work and their models should then show what the units of work are and to which roles they are entrusted. During their execution, the units of work are assigned to the appropriate participants through their work lists.

However, the organization of the work lists is not independent of the approach adopted for the representation of the business processes. With activity-centric notations, such as BPMN, the entries of the work lists draw on the tasks defined in the process models and then their labels follow the pattern "task-name info" where info stands for information taken from the task parameters. By clicking on an entry, a participant may perform the corresponding task through the graphical interface provided by the implementation.

In artifact-centric notations, instead, the focus is on the artifacts and then it is natural to structure the work lists on the basis of the artifacts that their owners are in charge of. This section discusses the issue with reference to the examples illustrated in the previous sections.

The content of work lists is a kind of viewpoint that the participants in the process are provided with on the artifacts they are in charge of. The viewpoint includes the options currently available and it changes on the basis of the decisions taken by the participants. One of the challenges is the identification of the most expressive technique to help participants work in this way. The Acta approach suggests replacing textual entries with a more structured representation stressing the distinction between master artifacts and subordinate ones.

An example is given in Fig.4 with reference to the purchase requisition process. A work list for a certain account manager is presented; Fig.4a shows the content of the work list before task fulfill2 is performed and Fig.4b shows it immediately after.

The work list is organized in two columns, the left one referring to master artifacts and the right one to subordinate artifacts. The master column includes two purchase requisitions, pr2 in the initial state and pr1 in the pending state; pr1 and pr2 represent identifiers. Each entry contains three major fields: the identifier, the life cycle state and the options available. Artifact pr2 is in the initial state and hence there are no subordinate items; the performer may select one option out of three. On the contrary, purchase requisition pr1 has three subordinate requests for quote, i.e., rfq1 and rfq3 in state fulfilled and rfq2 in state rejected. For the last one, there is no option available, while the others may be rejected or selected.

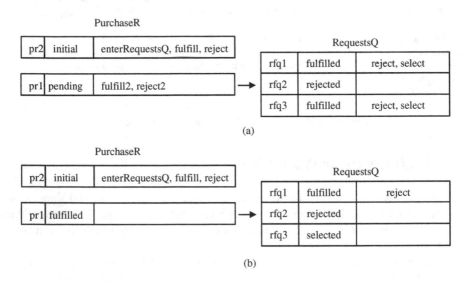

Fig. 4. An example of work list before (a) and after (b) fulfill2 is performed

The select option appears when an artifact, such as rfq1 or rfq3, may be involved in a selective correlation through a subordinate transition. Since a subordinate transition can only take place in conjunction with the corresponding master transition, the select option enables the performer to mark a subordinate artifact before carrying out the spanning task. As a matter of fact, if the account manager wants to fulfill purchase requisition pr1 with the help of request for quote rfq3, first they select rfq3 and then they choose the fulfill2 option on pr1. At the end of the task, the state of pr1 is fulfilled and no options are available because this state is handled by the customer who issued the purchase requisition. The request for quote selected, i.e., rfq3, is in the selected state and no options are available because the output transitions of this state are directly correlated with their master transitions; when the master artifact is acted on, the subordinate artifacts are acted on as well. For the request for quote rfq1, the select option is no longer available as the master artifact is no longer in the pending state.

7   Related Work

According to Sanz [17], the roots of the artifact-centric perspective can be found in past research on entity-based dynamic modeling, but only in recent years, the core ideas have permeated the discipline of Business Process Management.

Term artifact has been introduced in [15] to designate a concrete and self-describing chunk of information that business people use to run a business. The artifact types and their life cycles come from experience and show how the actual entities evolve over time: the business activities, which are responsible for the state transitions, are introduced in a subsequent step of analysis along with the business rules governing their execution.

In the case-handling approach [1], a process is meant to take care of a specific entity type (e.g., an insurance claim), called the process case: the purpose is to improve the flexibility of the control flow as the process evolution depends on the state of the case and not only on the tasks performed [10].

The BALSA framework [7] builds on the notion of artifact and adds services, which encapsulate units of work acting on one or more artifacts, and associations, which specify various kinds of constraints for the services.

In the artifact-centric approach, there are three major issues to cope with, i.e., structure, dynamics and coordination. Structure is about the properties (attributes and associations) of the artifacts involved, dynamics encompass the artifact life cycles and coordination is concerned with the synchronization of the life cycles. Such issues are dealt with in various ways, ranging from separate models to holistic ones.

The Guard-Stage-Milestone (GSM) approach [8] is a holistic technique: the major building blocks are the artifacts, which contain informational aspects (attributes and associations), life cycles and coordination items (events and rules). The drawback is the difficulty of understanding the propagation of the events between the life cycles. For example, an activity performed on an entity can produce an event that is targeted at another entity and triggers an activity affecting this entity. The second activity then turns out to be correlated with the first one.

In ArtiNet [9] and Chant [3], the life cycles are integrated in one model and coordination is obtained with transitions operating on two or more artifacts. Models are monolithic in that they are based on Petri nets where places represent artifact states and tokens denote artifacts.

In other approaches, where the relationships between artifact types are explicitly defined, coordination takes advantage of them, in particular of hierarchical relationships. Hierarchical structures, such as those related to physical systems, are addressed by COREPRO [14], which provides specific means to achieve mutual synchronization between the state of a compound object and those of its components. The approach presented in [12] is aimed at automatically generating a process model from the entity life cycles provided that the synchronization points are manually identified beforehand.

The PHILharmonicFlows approach [11] is based on micro processes and macro processes; the former define the life cycles of the artifacts and the latter provide a coordination mechanism consisting of macro steps and macro transitions. A macro step is associated with an artifact type and a particular state of its life cycle; at run time, it refers to the artifacts being in that state. A macro transition activates an output macro step only when the artifacts collected in the input macro steps satisfy certain conditions, which are related to the structure of the artifacts (in particular to the associations). In Acta, coordination is carried out by the spanning tasks, which are included in the life cycles and then no separate coordination model is needed.

With the Proclets framework [13], life cycles can be defined in separate building blocks, called Proclets, equipped with ports through which they can send and receive messages. The drawback is to address coordination with notions, such as messages and send/receive operations, which are too close to the programming domain.

8   Conclusion and Future Work

The artifact-centric approach is a promising viewpoint on business processes in that it places emphasis on the life cycles of the major business entities involved. However, it must face a number of challenges, such as the coordination of the life cycles and the handling of tasks spanning two or more life cycles.

This paper has analyzed these challenges on the basis of a new notation named Acta and with the help of two motivating examples. The main contribution of this paper is the notion of correlated transitions along with their classification in three major kinds, i.e., generative, selective and direct correlations. This notion is grounded on the distinction between master transitions and subordinate ones; the corresponding artifacts are called master artifacts and subordinate ones, respectively. Tasks spanning two or more life cycles determine a group of correlated transitions operating on one master artifact and on a number of subordinate artifacts. A spanning task is defined in the life cycle of the master artifact and its name appears in the labels of the subordinate transitions.

The artifact-centric approach has greater potential than the activity-centric one for coping with human-centric processes in which different courses of action are possible and the choices depend on human decisions. For this reason, it is essential to work out new structures for the work lists. This paper has illustrated one in which the artifacts are shown along with the options compatible with their states and the spanning tasks take advantage of a visual representation emphasizing the connections between the master artifacts and the subordinate ones.

The Acta notation is a proof of concept and current work is devoted to the definition of a suitable support environment.

Acknowledgements. The author wishes to thank the anonymous reviewers for their helpful comments.

References

1. van der Aalst, W.M.P., Weske, M., Grünbauer, D.: Case handling: a new paradigm for business process support. Data & Knowledge Engineering 53, 129–162 (2005)
2. BPMN, Business Process Model and Notation, V.2.0, http://www.omg.org/spec/BPMN/ (retrieved May 13, 2013)
3. Bruno, G.: Combining information and activities in business processes. In: Liddle, S.W., Schewe, K.-D., Tjoa, A.M., Zhou, X. (eds.) DEXA 2012, Part II. LNCS, vol. 7447, pp. 481–488. Springer, Heidelberg (2012)
4. Chao, T., et al.: Artifact-Based Transformation of IBM Global Financing. In: Dayal, U., Eder, J., Koehler, J., Reijers, H.A. (eds.) BPM 2009. LNCS, vol. 5701, pp. 261–277. Springer, Heidelberg (2009)
5. Davenport, T.H., Short, J.E.: The new industrial engineering: information technology and business process redesign. Sloan Management Review, 11–27 (Summer 1990)
6. Dumas, M., van der Aalst, W.M.P., ter Hofstede, A.H.M.: Process-Aware Information Systems: bridging people and software through process technology. Wiley, New York (2005)
7. Hull, R.: Artifact-centric business process models: Brief survey of research results and challenges. In: Meersman, R., Tari, Z. (eds.) OTM 2008, Part II. LNCS, vol. 5332, pp. 1152–1163. Springer, Heidelberg (2008)
8. Hull, R., et al.: Introducing the Guard-Stage-Milestone approach for Specifying Business Entity Lifecycles. In: Bravetti, M., Bultan, T. (eds.) WS-FM 2010. LNCS, vol. 6551, pp. 1–24. Springer, Heidelberg (2011)
9. Kucukoguz, E., Su, J.: On lifecycle constraints of artifact-centric workflows. In: Bravetti, M., Bultan, T. (eds.) WS-FM 2010. LNCS, vol. 6551, pp. 71–85. Springer, Heidelberg (2011)
10. Künzle, V., Reichert, M.: Towards object-aware process management systems: Issues, challenges, benefits. In: Halpin, T., Krogstie, J., Nurcan, S., Proper, E., Schmidt, R., Soffer, P., Ukor, R. (eds.) BPMDS 2009 and EMMSAD 2009. LNBIP, vol. 29, pp. 197–210. Springer, Heidelberg (2009)
11. Künzle, V., Reichert, M.: PHILharmonicFlows: towards a framework for object-aware process management. Journal of Software Maintenance and Evolution: Research and Practice 23(4), 205–244 (2011)
12. Küster, J.M., Ryndina, K., Gall, H.C.: Consistency of business process models and object life cycles. In: Kühne, T. (ed.) MoDELS 2006. LNCS, vol. 4364, pp. 80–90. Springer, Heidelberg (2007)
13. Mans, R.S., et al.: Proclets in healthcare. Journal of Biomedical Informatics 43(4), 632–649 (2010)
14. Müller, D., Reichert, M., Herbst, J.: Data-driven modeling and coordination of large process structures. In: Meersman, R., Tari, Z. (eds.) OTM 2007, Part I. LNCS, vol. 4803, pp. 131–149. Springer, Heidelberg (2007)
15. Nigam, A., Caswell, N.S.: Business artifacts: an approach to operational specification. IBM Systems Journal 42(3), 428–445 (2003)
16. OCL, Object Constraint Language, V.2.3.1, http://www.omg.org/spec/OCL/ (retrieved May 13, 2013)
17. Sanz, J.L.C.: Entity-centric operations modeling for business process management - A multidisciplinary review of the state-of-the-art. In: 6th IEEE International Symposium on Service Oriented System Engineering, pp. 152–163. IEEE Press, New York (2011)

Exploring Data Locality
for Clustered Enterprise Applications

Stoyan Garbatov and João Cachopo

INESC-ID Lisboa / Instituto Superior Técnico,
Universidade Técnica de Lisboa, Rua Alves Redol, 9
1000-029 Lisboa, Portugal
stoyangarbatov@gmail.com, joao.cachopo@ist.utl.pt

Abstract. Exploring data locality is crucial to achieve good performance on a distributed system. For many complex, constantly evolving applications, relying on programmers to write their code so as to explore data locality results often in sub-par performance. We propose an automatic approach for dealing with this problem. Instead of expecting programmers to identify data locality, the solution developed here relies on a stochastic analysis of the data-access patterns exhibited by the application at run-time. The analysis makes it possible to correlate not only domain data but application functionality as well. This information is used to explore data locality in clustered enterprise applications by combining two orthogonal and complementary approaches. The first approach reduces the memory foot-print by using a more compact in-memory representation for the application's domain classes and, furthermore, by delaying the loading of less frequently accessed data. The second approach generates a new request distribution policy. It employs the Latent Dirichlet Allocation partitioning algorithm, generating sub-sets of highly correlated application functionality. Every cluster node is responsible for processing requests belonging to a single sub-set. The combination of these approaches allows cluster nodes to make better use of their memory, thereby increasing the computational efficiency of the system. The work has been validated on the TPC-W benchmark, demonstrating significant performance improvements.

Keywords: heap management; in-memory object representation; persistence; clustered web servers; load balance; locality awareness; Latent Dirichlet Allocation; scalability; performance.

1   Introduction

Nowadays, it is well-known that the performance of any computational system is strongly dependent on its data locality. The "principle of locality", which was introduced by Denning [7], has two basic variants, temporal and spatial. Over short periods of time, a program distributes its memory references non-uniformly over its address space, but the portions of the address space that are favoured remain largely the same for long periods of time. Temporal locality implies that the information that will be in use in the near future is likely to be already in use. Spatial locality states

H. Decker et al. (Eds.): DEXA 2013, Part I, LNCS 8055, pp. 224–238, 2013.

that the portions of the address space that are in use consist of a small number of individually contiguous segments of that address space. As a consequence, locality of space denotes that the referenced locations of the program in the near future are likely to be near the currently referenced locations.

An application can be qualified as having good data locality if it operates in such a way that whenever it is executing a given functionality, all the relevant and necessary data for its successful completion is in an easily accessible location, with minimal presence of unnecessary data intermingled with it. While it is easy to define good locality in such a way, achieving it in practice is quite complicated. The first difficulty resides in identifying the working set of a given functionality, prior to executing it. The expression "working set" is employed here to designate the minimal set of application data without which certain functionality cannot be executed in its completeness. The non-determinism present in most systems makes this even harder. The second major difficulty is to guarantee that the necessary data is in an easily-accessible location. For a system where application data originates from a persistence layer (such as data-base) the trivial solution would be to load all existing data into memory and it would be "easily-available" for any operation. Unfortunately, having all data in memory does not automatically lead to good locality. This is because, with respect to any single operation, the large volume of unnecessary data that is interweaved between the relevant pieces of information, leads to a slow-down in the process of accessing all the material information. Good locality would only be achieved if the necessary data were placed spatially close to one another, to minimize access costs.

These deliberations lead to the following conclusion. The direction that should be followed, when attempting to achieve better locality, is to develop an automatic solution capable of analyzing and predicting with accuracy a target application's behaviour, in terms of the data accesses it performs. The solution should explore the correlation that has been identified to exist between data and functionality, when deciding upon the appropriate measures that should be taken, to improve the target's locality and, subsequently, its performance and computational efficiency.

Taking this into account, our work presents a solution focused on improving the performance and computational efficiency of clustered Java enterprise applications. The solution developed here can be applied, without any modifications (save trivial ones), to any large-scale distributed Java application where domain data originates from an external source (such as a persistence layer) that has fine granularity control over the data that is being loaded.

The solution consists in two complementary approaches for achieving the goals of this work. Both approaches make use of stochastic models for analyzing and predicting the behaviour of the target system, with regards to the data accesses it performs at run-time.

The first approach reduces the memory footprint of target applications' domain data. At compile time, it identifies the most compact and efficient in-memory representation of domain objects, while delaying, at run-time, the loading of data with low probability of being accessed.

The second approach consists in a request routing algorithm that explicitly takes into account the correlation between the working-sets of incoming requests, distributing them in a way that attempts to maximize the data locality of operations performed at server nodes. With this algorithm, server nodes are responsible for processing only a particular subset of request types: Requests are partitioned into disjoint groups, based on their working-sets. The composition of the working-sets of request types is established by using automatic data access pattern analysis and prediction routines. The Latent Dirichlet Allocation partitioning algorithm is used to maximize the correlation between the working sets of all requests placed in a particular group. This is done to improve the locality of all operations that are to be executed on a given node.

It is important to note that the present article is a continuation of the studies presented in [14] and [15]. The main contribution of the current work resides in the incorporation of the two previously developed approaches and a more extensive and comprehensive evaluation of the benefits of their usage.

The article is organized as follows. Section 2 discusses related work. Section 3 describes our new proposal in detail. Section 4 presents the benchmark that we used to evaluate the solution and discusses the results obtained. Finally, Section 5 presents some concluding remarks.

2   Related Work

Accounting for the two complementary approaches developed in this work, for the purpose of improving a clustered web server system's performance and computational efficiency, the relevant related works for each shall be discussed separately.

2.1   Memory Management

Acceptable application performance cannot be expected without proper memory management. Some research works, such as [17], [6] and [3] focus on analyzing the conditions under which application data is allocated and manipulated in the heap. Their goal is to devise ways for improving the memory management of the data.

Jones and Ryder [17] performed a study on Java objects' lifetimes. They suggest that the lifetimes of allocated instances belong to a reduced number of small ranges. Furthermore, the authors state that it is possible to form highly cohesive groups of allocation sites, based on the lifetime length of their allocated objects. They demonstrate that objects associated to identically grouped allocation sites usually only exist during certain phases of an application's execution. The authors reach the conclusion that to perform an accurate prediction of an object's lifetime distribution, its necessary to take into account an additional stack level, besides the allocation site itself.

Bhattacharya et al. [3] proposed an approach that decreases the loss in performance observed whenever there is memory bloat caused by excessive generation of short-duration objects when executing loops. They identify which objects can be re-used and, subsequently, apply a source-to-source transformation, which allows the previously determined objects to be employed in a more memory-efficient manner.

From the works considered so far, a few points can be highlighted. Overly large heaps are prone to causing serious performance degradation. Two lines of action have been identified for dealing with this issue. One seeks to keep the memory footprint small by preventing the loading of unnecessary data or by delaying its loading to the moment when it is needed. The other consists in keeping the heap compact by de-allocating no-longer-necessary data and recycling its memory. Several features, essential for the operation of both lines of action, include performing automatic analysis of the target application's behaviour and its precise prediction to allow for dynamic and adaptive identification of the most appropriate approach that should be employed.

2.2   Cluster Request Distribution

Scalability is a major property for web systems that have to deal, on a regular basis, with heavy workloads caused by large volumes of traffic and significant variations in the number of clients. This has led to extensive research performed in this domain. This section's related work discussion shall be limited to research covering request distribution for clustered web systems. Cardellini et al. [5] performed a comprehensive study of locally clustered systems, whereas Amza et al. [2] evaluated transparent scaling approaches tailored for dynamic content systems.

Pai et al. [18] introduced the concept of locality-aware request distribution (LARD). The main idea behind this concept is that for a clustered web server system to have good scalability and efficiency, the load-balancing policy should take into account the content associated with incoming requests and redirect them so that the data locality of the server node responsible for processing them is improved. A hash function is used to partition all system functionality, while the load balancer ensures that every server node only processes requests associated to a given partition. This approach decreases the working-sets of all nodes to a portion of the system working-set, allowing for improved data locality and scalability.

Zhang et al. [21] performed a lengthy simulation study of AdaptLoad. AdaptLoad is a load-balancing policy with self-adjusting capabilities that observes and takes into consideration workload variations to adjust its control parameters. When this load-balancing policy is employed, server nodes receive only requests that have similar processing times. This is done with the goal of minimizing the overall task slowdown, by executing separately requests with different sizes.

Amza et al. [1] presented a new conflict-aware scheduling technique for database back-ends of dynamic content site applications in cluster environments that provides throughput scalability and one-copy serializability. The authors employ information about the application data accessed during transactions to develop a conflict-aware scheduler that provides one-copy serializability and reduces the rate of conflicts.

Elnikety et al. [8] proposed a memory-aware load balancing method for dispatching transactions to replicas in systems with replicated databases. The algorithm uses information about the data manipulated in transactional contexts, with the aim of assigning transactions to replicas so as to guarantee that all necessary data for their execution is in memory, thereby reducing disk I/O.

Zhong et al. [22] studied the improvements that can be obtained with data placement, when accounting for the correlation within data. The authors proposed a polynomial-complexity algorithm for calculating object placement that achieves a close to optimal solution, in terms of minimizing communication costs. Further optimizations of the algorithm consist in focusing on a small set of higher importance objects.

Considering the research works discussed in this section, it is possible to draw several conclusions. Request distribution policies that emphasize data locality instead of focusing solely on load distribution appear to offer much better scalability and performance gains. Uniform load distribution is still important and should be used to complement locality aware policies, without taking precedence over them. Despite all the research accomplished in the area of locality-aware load distribution, existing solutions have (at least) one of the following shortcomings:

- While seeking to improve data locality, they do not take explicitly into account the access patterns performed while processing requests, either because they lack a proper analysis of data usage or because they expect that locality emerges "naturally" when requests are distributed among server nodes, without accounting for the data access patters performed in their contexts.
- Data access pattern analysis is performed manually.
- The partitioning of requests/functionality among server nodes is performed manually.

3   System Description

The solution presented in this paper explores the data locality subjacent in the operation of target applications for the purpose of improving their performance and computational efficiency. This is achieved through two orthogonal approaches, based on [14] and [15]. The first seeks to reduce the memory footprint of the application, whilst the second improves its data locality by accounting for the correlation between functionality and data accessed within its context. A brief description of the implementation of each of these approaches shall be presented next.

3.1   Memory Footprint Reduction

The first part of the solution presented here employs two complementary techniques for reducing the memory footprint of a Java application's domain data, while, simultaneously, improving performance when the application is placed under sub-optimal memory availability conditions. This was performed in the context of the Fénix Framework [9], which allows the development of Java applications with a transactional and persistent domain model.

One of the technique uses a carefully selected in-memory layout for instances of domain classes to achieve a compact representation with minimal memory overhead. The second technique consists in delaying the loading of domain data until the moment when it is effectively needed, as opposed to eager loading. This strategy seeks to decrease the upper limit of the effectively used heap, by avoiding the loading of domain data that is never to be accessed during the lifetime of the application.

Both techniques make informed decisions about the actions that are to be taken. These decisions are based on predictions made by stochastic models of the application's data access patterns. A comprehensive description and discussion of these models, which have been developed previously, can be found in [16, 11, 10, 12]. Their implementations have been evaluated and demonstrated to generate correct and highly precise predictions about the effectively accessed domain data throughout the execution contexts of an application. In the context of the work presented here, these methods supply the access probabilities of all domain classes and their fields.

An important concept when discussing object layouts is that of a "Box". A Box is responsible, among other things, for holding the persistent state of domain objects, as well as for loading it from persistence, whenever it is needed.

Before describing the new domain object layout DynamicL, the two previously existing schemas OneBoxPerSlotL and OneBoxPerObjectL, will be considered. For OneBoxPerSlotL, every object attribute is contained by a single Box. At runtime, when a domain object is referenced for the first time, it is initialized as a thin wrapper containing the ObjectId (a unique system-wide identification of that instance). The first time that any of an object's attributes are accessed, all Boxes of that object are initialized and their contents loaded from the persistence layer, in a single round-trip.

For the OneBoxPerObjectL layout, every domain object has all of its persistent state kept in a single Box. The behaviour, when referencing an instance or accessing any of its attributes for the first time, is identical to that of the OneBoxPerSlotL.

By analysing these layout styles, a few pertinent points were identified. If a compact object memory representation is to be achieved, then domain classes should have as few as possible Boxes to minimize the induced memory overhead. Furthermore, both approaches load significant volumes of persistent data into memory, even though most of it might not end up being necessary for the operation of the application.

The DynamicL layout was developed to account for these considerations. The approach is adaptive since the actual layout configuration is chosen at compile-time, based on the data access pattern analysis provided by the stochastic models. This makes it possible for domain data to have different layouts from one deployment to another, if a significant change in the application behaviour is detected.

DynamicL code generation proceeds as follows. For every domain class, its attributes are split into a high access probability set (HighP) and a low access probability set (LowP). For a given domain object, the code generator places HighP fields in individual Boxes, while all LowP fields are wrapped by a single Box.

Even though HighP fields are assigned individual Boxes, the overhead caused by these boxes, when compared against OneBoxPerObjectL is negligible, because, in most applications, only a fraction of data is responsible for the majority of accesses performed. It is rare to have a class with more than a few HighP fields. As such, DynamicL is close to the optimal solution, in terms of the number of Boxes per instance.

With DynamicL, when an instance is referenced for the first time, the procedure is identical to the other approaches. On the other hand, when a field that has not been loaded yet is accessed, the associated Box is initialized and all fields held by that Box are loaded from persistence. Only the field(s) within that Box are loaded, regardless of the state in which the remainder of instance Boxes might be. This measure was taken to minimize the loading of unnecessary data from persistence. It is still possible for unnecessary data to be loaded, but this is much less likely to happen.

It should be noted that a lazy version of the OneBoxPerSlotL is not a viable solution. The overhead of a single Box per field cannot be offset by delaying/avoiding the loading of low access probability data. The resulting footprint would be between that of the eager OneBoxPerSlotL and OneBoxPerObjectL.

3.2   Request Distribution

The clustering solution has been implemented in three modules – an access pattern analysis module, an optimal clustering module, and a request distribution module. The first module identifies the composition of the working-sets of all request types supported by the target system. In practice, this is achieved by employing the stochastic models referred to in the previous section.

The second module, which has been described in more detail in [13], calculates the optimal clustering of the target's functionality (request types) and domain data (working-set composition), based on the access pattern behaviour observed at runtime.

The algorithm used for partitioning the request types is Latent Dirichlet Allocation (LDA) [4]. LDA is the current state-of-the-art in multivariate clustering algorithms. By providing the request types and their respective working-sets as input to LDA, the algorithm groups requests into clusters. LDA seeks to maximize the (positive) correlation between elements placed in the same cluster and, since the request types are characterized by their working-sets, this leads to clusters being composed of request types that have very similar working-sets.

There are several control parameters (among which is the number of clusters into which the elements should be grouped) that are supplied as input to LDA. As such, LDA can only guarantee that the cluster composition maximizes the correlation between elements for the control parameters given. There is no way to know, a priori, what parameter values lead to the best results. Intuitively, good clusters have the property that their elements are (conceptually) close one another and far from elements of other clusters. The Silhouette technique [19] captures this notion and provides a numerical value of how good a particular clustering is. By calculating the average Silhouette values for the clusters generated by LDA, across all possible combinations of control parameter values, within their valid ranges, it was possible to identify the control values that lead to the highest quality results.

The request distribution module, which enforces the new distribution policy, was implemented as a software switch (Layer 7 of the OSI stack). The switch is the sole entry point through which all incoming requests pass, before being redirected to a server node. It is the only visible component of the system, from client point of view, hiding the multiplicity of application server nodes.

When a request arrives at the switch, its type is determined and then used to identify the cluster to which it belongs. Afterwards, the request is forwarded to a node that only deals with requests of that particular group. Once the request has been processed, the result is returned to the switch, and then back to the client.

Servers dealing only with a particular subset of request types allows them to have working-sets that are much smaller than the system working-set, not only because a subset of all functionality is processed there, but also because the requests were specifically selected to maximize the similarity of their working sets. This allows for a more efficient use of computational resources (e.g. smaller heaps, less garbage collection, etc) and better data locality, leading ultimately to improved system performance.

4   Results

The TPC-W benchmark [20] was used to validate the work presented here. It specifies an e-commerce workload simulating the activities of a retail store website, where emulated users can browse and order products. The main metric is the WIPS – web interactions per second that the system under test can sustain. The benchmark is characterised by a series of input parameters. One of these is the type of workload, which defines the percentage of read and write operations, that is to be simulated by the emulated browsers (EB). The workload types considered here are Type1 (95% read and 5% write); Type2 (80% read and 20% write); Type3 (50% read and 50% write).

The analysis of the system was performed with the benchmark executing in Type2 mode. The same profiling results (from Type2) were employed for all 3 workload configurations (Type1, Type2, and Type3) in the performance testing phase.

The remaining parameters are: number of EBs - 10; ramp-up - 180s; measurement - 600s; ramp-down - 60s; number of book items in the database - 1k, 10k, and 100k; think time - 0, ensuring that the EBs do not wait before making a new request.

Measurements were made with the benchmark running in two different configuration scenarios. The first configuration (Alpha) is a single machine equipped with 2x Intel Xeon E5520 (total of 8 physical cores with hyper-threading running at 2.26 GHz) and 24 GB of RAM. Alpha's operating system is Ubuntu 10.04.3. The second configuration (Beta) is a cluster of five machines with identical hardware specifications. Each machine has 2x Intel Xeon E5506 (a total of 8 physical cores running at 2.13GHz) and 2 GB of RAM. The operating system is Ubuntu 10.04.4 LTS. The average latency of any inter-node communication was measured to be 0.3ms.

For both configurations, the JVM is Java(TM) SE Runtime Environment (build 1.6.0 22-b04), Java HotSpot(TM) 64-Bit ServerVM (build 17.1-b03, mixed mode). The benchmark server replicas, as well as the load-balancer, were run on top of Apache Tomcat 6.0.24, with the options "-Xshare:off -Xms64m -Xmx$(heapSize)m -server -XX:+UseConcMarkSweepGC -XX:+AggressiveOpts".

4.1   Single Machine Configuration (Alpha)

For the single machine configuration, the client (emulated browser generator), the load-balancer and all server replicas were run on the same physical machine. The load-balancer and all replicas were deployed on individual Tomcat instances.

While evaluating the request distribution part of this work on a single physical machine might seem somehow forced, there were two reasons for doing so. Having all server nodes on the same physical machine simulates a scenario where the system is heavily loaded, making it possible to appreciate the effects of less-than-optimal resource availability (CPU and RAM). This aspect was expanded by having the Tomcat instances where the benchmark replicas were deployed operate under five different maximum heap size configurations - 512MB, 640MB, 768MB, 896MB and 1024MB. Furthermore, with the emergence of cloud-computing, it is no longer that much of a stretch to imagine several virtual servers running on top of a single physical machine, depending on the resource availability and allocation.

It should be noted that, for Alpha, a total of 540 distinct benchmark and server configurations were evaluated (2 layout policies, 3 request distribution policies, 2 cluster sizes, 5 max heap sizes, 3 workload types and 3 data-base sizes). Additionally, every single configuration was executed 4 times, independently of previous runs, to provide a more representative view of the behaviour of the system under test. Taking all this into account and the fact that a single execution of the benchmark takes approximately 15min (14min benchmark execution and 1min for Tomcat reboot, benchmark redeploy and database refresh), the results for Alpha took 540h to generate.

The two layout policies under which the benchmark performance was evaluated are the DynamicL and the OneBoxPerObjectL. OneBoxPerSlotL results are omitted because they are worse than the ones provided by any of the other two alternatives.

The first of the three request distribution policies to be evaluated is the one discussed in Section 3.2, which shall be called the LDA policy. The second policy corresponds to an idealized locality-aware request distribution (LARD), where each node is responsible for processing a subset of request types, and there are no intersections among the sets assigned to different server nodes. This policy is idealized because it assumes prior knowledge of the composition of the workload, in terms of the relative proportions of incoming request types, as well as the average time that requests of a given type take to be processed. The policy attempts to achieve the most uniform load distribution possible, whilst keeping every server node dedicated to processing a fixed subset of request types. The third policy employs a classic (non-weighted) round-robin (RR) approach for distributing incoming requests among existing server nodes.

The different solutions are evaluated for 3 and 4 server replicas. Only these server replication values were considered because it has been demonstrated in [13] that the optimal number of unique and non-intersecting sub-sets of functionality into which the TPC-W's request types can be subdivided into, resides in between 3 and 4.

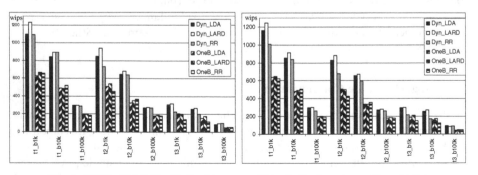

Fig. 1. WIPS, 1024MB, Alpha, 3 server replicas (left), 4 server replicas (right)

The WIPS achieved through the 6 possible combinations of layout and request distribution policies, when the servers are allocated 1024MB of heap, for 3 and 4 server replicas, can be seen in Fig. 1 left and right, respectively. Each group of bars corresponds to a particular benchmark configuration, where t1, t2, and t3 indicate workload type, and b1k, b10k, and b100k indicate the size of the database used.

By analyzing the results, it can be said that all configurations with the DynamicL layout display significantly better throughput than their OneBoxL counterparts. In terms of the request distribution policies, the RoundRobin policy is rather consistently outperformed by the other two policies. The LDA and LARD policies display similar throughput, even though, for most configurations, the LARD version is slightly better.

The relative throughput gains obtained when comparing DynamicL_LDA against the remaining 5 policy configurations can be seen in Fig. 2 left and right for 3 and 4 server node configurations, respectively. The relative gain has been calculated as $\left(\left(T_{Dyn\_LDA} - T_{OLP\_RDP}\right)/T_{OLP\_RDP}\right) \times 100,\%$ where T stands for throughput and OLP and RDP indicate the object layout and request distribution policies against which the Dyn_LDA is being compared. The average throughput grains are as follows: Dyn_LARD -4.89% (3), -4.68(4); Dyn_RR 16.66%(3), 18.49(4); OneB_LDA 61.66%(3), 65.01%(4); OneB_LARD 60.72%(3), 61.67%(4); OneB_RR 82.29%(3), 81.31(4).

If we consider the effect over throughput variation, when one of the policies is changed, while keeping the other fixed (e.g. Dyn_LARD to OneB_LARD, or OneB_LDA to OneB_RR), the results are as follows. The average throughput gain of Dyn vs OneB is 62.52%; LDA vs LARD is -3.07% and LDA vs RR is 14.96%.

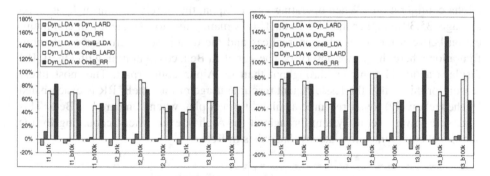

Fig. 2. Throughput gain, Dyn_LDA vs All, Alpha, 3 servers (left), 4 servers (right)

Based on these results, several remarks can be made. Employing the DynamicL object representation policy offers significant and unambiguous performance gains, independently of the other configuration parameters. In terms of request distribution policy, the LDA clearly offers better throughput than RoundRobin, unfortunately it falls slightly behind the throughput that LARD displays. Nevertheless, if we account for the fact that this particular LARD implementation is an idealized approach that makes use of perfect prior knowledge, about the target system's behaviour, then it is quite positive that the performance offered by the LDA policy is so similar to an approach that would be otherwise impossible to achieve in practice.

4.2   Cluster Configuration (Beta)

For the cluster configuration, the client and load-balancer were deployed on the same node, while having a separate node for each of the 3/4 server replicas. The data-base was operating on yet another node, different from the ones listed so far.

For Beta, 108 distinct benchmark and server configurations were evaluated (2 layout policies, 3 request distribution policies, 2 cluster sizes, 1 heap size - 1024MB, 3 workload types and 3 data-base sizes). Every configuration was evaluated 2 times, independently of previous executions, leading to a total of 54h execution time.

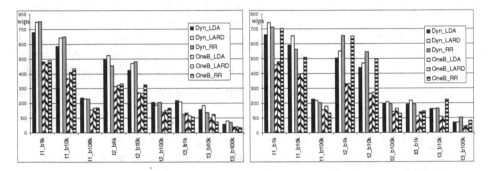

Fig. 3. WIPS, 1024MB, Beta, 3 server replicas (left), 4 server replicas (right)

The average throughput achieved for Beta (see Fig. 3) is lower than the one for the Alpha configurations. When operating under Alpha, the benchmark throughput is, on average, 35.39% higher than for Beta. This is mainly due to the latency in communication between the load-balancer, replicas and the data-base. There is a single configuration where this is not observed. The 4 replica Beta configuration with OneB_RR displays about 21% higher throughput than its Alpha counterpart. The most likely reason for this is that the less efficient resource usage of the OneB_RR is exacerbated by the (relatively) lower resource availability in Alpha, when compared to Beta. This leads to a lower overall system throughput (in Alpha), even when accounting for the additional communication latency that exists in Beta.

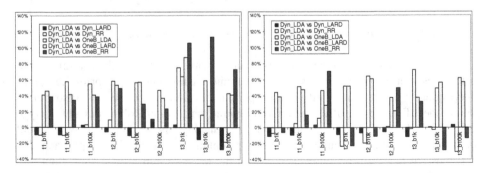

Fig. 4. Throughput gain, Dyn_LDA vs All, Beta, 3 servers (left), 4 servers (right)

A comparison of Dyn_LDA against the other 5 policy configurations can be seen in Fig. 4. The average throughput variations are the following: Dyn_LARD -6.55% (3), -4.87% (4); Dyn_RR 6.06% (3), -7.18% (4); OneB_LDA 53.53% (3), 53.58% (4); OneB_LARD 48.01% (3), 44.49% (4); OneB_RR 56.36% (3), 9.83% (4).

The results obtained when only the object layout or the request distribution policy is changed, while the other remains fixed, are considered next. Using Dyn instead of OneB leads to an average of 47.91% increase in throughput. The average throughput of LDA is 5.21% lower than LARD and 6.80% lower than RR.

Some considerations about the results observed thus far are necessary. The benefits of using the DynamicL policy for Beta (47.91%) are significant, yet, they are nevertheless lower than the ones achieved for Alpha (66.06%). The relation between the LDA request policy and LARD remains mostly the same (-4.23% for Alpha and -5.21% for Beta), however the ratio between LDA and RR changed significantly (from 9.71% for Alpha to -6.80% for Beta). Furthermore, it should be noted that these variations are not so much due to a decrease in the DynamicL and LDA throughput, when moving from Alpha to Beta, so much as due to an increase in the throughput when the benchmark operates with OneBoxPerObject and RR.

This phenomena is explainable with the fact that both DynamicL and LDA focus on improving the locality of data with which the target application operates. This makes it possible for the application to operate longer without performance handicaps when operating with limited computational resources (e.g. less CPU time and free memory). On the other hand, the less efficient resource-management of OneBoxPerObjectL and RR lead any system that is employing them to require higher resource availability levels to operate without suffering sub-optimal performance.

All this leads to the conclusion that any system capable of employing the dual-approach solution (DynamicL + LDA request distribution) presented with this work will be able to remain operating at normal performance levels for longer, when faced with increasingly restricted availability of resources.

4.3   Further Discussion

Even though it has been repeatedly stated that the DynamicL approach reduces the memory footprint of its target application, as well as minimizing the garbage-collection activity, no direct evidence has been given so far to back these statements. The percentage of CPU usage, as well as the intensity of GC activity (as a function of % of CPU employed for it) for a particular TPC-W configuration (1024MB max heap, type 2 workload, 100k books in data-base) can be observed in Fig. 5 (left) for the OneBoxPerObjectL, and in Fig. 5 (right) for the DynamicL. It should be noted that these results were obtained with the benchmark operating with a single instance (no server replication, nor load-balancing). As can be seen from these results (and without pretending to be exhaustive in their coverage), when the TPC-W is operating with the OneBoxPerObjectL, the GC activity remains steady in the 20% range of CPU occupation, while for the DynamicL approach, there are only a few short bursts of GC, during a corresponding period of time (with identical length).

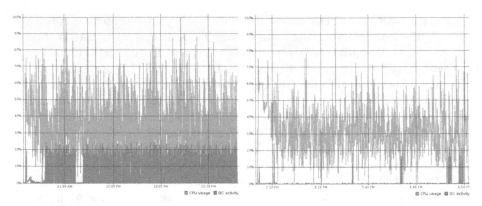

Fig. 5. CPU/CG activity, 1024MB, t2, b100k, OneBoxPerObjectL (left) DynamicL (right)

Last but not least, Fig. 6 presents the average size of the effectively used server heaps for a particular TPC-W configuration, when operated with 4 replicas, on Alpha. The clouds of dots reflect the average heap size of the four replicas (without considering the load-balancer). On the average, the LDA heap is 5.30% bigger than the LARD heap and 33.34% smaller than the RR. If linear regression is applied on the clouds-of-dots, it is possible to conclude that, at least for this particular configuration, the heap growth rate for the LDA approach is 45.52% lower than the growth rate exhibited by the LARD, and 67.02% lower than the RR heap growth rate.

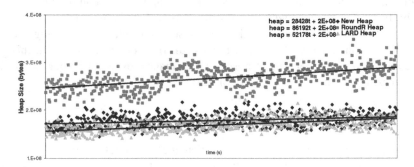

Fig. 6. Heap size at run-time, 4 serv, 640MB, t3, b10k, Alpha

5   Conclusions

The article presented here proposed an automatic solution for improving the efficiency and performance of enterprise Java distributed-system applications, by exploring their data-locality. The solution incorporates two previously developed approaches (an object representation layout policy named DynamicL and a request distribution policy based on Latent Dirichlet Allocation), expanding them, thus, in their applicability. The system has been validated on the TPC-W benchmark, in two distinct hardware configurations.

The results demonstrate that the usage of the solution developed here allows for significant improvements. These cover an increase in the efficiency of the usage of computational resources, which allows the target system to remain operating with optimal performance even when subjected to increasingly restrictive resource availability constraints.

Furthermore, when compared to previously existing solutions, the newly developed one demonstrates expressive performance gains. On a single heavily loaded machine, the individual approaches that compose the solution are capable of offering up to 9.71% (LDA request distribution) and 66.06% (DynamicL object layout) increase in throughput respectively. However, the integrated solution allows for an increase in throughput of up to 77.56%, which is actually better than the sum of its parts.

When evaluating performance in a clustered environment, the individually applied approaches can provide up to 6.80% and 47.91%, whereas the integrated solution displays gains of up to 56.36%.

Based on all the results, it is possible to observe that the integrated solution provides a magnifying effect over the performance achieved by the two approaches that it employs.

Acknowledgments. This work was partially supported by national funds through FCT – Fundação para a Ciência e a Tecnologia, under project PEst-OE/EEI/LA0021/2013, as well as by FCT (INESC-ID multiannual funding) PIDDAC Program funds and by the Specific Targeted Research Project (STReP) Cloud-TM, which is co-financed by the European Commission through the contract no. 257784. The first author has been funded by the Portuguese FCT under contract SFRH/BD/64379/2009.

References

1. Amza, C., Cox, A.L., Zwaenepoel, W.: Conflict-aware scheduling for dynamic content applications. In: Proceedings of the 4th conference on USENIX Symposium on Internet Technologies and Systems, vol. 4, pp. 6–20. USENIX Association (2003)
2. Amza, C., Cox, A.L., Zwaenepoel, W.: A comparative evaluation of transparent scaling techniques for dynamic content servers. In: Proceedings of the 21st International Conference on Data Engineering (ICDE 2005), pp. 230–241. IEEE (2005)
3. Bhattacharya, S., Nanda, M.G., Gopinath, K., Gupta, M.: Reuse, Recycle to De-bloat Software. In: Mezini, M. (ed.) ECOOP 2011. LNCS, vol. 6813, pp. 408–432. Springer, Heidelberg (2011)
4. Blei, D., Ng, A., Jordan, M.: Latent dirichlet allocation. Journal of Machine Learning Research 3, 993–1022 (2003)
5. Cardellini, V., Casalicchio, E., Colajanni, M., Yu, P.: The state of the art in locally distributed Web-server systems. ACM Computing Surveys (CSUR) 34(2), 263–311 (2002)
6. Chis, A.E., Mitchell, N., Schonberg, E., Sevitsky, G., O'Sullivan, P., Parsons, T., Murphy, J.: Patterns of Memory Inefficiency. In: Mezini, M. (ed.) ECOOP 2011. LNCS, vol. 6813, pp. 383–407. Springer, Heidelberg (2011)
7. Denning, P.J., Schwartz, S.C.: Properties of the working-set model. Communications of the ACM 15(3), 191–198 (1972)

8. Elnikety, S., Dropsho, S., Zwaenepoel, W.: Tashkent+: Memory-aware load balancing and update filtering in replicated databases. ACM SIGOPS Operating Systems Review 41(3), 399–412 (2007)

9. Fernandes, S., Cachopo, J.: Strict serializability is harmless: a new architecture for enterprise applications. In: Proceedings of the ACM International Conference on Object-Oriented Programming Systems, Languages and Applications, Portland, Oregon, USA, pp. 257–276. ACM (2011)

10. Garbatov, S., Cachopo, J.: Importance Analysis for Predicting Data Access Behaviour in Object-Oriented Applications. Journal of Computer Science and Technologies 14(1), 37–43 (2010)

11. Garbatov, S., Cachopo, J.: Predicting Data Access Patterns in Object-Oriented Applications Based on Markov Chains. In: Proceedings of the Fifth International Conference on Software Engineering Advances (ICSEA 2010), Nice, France, pp. 465–470 (2010)

12. Garbatov, S., Cachopo, J.: Data Access Pattern Analysis and Prediction for Object-Oriented Applications. INFOCOMP Journal of Computer Science 10(4), 1–14 (2011)

13. Garbatov, S., Cachopo, J.: Optimal Functionality and Domain Data Clustering based on Latent Dirichlet Allocation. In: Proceedings of the Sixth International Conference on Software Engineering Advances (ICSEA 2011), Barcelona, Spain, pp. 245–250. ThinkMind (2011)

14. Garbatov, S., Cachopo, J.: Decreasing Memory Footprints for Better Enterprise Java Application Performance. In: Liddle, S.W., Schewe, K.-D., Tjoa, A.M., Zhou, X. (eds.) DEXA 2012, Part I. LNCS, vol. 7446, pp. 430–437. Springer, Heidelberg (2012)

15. Garbatov, S., Cachopo, J.: Explicit use of working-set correlation for load-balancing in clustered web servers. In: Proceedings of the Seventh International Conference on Software Engineering Advances (ICSEA 2012), Lisbon, Portugal (2012) (in print)

16. Garbatov, S., Cachopo, J., Pereira, J.: Data Access Pattern Analysis based on Bayesian Updating. In: Proceedings of the First Symposium of Informatics (INForum 2009), Lisbon, Paper 23 (2009)

17. Jones, R.E., Ryder, C.: A study of Java object demographics. In: Proceedings of the 7th International Symposium on Memory Management, Tucson, AZ, USA, pp. 121–130. ACM (2008)

18. Pai, V., Aron, M., Banga, G., Svendsen, M., Druschel, P., Zwaenepoel, W., Nahum, E.: Locality-aware request distribution in cluster-based network servers. In: Proceedings of the Eighth International Conference on Architectural Support for Programming Languages and Operating Systems, San Jose, California, United States, pp. 205–216. ACM (1998)

19. Rousseeuw, P.J.: Silhouettes: a graphical aid to the interpretation and validation of cluster analysis. Journal of Computational and Applied Mathematics 20, 53–65 (1987)

20. Smith, W.: TPC-W: Benchmarking An Ecommerce Solution. Intel Corporation (2000)

21. Zhang, Q., Riska, A., Sun, W., Smirni, E., Ciardo, G.: Workload-aware load balancing for clustered web servers. IEEE Transactions on Parallel and Distributed Systems 16(3), 219–233 (2005)

22. Zhong, M., Shen, K., Seiferas, J.: Correlation-Aware Object Placement for Multi-Object Operations. In: Proceedings of the 2008 the 28th International Conference on Distributed Computing Systems, pp. 512–521. IEEE Computer Society (2008)

A Framework for Data-Driven Workflow Management: Modeling, Verification and Execution

Nahla Haddar, Mohamed Tmar, and Faiez Gargouri

University of Sfax, B.P. 1069, 3029 Sfax, Tunisia
nhaddar@ymail.com, {mohamed.tmar,faiez.gargouri}@isimsf.rnu.tn

Abstract. In recent years, many data-driven workflow modeling approaches has been developed, but none of them can insure data integration, process verification and automatic data-driven execution in a comprehensive way. Based on these needs, we introduced, in previous works, a data-driven approach for workflow modeling and execution. In this paper, we extend our approach to ensure a correct definition and execution of our workflow model, and we implement this extension in our Framework *Opus*.

Keywords: Data-driven workflow management Framework, Petri nets, Relational algebra, Workflow analysis and verification, Soundness property.

1 Introduction

In a competitive environment continually evolving, companies are recognized the need to manage their business processes in order to align their information systems, more and more quickly, in a process-oriented way. In this context, workflow management systems (WMS) offer promising perspectives for modeling, processing and controlling processes. In the most common WMS, only the control flow [1] is completely included [1]. In Fact, during process execution, a process-oriented view (e.g. worklists) is provided to end-users. However, the behavior of an activity during its execution is out of the control of the WMS [2]. As almost all processes are related to data, such as the costs of the ordered products, the addresses information for delivery, etc., the main goal from using a WMS is to automate, as possible, the manipulation of data in business processes, and to restrict as possible the manual tasks performed by human actors.

Regarding the existing literature, the need for modeling processes that combine data and control flow has been widely studied. Most of them are inspired from Petri nets (P-nets) formalism, such as the approaches proposed in [3–7]. Thus, to enhance earlier approaches that have mainly focused on process activities and largely overlooked the data, we previously extended, in [8,9], the P-nets

---

[1] Control flow: is a set of synchronized activities representing the business process functions, and a set of ordering constraints defining their execution sequence [1].

H. Decker et al. (Eds.): DEXA 2013, Part I, LNCS 8055, pp. 239–253, 2013.

formalism by data operations inspired from the relational algebra to model data-driven workflows. This extension improves the generation of business functions from the process definition, without the need for a programmer, and provides advanced abilities for the information system (IS) and the users' integration. In this paper, we extend the modeling method of the proposed approach [8, 9] by some rules, to ensure the consistency of the process model during runtime. We also provide a technique to verify a released notion of the classical soundness property.

The remainder of the paper is organized as follows: we present the related work in Sect. 2 and we continue by introducing our workflow model in Sect. 3. In Sect. 4, we present the definition of the firing rules that ensure a uniform execution of all the process activities. Then we discuss, in Sect. 5, the technique provided to ensure the analysis and the verification of our workflow model. In Sect. 6, we present our Framework *Opus*. Section 7 concludes.

2   Related Work

The need for modeling processes integrating data has been recognized by several authors [3–7, 10, 11]. The Case Handling Paradigm [11] aims to coordinate activities which are presented as forms in relation to atomic data elements. The problem here is that data may be omitted or activities are unwittingly ignored or executed many times. In addition, more than one user can handle the same case simultaneously, which damages the data coherence.

The PHILharmonicFlows system [2] provides a comprehensive approach that combines object behavior based on states with data-driven process execution. In fact, it allows the control of activities by presenting them as form-based [2] and black-box activities [3]. The proper execution as well as termination of processes at runtime is further ensured by a set of correctness rules [12]. But, the execution of form-based activities increases the rate of errors that may be caused by the manual seizure performed by human actors, even if the seized data values respect the data types requested by the form' input fields.

Many extensions of P-nets in which tokens carry data have been defined in the literature. The workflow nets based on colored P-nets (WFCP-nets) [10] consider a P-net color as an abstraction of data objects and flow control variants. The execution of a WFCP-net depends on the interpretation of its arc expressions and guard expressions, which describe the business rules. Besides, the verification methods of workflow nets [13] are adapted to WFCP-nets. The weakness of this approach is that the process specification consists of a graphical part and a WF script part. The latter is a hard-coding process logic that describes the data elements and the behavior of activities. So, resulting applications are both complex to design and costly to deploy, and even simple process changes require

---

[2] Form-based activities: provide input fields for writing and data fields for reading selected attributes of data object instances [2].

[3] Black-box activities: allow the integration of advanced functionalities (e.g. sending e-mails) [2].

costly code adaptations and testing efforts. Another extension of P-nets is the workflow nets with data (WFD-nets) [7], in which transitions can read from or write to some data elements. This extension does not provide a support for executing process models. However, it defines algorithms to verify a soundness property that guarantees the proper termination of a WFD-net and that only certain transitions are not dead.

None of the existent approaches considers data integration, process verification and data-driven execution issues in a comprehensive way. Thus, in this area a comprehensive approach for supporting these three issues is still missing.

3   Our Data-Driven Workflow Model

As described by the most modeling approaches, a process is defined, in a higher level of abstraction, as a set of synchronized activities performed by roles according to the available data. If we stop at this level, we will not be able to generate the process functions from the process model definition, and activities will behave as a black box in which data are managed by invoked application components. To attempt the lowest level of abstraction, we propose to split each activity, in a process, to tasks applied on data. Each task consumes data to produce others. Thus, each activity is presented as a set of data-driven tasks. Each task consumed data provenance can be either the IS, or data produced by other tasks. But, in some cases, to complete the processing of a task, data can be seized by a role. Accordingly, to enable tasks to generate new data from old ones and import data from the IS, data have to be well structured. We introduced this approach in [8, 9], in a formal way, as a data-driven modeling approach based on combination between structured tokens P-nets and relational algebra.

3.1   Data Structure

According to [8, 9], we define each handled data as a data structure; i.e, a pair $s = (C, D)$, where C is a list of attributes and D is a list of data tuples. Each tuple is an ordered list of attributes values, formally defined by:

$$C = (c_1, c_2 \ldots c_n), \ D = \{(d_{1_1}, d_{1_2} \ldots d_{1_n}), (d_{2_1}, d_{2_2} \ldots d_{2_n}) \ldots (d_{m_1}, d_{m_2} \ldots d_{m_n})\}.$$

Where n (resp. m) is the number of attributes (resp. tuples) in s.

Each attribute $c_j = (\alpha_j, \beta_j)$ is a pair characterized by an attribute identifier α_j and an attribute type β_j, such as: $\forall j \in \{1, 2 \ldots n\}$, $\beta_j \in \{Int, Float, Char, String, Date, Boolean \ldots\}$, and $\forall i \in \{1, 2 \ldots m\}$, an attribute value d_{i_j} is a specific valid value for the type β_j of the attribute c_j.

At modeling step, the designer has just to define the data structure attributes and the values types put up with each one. At runtime, the different data structure tuples comprise varying values according to each attribute type.

3.2   Process Structure

The workflow process is defined as a P-net representing the work, where a place corresponds to a *data structure* that contains *structured tokens* (tuples) and a transition corresponds to a *task*. A workflow is then a quadruplet [8, 9] $WF = (S, T, Pre, Post)$, where:

$S = \{s_1, s_2 \ldots s_{|S|}\}$ is a finite set of data structures,

$T = \{t_1, t_2 \ldots t_{|T|}\}$ is a finite set of tasks inspired from the relational algebra,

$Pre : S \times T \to \mathbb{N}$ is the pre-incidence matrix, such as, $\forall i \in \{1, 2 \ldots |S|\}$ and $j \in \{1, 2 \ldots |T|\}$, $Pre(s_i, t_j)$, is the edge between a data structure s_i and a task t_j weight, representing the number of tokens consumed by t_j in order to be firable, i.e. executable.

$Post : T \times S \to \mathbb{N}$ is the post-incidence matrix. Due to the dynamic of the relational algebra, we cannot be limited to a static post-incidence matrix thus, $\forall i \in \{1, 2 \ldots |S|\}$ and $j \in \{1, 2 \ldots |T|\}$, $Post(t_j, s_i) \in [Post_{Min}(t_j, s_i), Post_{Max}(t_j, s_i)]$. Where: $Post_{Min}(t_j, s_i)$ (resp. $Post_{Max}(t_j, s_i)$), is the edge between a task t_j and a data structure s_i minimal (resp. maximal) weight, representing the minimal (resp. maximal) number of t_j produced tokens.

3.3   Data Operations

A task can be viewed as a data operation applied on data structure tokens to produce others. Therefore, we have inspired from the relational algebra to define the behavior of operations. We presented these operations in [9]. So, in this paper, we detail in Appendix A, only operations that we will use to demonstrate the new extensions of the model. Noting that the definition of the *Add_Tuples* operation, presented in Table 1, is an extended version of its definition in [9]. In fact, in this version, we allow to inserts all a data structure tuples in another data structure, instead of inserting only a single tuple [9].

3.4   Workflow Example

The customer solvency check role (SCRole) evaluates the received orders, and sends them to the inventory check. After the evaluation, either an order is rejected, or sent to shipping and billing. As illustrated by Fig. 1, considering that s_8, s_{13}, and s_{19} present tables from the IS, we restrict our example to the inventory check role (ICRole) sub-process, which performs the following activities:

Select the ordered products: t_7 extends s_8 (which contains all the products data) by the *ord_qtity* attribute, in order to allow the ICRole to enter the ordered quantities. Then, t_8 selects from s_9 only tuples having an ordered quantity value higher than zero and lower than the stocked product quantity.

Verify the products availability: t_9 checks s_{10} content. If it contains one or more tokens, t_9 will reproduce s_{11} token in s_{12}, otherwise, it will end the process.

Create a new order: according to the decision of t_9, if there are available products, t_{10} will add a new order in s_{13}.

Create the new order lines: the ICRole enters the new order identifier and t_{13} saves it in s_{17}. Then, t_{14} will create the new order lines and finally, t_{15} will save the resulting structured s_{18} tokens in s_{19}.

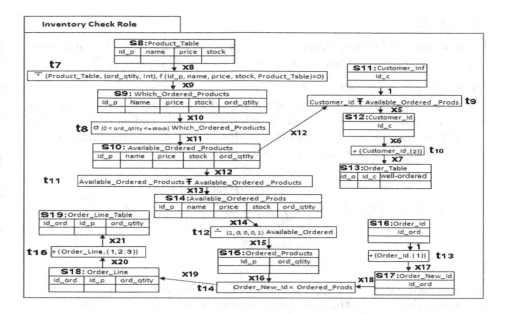

Fig. 1. The inventory check role sub-process [9] (modified version)

We deduce the *Pre* and *Post* matrix of the example in Fig. 1.

$$
Pre = \begin{array}{c} \\ s_8 \\ s_9 \\ s_{10} \\ s_{11} \\ s_{12} \\ s_{13} \\ s_{14} \\ s_{15} \\ s_{16} \\ s_{17} \\ s_{18} \\ s_{19} \end{array}
\begin{array}{ccccccccc}
t_7 & t_8 & t_9 & t_{10} & t_{11} & t_{12} & t_{13} & t_{14} & t_{15} \\
x_8 & 0 & 0 & 0 & 0 & 0 & 0 & 0 & 0 \\
0 & x_{10} & 0 & 0 & 0 & 0 & 0 & 0 & 0 \\
0 & 0 & x_{12} & 0 & x_{12} & 0 & 0 & 0 & 0 \\
0 & 0 & 1 & 0 & 0 & 0 & 0 & 0 & 0 \\
0 & 0 & 0 & x_6 & 0 & 0 & 0 & 0 & 0 \\
0 & 0 & 0 & 0 & 0 & 0 & 0 & 0 & 0 \\
0 & 0 & 0 & 0 & 0 & x_{14} & 0 & 0 & 0 \\
0 & 0 & 0 & 0 & 0 & 0 & 0 & x_{16} & 0 \\
0 & 0 & 0 & 0 & 0 & 0 & 1 & 0 & 0 \\
0 & 0 & 0 & 0 & 0 & 0 & 0 & x_{18} & 0 \\
0 & 0 & 0 & 0 & 0 & 0 & 0 & 0 & x_{20} \\
0 & 0 & 0 & 0 & 0 & 0 & 0 & 0 & 0
\end{array}
$$

Following the definition of tasks in Appendix A, the number of tokens produced by each task in Fig. 1, are defined as follows: $x_5 \in [0, 1]$, $x_6 = 1$ (i.e. there is only a single customer identifier), $x_7 = |D_{13}| + x_6$, $x_9 = x_8$, $x_{11} \in [0, x_{10}]$, $x_{13} \in [0, x_{12}]$, $x_{15} \in [0, x_{14}]$, $x_{17} = |D_{17}|+1 = 1$ (because $D_{17} = \emptyset$), $x_{19} = x_{16} \times x_{18}$, $x_{21} = |D_{19}| + x_{20}$. Accordingly, the *Post* matrix is defined by:

$$
Post = \begin{array}{c} \\ s_8 \\ s_9 \\ s_{10} \\ s_{11} \\ s_{12} \\ s_{13} \\ s_{14} \\ s_{15} \\ s_{16} \\ s_{17} \\ s_{18} \\ s_{19} \end{array}
\begin{array}{ccccccccc}
t_7 & t_8 & t_9 & t_{10} & t_{11} & t_{12} & t_{13} & t_{14} & t_{15} \\
0 & 0 & 0 & 0 & 0 & 0 & 0 & 0 & 0 \\
x_8 & 0 & 0 & 0 & 0 & 0 & 0 & 0 & 0 \\
0 & [0, x_{10}] & 0 & 0 & 0 & 0 & 0 & 0 & 0 \\
0 & 0 & 0 & 0 & 0 & 0 & 0 & 0 & 0 \\
0 & 0 & [0, 1] & 0 & 0 & 0 & 0 & 0 & 0 \\
0 & 0 & 0 & |D_{13}| + x_6 & 0 & 0 & 0 & 0 & 0 \\
0 & 0 & 0 & 0 & [0, x_{12}] & 0 & 0 & 0 & 0 \\
0 & 0 & 0 & 0 & 0 & [0, x_{14}] & 0 & 0 & 0 \\
0 & 0 & 0 & 0 & 0 & 0 & 0 & 0 & 0 \\
0 & 0 & 0 & 0 & 0 & 0 & 1 & 0 & 0 \\
0 & 0 & 0 & 0 & 0 & 0 & 0 & x_{16} \times x_{18} & 0 \\
0 & 0 & 0 & 0 & 0 & 0 & 0 & 0 & |D_{19}| + x_{20}
\end{array}
$$

3.5   Marking

The marking $M^{[\theta]}$ defines the state of the process described by WF at a given time $\theta \in \{0, 1, 2\ldots\}$. Thus, $\forall i \in \{1, 2\ldots|S|\} : M^{[\theta]} = \left(m_1^{[\theta]}\ m_2^{[\theta]}\ \ldots\ m_{|S|}^{[\theta]}\right)$, where $m_i^{[\theta]} \in \mathbb{N}$ is the number of tuples in s_i.

The initial marking $M^{[0]}$ defines the state of WF at time $\theta = 0$, in which only root nodes of a WF process can be initiated by a finite number of tokens. The evolution of the markings, in the other nodes, results due to the firing of WF tasks. A valid initial marking must follow (1) [8,9].

$$\forall j \in \{1,2\ldots|S|\}, m_j^{[0]} = \begin{cases} \max\limits_{k \in \{1,\ 2\ldots|T|\}} Pre(s_j,\ t_k) \\ if\ \forall l \in \{1,2\ldots|T|\} Post_{Max}(s_j,\ t_l) = 0, \\ 0\ \text{otherwise}. \end{cases} \quad (1)$$

3.6   Synthesis

Our proposed data-driven approach allows for a comprehensive integration between data flow and control flow, which ensures a successful data driven execution of the workflow. Indeed, data integration is granted through data structures that can handle various data elements types. Furthermore, data manipulation is enable through data operations that can read, write and generate new data elements without any risk of simulating a WF process in which data can be lost. This is granted through the dynamic behavior of the relational data operations, which entails a generalization of the static post-incidence matrix of the classical P-nets. In the next section, we present how we improve our approach by the application of some firing rules. These latter grant a uniform definition of a WF process and introduce the basic notions of our verification method, that ensures a valid WF process execution.

4   Firing Rules

To ensure the process consistency during runtime, we improve the modeling approach described in [8, 9] by adding some firing rules. The latter indicate under which conditions a task may fire, and what the effect of the firing on the marking is.

1. Assuming that t_i, $t_j \in T$, are two successive tasks in a WF, and $s \in S$ is an output data structure of t_i and an input data structure to t_j. Thus, tokens produced by t_i will be automatically consumed by t_j:

$$\forall t_i,\ t_j \in T, \exists\ s \in S\ |\ <t_i,\ t_j> \Rightarrow\ pre(t_j,\ s) = post(t_i,\ s)\ . \quad (2)$$

2. Assuming that δ is the function calculating the possible markings resulting of the firing of a task $t_i \in T$ from a marking M. So, $\forall M \in \mathbb{N}^{|S|}$, $t_1, t_2\ldots t_k \in T$:

$$\delta(\{M_1,\ M_2\ldots M_n\},\ t_1\ t_2\ldots\ t_k) = \bigcup\limits_{M \in \{M_1,\ M_2\ldots M_n\}} \delta(\{M\},\ t_1\ t_2\ldots\ t_k) \quad (3)$$

Where: $\delta(\{M\}, t_1\, t_2 \ldots t_k) = \delta(\delta(\{M\}, t_1), t_2 \ldots t_k)$
$$= \delta(\delta(\delta(\{M\}, t_1), t_2), t_3 \ldots t_k)$$
$$= \ldots$$

The function $\delta(\{M\}, t_i)$ is defined as follows:

$$\delta(\{M\}, t_i) = \begin{cases} \varnothing \text{ if } M < Pre(., t_i), \\ \{M - Pre(., t_i) + x\} \text{ otherwise.} \end{cases} \qquad (4)$$

Where $x \in [Post_{Min}(., t_i), Post_{Max}(., t_i)]$, and $\delta(\varnothing, t_i) = \varnothing$.

We apply (3) and (4) on the example illustrated in Fig. 1, and we calculate the possible markings resulting from the firing sequence of tasks $< t_7\, t_8\, t_{11}\, t_{12}\, t_{13}\, t_{14}\, t_{15} >$:

$\{M^{[0]}\} = \{(x_8\ 0\ 0\ 0\ 0\ 0\ 0\ 0\ 0\ 0\ 0), (x_8\ 0\ 0\ 1\ 0\ 0\ 0\ 0\ 0\ 0\ 0), (x_8\ 0\ 0\ 0\ 0\ 0\ 0\ 0\ 1\ 0\ 0\ 0),$
$\qquad (x_8\ 0\ 0\ 1\ 0\ 0\ 0\ 0\ 1\ 0\ 0\ 0), (0\ 0\ 0\ 1\ 0\ 0\ 0\ 0\ 0\ 0\ 0\ 0), (0\ 0\ 0\ 1\ 0\ 0\ 0\ 0\ 1\ 0\ 0\ 0),$
$\qquad (0\ 0\ 0\ 0\ 0\ 0\ 0\ 0\ 1\ 0\ 0\ 0), (0\ 0\ 0\ 0\ 0\ 0\ 0\ 0\ 0\ 0\ 0\ 0)\}$

$\delta(\{M^{[0]}\}, t_7\, t_8\, t_{11}\, t_{12}\, t_{13}\, t_{14}\, t_{15}) = \delta(\delta(\{M^{[0]}\}, t_7), t_8\, t_{11}\, t_{12}\, t_{13}\, t_{14}\, t_{15})$

$= \delta(\{\{(0\ x_8\ 0\ 0\ 0\ 0\ 0\ 0\ 0\ 0\ 0), (0\ x_8\ 0\ 1\ 0\ 0\ 0\ 0\ 0\ 0\ 0), (0\ x_8\ 0\ 0\ 0\ 0\ 0\ 0\ 1\ 0\ 0\ 0),$
$\qquad (0\ x_8\ 0\ 1\ 0\ 0\ 0\ 0\ 1\ 0\ 0\ 0)\}, \varnothing, \varnothing, \varnothing, \varnothing\}, t_8\, t_{11}\, t_{12}\, t_{13}\, t_{14}\, t_{15})$

$= \delta(\delta(\{\{(0\ x_8\ 0\ 0\ 0\ 0\ 0\ 0\ 0\ 0\ 0\ 0), (0\ x_8\ 0\ 1\ 0\ 0\ 0\ 0\ 0\ 0\ 0), (0\ x_8\ 0\ 0\ 0\ 0\ 0\ 0\ 1\ 0\ 0\ 0),$
$\qquad (0\ x_8\ 0\ 1\ 0\ 0\ 0\ 0\ 1\ 0\ 0\ 0)\}, \varnothing), t_8), t_{11}\, t_{12}\, t_{13}\, t_{14}\, t_{15})$

$= \delta(\{\{(0\ 0\ [0, x_{10}]\ 0\ 0\ 0\ 0\ 0\ 0\ 0\ 0), (0\ 0\ [0, x_{10}]\ 1\ 0\ 0\ 0\ 0\ 0\ 0\ 0),$
$\qquad (0\ 0\ [0, x_{10}]\ 0\ 0\ 0\ 0\ 1\ 0\ 0\ 0), (0\ 0\ [0, x_{10}]\ 1\ 0\ 0\ 0\ 1\ 0\ 0\ 0)\}, \varnothing\}, t_{11}\, t_{12}\, t_{13}\, t_{14}\, t_{15})$

$= \ldots \quad = \{\{(0\ 0\ 0\ 0\ 0\ 0\ 0\ 0\ 0\ 0\ 0\ |D_{19}|+x_{20}), (0\ 0\ 0\ 1\ 0\ 0\ 0\ 0\ 0\ 0\ 0\ |D_{19}|+x_{20})\}, \varnothing\}$

3. According to (1), any loop in a WF model will cause a blocking state. In fact, if a task t_j is waiting for a data structure s tuples, and if these tuples are produced by a task t_i, which will never be executed in certain conditions: if his input data structures are not root nodes, and if $t_j \in < t_1\, t_2 \ldots t_i >$, i.e. the firing sequence leading to execute t_i, this case is identified as a deadlock. In addition, even if the input data structures of t_j are root nodes, the occured cycle may cause a livelock (i.e. a loop without progress). Thus, we require that each task $t \in T$ is fired, at most once, in a sequence of tasks starting from a marking M.

$$\forall i_1, i_2 \ldots i_k \in \{1, 2 \ldots |T|\}, \delta(\{M\}, t\, t_{i_1}\, t_{i_2} \ldots t_{i_k}\, t) = \varnothing . \qquad (5)$$

We explain this rule through the example illustrated through Fig. 3.

4. To keep the coherency of a WF process at runtime, whatever the firing sequence, starting from a marking M, the final marking has to be the same. Formally: $\forall i_1, i_2 \ldots i_k \in \{1, 2 \ldots |T|\}$ and $\forall j_1, j_2 \ldots j_k \in \{1, 2 \ldots |T|\}$, if $\delta(\{M\}, t_{i_1}\, t_{i_2} \ldots t_{i_k}) = \varnothing$ (resp. $\delta(\{M\}, t_{j_1}\, t_{j_2} \ldots t_{j_k}) = \varnothing$) then: $\delta(\{M\}, t_{i_1}\, t_{i_2} \ldots t_{i_{k-1}}) \neq \varnothing$ (resp. $\delta(\{M\}, t_{j_1}\, t_{j_2} \ldots t_{j_{k-1}}) \neq \varnothing$) is the final marking. In such case we have to get:

$$\delta(\{M\}, t_{i_1}\, t_{i_2} \ldots t_{i_{k-1}}) = \delta(\{M\}, t_{j_1}\, t_{j_2} \ldots t_{j_{k-1}}) . \qquad (6)$$

We elucidate this rule through the example illustrated in Fig. 1.

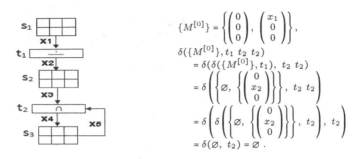

Fig. 2. Process model (a) with deadlock

So, we calculate the possible markings resulting from the firing sequence of tasks $< t_7\ t_8\ t_{11}\ t_{12}\ t_{13}\ t_{14}\ t_{15}\ t. >$, such as $t.$ refers to any task $\in T$ that is not in the above firing sequence:

$\delta(\{M^{[0]}\},\ t_7\ t_8\ t_{11}\ t_{12}\ t_{13}\ t_{14}\ t_{15}\ t.) = \delta(\delta(\{M^{[0]}\},\ t_7),\ t_8\ t_{11}\ t_{12}\ t_{13}\ t_{14}\ t_{15}\ t.)$

$= \dots$

$= \delta(\{\{(0\ 0\ 0\ 0\ 0\ 0\ 0\ 0\ 0\ 0\ 0\ 0\ |D_{19}| + x_{20}),\ (0\ 0\ 0\ 1\ 0\ 0\ 0\ 0\ 0\ 0\ 0\ 0\ |D_{19}| + x_{20})\},\ \varnothing\},\ t.) = \varnothing$

$\Rightarrow \{\{(0\ 0\ 0\ 0\ 0\ 0\ 0\ 0\ 0\ 0\ 0\ 0\ |D_{19}| + x_{20}),\ (0\ 0\ 0\ 1\ 0\ 0\ 0\ 0\ 0\ 0\ 0\ 0\ |D_{19}| + x_{20})\},\ \varnothing\}$ is the final marking. We also calculate the possible markings resulting from the firing sequence of tasks $< t_7\ t_8\ t_{11}\ t_{13}\ t_{12}\ t_{14}\ t_{15}\ t. >$:

$\delta(\{M^{[0]}\},\ t_7\ t_8\ t_{11}\ t_{13}\ t_{12}\ t_{14}\ t_{15}\ t.) = \delta(\delta(\{M^{[0]}\},\ t_7),\ t_8\ t_{11}\ t_{13}\ t_{12}\ t_{14}\ t_{15}\ t.)$

$= \dots$

$= \delta(\{\{(0\ 0\ 0\ 0\ 0\ 0\ 0\ 0\ 0\ 0\ 0\ 0\ |D_{19}| + x_{20}),\ (0\ 0\ 0\ 1\ 0\ 0\ 0\ 0\ 0\ 0\ 0\ 0\ |D_{19}| + x_{20})\},\ \varnothing\},\ t.) = \varnothing$

$\Rightarrow \{\{(0\ 0\ 0\ 0\ 0\ 0\ 0\ 0\ 0\ 0\ 0\ 0\ |D_{19}| + x_{20}),\ (0\ 0\ 0\ 1\ 0\ 0\ 0\ 0\ 0\ 0\ 0\ 0\ |D_{19}| + x_{20})\},\ \varnothing\}$ is the final marking. So, as t_{12} and t_{13} are two parallel tasks:

$\delta(\{M^{[0]}\}, t_7\ t_8\ t_{11}\ t_{12}\ t_{13}\ t_{14}\ t_{15}) = \delta(\{M^{[0]}\}, t_7\ t_8\ t_{11}\ t_{13}\ t_{12}\ t_{14}\ t_{15}).$

5. When the final marking has been reached, a WF process needs to be revived in order to be executed again. In other words, we have to ensure that $\forall M \in \mathbb{N}^{|S|}$, $i_1, i_2 \dots i_k \in \{1, 2 \dots |T|\}$:

$$\delta(\{M^{[0]}\},\ t_{i_1}\ t_{i_2} \dots t_{i_k}\ t_{i_1}) = \varnothing . \tag{7}$$

To do so, we extend WF by a restitution task $t_r \notin T$, i.e. t_r is not a data operation, it is a simple transition used to return from all the final states to the initial states. In such case:

$$\delta(\varnothing,\ t_r) = \{M^{[0]}\} . \tag{8}$$

5   Workflow Analysis

As introduced in [13], the classical soundness property grants that a process has always the possibility to terminate and all its tasks are coverable (i.e., can potentially be executed). Termination ensures that the workflow can, during its execution, neither get stuck (i.e., it is deadlock free) nor enter a loop that cannot be left (i.e., it is livelocks free), whereas coverable excludes dead tasks in the workflow. But, to ensure these criteria, the soundness property needs, to be verified, that the process has a single source place i and a single final place o.

Nevertheless, to reflect the reality of business processes, we allow a WF model to present initial and final states as needed and accordingly, it is not possible to detect this classical soundness property. So, we propose a released notion of soundness which ensures that there are no livelocks, or deadlocks, or dead tasks in a WF. In other words, we will verify the well-formedness property of a WF process. According to [14], a P-net is well-formed if it is live and bounded. We adopt this rule to WF, thus, the first step is to verify its liveness property.

5.1   Verification of the Liveness Property

We tackle this issue in [8,9], but by analyzing other process cases, we have been aware that the proposed technique is not enough to ensure the liveness property of a WF model. So, we improve it as follows:

Assuming that $\{M^{[0]}\}$ is the set of the possible initial markings, a WF model is live if and only if: $\forall t \in T,\ \exists\ t_1,\ t_2 \ldots t_n \in T\ |\ \delta(M^{[0]},\ t_1\ t_2 \ldots t_n\ t) \neq \varnothing \Rightarrow$ t is live.

To ensure the verification of this property, we proposed in [8, 9] a simple algorithm based on (9) and (10), which are defined as the following:

$$\text{Firable}(t) = \bigwedge_{\substack{i \in \{1,\ 2 \ldots |S|\} \\ Pre(s_i,\ t) \neq 0}} \text{Expectable}(s_i)\ . \tag{9}$$

$$\text{Expectable}(s) = \begin{cases} \text{true if } \forall i \in \{1,\ 2 \ldots |T|\},\ Post_{Max}(s,\ t_i) = 0\ . \\ \text{Firable}(t_i) \text{ if } \exists\ i \in \{1,\ 2 \ldots |T|\},\ \text{where } Post_{Max}(s,\ t_i) \neq 0\ . \end{cases} \tag{10}$$

We elucidate (9) and (10) through the example illustrated in Fig. 1.

$$\text{Firable}(t_{15}) = \bigwedge_{\substack{i \in \{8,9 \ldots 19\} \\ Pre(s_i,\ t_{15}) \neq 0}} \text{Expectable}(s_i) = \text{Expectable}(s_{18}) = \text{Firable}(t_{14})$$

$= \text{Expectable}(s_{15}) \wedge \text{Expectable}(s_{17}) = \text{Firable}(t_{12}) \wedge \text{Firable}(t_{13}) = \text{Expectable}(s_{14}) \wedge$
$\quad \text{Expectable}(s_{16}) = \text{Firable}(t_{11}) \wedge \text{true} = \text{Firable}(t_{11}) = \text{Expectable}(s_{10})$
$= \text{Firable}(t_8) = \text{Expectable}(s_9) = \text{Firable}(t_7) = \text{Expectable}(s_8) = \text{true}$

By using (9) and (10), we can verify that every task in a WF process is firable if its expected tokens can be provided by the evolution of the marking. However, this is not sufficient to verify its liveness proverty. In fact, if a WF model contains structural conflicts, there will be a part in the workflow that may not be executed. So, before applying (9) and (10), we have to start by verifying that the workflow does not contains structural conflicts.

Conflict Resolution: we assume that a WF model has a structural conflict, if it contains at least two tasks t_i and t_j having the same input data structure s_k, e.g., t_2 and t_5 sharing s_2 in the $Role$ 1 sub-process, t_9 and t_{11} sharing s_{10}

in the *Role* 2 sub-process. To resolve such conflicts [8,9], we extend the model by adding extra tasks $T^* = \{t_{clone_1}, t_{clone_2} \ldots t_{clone_l}\}$ such as l is the number of conflict tasks, and t_{clone} is a *clone* operation formally defined as follows: whether $s_k = (C_k, D_k)$, $t_{clone}(s_k, l) = \{s_{k_1}, s_{k_2} \ldots s_{k_l}\}$.

The extended model $WF_+ = (S_+, T_+, Pre_+, Post_+)$ such as: $S_+ = S$, $T_+ = T \cup T^*$, $Pre_+ = S \times T_+$, and $Post_+ = T_+ \times S$.

Blocking State Resolution: after extending WF, the process has to be verified to ensure that there are no deadlocks or livelocks. In fact, if we apply (9) and (10) directly on a WF_+, which contains deadlocks or livelocks, the equations will enter in an infinite loop, as the case of model (a) presented in Fig. 2:

$$\text{Firable}(t_2) = \bigwedge_{\substack{i \in \{1,\ 2,\ 3\} \\ Pre(s_i,\ t_2) \neq 0}} \text{Expectable}(s_i) = \text{Expectable}(s_2) \wedge \text{Expectable}(s_3)$$

$$= \text{Firable}(t_1) \wedge \text{Firable}(t_2) = \text{Expectable}(s_1) \wedge \text{Expectable}(s_2) \wedge \text{Expectable}(s_3)$$

$$= \text{true} \wedge \text{Firable}(t_1) \wedge \text{Firable}(t_2) = \text{Firable}(t_1) \wedge \text{Firable}(t_2) = \text{Expectable}(s_1) \wedge$$

$$\text{Expectable}(s_2) \wedge \text{Expectable}(s_3) = \text{true} \wedge \text{Firable}(t_1) \wedge \text{Firable}(t_2) = \ldots$$

The verification of deadlocks or livelocks is ensured by (5) defined by the firing rule 3, which prohibits the existence of loops in a WF model. If this rule is not verified, it means that the model contains deadlocks or livelocks and accordingly, it is not live.

5.2   Verification of the Boundedness Property

As we extended WF to WF_+, the boundedness property will be verified relatively to the extended model. If WF_+ is not bounded, it means that the workflow will contain at least one data structure having a number of tokens increasing infinitely with the evolution of the marking. To verify the boundedness property of a WF, we assume that if its WF_+ has no loop, it will be bounded. We prove this idea as follows: According to (2): $\forall t_i, t_j \in T, \exists s \in S \mid < t_i, t_j >$, $pre(t_j, s) = post(t_i, s)$, which ensures that the number of tokens produced by a task, in its output data structure, will be automatically consumed by the next task having, as input, the same data structure. Besides, according to (5): $\forall i \in \{1, 2 \ldots |T|\}$, $M \in \{M_1, M_2 \ldots M_n\}$, $\delta(\{M\}, t\ t_{i_1}\ t_{i_2} \ldots t_{i_k}\ t) = \varnothing$, which means that, there is no cycle in a WF model.

Consequently, $\{i \in \{1, 2 \ldots |T|\}, M > Pre(., t_i)\} = \varnothing$, and accordingly, in any case, the marking of a data structure will never be higher then the number of tuples requested by the task consuming this data structure tuples. Thus, $| \bigcup\limits_{\theta=0}^{+\infty} \delta(M^{[\theta]}, t_i)| < +\infty$, which means that, the set of possible markings $\delta(\{M^{[\theta]}\}, t_i)$ is a finite set $\forall t_i \in T$, and consequently, WF_+ is bounded.

6   Opus Framework

Opus Framework is implemented using Java Swing language with a set of Java library, namely, JGraph, UMLGraph, JTable. . . It consists of a number of components including a modeling editor, a workflow engine, and a verification module.

6.1 Opus Editor

The graphical modeling of workflow processes is ensured using *Opus editor*. The latter is equipped with a set of graphical interfaces to create profiles of roles performing the work, define data flow interactions between roles and the IS (e.g. ICRole receives the data structure *Customer_Inf* from SCRole and saves *Order_Table* and *Order_Line_Table* tokens in the IS), and finally, define the sub-process model related to each role work. It also provides to the designer a customized assistant for each operation in the process model, in order to help him to model the process structure.

6.2 Verification of the Workflow Model

Opus system is equipped with a *verification module* which ensures the analyses and the verification of the conceived workflow models, as described in Sect. 5. The verification result of the ICRole sub-process is illustrated in Fig. 3. We illustrate also the verification of model (a) (defined in Fig. 2), through Fig. 4.

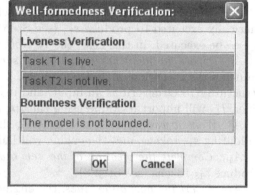

Fig. 3. ICRole sub-process verification (processing time 1.37 sec)

Fig. 4. Model (a) verification (processing time 0.24 sec)

6.3 Opus Engine

Opus engine follows up the data flow routing, simulates the processing of tasks according to its formal definition, considering the firing rules defined in Sect. 4, and invites each role to perform its tasks according to its feasibility and urgency. Furthermore, tokens of workflow initial states may be imported from the IS. And in the same way, tokens of final states may be stored in. For this purpose, *Opus* engine is equipped with the *IS Integration Module* that provides the *Import tool* (which imports tuples from a definite IS table to a definite data structure [9]), the *ImportId tool* (which imports the identifier of the last tuple inserted in a definite IS table, instead of being entered by a role), and the *Insert tool* (which stores a data structure tuples in a definite IS table. It is considered as an *Add_Tuples* operation such as t_{10} and t_{15} in Fig 1). We detail all the actions, performed either by the ICRole or by *Opus* engine, through Fig. 5.

A_1. *Starting the ICRole sub-process:* when the *ICRole* launches his process, the engine will present to him the data structure *Customer_Inf* received from *SCRole* and will ask him to instantiate the data structure *Product_Table*.

A_2. *Alimentation of the data structure Product_Table from the IS:* the *ICRole* imports tuples to *Product_Table* from the IS using the *Import* tool, and validates its tokens in order to execute t_7 (see Step 1).

A_3. *Seizure of the ordered quantities:* the *ICRole* seizes the ordered quantities relatively to the ordered products (OrdPs), in the resulting data structure of t_7, and validates his seizures (see Step 2).

A_4. *Select the OrdPs:* during the execution of t_8, the engine invites the *ICRole* to enter the selection property in order to select the OrdPs (see Step 3). Then, the *ICRole* saves the selection property for the next executions of t_8.

A_5. *Verify the availability of the OrdPs:* in this runtime example, $s_{10} \neq \emptyset$, so, t_{11} decides to send s_{14} to t_{12}, also t_9 decides to send s_{11} to t_{10}.

A_6. *Insert a new order:* t_{10} uses the *Insert tool* to insert a new order in the IS *Orders* Table (see Steps 5, 5.1, 5.2).

A_7. *Create the new order lines:* in parallel with t_9, t_{11} then t_{12} will be automatically executed to produce s_{15} (see Step 4-2). Then, t_{14} will be waiting for s_{17} to be executed. In this case, the *ICRole* has to launch the execution of t_{13}.

A_8. *Import the identifier of the last inserted order from the IS:* the *ICRole* launches the execution of t_{13} (see Step 6-1). The latter receives the empty data structure $S16\_Order\_Id$ as an input, and instead of seizing the new order identifier, t_{13} will import its value using the *ImportId* tool (see Step 6-2).

A_9. *Wake a waiting task:* at this level, the *ICRole* can turn to wake t_{14} by validating s_{15} tokens, and the engine will launch its execution (see Step 7).

A_{10}. *Complete the creation of the new order lines:* the engine executes t_{14} to produce s_{18} tokens (see Step 8).

A_{11}. *Insert the new order lines:* the engine executes t_{15} and asks the *ICRole* to choose the suitable IS table for the insertion (see Step 9-1), and to perform the matching between the data structure s_{18} and the chosen table (see Step 9-2). If these two steps are well done, the engine will properly end the workflow.

We can deduce, from the execution details, the presence of four types of actions: 9.1% of actions are based on *manual tasks* (i.e. tasks that are performed only by a role without the intervention of the engine, such as A_3), 27.27% of actions are based on *automatic tasks* (i.e. tasks that are performed only by the engine without the intervention of a role, such as A_5, A_7 and A_{10}), 18.18% of actions are based on *semi-automatic tasks* (i.e. tasks that are performed by the engine under control of a role, such as A_1 and A_9), and 45.45% of actions are based on *semi-automatic tasks only in the first execution (SA-FE)* (i.e. tasks that are semi-automatic tasks only in their first executions, but during their next executions, they will migrate to be automatic tasks, such as A_2, A_4, A_6, A_8 and A_{11}).

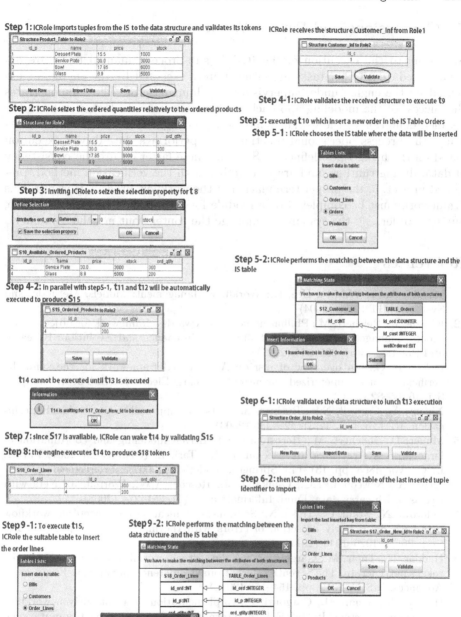

Fig. 5. Executing the ICRole sub-process

7   Conclusions and Future Work

According to the execution of the ICRole sub-process, manual tasks are extremely reduced. Other tasks are either purely executed by the engine, or executed by the engine under supervising of a human actor. That demonstrates the success of the modeling approach in execution issue. In fact, thanks to the detailed definition of the workflow model, *Opus* engine can interpret, automatically, the process operational functions and perform a data-driven execution based on the firing rules defined in Sect. 4. These latter ensure the consistency of data, during runtime, and grant, together with the verification method, presented in Sect. 5, the proper termination of the workflow process. However, this Framework must be completed by a module for documents generation (invoice, purchase order...): the system can manage the content but not the container.

References

1. van der Aalst, W.M.P., Hee, K.: Workflow Management: Models, Methods, and Systems. MIT Press (2004)
2. Künzle, V., Reichert, M.: Philharmonicflows: towards a framework for object-aware process management. Journal of Software Maintenance and Evolution: Research and Practice 23(4), 205–244 (2011)
3. Delzanno, G.: An overview of msr(c): A clp-based framework for the symbolic verification of parameterized concurrent systems. Electr. Notes Theor. Comput. Sci. 76, 65–82 (2002)
4. Nigam, A., Caswell, N.S.: Business artifacts: An approach to operational specification. IBM Syst. J. 42(3), 428–445 (2003)
5. Müller, D., Reichert, M., Herbst, J.: Data-driven modeling and coordination of large process structures. In: Meersman, R., Tari, Z. (eds.) OTM 2007, Part I. LNCS, vol. 4803, pp. 131–149. Springer, Heidelberg (2007)
6. Lazic, R., Newcomb, T.C., Ouaknine, J., Roscoe, A.W., Worrell, J.: Nets with tokens which carry data. Fund. Informaticae 88(3), 251–274 (2008)
7. Sidorova, N., Stahl, C., Trčka, N.: Soundness verification for conceptual workflow nets with data: Early detection of errors with the most precision possible. Inf. Syst. 36(7), 1026–1043 (2011)
8. Haddar, N., Tmar, M., Gargouri, F.: A data-driven workflow based on structured tokens petri net. In: The Seventh International Conference on Software Engineering Advances, ICSEA 2012, pp. 154–160 (2012)
9. Haddar, N., Tmar, M., Gargouri, F.: Implementation of a data-driven workflow management system. In: IEEE 15th International Conference on Computational Science and Engineering, CSE 2012, pp. 111–118. IEEE Computer Society (2012)
10. Liu, D., Wang, J., Chan, S.C.F., Sun, J., Zhang, L.: Modeling workflow processes with colored petri nets. Comput. Ind. 49(3), 267–281 (2002)
11. Aalst, W., Weske, M., Grünbauer, D.: Case handling: a new paradigm for business process support. Data Knowl. Eng. 53(2), 129–162 (2005)
12. Künzle, V., Reichert, M.: Philharmonicflows: Research and design methodology. Technical report, University of Ulm (May 2011)
13. Aalst, W.: Verification of workflow nets. In: Azéma, P., Balbo, G. (eds.) ICATPN 1997. LNCS, vol. 1248, pp. 407–426. Springer, Heidelberg (1997)

14. van der Aalst, W.M.P.: Workflow verification: Finding control-flow errors using petri-net-based techniques. In: van der Aalst, W.M.P., Desel, J., Oberweis, A. (eds.) Business Process Management. LNCS, vol. 1806, pp. 161–183. Springer, Heidelberg (2000)

Appendix A: Data Operations Definition

Table 1. Operations definition

Operation	Formal definition
Inner Product: performs the combination of a data structure tuples with those of another data structure. Noted: ×	$\forall\, s_j = (C_j, D_j)$, $s_k = (C_k, D_k)$, $C_j = (c_{j_1}, c_{j_2} \ldots c_{j_{n_j}})$, $C_k = (c_{k_1}, c_{k_2} \ldots c_{k_{n_k}})$, $s_i = s_j \times s_k = ((c_{j_1} \ldots c_{j_{n_j}}, c_{k_1} \ldots c_{k_{n_k}}), D_i)$ Where: $D_i = \bigcup \{(d_{j_l}1 \ldots d_{j_l}{}^{n_j}, d_{k_p}1 \ldots d_{k_p}{}^{n_k})\}$ $l \in \{1 \ldots n_j\}$, $p \in \{1 \ldots n_k\}$ Resulted tokens number: $|D_i| = |D_j| \times |D_k|$
Selection: selects the tuples of a data structure that meet the desired criteria. Noted: σ	Whether P is the selection property, $\forall\, s_j = (C_j, D_j)$, $s_i = \sigma_P s_j$ $\Leftrightarrow s_i = (C_j, \bigcup \{(e)\})$ $e \in D_j$ $P(e)$ Resulted tokens number: $|D_i| \in [0, |D_j|]$
Projection: selects the values of specific attributes in a data structure. Noted: $\doteq$	$s_j = (C_j, D_j), \forall (b_1 \ldots b_n) \in \{0,1\}^n$, $s_i = (C_i, D_i) = \doteq_{(b_1 \ldots b_n)} s_j$ Where c_i is a selected (resp. not selected) attribute, if $b_i = 1$ (resp $b_i = 0$). $\Leftrightarrow C_i = (c_{j_{j'_1}}, c_{j_{j'_2}} \ldots c_{j_{j'_q}})$, $D_i = \{(d_{j_{1_{j'_1}}}, d_{j_{1_{j'_2}}} \ldots d_{j_{1_{j'_q}}}),$ $(d_{j_{2_{j'_1}}}, d_{j_{2_{j'_2}}} \ldots d_{j_{2_{j'_q}}}) \ldots (d_{j_{m_{j'_1}}}, d_{j_{m_{j'_2}}} \ldots d_{j_{m_{j'_q}}})\}$. Such as: $q = \sum_{k=1}^n b_k$: is the number of attributes in the structure result, $j'_k = \min_{l = \{1,2 \ldots n\}} l$: refers to the projection attributes indices. $\sum_{p=1}^l b_p = k$ Resulted tokens number: $|D_i| \in [0, |D_j|]$
Substitution: changes the name of an attribute in a data structure. Noted: $\varnothing$	$\forall s_j = (C_j, D_j)$, $s_i = \varnothing(c_{j_k}, c, s_j) = ((c_{j_1} \ldots c_{j_{k-1}}, c, c_{j_{k+1}} \ldots c_{j_n}), D_j)$ Resulted tokens number: $|D_i| = |D_j|$
Permutation: allows to permute two columns in a data structure. Noted: $\overset{\frown}{\sim}$	$\forall s_i = (C_i, D_i), s_j = \overset{\frown}{\sim}(s_i, k, l)$, such as: $k, l \in \{1, 2 \ldots n\}$, $k < l$, $C_j = (c_{i_1} \ldots c_{i_{k-1}}, c_{i_l}, c_{i_{k+1}} \ldots c_{i_{l-1}}, c_{i_k}, c_{i_{l+1}} \ldots c_{i_n})$ $D_j = \{(d_{1_{i_1}} \ldots d_{1_{i_{k-1}}}, d_{1_{i_l}}, d_{1_{i_{k+1}}} \ldots d_{1_{i_{l-1}}}, d_{1_{i_k}}, d_{1_{i_{l+1}}} \ldots d_{1_{i_n}}),$ $(d_{m_{i_1}} \ldots d_{m_{i_{k-1}}}, d_{m_{i_l}}, d_{m_{i_{k+1}}} \ldots d_{m_{i_{l-1}}}, d_{m_{i_k}}, d_{m_{i_{l+1}}} \ldots d_{m_{i_n}})\}$. Resulted tokens number: $|D_i| = |D_j|$
Extension: Extends a structure scheme by adding an attribute c =(n, t) and applying a function f. Noted: τ	$\forall s_j = (C_j, D_j)$, $s_i = \tau(s_j, c, f)$, such as: $C_i = ((c_{j_1}, c_{j_2} \ldots c_{j_n}, c), D_i = \{(d_{j_{1_1}}, d_{j_{1_2}} \ldots d_{j_{1_n}}, f(d_{j_{1_1}}, d_{j_{1_2}} \ldots d_{j_{1_n}}, D_j)) \ldots (d_{j_{m_1}}, d_{j_{m_2}} \ldots d_{j_{m_n}}, f(d_{j_{m_1}}, d_{j_{m_2}} \ldots d_{j_{m_n}}, D_j))\}$ Resulted tokens number: $|D_i| = |D_j|$
Add_Tuples: inserts the tuples of a data structure in another one. Noted: +	$\forall s_j = (C_j, D_j)$, $s_k = (C_k, D_k)$, $D_j = \{(d_{j_{1_1}}, d_{j_{1_2}} \ldots d_{j_{1_n}}), (d_{j_{2_1}}, d_{j_{2_2}} \ldots d_{j_{2_n}}) \ldots (d_{j_{m_1}}, d_{j_{m_2}} \ldots d_{j_{m_n}})\}$ $D_k = \{(d_{k_{1_1}}, d_{k_{1_2}} \ldots d_{k_{1_h}}), (d_{k_{2_1}}, d_{k_{2_2}} \ldots d_{k_{2_h}}) \ldots (d_{k_{l_1}}, d_{k_{l_2}} \ldots d_{k_{l_h}})\}$, $b_1, b_2 \ldots b_n) \in \{1, 2 \ldots n\}$: refers to the positions of the added values in the resulting data structure. $s_j = +(s_k, (b_1, b_2 \ldots b_n))$ if $h < n$ then: $s_j = ((c_1, c_2 \ldots c_n), \{(d_{j_{1_1}}, d_{j_{1_2}} \ldots d_{j_{1_n}}), (d_{j_{2_1}}, d_{j_{2_2}} \ldots d_{j_{2_n}})$ $\ldots (d_{j_{m_1}}, d_{j_{m_2}} \ldots d_{j_{m_n}}), (d_{k_{1_{b_1}}}, d_{k_{1_{b_2}}} \ldots d_{k_{1_{b_h}}} \ldots d_{k_{1_n}}),$ $(d_{k_{2_{b_1}}}, d_{k_{2_{b_2}}} \ldots d_{k_{2_{b_h}}} \ldots d_{k_{2_n}}) \ldots (d_{k_{l_{b_1}}}, d_{k_{l_{b_2}}} \ldots d_{k_{l_{b_h}}} \ldots d_{k_{l_n}})\}$ if $h = n$ then: $s_j = ((c_1, c_2 \ldots c_n), \{(d_{j_{1_1}}, d_{j_{1_2}} \ldots d_{j_{1_n}}), (d_{j_{2_1}}, d_{j_{2_2}} \ldots d_{j_{2_n}})$ $\ldots (d_{j_{m_1}}, d_{j_{m_2}} \ldots d_{j_{m_n}}), (d_{k_{1_{b_1}}}, d_{k_{1_{b_2}}} \ldots d_{k_{1_n}}),$ $(d_{k_{2_{b_1}}}, d_{k_{2_{b_2}}} \ldots d_{k_{2_n}}) \ldots (d_{k_{l_{b_1}}}, d_{k_{l_{b_2}}} \ldots d_{k_{l_n}})\}$. Resulted tokens number: $|D_j| = |D_j| + |D_k|$
Control 1: Decides to continue or not the information flow routing, according to condition1. Noted: $\pm$	Condition 1: if s_i is the data structure expected by the next task, and s_j is the controlled data structure, then: $s_i = s_k \pm s_j = \begin{cases} s_k \ if \ s_j = \varnothing, \\ \varnothing \ otherwise. \end{cases}$ Resulted tokens number: $|D_i| \in [0, |D_k|]$
Control 2: Decides to continue or not the information flow routing, according to condition 2. Noted: $\mp$	Condition 2: if s_i is the data structure expected by the next task, and s_j is the controlled data structure, then: $s_i = s_k \mp s_j = \begin{cases} s_k \ if \ s_j \neq \varnothing, \\ \varnothing \ otherwise. \end{cases}$ Resulted tokens number: $|D_i| \in [0, |D_k|]$

Generic Top-k Query Processing
with Breadth-First Strategies

Mehdi Badr and Dan Vodislav

ETIS, ENSEA, University of Cergy-Pontoise, CNRS, France
firstname.lastname@u-cergy.fr

Abstract. Many algorithms for top-k query processing with ranking predicates have been proposed, but little effort has been directed toward genericity, i.e. supporting any type (sorted and/or random) or cost settings for the access to the lists of predicate scores. In previous work, we proposed BreadthRefine (BR), a generic algorithm that considers the current top-k candidates as a whole instead of focusing on the best candidate, then we compared it with specific top-k algorithms. In this paper, we compare the BR breadth-first strategy with other existing generic strategies and experimentally show that BR leads to better execution costs. To this end, we propose a general framework GF for generic top-k processing, able to express any top-k algorithm and present within this framework a first comparison between generic algorithms. We also extend the notion of θ-approximation to the GF framework and present a first experimental study of the approximation potential of top-k algorithms on early stopping.

Keywords: Top-k procesing query, ranking, multicriteria information retrieval.

1 Introduction and Related Work

We address the problem of top-k query processing, where queries are composed of a set of ranking predicates, each one expressing a measure of similarity between data objects on some specific criteria. Unlike traditional boolean predicates, similarity predicates return a relevance score in a given interval. The query also specifies an aggregation function that combines the scores produced by the similarity predicates. Query results are ranked following the global score and only the best k ones are returned.

Ranking predicates acquired an increasing importance in today's data retrieval applications, especially with the introduction of new, weakly structured data types: text, images, maps, etc. Searching such data requires content-based information retrieval (CBIR) techniques, based on predicates measuring the similarity between data objects, by using content descriptors such as keyword sets, image descriptors, geographical coordinates, etc. We consider here the case of *expensive* ranking predicates over data objects, whose specificity is that their evaluation cost dominates the cost of the other query processing operations.

The general form of the top-k queries that we consider is expressed in Figure 1. The query asks for the k best objects following the scores produced by m ranking predicates $p_1, ..., p_m$, aggregated by a monotone function $\mathcal{F}$. Figure 1 also presents a query example from a touristic application, where the visitor of a monument takes a

H. Decker et al. (Eds.): DEXA 2013, Part I, LNCS 8055, pp. 254–269, 2013.
© Springer-Verlag Berlin Heidelberg 2013

select * from Object o select * from Monument m
order by $\mathcal{F}(p_1(o), ..., p_m(o))$ order by $near(m.address, here())$ +
limit k $similar(m.photo, myPhoto)$ +
                                                   $ftcontains(m.descr, \text{'Renaissance sculpture'})$
                                          limit 1

Fig. 1. General form and example of top-k query

photo of a detail and searches for the "best" monument on three criteria: *near* to its current location, reproducing a *similar* detail, and *exposing Renaissance sculptures*.

As in this example, expensive ranking predicates come often from the evaluation of similarity between images, text, locations and other multimedia types, whose content is described by numerical vectors. This results in expensive search in highly dimensional spaces, based often on specific multidimensional index structures [3].

In many cases, predicates are evaluated by distant, specialized sites, that provide specific web services, e.g. map services evaluating spatial proximity, photo sharing sites allowing search of similar images, specialized web sites proposing rankings for hotels, restaurants, etc. Internet access to such services results into expensive predicate evaluation by distant, independent sites. Moreover, the control over predicate evaluation is minimal most of the time, reduced to the call of the provided web service.

For each query, a ranking predicate may produce a score for each object. In the following, we call a *source* the collection of scores produced by a ranking predicate for the set of data objects. The list of scores may be produced e.g. by the access to a local index structure that returns results by order of relevance. We consider here the general case, where the access to the scores of a source is limited to sorted and/or random access. This allows three possible types for a source S:

- *S-source*: sorted access only, through the operator *getNext(S)* returning the next highest score s and the corresponding object identifier o.

- *R-source*: random access only, through the operator *getScore(S, o)* returning the score of object o.

- *SR-source*: a source with both sorted and random access.

The general idea of a top-k algorithm is to avoid computing all the global scores, by maintaining a list of candidate objects and the interval $[L, U]$ of possible global scores for each of them. The monotonicity of the aggregation function ensures that further source accesses always decrease the upper bound U and increase the lower bound L. The algorithm stops when the score of the best k candidates cannot be exceeded by the other objects anymore. Figure 2 presents a possible execution for the example query in Figure 1. We suppose S_1 is an S-source, S_2 an SR-source, S_3 an R-source; scores are presented in descending order for S/SR sources and by object id for R-sources. Local scores belong to the $[0, 1]$ interval, so the initial global score interval is $[0, 3]$ for all objects. We note *candidates* the set of candidates and U_{unseen} the maximum score of unseen objects (not yet discovered). Initially, *candidates* = $\emptyset$ and $U_{unseen} = 3$.

A sorted access to S_1 retrieves (o_2, 0.4), so o_2's global score interval becomes $[0.4, 2.4]$. Also U_{unseen} becomes 2.4 because further scores in S_1 cannot exceed 0.4. Then, a sorted access to S_2 retrieves (o_3, 0.9). This adds a new candidate (o_3), lowers U_{unseen} to 2.3 (further S_2 scores cannot exceed 0.9), but also lowers the upper bound of

S_1 (S)	S_2 (SR)	S_3 (R)
$(o_2, 0.4)$	$(o_3, 0.9)$	$(o_1, 0.9)$
$(o_1, 0.3)$	$(o_1, 0.2)$	$(o_2, 0.7)$
$(o_4, 0.25)$	$(o_4, 0.15)$	$(o_3, 0.8)$
$(o_3, 0.2)$	$(o_2, 0.1)$	$(o_4, 0.6)$

Access	Retrieved	candidates	U_{unseen}
		$\emptyset$	3.0
S_1/S	$(o_2, 0.4)$	$\{(o_2, [0.4, 2.4])\}$	2.4
S_2/S	$(o_3, 0.9)$	$\{(o_2, [0.4, 2.3]), (o_3, [0.9, 2.3])\}$	2.3
S_2/R	$(o_2, 0.1)$	$\{(o_2, [0.5, 1.5]), (o_3, [0.9, 2.3])\}$	2.3
S_3/R	$(o_3, 0.8)$	$\{(o_2, [0.5, 1.5]), (o_3, [1.7, 2.1])\}$	2.3
S_2/S	$(o_1, 0.2)$	$\{(o_2, [0.5, 1.5]), (o_3, [1.7, 2.1]), (o_1, [0.2, 1.6])\}$	1.6

Fig. 2. Examples of sources and query execution for the query example

o_2 to 2.3, because the maximum score of S_2 is now 0.9. Next, a random access to S_2 for o_2 retrieves $(o_2, 0.1)$. This changes only the global score interval of o_2, etc. Execution ends when the minimum global score of o_3 exceeds both U_{unseen} and the maximum global score of all the other candidates, i.e. o_3 is the best (top-1) object.

Related Work and Contribution

A large spectrum of top-k query processing techniques [11] has been proposed at different levels: query model, access types, implementation structures, etc. We consider here the most general case, of simple top-k selection queries, with expensive access to sources, limited to individual sorted/random probes, without additional information about local source scores/objects, and out of the database engine.

This excludes from our context join queries [17,10] or interaction with other database operators for query optimization [13,10,12]. We consider sequential access only, parallel processing is out of the scope of this paper. We exclude also approaches such as TPUT [5], KLEE [16] or BPA [1], able to get several items at once, or disposing of statistical information about scores, or disposing also of the local rank. Algorithms such as LARA [14], that optimize the management of the candidate list, are orthogonal to our approach for expensive predicates, which focuses on source access.

In this context, top-k algorithms proposed so far fit with the general method presented in Figure 2 and propose their own heuristic for deciding the next access. However, most algorithms focus on specific source types and cost settings.

Algorithms such as *NRA*[7] (No Random Access) and *StreamCombine*[9] consider only S-sources. NRA successively consults all the sources in a fixed order, while StreamCombine selects at each step the next access based on a notion of *source benefit*.

Other algorithms consider only SR-sources. The best known is *TA*[7] (Threshold Algorithm), which consults sorted sources in a fixed order (like NRA), but fully evaluates the global score of each candidate through random access to the other sources. The algorithm stops when at least k global scores exceed U_{unseen}. *QuickCombine*[8], uses the same idea as StreamCombine to select the next sorted source to probe. *CA*[7] (Combined Algorithm) is a combination of TA with NRA that considers random accesses being h times more expensive than sorted ones. It reduces the number of random probes by performing h sorted accesses in each source before a complete evaluation of the best candidate by random probes.

Also supposing cost asymmetry, a third category of algorithms considers one cheap S-source (providing candidates) and several expensive R-sources. *Upper*[4,15] focuses on the candidate with the highest upper bound U and performs a random probe for it, unless $U < U_{unseen}$, in which case a sorted access is done. The choice of the next R-source to probe is based on a notion of source benefit, dynamically computed. *MPro*[6] is similar to Upper, but fixes for all the candidates the same order for probing the R-sources, determined by sampling optimization.

Surprisingly, little effort has been made towards generic top-k processing, i.e. adapted to any combination of source types and any cost settings. *NC*[19] (Necessary Choices) proposes a framework for generic top-k algorithms in the case of results *with complete scoring*, a strategy SR that favors sorted accesses, and a specific algorithm SR/G that uses sampling optimization to find the best fit with the source settings.

Approximate top-k processing has been considered in several approaches. A variant of the TA algorithm, called TA$_\theta$ [7], defines an approximation parameter $\theta > 1$ and the θ-approximation of the top-k result as being a set K of k objects such that $\forall x \in K$, $\forall y \notin K, \theta score(x) \geq score(y)$ (global and local scores are considered to belong to the [0,1] interval). To obtain a θ-approximation, TA$_\theta$ simply changes the threshold condition: the algorithm stops when at least k objects have a global score $\geq U_{unseen}/\theta$, i.e. TA$_\theta$ is equivalent to an early stopping of the TA algorithm.

Other approximation algorithms for top-k selection queries are proposed in [18], for S-source algorithms, or in the *KLEE* system [16] for top-k processing in distributed environments. Note that [18] is based on dropping candidates that have low probability to be in the top-k and provides probabilistic guarantees for the result, but requires knowledge about score distribution in sources.

In previous work, we have proposed *BR* (BreadthRefine) [2], a generic algorithm that uses a breadth-first strategy for top-k processing in a larger context than NC, i.e. incomplete scoring. The BR strategy considers the current top-k as a whole to be refined, while all the other proposed strategies focus on the best candidate. BR has been compared to algorithms of the three categories mentioned above and proved that it successfully adapts to their specific settings, with better cost.

This paper completes the BR approach with the following contributions:

- A general framework *GF* for generic top-k processing, that allows expressing any top-k algorithm in our context, thus providing a basis for comparative analysis.
- New, comparable generic variants of the BR, NC and CA algorithms with experimental comparison, showing that the BR strategy leads to better costs.
- A generalization of θ-approximation computing in the context of GF, and a first experimental study of the ability of generic top-k algorithms to produce good approximate results on early stopping, showing that the BR strategy has a better approximation potential.

The rest of the paper is organized as follows: the next section introduces the generic framework for top-k processing and compares in this context the BR algorithm with generic variants of NC and CA; Section 3 presents our approach for top-k approximation, then we report experimental results and end with conclusions.

2   Generic Top-k Framework and Algorithms

We propose *GF*, a *generic framework* for top-k processing (Figure 3). As in the example of Figure 2, GF considers a top-k algorithm as a sequence of accesses to the sources, that progressively discover scoring information about data objects. The input parameters are the query q and the set of sources S. Query q specifies the number k of results to return and the monotone aggregation function $\mathcal{F}$, while the set of sources S materializes the scores returned by the query's ranking predicates.

In GF, algorithms maintain *a set of candidates* (initially empty) with their interval scores, *the threshold* U_{unseen} (initialized with the aggregation of the maximum scores max_j of the sources), and possibly other local data structures.

Notations: For a candidate c, we note $[L(c), U(c)]$ its current interval of scores. We note $\mathcal{U}_k(\mathcal{L}_k)$ the current subset of k candidates with the best k upper (lower) bound scores[1]. We note U_k the current k-th highest upper bound score among the candidates, i.e. $U_k = min_{c \in \mathcal{U}_k}(U(c))$, respectively L_k the current k-th highest lower bound score. We note $\chi \in \mathcal{U}_1$ the candidate with the current best upper bound score.

One source access is performed at each iteration, the access type being decided by the *SortedAccessCondition* predicate. In the case of a sorted access, a source S_j is chosen by the **BestSortedSource** function, then is accessed through *getNext*. The returned object-score couple is used to update the threshold, the set of candidates and local variables. The retrieved object is added/updated in the *candidates* set and objects not yet retrieved in S_j update their upper bounds. Update also includes *the discarding of non-viable candidates*. A candidate c with $U(c) < L_k$ is called non-viable because it will never be in the top-k result since at least k candidates surely have better scores.

In the case of a random access, the **ChooseCandidate** function selects a candidate c, then **BestRandomSource** gives a random source to probe for it. After the random access through *getScore*, the *candidates* set and local variables are updated (among candidates, only c changes).

The end of the algorithm is controlled by the generic *StopCondition* predicate, which depends on the type of top-k result expected (e.g. with complete or incomplete scoring). The earliest end is obtained with predicate

$$StopCondition \equiv (|candidates| = k \land L_k \geq U_{unseen}) \qquad (1)$$

i.e. only k candidates are viable and there is no viable unseen object. It is simple to demonstrate that this condition is necessary and sufficient for a correct top-k result. Since this result may have incomplete scoring, additional conditions are necessary to ensure properties such as ordering or complete scoring of the results.

It is easy to see that any top-k algorithm in our context can be expressed in GF. Indeed, for a given query and set of sources, each algorithm is equivalent to the sequence of accesses to the sources it produces, which can be obtained with a sequence of decisions about the access type, the source and the candidate for random probes.

Given its ability to express any top-k algorithm, the GF framework is a valuable tool for comparing top-k strategies. In the following, we express in GF and compare three generic algorithms: a new variant of BR and new, generic and comparable variants of the NC and CA algorithms.

2.1 BreadthRefine

BreadthRefine (BR) [2] proposes a generic algorithm framework that can be instantiated to several variants. The main idea of the BR strategy is to maintain the set of current

---

[1] With random selection among candidates with the same score if necessary.

GF $(q, \mathcal{S})$

   candidates $\leftarrow \emptyset$; $U_{unseen} \leftarrow \mathcal{F}(max_1, ..., max_m)$; ... *//other local variables*

   repeat *//choice between sorted or random access*

      if *SortedAccessCondition()* **then** *//sorted access*

         $S_j \leftarrow$ **BestSortedSource()** *//choice of a sorted source*

         $(o, s) \leftarrow$ getNext(S_j) *//sorted access to the selected source*

         Update *candidates*, U_{unseen} and other local variables

      else *//random access*

         $c \leftarrow$ **ChooseCandidate()** *//choice of a candidate*

         $S_j \leftarrow$ **BestRandomSource**(c) *//choice of a random source*

         $s \leftarrow$ getScore(S_j, c) *//random access to the selected source*

         Update *candidates* and other local variables

      endif

   until *StopCondition()*

   return *candidates*

Fig. 3. The GF generic top-k framework

top-k candidates $\mathcal{U}_k$ as a whole, instead of focusing on the best candidate χ, which is the common approach.

The BR framework can be expressed in the more general GF framework by instantiating *SortedAccessCondition* and **ChooseCandidate** to realize the BR strategy.

The *SortedAccessCondition* in the BR strategy combines three conditions: $|candidates| < k$ **or** $U_{unseen} > U_k$ **or** *CostCondition()*. A sorted access is scheduled if (i) there are not yet k candidates, or (ii) an unseen object could belong to the current top-k $\mathcal{U}_k$ ($U_{unseen} > U_k$), or (iii) a generic *CostCondition* favors sorted access in the typical case where a random access is more expensive than a sorted one. Condition (ii) targets the decrease of U_{unseen} through sorted accesses and is the heart of the BR strategy for sorted sources: it maintains the whole current top-k free of unseen objects, while the common strategy is to consider only the best candidate ($U_{unseen} > U(\chi)$).

The BR strategy is completed by the **ChooseCandidate** function for refinement by random probes. All the existing algorithms facing this choice systematically select the best current candidate χ. The BR strategy maintains the k best candidates as a whole by first selecting the least refined candidate in $\mathcal{U}_k$.

BR considers top-k with incomplete scoring, thus *StopCondition* is given by (1).

BR-Cost*

Several instantiations of the BR framework have been proposed in [2]. The best one was *BR-Cost*, that fully implements the BR strategy and uses a *CostCondition* inspired from CA: if r is the ratio between the average costs of random and sorted accesses, then successive random probes must be separated by at least r sorted accesses.

In BR-Cost, **BestSortedSource** adopts the benefit-oriented strategy proposed by StreamCombine [9] for choosing a sorted source. The benefit of source S_j is $B_j = (\partial \mathcal{F} / \partial S_j) \times N_j \times \delta_j / C_s(S_j)$, where $(\partial \mathcal{F} / \partial S_j)$ is the weight of S_j in the aggregation function, N_j the number of candidates in $\mathcal{U}_k$ not yet seen in S_j, δ_j the expected decrease of the score in S_j and $C_s(S_j)$ the cost of a sorted access in S_j. Since $(\partial \mathcal{F} / \partial S_j)$ cannot be computed for any monotone function $\mathcal{F}$, we consider here, for simplicity, *only the*

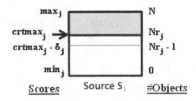

Fig. 4. Scores in a sorted source S_j

case of weighted sum, in which $(\partial \mathcal{F}/\partial S_j) = coef_j > 0$, where $coef_j$ is the coefficient corresponding to source S_j in the weighted sum.

BestRandomSource uses also a benefit-oriented strategy inspired from algorithms with controlled random probes such as Upper [4]; the benefit of source S_j is $B_j = coef_j \times (crtmax_j - min_j)/C_r(S_j)$, where $crtmax_j$ and min_j are respectively the current maximum score and the minimum score in S_j and $C_r(S_j)$ is the cost of a random probe in S_j. Note that $crtmax_j$ decreases in SR-sources (after sorted accesses), but remains constant (equal to max_j) in R-sources. Note also that $coef_j \times (crtmax_j - min_j)$ measures the reduction of the candidate's score interval size after a random probe in S_j.

We propose here *BR-Cost**, an improved variant of BR-Cost, using a different method for estimating r as a *ratio of benefits*. Roughly speaking, the benefit of an access to a source is the ratio between the refinement produced on the candidate score intervals and the cost of that access. This approach favors the comparison with the NC strategy.

Consider the case of a S-source S_j in Figure 4 at the moment when the current score is $crtmax_j$ and Nr_j objects have not been yet accessed. A sorted access to S_j refines the score of the retrieved object, but also produces a decrease δ_j of $crtmax_j$ that affects the upper bound of the remaining $Nr_j - 1$ objects. For the retrieved object, the width of the score interval decreases with $coef_j \times (crtmax_j - min_j)$. For each one of the remaining $Nr_j - 1$ objects, the upper bound decreases with $coef_j \times \delta_j$.

In conclusion, the benefit of a sorted access to S_j is:

$$B_s(S_j) = coef_j \times (crtmax_j - min_j + (Nr_j - 1) \times \delta_j)/C_s(S_j)$$

Benefit varies in time; if δ_j does not vary much, benefit globally decreases because $crtmax_j$ and Nr_j decrease. We approximate the average benefit by considering $\delta_j \approx (max_j - min_j)/N$, $crtmax_j \approx (max_j - min_j)/2$ and $Nr_j \approx N/2$:

$$\overline{B_s}(S_j) \approx coef_j \times (max_j - min_j)/C_s(S_j) \tag{2}$$

Benefit for a random access is computed in a similar way, but in this case only the score interval of the selected candidate changes. If S_j is a SR-source, the benefit, respectively the average benefit of a random access are:

$$B_{rs}(S_j) = coef_j \times (crtmax_j - min_j)/C_r(S_j)$$
$$\overline{B_{rs}}(S_j) \approx coef_j \times (max_j - min_j)/2C_r(S_j) \tag{3}$$

For a R-source $crtmax_j = max_j$ all the time, therefore

$$B_r(S_j) = \overline{B_r}(S_j) = coef_j \times (max_j - min_j)/C_r(S_j) \tag{4}$$

The global benefit SB (RB) of processing sorted (random) accesses is defined as the sum of average benefits of the sources allowing this kind of access.

$$SB = \sum_{S_j \in \mathcal{S}_S \cup \mathcal{S}_{SR}} \overline{B_s}(S_j) \qquad RB = \sum_{S_j \in \mathcal{S}_R} \overline{B_r}(S_j) + \sum_{S_j \in \mathcal{S}_{SR}} \overline{B_{rs}}(S_j)$$

where $\mathcal{S}_S$, $\mathcal{S}_R$ and $\mathcal{S}_{SR}$ are respectively the disjoint sets of S-, R- and SR-sources. In conclusion, the access ratio r used by BR-Cost* is:

$$r = SB/RB = \frac{\sum_{S_j \in \mathcal{S}_S \cup \mathcal{S}_{SR}} \frac{A_j}{C_s(S_j)}}{\sum_{S_j \in \mathcal{S}_R} \frac{A_j}{C_r(S_j)} + \sum_{S_j \in \mathcal{S}_{SR}} \frac{A_j}{2C_r(S_j)}} \qquad (5)$$

where $A_j = coef_j \times (max_j - min_j)$ is the amplitude of the interval produced by S_j in the aggregated score.

2.2 Necessary Choices

As mentioned above, *Necessary Choices* (NC) [19] was the first proposal for a generic algorithm, yet constrained to the case of complete top-k scoring. In this context, NC identifies *necessary accesses* at some moment, as being those for candidates in $\mathcal{U}_k$.

In this framework, NC proposes an algorithm *SR/G* that favors sorted against random accesses for each candidate. SR/G is guided by two parameters: $D = \{d_1, ..., d_m\}$, which indicates *a depth of sorted access* in each S- or SR-source, and H, which indicates *a fixed order of probes* in the random (R and SR) sources for all the candidates. The meaning of D is that sorted access to a source S_j where $crtmax_j \geq d_j$ has always priority against random probes.

Among all the possible couples (D, H), SR/G selects the optimal one by using sampling optimization. The optimization process converges iteratively: for some initial H one determines the optimal D, then an optimal H for this D, etc.

Despite its genericity, NC is hardly comparable with BR. In the context of incomplete top-k scoring adopted by BR, NC's analysis of necessary accesses is no longer valid. Source sampling used by SR/G is not always possible and does not guarantee similar score distribution. We propose here a variant of SR/G, adapted to the context of BR by considering incomplete scoring and a heuristic approximation of (D, H) inspired by BR-Cost*. The intention is to compare *the strategies* proposed by BR-Cost* and SR/G in a context as similar as possible.

The SR/G variant we propose is expressed in the GF framework as follows:

- Besides D and H, a local variable keeps the *best candidate*, i.e. the candidate in $\mathcal{U}_k$ with incomplete scoring having the highest upper bound. SR/G realizes a first sorted access to some source; the best object is initialized with this first retrieved object and updated after each iteration. Note that at least one object in $\mathcal{U}_k$ has incomplete scoring if the *StopCondition* has not been yet reached.
- *SortedAccessCondition* returns true if the set of sorted sources in which the best candidate has not been yet retrieved and where $crtmax_j \geq d_j$ is not empty.
- **BestSortedSource** returns one of the sources in this set.
- **ChooseCandidate** returns the best candidate.
- **BestRandomSource** returns the first random source not yet probed for the best candidate, following the order defined by H.
- *StopCondition*, for incomplete scoring, is given by (1).

We propose an heuristic approximation of D and H, based on the notion of source benefit used for BR-Cost*. For H we consider the descending order of the random source benefit computed with (3) and (4). Estimation of D is based on three hypotheses:

1. The number of sorted accesses to a source must be proportional to the source benefit given by (2).
2. Sorted accesses until depth d_j in each source should produce a decrease of the threshold enough for discriminating the top-k result, which is at least until $U_{unseen} = R_k$, where R_k is the k-th highest real score of an object.
3. If $n_j = N - Nr_j$ is the number of sorted accesses in S_j for reaching depth d_j (see Figure 4), the relation between n_j and d_j depends on the score distribution in sources, generally unknown and approximated here with uniform distribution.

If we note $\Delta_j = max_j - d_j$ the score decrease to reach depth d_j, we obtain:
1. $\forall j, n_j = C \times \overline{B_s(S_j)}$, where C is a constant.
2. $U_{max} - R_k = \sum coef_j \times \Delta_j$, where $U_{max} = \mathcal{F}(max_1, ..., max_m)$ is the highest possible aggregated score.
3. $\forall j, n_j/N = (max_j - d_j)/(max_j - min_j)$.

Resolving this equation system produces the following estimation for the depth:

$$d_j = max_j - \frac{A_j^2}{coef_j \times C_s(S_j)} \times \frac{U_{max} - R_k}{\sum_j A_j^2/C_s(S_j)} \tag{6}$$

Real score R_k may be estimated by various methods. This is not important in our experimental evaluation, since we precompute the R_k value, hence considering the best case for SR/G.

2.3 Combined Algorithm

Although Combined Algorithm (CA) [7], limited to SR-sources, is not a generic algorithm, it was a first attempt towards genericity, by proposing to combine NRA and TA strategies to adapt to the case of different costs for random and sorted access.

We propose here *CA-gen*, a generic variant of CA adapted to any source types. Like for CA, if r is the ratio between the average costs of random and sorted access, each sorted (S- and SR-) source is accessed r times, before performing all the random probes for the best candidate in $\mathcal{U}_k$ with incomplete scoring in random sources.

The cycle of r sorted accesses in each source can be simulated in GF with local variables indicating the next source to access and the number of accesses already performed in the cycle. Then *SortedAccessCondition* returns true if the cycle is not yet finished and **BestSortedSource** simply returns the next source. **ChooseCandidate** returns the best candidate, as defined above and **BestRandomSource** returns the first random source not yet probed for the best candidate. If no such source exists, the cycle stops. *StopCondition*, for incomplete scoring, is given by (1).

3   Approximation by Early Stopping

Top-k processing in our context is usually expensive because of predicate evaluation, therefore reducing the execution cost by accepting approximate results is a promising approach. We adopt the method proposed by TA_θ [7], based on relaxing the threshold condition in TA with a factor $\theta > 1$, i.e. the algorithm stops when the score of at least k candidates exceeds U_{unseen}/θ. This produces a θ-approximation of the top-k result, i.e. a set K_a of k objects such that $\forall x \in K_a, \forall y \notin K_a, \theta \times score(x) \geq score(y)$.

Note that this method is equivalent to an *early stopping* of the exact algorithm, i.e. TA and TA_θ have the same execution until the end of TA_θ, which occurs first.

We generalize here the TA_θ approach in the case of incomplete scoring within the GF framework and thus enable a comparison between various top-k algorithms.

Note that TA_θ considers that all source scores belong to the $[0, 1]$ interval. In the general case, in order to preserve the meaning of θ-approximations, we simply consider that scores in source S_j belong to a $[0, max_j]$ interval.

Consider an approximate solution K_a composed of k candidates with possibly incomplete scoring at some point during the execution of the algorithm in the GF framework. Then the condition for detecting K_a as a θ-approximation of the top-k result is given by the following theorem.

Theorem 1. *An approximate solution K_a composed of k objects with incomplete scoring is surely a θ-approximation of the top-k result iff $\theta \times min_{c \in K_a}(L(c)) \geq max_{c \notin K_a}(U(c))$*

Proof. Since $L(c) \leq score(c) \leq U(c)$, then $\forall x \in K_a, score(x) \geq L(x) \geq min_{c \in K_a}(L(c))$ and $\forall y \notin K_a, max_{c \notin K_a}(U(c)) \geq U(y) \geq score(y)$. If the theorem condition holds, then $\forall x \in K_a, y \notin K_a, \theta \times score(x) \geq score(y)$, i.e. K_a is a θ-approximation.

Consider now $x = argmin_{c \in K_a}(L(c))$ and $y = argmax_{c \notin K_a}(U(c))$. If the theorem condition does not hold for K_a, then $\theta \times L(x) < U(y)$, so it is possible that $\theta \times score(x) < score(y)$, i.e. K_a may not be a θ-approximation.

In the GF context, algorithms manage only the set of candidates discovered in sorted sources. Considering $K_a \subset candidates$, the stop condition (1) becomes:

$$\theta \times min_{c \in K_a}(L(c)) \geq max(U_{unseen}, max_{c \in candidates - K_a}(U(c))) \qquad (7)$$

The difference with Theorem 1 is that here U_{unseen} gives the upper bound score for all the objects not yet discovered and thus not members of *candidates*.

Theorem 2. *Eliminating non-viable candidates does not affect the stop condition (7).*

Proof. Suppose that at some moment a non viable candidate x affects the stop condition. Since $x \notin K_a$, x can only impact the right side of the inequality and only if $U(x) > U_{unseen}$ and $U(x) > U(y), \forall y \in candidates - K_a$. But $U(x) < L_k$ (x non-viable), so all the objects in $candidates - K_a$ are non-viable and $L_k > U_{unseen}$, which in accordance to (1) means that the exact top-k has been already found.

To estimate the precision of an approximate solution, we propose a distance measure based on two principles: (i) order and final scores of elements in the top-k solution are not important, and (ii) only wrong elements in the approximate solution affect precision.

Distance is measured by the difference between the real scores of wrong elements and R_k, the k-th score in the exact solution, normalized to the $[0, 1]$ interval by dividing it by R_k. Indeed, R_k is the maximum possible distance to R_k, since the lowest possible global score is 0. The distance between an element $o \in K_a$ and the real result K, respectively between K_a and K are defined as follows:

$$\text{dist}(o, K) = \begin{cases} \frac{R_k - score(o)}{R_k}, & \text{if } o \notin K \\ 0, & \text{if } o \in K \end{cases}, \qquad \text{dist}(K_a, K) = \frac{1}{k} \sum_{o \in K_a} \text{dist}(o, K) \quad (8)$$

We measure *the precision* of an approximate solution K_a as being $1 - dist(K_a, K)$.

The relation between our distance measure and θ-approximations is given by the following theorem.

Theorem 3. *If K_a is a θ-approximation of the real solution K, then $dist(K_a, K) \leq \theta - 1$.*

Proof. If $K_a = K$ then $dist(K_a, K) = 0$ and the inequality is true. Otherwise, considering $x \in K - K_a$, then $score(x) \geq R_k$. K_a being a θ-approximation of K, $\forall o \in K_a, \theta \times score(o) \geq score(x) \geq R_k$, so $R_k - score(o) \leq (\theta - 1)score(o)$.

In conclusion, $dist(K_a, K) = \frac{1}{k} \sum_{o \in K_a} dist(o, K) = \frac{1}{k} \sum_{o \in K_a - K} \frac{R_k - score(o)}{R_k} \leq \frac{1}{k} \sum_{o \in K_a - K} \frac{(\theta - 1)score(o)}{R_k} = \frac{\theta - 1}{kR_k} \sum_{o \in K_a - K} score(o) \leq \frac{\theta - 1}{kR_k} kR_k = \theta - 1$

We propose here a comparative study of the approximation potential of generic top-k algorithms. We draw cost-distance curves for these algorithms and compare their shapes. A point on the cost-distance curve indicates the precision of the approximate result on early stopping at that moment/cost. Since arbitrary early stopping comes with no guarantees on the precision of the approximate result, we also produce θ-approximations in each case and compare costs for measured and guaranteed precision.

4   Experiments

We experimentally compare the BR strategy with that of the other generic algorithms in terms of *execution cost* and of *approximation potential*.

Data Sets and Parameters
We use synthetic sources, independently generated as lists of scores in the $[0, 1]$ interval for the N objects, then organized for S, R or SR access, depending on the source type. We consider two types of score distribution in a source: uniform or exponential $(p(x) = \lambda e^{-\lambda x})$, for $\lambda = 1$ and restricted to the $[0, 1]$ interval. Exponential distribution illustrates S-sources where scores have fast decrease at the beginning, potentially more discriminant than sources with uniform distribution.

We measure the execution costs for each algorithm as the sum of costs of all the source accesses. We consider that all the sorted (random) accesses have the same cost C_s (C_r). Each result in the experiments is the average of 10 measures over different randomly generated sources. We consider weighted sum as the aggregation function, with coefficients randomly generated for each of the 10 measures.

The following parameters are considered in the experiments: $N = 10000$ database objects, $k = 50$, 6 sources of each type, and the most common cost settings, with random accesses more expensive than sorted ones ($C_r = 10$, $C_s = 1$). Two configurations for data distribution in sources are considered: *uniform* for all the sources or *mixed*, i.e.

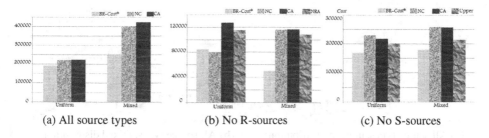

| (a) All source types | (b) No R-sources | (c) No S-sources |

Fig. 5. Execution cost comparison

exponential distribution for half of the sorted sources (3 S-sources and 3 SR-sources) and uniform for the other sources.

Comparison of the Execution Cost
We compare the execution cost of BR-Cost* with the NC variant and CA-gen in three configurations of source types: no R-sources, no S-sources, all source types. We also add to the comparison the reference non-generic algorithms compatible with that setting. In each configuration, uniform and mixed data distribution are considered.

All source types (Figure 5.a). BR-Cost* behaves visibly better (10%) than both NC and CA-gen for uniform distribution, while the difference becomes important for mixed distribution: approximatively 37% better than NC and 40% better than CA-gen.

No R-sources (only S and SR). Note that here cost and source settings are in favor of algorithms that realize only sorted access (NRA) or strongly favors them (NC). Figure 5.b shows that in the uniform distribution case BR-Cost* and NC are the best, with very close costs, much better than CA-gen (around 33%), which is even outperformed by NRA. For mixed distribution, BR-Cost* is clearly much better than NC and CA-gen (almost 60%), which are outperformed by NRA.

No S-sources (only R and SR). Figure 5.c shows that in all the cases BR-Cost* outperforms the other algorithms and that NC and CA-gen are less adapted to this setting, performing worse than Upper. The benefit of using BR-Cost* is bigger in the mixed distribution case (around 28% better than NC and CA-gen) compared to uniform distribution (24%). Compared to Upper, the benefit is similar in both cases, around 15%.

In conclusion, BR-Cost* successfully adapts to various source types and data distribution settings, and outperforms not only the other generic approaches, but also specific algorithms designed for that case. We also note a weakness for the other generic strategies in one of the studied cases: no S-sources for NC and no R-sources for CA-gen. Paradoxically, mixed distribution does not improve cost in most cases; we guess that discriminant distributions are counter-balanced here by the lack of correlation between sources and by their relatively high number.

Approximation Potential
We measure the potential of approximation by early stopping of the generic top-k algorithms by drawing their cost-distance curves. For space reasons, only the case of all source types is presented here.

Distance between approximate and real solution, computed with Formula (8), is measured every 2000 cost units during the algorithm's normal execution and a curve relying

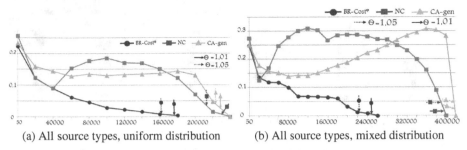

(a) All source types, uniform distribution   (b) All source types, mixed distribution

Fig. 6. Approximation with the best k upper bound scores, all source types

these points is extrapolated. Each point on the curve represents the distance between the approximate solution and the real one if the algorithm stops at that moment. A curve "below" another one indicates a better approximation potential.

The form of the curve also indicates *approximation stability*. A monotone descending curve means stable approximation, that improves if execution continues, while non-monotony indicates an algorithm badly adapted for approximation by early stopping.

For each cost-distance curve we measure the end point that corresponds to a θ-approximation obtained with the stop condition (7). We consider two values, $\theta = 1.05$ and $\theta = 1.01$, that correspond to a guaranteed distance of 0.05, respectively 0.01 (see Theorem 3), i.e. a precision of 95%, respectively 99%. We compare the position of these points with that of the intersection between the curve and the corresponding distance.

We consider two cases for the approximate solution. The first one is the set $\mathcal{U}_k$ of k candidates with the highest upper bound. This is a natural choice for the approximate solution, since $\mathcal{U}_k$ is the set of candidates that top-k algorithm focus on during execution. More precisely, all the algorithms proposed so far base their strategies on $\mathcal{U}_k$, either for deciding a sorted access, or for the choice of a candidate for random probes. Intuitively, candidates with high upper bounds must be "refined" because their upper bound make them potentially belong to the final top-k. The algorithm *must* decide if they really belong to the result or not - if not, the algorithm cannot end without refining the candidate's score to make it non-viable.

The second proposal for an approximate solution is the set of k candidates with the highest lower bound $\mathcal{L}_k$. Intuitively, belonging to $\mathcal{L}_k$ means that the candidate was already refined with good scores in some sources. This may be a good indication that the candidate belongs to the final top-k, better than for $\mathcal{U}_k$ where high upper bounds may be the result of little refinement, thus with high uncertainty.

Approximation with $\mathcal{U}_k$

Figure 6 presents the cost-distance curves for uniform and mixed data distributions. Final costs for algorithms may be less visible, they are already indicated in Figure 5.

For uniform distribution (Figure 6.a), BR-Cost* approximation distance quickly decreases and the algorithm has clearly better approximation properties than CA-gen (much higher distance, only decreasing at the end) or NC (totally unstable). Mixed distribution (Figure 6.b) accentuates the problems of NC and CA-gen (which becomes unstable), while BR-Cost* keeps a good curve shape. However, θ-approximation significantly reduces the cost saving for BR-Cost*, e.g. for θ=1.05 algorithm stops

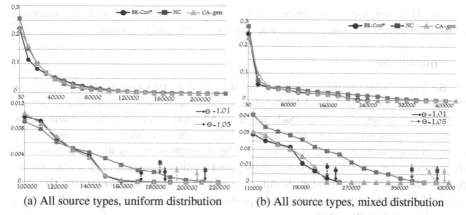

(a) All source types, uniform distribution   (b) All source types, mixed distribution

Fig. 7. Approximation with the best k lower bound scores, all source types

at cost 160 000, while the corresponding distance of 0.05 is already reached at cost 70 000.

In conclusion, BR-Cost* has clearly the best approximation potential with $\mathcal{U}_k$ among the generic algorithms, with good properties for the different data distributions. The other generic algorithms are badly adapted to approximation with $\mathcal{U}_k$: NC and CA-gen are systematically unstable. We guess that the good approximation properties of BR-Cost* come from its breadth-first strategy. Handling the current top-k $\mathcal{U}_k$ as a whole, instead of focusing on the best candidate only, produces a more stable evolution of $\mathcal{U}_k$ toward the final solution. The price to pay for guaranteed precision in θ-approximations is important for algorithms with good approximation curves - we notice a significant difference with the potential cost at the same precision. Comparison with specific algorithms (not presented here for space reasons) suggests that the cost of θ-approximations is dependent on the total cost of the algorithm: for algorithms with very close cost-distance curves, higher total costs systematically lead to higher θ-approximation costs.

Approximation with $\mathcal{L}_k$

For both uniform (Figure 7.a) and mixed distribution (Figure 7.b), the curves for all the algorithms are very close, stable, with good approximation potential. The sub-figure for each case presents, besides the curves, a zoom on the final part of the execution, where curves are very close. BR-Cost* and CA-gen have slightly better curves than NC, the difference being visible in the mixed distribution case and on the final part of the uniform case. Comparison of θ-approximations follows the conclusion of the previous point, algorithms with better execution costs produce better θ-approximations, i.e. BR-Cost* is the best, while CA-gen and NC are very close. We notice that cost-distance curves with $\mathcal{L}_k$ are better than those with $\mathcal{U}_k$ in all the cases. This also leads to an increased difference between the cost with θ-approximation and the potential one.

In conclusion, we notice that approximation with $\mathcal{L}_k$ has better quality than with $\mathcal{U}_k$ for all the algorithms. Compared with the $\mathcal{U}_k$ case, approximation is always stable with $\mathcal{L}_k$ and has better precision at the same execution cost. Even if BR-Cost* has globally the best properties, the approximation potential of generic algorithms is very close in this case. However, θ-approximations are not improved by $\mathcal{L}_k$ and lead to an increased difference between the potential cost and that for guaranteed precision.

5   Conclusion

The BR breadth-first strategy adapts itself very well to various source type configurations and data distributions, leading to better execution cost than the other generic or specific strategies. Also, it globally has the best approximation potential among the generic strategies, with a clear advantage in the $\mathcal{U}_k$ approximation case. However, $\mathcal{L}_k$ approximation produces better results for all the algorithms and greatly reduces the differences between them. We noticed that θ-approximation is weakly correlated with the approximation potential and significantly depends on the total execution cost. This cancels the difference between $\mathcal{U}_k$ and $\mathcal{L}_k$ approximation and favors again the BR strategy that produces better total costs.

References

1. Akbarinia, R., Pacitti, E., Valduriez, P.: Best position algorithms for top-k queries. In: VLDB, pp. 495–506 (2007)
2. Badr, M., Vodislav, D.: A general top-k algorithm for web data sources. In: Hameurlain, A., Liddle, S.W., Schewe, K.-D., Zhou, X. (eds.) DEXA 2011, Part I. LNCS, vol. 6860, pp. 379–393. Springer, Heidelberg (2011)
3. Böhm, C., Berchtold, S., Keim, D.A.: Searching in high-dimensional spaces: Index structures for improving the performance of multimedia databases. ACM Comput. Surv. 33(3), 322–373 (2001)
4. Bruno, N., Gravano, L., Marian, A.: Evaluating top-k queries over web-accessible databases. In: ICDE, p. 369 (2002)
5. Cao, P., Wang, Z.: Efficient top-k query calculation in distributed networks. In: PODC, pp. 206–215 (2004)
6. Chang, K.C.-C., Hwang, S.W.: Minimal probing: supporting expensive predicates for top-k queries. In: SIGMOD Conference, pp. 346–357 (2002)
7. Fagin, R., Lotem, A., Naor, M.: Optimal aggregation algorithms for middleware. J. Comput. Syst. Sci. 66(4), 614–656 (2003)
8. Güntzer, U., Balke, W.-T., Kießling, W.: Optimizing multi-feature queries for image databases. In: VLDB, pp. 419–428 (2000)
9. Güntzer, U., Balke, W.-T., Kießling, W.: Towards efficient multi-feature queries in heterogeneous environments. In: ITCC, pp. 622–628 (2001)
10. Ilyas, I.F., Aref, W.G., Elmagarmid, A.K.: Supporting top-k join queries in relational databases. VLDB J. 13(3), 207–221 (2004)
11. Ilyas, I.F., Beskales, G., Soliman, M.A.: A survey of top-k query processing techniques in relational database systems. ACM Comput. Surv. 40(4) (2008)
12. Li, C., Chang, K.C.-C., Ilyas, I.F.: Supporting ad-hoc ranking aggregates. In: SIGMOD Conference, pp. 61–72 (2006)
13. Li, C., Chang, K.C.-C., Ilyas, I.F., Song, S.: Ranksql: Query algebra and optimization for relational top-k queries. In: SIGMOD Conference, pp. 131–142 (2005)
14. Mamoulis, N., Cheng, K.H., Yiu, M.L., Cheung, D.W.: Efficient aggregation of ranked inputs. In: ICDE, p. 72 (2006)
15. Marian, A., Bruno, N., Gravano, L.: Evaluating top-k queries over web-accessible databases. ACM Trans. Database Syst. 29(2), 319–362 (2004)

16. Michel, S., Triantafillou, P., Weikum, G.: Klee: A framework for distributed top-k query algorithms. In: VLDB, pp. 637–648 (2005)
17. Natsev, A., Chang, Y.-C., Smith, J.R., Li, C.-S., Vitter, J.S.: Supporting incremental join queries on ranked inputs. In: VLDB, pp. 281–290 (2001)
18. Theobald, M., Weikum, G., Schenkel, R.: Top-k query evaluation with probabilistic guarantees. In: VLDB, pp. 648–659 (2004)
19. Hwang, S.W., Chang, K.C.-C.: Optimizing top-k queries for middleware access: A unified cost-based approach. ACM Trans. Database Syst. 32(1), 5 (2007)

Evaluating Spatial Skyline Queries on Changing Data

Fabiola Di Bartolo and Marlene Goncalves

Universidad Simón Bolívar, Departamento de Computación, Apartado 89000
Caracas 1080-A, Venezuela
{fbartolo,mgoncalves}@usb.ve

Abstract. The amount of data being handled is enormous these days. To identify relevant data in large datasets, Skyline queries have been proposed. A traditional Skyline query selects those points that are the best ones based on user's preferences. Spatial Skyline Queries (SSQ) extend Skyline queries and allow the user to express preferences on the closeness between a set of data points and a set of query points. However, existing algorithms must be adapted to evaluate SSQ on changing data; changing data are data which regularly change over a period of time. In this work, we propose and empirically study three algorithms that use different techniques to evaluate SSQ on changing data.

Keywords: Skyline queries, Spatial Skyline queries, changing data.

1 Introduction

Wireless sensor networks (WSN) are commonly used to monitor control variables of diverse applications, such as environmental, medical, etc. Monitoring systems over WSN are able to handle in real time the data generated by sensors. Sensor data are highly changing, i.e., new observations are recorded even while using the monitoring systems. Usually, very large datasets are collected by sensors since sensor data regularly change over a period of time. In this sense, individuals and communities require to analyze a broad range of data; therefore, techniques to filter relevant data must be applied. To identify relevant data in large datasets, Skyline queries have been proposed.

A traditional Skyline query selects those points that are the best ones according to multiple user's criteria. Spatial Skyline Queries (SSQ) are an extension of Skyline queries; they allow to express preferences on the closeness between a set of data points and a set of query points [5].

In this work, we study SSQ on changing data. Suppose an online mobile recommendation system that is able to suggest the best restaurants to be booked by a customer who would like to have a lunch with 5 friends from 2 p.m. until 4 p.m. The customer wants to find a restaurant close to each friend's location with a low serving time and a high overall score. Based on the customer criteria, the result must include the restaurants with the smallest distance from each friend's location, the lowest serving time and the highest overall score. To identify the best restaurants, a Skyline query may be evaluated. A restaurant belongs to the Skyline set if and only if there is no other restaurant which is nearer to each friend's location, and also it has a lower serving time and a higher overall score. In this example, the data is changing, e.g., the friends may move

H. Decker et al. (Eds.): DEXA 2013, Part I, LNCS 8055, pp. 270–277, 2013.

continuously or the restaurant availability may change frequently. Therefore, the Skyline set changes depending on each friend's location and/or restaurant avalaibility. If the friends move a significant distance, some restaurants must be discarded because they are further away or new restaurants must be included because of their closeness. Additionally, availability of a restaurant may change since some customers finished eating, cancelled or made new reservations. Thus, the recommendations or the Skyline must be continuously refreshed according to changes of data. Particularly, availability is a changing non-spatial datum and location is a moving query point.

On the other hand, time complexity for answering SSQ is higher than evaluating traditional Skyline queries due to calculation of multiple distances to the query points. Evaluation time is even higher in SSQ on changing data because of distance calculation and Skyline recalculation. Our contribution is to provide solutions to the problem of evaluating SSQ on changing data. To the best our knowledge, algorithms for SSQ do not assume non-spatial data changes [6] and Skyline algorithms for changing non-spatial data only allow one query point [2–4, 7, 8]. We propose three algorithms that use different techniques to calculate the Skyline Spatial when data are changing. The first algorithm, AB2S, is a naive solution which calculates the Skyline Spatial set using the BNL algorithm [1] at each change of the data. The second algorithm, VC2S+, is an adaptation of VCS^2 [6] that allows changing data. The third algorithm, CD2S, is able to return the Skyline Spatial based on changing characteristics of the data and also prunes the search space using spatial properties in order to avoid exhaustive searches. We have empirically studied the performance of the three algorithms proposed to evaluate SSQ on changing data. In the best case, our experimental study shows that average execution time of CD2S was approximately one-third of the time taken by AB2S and VC2S+.

The paper is composed of four sections. Section 2 describes the three algorithms proposed to evaluate SSQ on changing data. Section 3 reports the results of our experimental study. Finally, Section 4 contains the conclusions and future work.

2 Proposed Algorithms

We propose three algorithms to evaluate SSQ on changing non-spatial data: AB2S, VC2S+ and CD2S. Particularly, AB2S and VC2S+ are adaptations of existing algorithms [1, 6] which consider changes of the non-spatial data and a moving query point.

All algorithms receive a Spatial Skyline query composed by multiple query points and non-spatial dimensions, one changing non-spatial dimension, and one moving query point. They use the Euclidean distance function to compute closeness between points.

The Spatial Skyline on changing data may transform its state according to the following: 1) *Initial Configuration*: Corresponds to the original values of the data. There is not any change of data; 2) *State Change Configuration*: New data state after a non-spatial dimension changes its value in at least one registry; and 3) *Movement Change Configuration*: New data state after a moving query point has changed.

2.1 Naive Algorithm

Adapted BNL for Spatial Skyline (AB2S) is a naive algorithm based on BNL (Block Nested Loop) [1]. The AB2S algorithm does not have methods to update the Spatial

Skyline according to changes of a moving query point. The Spatial Skyline is built in any configuration to identify the correct points. AB2S starts on the *Initial Configuration* which finds the candidates, computes the distances from each candidate to the query points and calculates the Spatial Skyline on the candidates. A point is candidate if its changing non-spatial dimension value is equal to a given state, e.g, the restaurant state is available. In the *State Change Configuration*, AB2S updates the current candidates. The AB2S algorithm determines the points that change its state. Those points in a given state are now candidates. For each candidate that has not had a given state before, its distance to the query points is calculated. Otherwise, for those candidates that already had a given state at some time, their distance to the moving query point are updated. Also, AB2S deletes invalid candidates, e.g., points that are no longer available. Finally, dominance[1] checks are done to update the Skyline set. In the *Movement Change Configuration*, AB2S only updates distances from each candidate to the moving query point. With the new distances the Spatial Skyline is calculated through dominance checks. In any configuration, AB2S performs the BNL algorithm on the found candidates.

2.2 Adaptation of VCS^2

We adapted VCS^2[6] to evaluate SSQ on changing non-spatial data. This adapted algorithm is named VC2S+. VC2S+ calculates the non-spatial Skyline or the Skyline on non-spatial dimensions, and builds the Voronoi diagram using the candidates; incorrect results may be produced if the unavailable points are included in the Voronoi diagram. The main disadvantage of this adaptation is that the Voronoi diagram and the Delaunay graph must be reconstructed in the *State Change Configuration*.

VC2S+ starts on the *Initial Configuration* which finds the candidates, constructs the Delaunay graph and proceeds to calculate the Non-Spatial Skyline on the candidates. Subsequently, VC2S+ computes the Spatial Skyline using the VS^2 algorithm [6].Thus, the Delaunay graph scanning begins. Some dominance checks may be avoided, e.g., a point belonging to the convex hull is a Skyline point, and therefore, it is not necessary to check if it is dominated.

In the *State Change Configuration*, VC2S+ re-initializes data structures for the Skyline, the Spatial Skyline, the Non-Spatial Skyline, and the Delaunay graph. The candidates are updated depending on state changes of the points. The Non-Spatial Skyline is computed, and the search region and Delaunay graph are recalculated considering updated candidates. Using the Delaunay graph, the Spatial Skyline is identified. Lastly, the final Skyline is built merging the Non-Spatial Skyline and the Spatial Skyline.

In the *Movement Change Configuration*, VC2S+ saves the previous convex hull polygon, and creates the convex hull that was modified by the movement of the query point. The algorithm identifies a movement pattern [6] and updates the Skyline set in almost all the patterns. Also, the distances from each candidate to the moving query point are modified. The new search region is calculated. Finally, VC2S+ returns the final Skyline after updating and merging the Non-Spatial Skyline and the Spatial Skyline.

---

[1] A point A dominates a point B if it is better or equal in all criteria, and is better in at least one criterion.

2.3   Continuous Dynamic Spatial Skyline Algorithm

The Continuous Dynamic Spatial Skyline (CD2S) algorithm identifies the Spatial Skyline in terms of the changing features of the data. In *Initial Configuration*, CD2S initializes a d-dimensional matrix to storage the state value of each point in the dataset P^d. Then, it fills the lists of query points and non-spatial dimensions with the coordinates and values indicated by the user. Also, the search region and the convex hulls are created. One of the convex hulls is composed of all the query points, and it is named Global Convex Hull while the Static Convex Hull comprises the query points excluding the moving query point. Initially, the search region is the minimum bounding rectangle that includes all the points in P^d.

To explain the algorithm, we assume that the user preferences are expressed over points that have a particular Boolean state, for instance the value *True* or *1*. Therefore, points whose state value is *True* are the candidates. To find these candidates in the matrix, the algorithm has to read the row value (or a piece of the row) as a number generated by the bits contained there. Only the rows or row pieces that have a value greater than zero are going to be verified to select the candidates. Then, CD2S looks for the candidates inside the search region and save them in a list. If the candidate is inside the Global Convex Hull, it is directly marked as Skyline. In this step, the Skyline is separated into two groups: Static Skyline and Dynamic Skyline. The first group is the set of candidates inside the Static Convex Hull, and the second group is the set of candidates inside the Global Convex Hull and outside the Static Convex Hull. While each point is marked as Skyline, if there are no preferences on non-spatial dimensions, the search region is reduced by the intersection with the dominance region of the point. In presence of non-spatial dimensions, similar as the VC2S+ algorithm, the Non-Spatial Skyline is calculated checking dominance between candidates considering only the non-spatial dimensions. In this case, the search region is composed of the union of the dominance regions of the Non-Spatial Skyline points. Next, the remaining Skyline is calculated, the Outer Skyline, by means of the dominance checks of the candidates inside the search region and outside the Global Convex Hull against previous Skyline.

In the *State Change Configuration*, when a change ocurrs CD2S updates the Skyline set. First, it stores the previous search region in a variable. Then, the matrix is synchronized with the new state values of the points and the points with a state value not required by the user (e.g. *False* or *0*) are removed from the candidate list and the Skyline set. If none was deleted from the Skyline in the previous step, then the candidate list is temporarily stored in another variable and the current candidate list is reinitialized. The reason is that old candidates which are not in the Skyline set, they will still be dominated by the current Skyline. Nevertheless, if any Skyline point was deleted, then the candidate list is held because old candidates can now be Skyline points and, the new search region is created considering only the Skyline points that were not deleted. The points whose state has changed to the value required and also are inside the search region, are added to the candidate list and analyzed to be included in the Skyline. If a candidate is inside the Global Convex Hull, the point is directly added to the Skyline set, but some skyline points outside the Global Convex Hull could be dominated by this new point; therefore, dominated points have to be deleted from the Skyline.

If some skyline point was deleted, it is necessary to find the candidates outside the old search region and inside the new one. These points have to be added to the candidate list, also this list is revised to remove points that are not in the new search region. Through the candidate list and excluding the added objects, the remaining Skyline set is obtained

For each position change of the moving query point, in the *Movement Change Configuration*, CD2S updates the Skyline set. The algorithm applies the same pattern identification as VCS^2 [6] to evaluate the Skyline depending on the case. In this step, the new Global Convex Hull is calculated, but the Static Convex Hull always is the same. Also, the region where the Skyline is not affected by the movement is generated; this region is named the Invariant Sector.

If there is any non-spatial dimension, the distances of the Non-Spatial Skyline points to the moving query point are updated and the new search region is obtained (this Skyline does not change) The Dynamic Skyline points and Outer Skyline points are analyzed to delete dominated points; also, the non-skyline points in the candidate list inside the new search region are checked to add new points to the Dynamic Skyline and Outer Skyline. Before a point may be compared, its distance to the moving query point has to be updated.

3   Experimental Study

We empirically study the three algorithms: AB2S, VC2S+, and CD2S. We synthetically generated several datasets and Spatial Skyline queries by means of a Java program. 200 datasets comprise points with a spatial location, three non-spatial dimensions, and a Boolean non-spatial dimension. Dataset sizes vary between 10,000 and 100,000 points. Spatial Skyline Queries contain from 5 to 18 query points, from 0 to 3 non-spatial dimensions and a Boolean non-spatial dimension. In all experiments, we measured the total execution time and the number of dominance checks or comparisons.

Using these datasets and random change configurations, 340,000 simulations for the three algorithms have been fulfilled. The simulations were generated in groups of 50. The first simulation corresponds to the *Movement Change Configuration*, and the others are continuous transformations on the data based on previously generated configurations. Three parameters were used in the simulations: type of change (mode), percentage of points whose changing non-spatial dimension was updated (%stC), maximum movement distance in meters for the moving query point (maxD). Also, for each dataset was used a percentage from the total area occupied by the points in P which was used for the initial position of the query points in Q (%area). Finally, the three algorithms were implemented in Java and were executed on a *Intel Q6600 CPU* with 4 GB RAM and disk of 100 GB.

3.1   Experiment Settings

We conduct five experiments. The first experiment shows the performance on 10,000 points of the three algorithms when the number of query points and non-spatial dimensions vary; the parameters are: mode=All and %stC=random in two variants: #1 maxD=10 and #2 maxD=20. The second experiment considers the movement of one query point for datasets of 10,000 and %area=10 in five variants: #1 maxD=5, #2 maxD=10, #3 maxD=15, #3 maxD=20 and #4 maxD=25. The third experiment reports the performance of the algorithms when the non-spatial dimension change its state on datasets of 10,000 points with %area=10 in five variants: #1 %stC=10, #2 %stC=30, #3 %stC=rand, #3 %stC=70 and #4 %stC=90. The fourth experiment shows the impact of a variation of %area for datasets of 50,000 in three variants: #1 %area=10-maxD=10, #2 %area=25-maxD=10 and #3 %area=50-maxD=10. Finally, the fifth experiment presents the results for datasets of 100,000 using queries from 12 to 18 query points and the parameter values %stC=random, maxD=50 and %area=10 in two variants: #1 any non-spatial dimension and #2 one non-spatial dimension.

3.2   Query Processing Performance

In this section, we compare our proposed algorithm CD2S against the AB2S and VC2S+ algorithms. In the first experiment, CD2S was the best algorithm in terms of execution time. In fact the average execution time of CD2S is almost the 60% and 50% of the time of AB2S and VC2S+. We can also observe that the average execution time of CD2S is closer to the best execution time taken by any of the three algorithms while

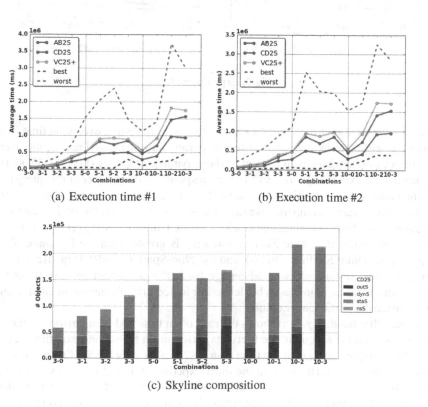

(a) Execution time #1   (b) Execution time #2

(c) Skyline composition

Fig. 1. Experiment I

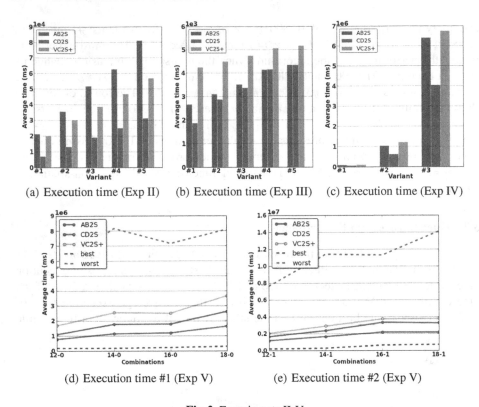

(a) Execution time (Exp II)   (b) Execution time (Exp III)   (c) Execution time (Exp IV)

(d) Execution time #1 (Exp V)   (e) Execution time #2 (Exp V)

Fig. 2. Experiments II-V

the average execution time of VC2S+ is the closest to the worst execution time. For 50 continuous simulations generated using the parameters %stC and maxD, the average execution time increases with respect to these parameter values (Figs. 1(a) and 1(b)). The horizontal axis of Figs. 1(a) and 1(b) corresponds to the number of the query points and the number of non-spatial dimensions. The execution time increases because of the number of dominance comparisons also increases. Additionally, the skyline cardinality increases too. Moreover, since the Static Convex Hull area is bigger when there are more query points, the Static Skyline (staS) is greater than the Dynamic Skyline (dynS), the Outer Skyline (outS) and the Non-Spatial (nsS) (Fig. 1(c)). Finally, a bigger Static Skyline set is an advantage for CD2S algorithm because it holds the Skyline set unless an point has to be deleted or added due to a new state change on the points inside the Static Convex Hull.

On the other hand, Fig. 2(a) shows the execution time and comparisons of the three algorithms from the second to the fourth experiment. In the second experiment, CD2S and VC2S+ require less execution time than AB2S to identify the Skyline when a query point is in movement. However, in the third experiment, Fig. 2(b) shows that the execution time of CD2S and AB2S is better than VC2S+ when a state change is produced. For a medium dataset in both change types, the fourth experiment reports that the execution time of VC2S+ is worse than CD2S and AB2S. This is because of the number of dominance comparisons and the construction of the Voronoi diagram (Fig. 2(c)). The best performance of CD2S compared to the others is in the fifth experiment #1. The average execution time taken by CD2S is 33% and 35% of the time of VC2S+ and AB2S, and its number of dominance comparisons is at least one order of magnitude smaller.

Additionally, we can observe in the Figures 2(d)) and 2(e) that the SSQ evaluation with a non-spatial dimension is more costly than SSQ answering with a spatial dimension. This is because the Skyline set is less selective.

4 Conclusions

In this paper, we have proposed three algorithms to evaluate SSQ on changing data. In this work, changes on the query points and the non-spatial data are allowed. We have adapted two state-of-the-art algorithms to consider changing data. Also, we have defined a new algorithm named CD2S which is able to reduce dominance comparisons and decrease the execution time. Our experimental study showed that CD2S reduces the number of dominance comparisons and execution time because it avoids rebuilding the whole Skyline each time that data are updated. However, only step wise data changes are allowed. As a future work, we plan to extend our algorithms for continuous changes. Also, we will perform experiments using real-world datasets and we will study theoretical time complexity of the the different algorithms in terms of $O-$notation.

References

1. Borzsonyi, S., Kossmann, D., Stocker, K.: The skyline operator. In: IEEE Conf. on Data Engineering, pp. 421–430 (2001)
2. Chen, N., Shou, L.-D., Chen, G., Gao, Y.-J., Dong, J.-X.: Prismo: predictive skyline query processing over moving objects. Journal of Zhejiang University Science C 13, 99–117 (2012)
3. Huang, Y.-K., Chang, C.-H., Lee, C.: Continuous distance-based skyline queries in road networks. Information Systems 37(7), 611–633 (2012)
4. Kodama, K., Iijima, Y., Guo, X., Ishikawa, Y.: Skyline queries based on user locations and preferences for making location-based recommendations. In: Proceedings of the 2009 International Workshop on Location Based Social Networks, LBSN 2009, pp. 9–16. ACM, New York (2009)
5. Sharifzadeh, M., Shahabi, C.: The spatial skyline queries. In: VLDB 2006: Proceedings of the 32nd International Conference on Very Large Data Bases, pp. 751–762. VLDB Endowment (2006)
6. Sharifzadeh, M., Shahabi, C., Kazemi, L.: Processing spatial skyline queries in both vector spaces and spatial network databases, vol. 34, pp. 14:1–14:45. ACM, New York (2009)
7. Shen, H., Chen, Z., Deng, X.: Location-based skyline queries in wireless sensor networks. In: NSWCTC 2009: Proceedings of the 2009 International Conference on Networks Security, Wireless Communications and Trusted Computing, pp. 391–395. IEEE Computer Society, Washington, DC (2009)
8. Zheng, B., Lee, K.C.K., Lee, W.-C.: Location-dependent skyline query. In: 9th International Conference on Mobile Data Management, MDM 2008, pp. 148–155 (April 2008)

SONIC: Scalable Multi-query OptimizatioN through Integrated Circuits

Ahcène Boukorca[1], Ladjel Bellatreche[1], Sid-Ahmed Benali Senouci[2], and Zoé Faget[1]

[1] LIAS/ISAE-ENSMA, Futuroscope, Poitiers, France
(boukorca,bellatreche,zoe.faget)@ensma.fr
[2] Mentors Graphics 38330 Montbonnot-Saint-Martin, France
sid-ahmed_senouci@mentor.com

Abstract. In the first generation of databases, query optimizers were designed to optimize individual queries. Due to the predefined number of tables of a given database, the probability to have interaction between queries is high. As a consequence, optimizers propose solutions for multi-queries optimization. Getting this optimization is known as NP-hard problem. To ensure a scalable solution, we borrow techniques used in the electronic design automation (EDA) domain. In this paper, we first make an analogy between the multi-query optimization problem and the EDA domain. Secondly, we propose to model our problem with hypergraphs massively used to design and test integrated circuits. Thirdly, we use our results to materialize views. Finally, experiments are conducted to show the scalability of our approach.

1 Introduction

The age of extremely large data is now a reality. This situation consolidates the traditional position of the database technology as a support for storing, managing, and accessing it in an efficient way. The database community made several attempts on query optimization since the early 70s. Several algorithms and systems have been proposed, such as the *System-R project* and its ideas have been largely incorporated in many commercial optimizers. The difficulty of query optimization is mainly related to the fact that each SQL query corresponding to a select-project-join query in the relational algebra may be represented by many query trees. The leaves of each query tree represent base relations and non-leaf nodes are algebraic operators such as selections, projections, unions or joins. An intermediate node indicates that the corresponding operator is applied on the relations generated by its children, the result of which is then sent further up. Thus, the edges of a tree represent data rows from bottom to top, i.e., from the leaves which correspond to data in the database to the root which is the final operator producing the query answer. For a complicated query, the number of all possible query trees may be enormous due to many algebraic laws that hold for relational algebra: commutative and associative laws of joins, laws involving selection and projection push down along the tree, etc.

H. Decker et al. (Eds.): DEXA 2013, Part I, LNCS 8055, pp. 278–292, 2013.
© Springer-Verlag Berlin Heidelberg 2013

To choose the optimal solution for a given query tree, a database optimizer may employ one of two optimization techniques: the rule-based optimization approach ($\mathcal{RBA}$) and the cost based optimization approach ($\mathcal{CBA}$). The $\mathcal{RBA}$ is the oldest one. It is simple since it is based on a set of rules concerning for instance the implementation algorithms of the join, the use of an index or not, whether the choice of the external relation is a nested loop, and so on. The optimizer chooses an execution plan based on the available access paths and their ranks. For instance, Oracle's ranking of the access paths is heuristic. If there is more than one way to execute a query, then the $\mathcal{RBA}$ always uses the operation with the lower rank. Usually, operations of lower rank execute faster than those associated with constructs of higher rank. In the $\mathcal{CBA}$, the optimizer estimates the cost of each possible execution plan[1] by applying heuristic formulas using a set of statistics about the database (sizes of tables, indexes, tuple length, selectivity factors of join and selection predicates, sizes of intermediate results, etc.) and hardware (size of buffer, page size, etc.). For each execution plan, the query optimizer does the following tasks: (1) the selection of an order and grouping for associative-and-commutative operations such as joins, unions and intersection, (2) the choice of implementation algorithms for different algebraic operations: for example, selections may be implemented using either a sequential scan or an index scan. Join operation may be implemented in different ways: nested loop, sort merge join and hash join (see Figure 1), (3) the management of additional operators like group-by, sorting, etc. The cost-based optimizer chooses the plan that has the lowest cost using advanced techniques (e.g., programming approaches as in System R). An important point to be mentioned is that the quality of the $\mathcal{CBA}$ depends strongly on the recency of the statistics. Deciding which statistics to create is a difficult task [1].

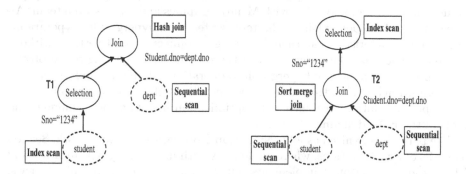

Fig. 1. Two different plans for the same query

---

[1] A query execution plan is a set of steps used to access data in relational databases. Figure 1 gives an example of an execution plan, where dashed circles and solid circles represent base tables and algebraic operations, respectively. Other execution plans might be generated.

The fact that a database is composed of a set of predefined tables indicates that there is a great chance that queries span common tables. This phenomenon is known as query interaction. In the 80s, several studies exploiting this interaction have been elaborated (for more details see [2]). Business intelligence technology and scientific applications use relational data warehouse technology to model their applications, thus increasing even more the focus on query itneraction [3,4]. A data warehouse ($\mathcal{DW}$) is usually modeled with a relational schema (star schema, snow flake schema). A star schema consists of one or several fact tables related to multiple dimension tables via foreign key joins. Dimension tables are relatively small compared to the fact table and rarely updated. They are typically denormalized so as to minimize the number of join operations required to evaluate a query. The typical queries on the star schema are called star join queries. They are characterized by: (i) a multi-table join among a large fact table and dimension tables and (ii) each of the dimension tables involved in the join operation has multiple selection predicates. In such context, an intermediate results benefit several queries. This phenomenon is known as multi-query optimization. Instead of considering each query plan individually, they are merged based on the common nodes in order to increase query interaction. Getting the optimal merged query plan is NP-hard problem. Several algorithms were proposed to deal with this problem, of mainly three types: dynamic programming algorithms [5,6], genetic and randomized algorithms [7]. A model for selecting the optimal merged query plan has to be established to facilitate the use of relevant algorithms. Existing works dealing with this problem use a cost model to quantify the quality of the obtained solution [6]. The major problem with these solutions is their scalability.

To reduce this complexity, we propose to use graph theory algorithms which have been proved of high interest in the electronic design automation domain, since electronic circuits manipulate a large number of logical ports. This paper is the fruit of the collaboration with Mentor graphics and our research team. An integrated global query plan can be represented by a hypergraph. Hypergraphs have the characteristic that their hyperedges connect several vertices without any requirement on the order between these vertices. Projecting our problem, we map vertices (nodes) of hypergraph to operations of queries, and hyperedges to queries. Each hyperedge connects the operations of the query. In this paper, we propose a solution for multi-query optimization based on the techniques and tools of the EDA domain.

The paper is organized as follows: Section 2 presents the related work. Section 3 details our contributions, where an analogy with multi-query optimization and EDA is given. Section 4 implements our results to select materialized views. Section 5 concludes the paper by summarizing the main results and suggesting future work.

2   Related Work

The multi-query optimization problem has been widely studied for traditional databases [2] and more recently for semantic databases [8]. The difficulty of this problem is the identification of relevant intermediate nodes. The tree structure was widely used to model the interaction between queries [9]. Using query interaction, the merged trees, each one representing a given query of the workload, gives raise to a global graph. This graph has been used as a data structure for several optimization problems such as materialized views [10,11,12]. Yang et al. [6] is one of the sole work in the context of $\mathcal{DW}$ that deals with two interdependent problems: (a) constructing an optimal and unified plan of queries and (b) materializing views using that plan. The authors propose a cost-driven approach that devided into two steps: (1) the construction of the unified plan and (2) the selection of intermediate nodes that may represent the future materialized views. We detail these two steps below.

1. Construction of the unified plan. This step is performed following the bottom up scenario. Initially, the authors select the optimal tree of each query using a cost model. Once the optimal trees are identified, common expressions (shared nodes) between these trees are obtained. These nodes are used to merge all query trees into a single graph, called *Multi-Views Processing Plan (MVPP)*. This plan has four levels: at level 0, we find the leave nodes representing the base tables of the $\mathcal{DW}$. At level 1, we find nodes which represent the results of unary algebraic operations such as selection and projection. At level 2, we find nodes representing binary operations such as join, union, etc. The last level represents the results of each query. Each intermediate node of the graph is tagged with the cost of each operation and its maintenance cost. Two algorithms are proposed for selecting the best MVPP. To reduce the *search space* of potential best MVPP, the authors first ignore unary operations and consider only joins. Once the best plan is obtained, they introduce the selections and projection. The first algorithm, called "A feasible solution", generates all possible MVPP. By the use of a cost model, the plan with the minimum cost will be chosen. This algorithm is costly in terms of computation. To simplify the previous algorithm, a second algorithm is proposed based on 0-1 integer programming. The different steps of this algorithm are:

   - identification for each query $Q_i(1 \leq i \leq k)$ of a set P of l $(l > 1)$ join plan trees, $P = \{p_1, \cdots, p_l\}$;
   - identification for each join plan tree $p_j \in P$; all sub-trees are identified, where each one is called a join pattern. Let $S = \{s_1, .., s_m\}$ be the set of all patterns;
   - construction of a *query-join-plan-tree* usage matrix $\mathcal{A}$, where each element a_{ij} takes value 1 if the query Q_i can be answered using the join plan tree p_j, else 0;
   - construction of a *contained* usage matrix $\mathcal{A}$, where each element b_{ij} takes value 1 if the join pattern s_j is contained in the join tree p_i, else 0;

The problem of selecting an optimal MVPP is reduced to the selection of a subset $\{p_1, .., p_l\}$ of l join plan trees that minimizes the total query processing cost x_0, whith $x_0 = \sum_{i=1}^{m} Ecost(s_i) * (\sum_{j=1}^{l} b_{ij} * x_j)$. In this formula, each query uses one and only one p_i, $x_j=0$ if no query use the join plan tree p_j, else $x_j 1$, and $Ecost(s_i)$ is the cost estimation of the join pattern s_i.

The main drawback of this approach is its high complexity due to the generation of all possible plans [13].

2. View Selection. This algorithm is performed in two steps: (1) generation of materialized views candidates which have positive benefit between query processing and view maintenance. This benefit corresponds to the sum of query processing using the view minus the maintenance cost of this view, (2) only candidate nodes with positive benefit are selected to be materialized.

The authors conducted a simplistic implementation using set of 5 queries. The obtained results cannot be generalized. To summarize, the existing studies on multiple query optimization suffer from scalability. To offer efficient algorithms, we propose an analogy between the problem of selecting MVPP and electronic circuit design.

3   Analogy between MVPP Generation and EDA

The multi-query optimization problem ($\mathcal{MQOP}$) is formalized as follows: given a workload of queries to be optimized, the $\mathcal{MQOP}$ consists in finding the best execution order of different operations of queries such as the total cost of the workload is minimized. It is well known that, for a given query, multiple execution plans are possible depending on the combination of intermediate operations of selection, join, projection, etc. Even if the query result is the same, the costs of two different execution plans can vary in terms of Input/Output. The optimal plan for a given query can be obtained by testing all plans. Combining different optimal plans of queries does not guarantee the optimality, and, conversely, the use of non-optimal individual plans can lead to better results [14,6]. Our answer to this problem is based on an analogy between the MVPP and an electronic circuit. As shown in the following example, a given MVPP can easily be mapped to an electronic circuit.

Example 1. Consider the star schema benchmark (SSB)[2]. The $\mathcal{DW}$ contains a fact table *Lineorder*, and four dimension tables: *Customer, Supplier, Part* and *Dates*. On top of this schema a set of 30 queries is executed. Figure 2 describes the MVPP of those 30 queries. Note that a MVPP contains four level of nodes (cf. Introduction). We designed the MVPP as an electronic circuit, where the intermediate results become electronics ports (AND, OR, XOR) as shown in Figure 3. This analogy allows us to borrow optimization techniques and tools defined in electronic circuits domain to handle very large MVPP.

4   Constructing the MVPP Using Hypergraphs

In this section we present our approach to generate the MVPP using all join nodes at the same time. Instead of using individual query plans, our approach

---

[2] http://www.cs.umb.edu/poneil/StarSchemaB.pdf

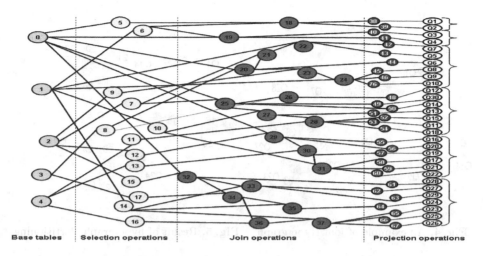

Fig. 2. An example of MVPP of 30 queries

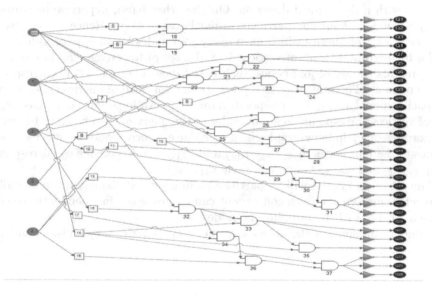

Fig. 3. Electronic circuit corresponding to the MVPP

follows a top down scenario, where all queries are considered simultaneously. We use the graph data structure to represent the different join plans.

The idea of our approach is to group join nodes in several connected components, where each component contains the nodes that can interact with them, so that these nodes can be reused. This allows us to use graph theory algorithms which have been proven useful in terms of efficiency and performance [15]. The generated join nodes have no order between them, so it is impossible to represent those nodes by a graph since a graph requires that an edge connects two

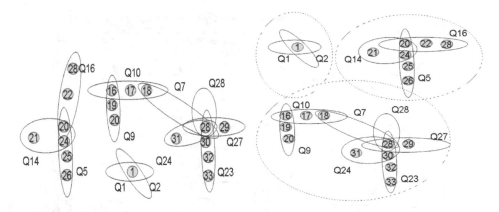

Fig. 4. An example of join Hypergraph   **Fig. 5.** Result of hypergraph Partitioning

vertices with a determined direction. On the other hand, hypergraphs connect many vertices without any particular order between them. Hence, we choose to represent a set of queries which have many join nodes by an hypergraph.

An hypergraph is a set of vertices V and a set of hyperedges E. In our case, V represents the set of join nodes, such that for each vertex $v_i \in V$ corresponds a join node n_j. The same way, E represents the workload of queries Q, such that for each hyperedge $e_i \in E$ corresponds a query q_j. An hyperedge e_i connecting a set of vertices corresponds to a join nodes of the query q_j. As shown in Figure 4, an example of a join hypergraph, the hyperedge e_{23} corresponds to query q_{23} and connects the join nodes nj_{28}, nj_{30}, nj_{32} and nj_{33}. The hyperedge e_2 corresponds to query q_2 and connects one join node nj_1.

The set of join nodes will be partitioned into several disjoint sub-sets, called *connected components*. Each component can be processed independently to generate a local MVPP by ordering the nodes.

Table 1 summarizes the correspondence between the *graph* vision and the *query* vision.

Table 1. correspondence graph -Query

Vision *hypergraph*	Vision of *query*
$V = \{v_i\}$ set of vertices	$\{n_j\}$ set of join nodes
Hyperedge e_j	Query q_j
Section (sub-hypergraph)	Connect component
Hypergraph HA	Workloads of queries Q
Oriented graph	Processing Plans

4.1   Hypergraph Partitioning

In this section, we adapt an existing algorithm derived from graph theory to aggregate the join nodes on connected components. First, we used *HMETIS*[3] programs to partition an hypergraph into several hypergraphs [16]. *HMETIS* is a free software library developed by the *Karypis* laboratory. This software allows parallel or serial partitioning. *HMETIS* algorithms are based on the multilevel hypergraph partitioning schemes, their principle is to divide an hypergraph into k partitions (new hypergraphs) such that the number of hyperedges cut is minimal [15,17,18,16]. More precisely, the main steps of the multilevel partitioning algorithms are: (1) A random division of the set of vertices into two disjoints subsets called sections. (2) An adjustment operation to minimize the number of the hyperedges cutting[4]. This operation will be repeated until no improvement is further possible. (3) Each bisection will be divided the same way until k partitions are formed. However, in our context, the exact number of partitions to do is unknown. We want to get all possible disjoint partitions (connected components). To this end, we adapt the original algorithm to our problem by changing the stopping criterion, so that the algorithm stops only when there is no possibility to partition without cutting hyperedges. Precisely, the algorithm works as follows: (1) The set of vertices will be divided if and only if the number of hyperedges cutting is null. (2) Each bisection result of the hypergraph partitioning will be divided in the same way until there is no more divisible hypergraph. The result of the hypergraph partitioning is a set of small disjoint hypergraphs. Each small hypergraph will be separately processed to generate the MVPP.

Figure 5 presents the join hypergraph partitioning of the hypergraph shown in Figure 4, where three connected components corresponding to three new hypergraphs are obtained.

4.2   From Hypergraph to Graph

In the previous section, we have presented the join hypergraph construction and its partitioning into several disjoints hypergraphs. The remaining step to generate the MVPP is to find an order between join nodes in each small hypergraph, i.e. to transform each hypergraph into an oriented graph.

A characteristic of OLAP queries is that all join operations involve the fact table. Figure 6a shows a join tree plan of an OLAP query. Figures 6b and c show that the join nodes graph in the MVPP for one connected component always has a start node called *pivot* which corresponds to the first join operation.

To generate the MVPP, the hypergraphs generated by the initial join hypergraph partitioning must be transformed into oriented join graphs *graph*. Adding an arc to a graph corresponds to establishing an order between two join nodes. We start by the node which maximizes the reuse benefit. As shown in Algorithm 1, the number of reuses of a node is the number of hyperedges which connect the

---

[3] http://glaros.dtc.umn.edu/gkhome/metis/hmetis/overview
[4] We said an hyperedge is cut if it connects vertices from both sections

Algorithm 1. findPivot(HyperGraph G)

1: $benefitemax \leftarrow 0$;
2: **for all** $v_i \in V$ **do**
3:     $nbr \leftarrow nbrUse(v_i)$; {get the number of hyperedges which use the vertex v_i; the reuse number is (nbr-1)}
4:     $cost \leftarrow costProcessing(v_i)$; {get the processing cost of the vertex (node) v_i, in the graph G}
5:     $benefit \leftarrow (nbr - 1) * cost - cost$ {get the reuse benefit of the vertex v_i }
6:     **if** $benefit \prec benefitemax$ **then**
7:         $benefitemax \leftarrow benefit$;
8:         $pivot \leftarrow v_i$;
9:     **end if**
10: **end for**
11: **return** $pivot$

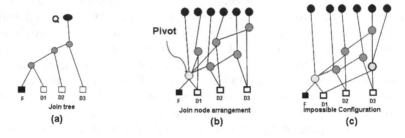

Fig. 6. Possible join nodes arrangement in MVPP

vertex corresponding to this node. The node which maximizes the benefit will be added to the graph and will be deleted from the hypergraph. This operation will be repeated until all nodes are added to the graph.

Algorithm 2 describes the transformation steps of an hypergraph into an oriented graph. (1) We start by finding the *pivot* node, which has the maximum benefit. (2) If the *pivot* is used by all hyperedges (ie. all queries), then the *pivot* is added to the graph and deleted from the hypergraph. When deleting the *pivot*, all hyperedges that do not have a vertices are deleted. (3) If the *pivot* is not used by all hyperedges, then the *pivot* is added to the graph and the hypergraph is partitioned into two disjoint hypergraphs *G1* and *G2* by Algorithm 3. *G1* includes all hyperedges using the *pivot* and *G2* has the other hyperedges. Both hypergraphs will be transformed the same way into oriented graphs *gráph*. Figure 7 shows the transformation steps of an hypergraph into an oriented graph.

Algorithm 1 allows to find the vertex (node) which is the pivot in the hypergraph G. This pivot corresponds to the node which has the best possible benefit of intermediate results reuse. The benefit is the number of reuse multiplied by the processing cost minus the cost processing to generate the intermediate result.

Algorithm 3 allows partitioning a hypergraph into two disjoint hypergraphs using a node as a *pivot*. The first hypergraph contains the hyperedges that

Algorithm 2. transformHyperGraph(Hypergraph G)

1: **while** V not empty **do**
2:    $pivot \leftarrow findPivot(G)$; {Algorithm 1}
3:    **if** $pivot \in all(e \in E)$ **then**
4:       $addVertexToGraph(pivot)$;{add the vertex $pivot$ to the graph $graph$ }
5:       $deleteVertex(pivot)$;{delete the vertex $pivot$ from V of G }
6:       **for all** $e_i \in E$ **do**
7:          **if** No vertice in e_i **then**
8:             $deleteHyperEdge(e_i)$;{delete all hyperedges which have no a vertex}
9:          **end if**
10:      **end for**
11:   **else**
12:      $partitionPivot(pivot, G1, G2)$; {Algorithm 3}
13:      $addVertexToGraph(pivot)$;
14:      $transformHyperGraph(G1)$; {re-transformation}
15:      $transformHyperGraph(G2)$; {re-transformation}
16:   **end if**
17: **end while**

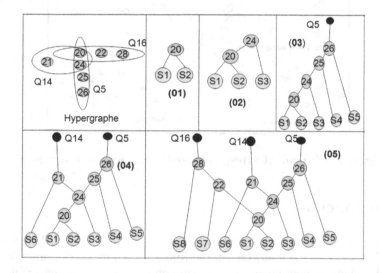

Fig. 7. Transformation steps of a join hypergraph to an oriented graph

use the *pivot*, and the second hypergraph contains the other hyperedges. If the second hypergraph contains vertices which are used by the first hypergraph, we duplicate the common vertices into the second hypergraph. Figure 8 shows an example of the result of hypergraph partitioning with the *pivot* nj_{28}. In this example, we note that node nj_{18} is used by both hypergraphs. Hence, this node is duplicated into node nj_{33} in the second hypergraph.

Algorithm 3. partitionPivot(Vertex *pivot*, HyperGraph *G1*, HyperGraph *G2*)

1: **for all** $e_i \in E$ **do**
2:    **if** *pivot* $\in e_i$ **then**
3:       $addToGraph(e_i, G1)$;{add the hyperedge e_i to the first hypergraph G1 }
4:    **else**
5:       $addToGraph(e_i, G2)$;{add the hyperedge e_i to the second hypergraph G2 }
6:    **end if**
7: **end for**
8: **for all** $v_i \in G2.V$ **do**
9:    **if** $v_i \in G1.Vi$ **then**
10:       $putNewID(v_i, G2)$;{Change the ID of the node to avoid conflict nodes into a global MVPP }
11:    **end if**
12: **end for**

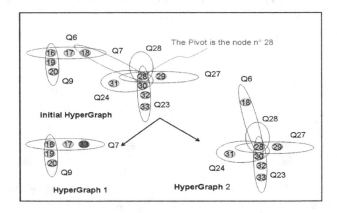

Fig. 8. An Example of hypergraph partitioning with node as a *pivot*

5   Experimentation

To validate our proposal, we developed a simulator tool in Java Environment. This tool consists in the following six modules. (1) An extraction module to get all selection, join, projection and aggregation nodes from a query as a SQL string. (2) A construction module which generates the join hypergraph of all queries. (3) An hypergraph partitioning module, which partitions a hypergraph into several hypergraphs without cutting hyperedges. (4) A transformation module, which transforms a hypergraph into an oriented graph by using a function maximizing the benefit of intermediate results reuse. (5) An aggregate module which assembles all selection, join, projection and aggregation nodes to generate the final MVPP. (6) A display module which displays the results of the generated MVPP using the *Cytoscape*[5] plug-in. This module can display the processing plan of a component as shown in Figure 10, or the individual query processing plan.

---

[5] http://www.cytoscape.org

Table 2. Execution time in milliseconds to generate a MVPP

Number of queries	Hypergraph	Feasible solution	0-1 integer programming
5	71	82	69
20	88	514	225
30	104	933	510
50	108	2258	781
60	114	4358	1224
499	283	25654	8604

Another module is developed to implement *Yang*'s algorithm [6] considering their two algorithms: feasible solution and 0-1 integer programming (cf. Related Work).

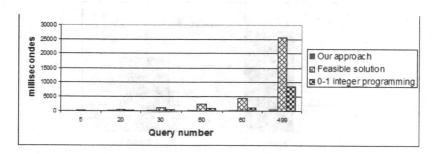

Fig. 9. The execution time to generate MVPP

A series of tests was applied on our approach module and *Yang*'s module. In each test, we change the number of queries as input to monitor the behavior of each algorithm. As shown in Table 2, our algorithm is more effective in terms of execution time compare to *Yang*'s algorithms which increase exponentially with the number of queries.

Figure 10 shows a graphic representation of three MVPP of three different connected components with *Cytoscape* tool.

We now evaluate the impact of our obtained MVPP on the problem of selecting materialized views. During the MVPP generation we have, for each connected component, a pivot join node which corresponds to the node that has the maximum reuse benefit relative to the node construction cost. As a consequence, each pivot node is materialized. Therefore for each connected component a join node will be materialized. Our materialization procedure is compared against Yang et al.'s *feasible solution* [6].

To estimate the cost of query processing in our tests, we used a cost model developed in [3]. This model estimates the number of Inputs/Outputs pages required for executing a given query.

To perform this experiment, we developed a simulator tool using Java Environment. The tool can automatically extract the data warehouse's meta-data

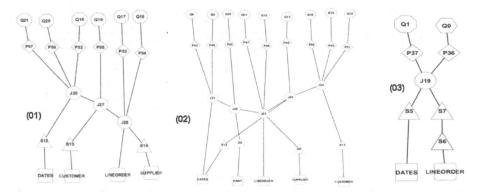

Fig. 10. MVPP of components

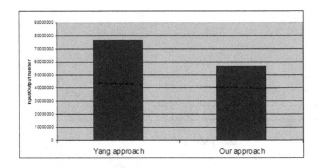

Fig. 11. Query processing cost using materialized views

characteristics. The data warehouse used in our test is SSB (Start Schema Benchmark) [19]. Its size is 1Go, with a fact table *Lineorder* of 6 000 000 tuples and four dimension tables: *Part, Customer, Supplier* and *Dates.* 30 queries are considered to conduct our evaluation. These queries cover the most types of OLAP queries[6].

The obtained results are described in Figure 11. They show that our method outperforms Yang's algorithm.

6   Conclusion

In this paper, we first identified the limit of the existing algorithms dealing with the problem of generating an optimal MVPP and its consequence on selecting materialized views. The use of graph theory techniques massively used to design and test electronic circuits allows us to alleviate the challenge of scalability of MVPP generation. Advanced algorithms to generate the best MVPP for a given

---

[6] The different types are :(a) 1-Join (10 queries), (b) 2-Joins (6 queries), (c) 3-Joins (11 queries) and (d) 4-Joins (3 queries).

workload are given. These algorithms are based on partitioning principle used in EDA domain. The experimental results prove the effectiveness of our approach even when applied to a large workload of queries. We also showed the impact of using our generated MVPP for selecting materialized views.

Currently, we are working on considering the dynamic aspects of generating MVPP and materialized views.

References

1. Bruno, N., Chaudhuri, S.: Efficient creation of statistics over query expressions. In: Proceedings of the International Conference on Data Engineering (ICDE), pp. 201–212 (2003)
2. Sellis, T.K.: Multiple-query optimization. ACM Transactions on Database Systems 13(1), 23–52 (1988)
3. Kerkad, A., Bellatreche, L., Geniet, D.: Queen-bee: Query interaction-aware for buffer allocation and scheduling problem. In: Cuzzocrea, A., Dayal, U. (eds.) DaWaK 2012. LNCS, vol. 7448, pp. 156–167. Springer, Heidelberg (2012)
4. Ahmad, M., Aboulnaga, A., Babu, S., Munagala, K.: Interaction-aware scheduling of report-generation workloads. VLDB Journal 20(4), 589–615 (2011)
5. Toroslu, I.H., Cosar, A.: Dynamic programming solution for multiple query optimization problem. Information Processing Letters 92(3), 149–155 (2004)
6. Yang, J., Karlapalem, K., Li, Q.: Algorithms for materialized view design in data warehousing environment. In: Proceedings of the International Conference on Very Large Databases (VLDB), pp. 136–145. Morgan Kaufmann Publishers Inc., San Francisco (1997)
7. Ioannidis, Y.E., Kang, Y.C.: Randomized algorithms for optimizing large join queries. In: Garcia-Molina, H., Jagadish, H.V. (eds.) ACM SIGMOD, pp. 312–321 (1990)
8. Le, W., Kementsietsidis, A., Duan, S., Li, F.: Scalable multi-query optimization for sparql. In: Proceedings of the International Conference on Data Engineering (ICDE), pp. 666–677. IEEE (2012)
9. ElMasri, R., Navathe, S.B.: Fundamentals of Database Systems. Benjamin Cummings, Redwood City (1994)
10. Gupta, H.: Selection and maintenance of views in a data warehouse. Ph.d. thesis, Stanford University (September 1999)
11. Yang, J., Karlapalem, K., Li, Q.: A framework for designing materialized views in data warehousing environment. In: ICDCS, p. 458 (1997)
12. Baralis, E., Paraboschi, S., Teniente, E.: Materialized view selection in a multidimensional database. In: Proceedings of the International Conference on Very Large Databases (VLDB), pp. 156–165 (August 1997)
13. Galindo-Legaria, C.A., Grabs, T., Gukal, S., Herbert, S., Surna, A., Wang, S., Yu, W., Zabback, P., Zhang, S.: Optimizing star join queries for data warehousing in microsoft sql server. In: Proceedings of the International Conference on Data Engineering (ICDE), pp. 1190–1199 (2008)
14. Gupta, A., Sudarshan, S., Viswanathan, S.: Query scheduling in multi query optimization. In: Proceedings of the International Database Engineering & Applications Symposium (IDEAS), pp. 11–19. IEEE Computer Society, Washington, DC (2001)

15. Karypis, G., Aggarwal, R., Kumar, V., Shekhar, S.: Multilevel hypergraph partitioning: applications in vlsi domain. IEEE Transactions on Very Large Scale Integration Systems 7(1), 69–79 (1999)
16. Karypis, G., Kumar, V.: Multilevel k-way hypergraph partitioning. In: ACM/IEEE Design Automation Conference (DAC), pp. 343–348. ACM, New York (1999)
17. Karypis, G., Aggarwal, R., Kumar, V., Shekhar, S.: Multilevel hypergraph partitioning: Application in vlsi domain. In: ACM/IEEE Design Automation Conference (DAC), pp. 526–529 (1997)
18. Selvakkumaran, N., Karypis, G.: Multiobjective hypergraph-partitioning algorithms for cut and maximum subdomain-degree minimization. IEEE Transactions on Computer-Aided Design of Integrated Circuits and Systems 25(3), 504–517
19. O'Neil, P., O'Neil, B., Chen, X.: Star schema benchmark (2009)

XML Schema Transformations

The ELaX Approach

Thomas Nösinger, Meike Klettke, and Andreas Heuer

Database Research Group
University of Rostock, 18051 Rostock, Germany
(tn,meike,ah)@informatik.uni-rostock.de

Abstract. In this article the transformation language ELaX (Evolution Language for XML-Schema) for modifying existing XML Schemas is introduced. This domain-specific language was developed to fulfill the crucial need to handle modifications on an XML Schema and to express such modifications formally. The language has a readable, simple, base-model-oriented syntax, but it is able to also express more complex transformations by using add, delete and update operations. A small subset of operations of the whole language is presented and illustrated partially by dealing with a real life XML Schema of the WSWC (Western States Water Council). Finally, the idea of integrating an ELaX interface into an existing research prototype for dealing with the co-evolution of corresponding XML documents is presented.[1]

1 Introduction

The eXtensible Markup Language (XML) is one of the most popular formats for exchanging and storing information in heterogeneous environments. To assure that well-defined XML documents can be understood by every participant (e.g. user or application) it is necessary to introduce a document description. XML Schema [3] is one commonly used standard for dealing with this problem. An XML document is called valid, if it fulfills all restrictions and conditions of an associated XML Schema.

After an XML Schema has been specified and widely used for XML documents, the requirements against the information contained in those documents may change. Therefore, the XML Schema may need to adapted as well. This may concern every possible structure within the XML Schema definition (XSD). The occurring problem is: how can adaptions be described and formulated under consideration of the underlying XML schema definition in a descriptive, intuitive and easy-understandable way? The definition of a schema update language is absolutely necessary; we introduce such a language in this paper.

A further issue in the overall context of exchanging information is the validity of XML documents. The resulting problem of modifying an XML Schema is,

---

[1] The comprehensive version is available as a technical report at:
www.ls-dbis.de/elax

H. Decker et al. (Eds.): DEXA 2013, Part I, LNCS 8055, pp. 293–302, 2013.

existing XML documents, which were valid against the former XML Schema, have to be adapted as well (co-evolution). The standardized description of the adaptions (e.g. a sequence of operations of an update language) is an essential prerequisite for the co-evolution. The new evolution language ELaX (Evolution Language for XML-Schema) is our answer to the above mentioned problems.

This paper is organized as follows. **Section 2** introduces a running example, which defines a realistic scenario for the use of ELaX. **Section 3** gives the necessary background of XML Schema and corresponding concepts. **Section 4** and **section 5** present our approach, by first specifying the basic statements of ELaX and then showing how our approach can contribute to the scenario discussed in **section 2**. In **section 6** we describe the practical use of ELaX in our prototype, which was developed for handling the co-evolution of XML Schema and XML documents. In **section 7** we discuss some related transformation language for XML Schema. Finally, in **section 8** we draw our conclusions.

2   Running Example

Information exchange specifications usually provide some kind of XML Schema which contains information about allowed structures, constraints, data types and so on. One example is the WSWC (Western States Water Council), an organization which accomplishes effective cooperation among 18 western states in the conservation, development and management of water resources. Another purpose is the exchange of views, perspectives and experiences among member states - summarized the exchange of information. The expected format for XML data exchange is defined in a set of XML Schemas, one report is presented in the following. Due to space limitations, the chosen example is a simple one.

```
<?xml version="1.0" encoding="utf-8"?>
<xsd:schema xmlns:WC="http://www.exchangenetwork.net/schema/WC/0"
    xmlns:xsd="http://www.w3.org/2001/XMLSchema"
    targetNamespace="http://www.exchangenetwork.net/schema/WC/0"> [..]
    <xsd:include schemaLocation="WC_ReportingUnit_v0.2.xsd"/>
    <xsd:annotation/>
    <xsd:element name="Report" type="WC:ReportDataType">[..]</xsd:element>
    <xsd:complexType name="ReportDataType">
      <xsd:annotation/>
      <xsd:sequence>
        <xsd:element ref="WC:ReportIdentifier"/>[..]
        <xsd:element ref="WC:ReportName" minOccurs="0"/>[..]
        <xsd:element ref="WC:ReportingUnit" maxOccurs="unbounded"/>
      </xsd:sequence>
    </xsd:complexType>
</xsd:schema>
```

Fig. 1. The original WSWC report XML Schema

The original XML Schema is illustrated in figure 1. According to the complex type ("ReportDataType") a report (the component "Report") is a sequence of an obligatory report identifier ("WC:ReportIdentifier"), an optional name

("WC:ReportName") and a set of reported units ("WC:ReportingUnit"). The element declarations not given in this schema are specified in external XML Schemas, which are represented by the "xsd:include" component.

This original report has been adapted; the result is the XML Schema presented in figure 2 (changed and added parts are highlighted by rectangles).

```
<?xml version="1.0" encoding="utf-8"?>
<xsd:schema xmlns:WC="http://www.exchangenetwork.net/schema/WC/0"
    xmlns:xsd="http://www.w3.org/2001/XMLSchema"
    targetNamespace="http://www.exchangenetwork.net/schema/WC/0"> [..]
    <xsd:include schemaLocation="WC_ReportingUnit_v0.2.xsd"/>
    <xsd:annotation/>
    <xsd:element name="Report" type="WC:ReportDataType">[..]</xsd:element>
    <xsd:complexType name="ReportDataType">
        <xsd:annotation/>
        <xsd:sequence>
            <xsd:element ref="WC:ReportIdentifier"/>[..]              3
            <xsd:element ref="WC:ReportName" minOccurs="1"/>[..]
            <xsd:element ref="WC:ReportingUnit" maxOccurs="unbounded"/>
        </xsd:sequence>
    </xsd:complexType>                                             1
    <xsd:complexType name="ReportListDataType">
        <xsd:sequence>
            <xsd:element ref="WC:Report" minOccurs="0" maxOccurs="10"/>
        </xsd:sequence>
    </xsd:complexType>                                             2
    <xsd:element name="ReportList" type="WC:ReportListDataType"></xsd:element>
</xsd:schema>
```

Fig. 2. The adapted WSWC report XML Schema

The modifications have the purpose to summarize information in only one report that otherwise would be spread over multiple small reports. The question is how can these modifications be described formally? Before presenting one possibility (i.e. ELaX), some background information and notations are presented in the following chapter.

3 Technical Background

The XML Schema *abstract data model* consists of different components or node types. Basically, these are: type definition components (simple and complex types), declaration components (elements and attributes), model group components, constraint components, group definition components and annotation components [1]. Additionally, the *element information item* serves as an XML representation of these components. The *element information item* defines which content and attributes can be used in an XML Schema. Table 1 gives an overview about the most important components and their representation.

The <include>, <import>, <redefine> and <overwrite> items are not explicitly given in the *abstract data model* (N.N. - Not Named), but they are important components for embedding externally defined XML Schemas. In the remaining parts of the paper, we will summarize them under the node type

Table 1. Abstract Data Model and XML representation

Abstract Data Model	Element Information Item
declarations	<element>, <attribute>
group-definitions	<attributeGroup>
model-groups	<all>, <choice>, <sequence>, <any>, <anyAttribute>
type-definitions	<simpleType>, <complexType>
N.N.	<include>, <import>, <redefine>, <overwrite>
annotations	<annotation>
constraints	<key>, <unique>, <keyref>, <assert>, <assertion>
N.N.	<schema>

"module". The <schema> item is the document root element of any W3C XML Schema. It is a container for all the declarations and definitions.

Analyzing the possibilities of specifying declarations and definitions leads to four different modeling styles of XML Schema: *Russian Doll*, *Salami Slice*, *Venetian Blind* and *Garden of Eden* [8]. These modeling styles mainly influence the re-usability of element declarations or defined data types but also the flexibility of an XML Schema in general. The scope of element and attribute declarations as well as the scope of type definitions is global iff the corresponding node is specified as a child of the <schema> item and can be referenced (e.g. by knowing the name and namespace). In contrast, locally specified nodes are not directly defined under the <schema> item, therefore the re-usability is low.

An XML Schema in the *Garden of Eden* style just contains global declarations and definitions. If the requirements against exchanged information change and the underlying schema has to be adapted then this modeling style is the most suitable one. That is, because all components can be easily identified by knowing the *QNAME* (qualified name). A qualified name is a colon separated string of the target namespace of the XML Schema followed by the name of the declaration or definition. The name of a declaration and definition is a string of the data type *NCNAME* (non-colonized name), a string without colons. Due to the characteristics of the *Garden of Eden* style, it is our basic modeling style.

4   XML Schema Transformation Language

In order to handle modifications on an XML Schema and to express these modifications formally, an adequate transformation language is absolutely essential. Therefore, we developed ELaX (Evolution Language for XML-Schema) which lets the user describe modifications in a simple, easily-understandable and explicit manner. The following criteria were important through the development process, parts were already mentioned above:

1. Consideration of the underlying data model (i.e. the *abstract data model* and *element information item* of the XML Schema definition)

2. Adequate and complete realization of the common operations ADD, UP-DATE, DELETE
3. Definition of an descriptive and readable interface for creating, changing and deleting XML Schema
4. Intuitive and simple syntax of operation steps

The *abstract data model* and the *element information item* consist of different node types, which have to be adapted (see section 3). On these node types the operations ADD, DELETE and UPDATE have to be executed; the first ELaX statements are in an EBNF (Extended Backus-Naur Form) like notation:

$$elax ::= ((< add > | < delete > | < update >) ";") + \, ; \qquad (1)$$

$$add ::= "add" \, (< addannotation > | < addattributegroup >$$
$$| < addgroup > | < addst > | < addct > | < addelement > \qquad (2)$$
$$| < addmodule > | < addconstraint >) \, ;$$

$$delete ::= "delete" \, (< delannotation > | < delattributegroup >$$
$$| < delgroup > | < delst > | < delct > | < delelement > \qquad (3)$$
$$| < delmodule > | < delconstraint >) \, ;$$

$$update ::= "update" \, (< updannotation > | < updattributegroup >$$
$$| < updgroup > | < updst > | < updct > | < updelement > \qquad (4)$$
$$| < updmodule > | < updconstraint > | < updschema >) \, ;$$

An ELaX statement always starts with "add", "delete" or "update" followed by one of the alternative components for modifying the different node types. Every component of rule (1) can optionally be repeated one or more times (i.e. "+"), consequently an encapsulation or ordered sequence of operations is possible. The operations are separated by ";". By using the rules (1), (2), (3) and (4), complex modifications of an XML Schema can be expressed formally. In the following subsections, the statements for adding (<addelement>), deleting (<delelement>) and updating (<updelement>) elements are presented.[2]

4.1 Adding Elements

According to the *Garden of Eden* modeling style, elements are either defined as element declarations in the global scope of an XML Schema or as references to such declarations. Furthermore, it is possible to define wildcards. The following statements realize the add operation for elements:

$$addelement ::= \quad < addelementdef > | < addelementref >$$
$$| < addelementwildard > \, ; \qquad (5)$$

---

[2] The statements not presented are available at: www.ls-dbis.de/elax

$$
\begin{aligned}
addelementdef ::=\ & "element"\ "name"\ ncname\ "type"\ qname \\
& (("default"|"fixed")\ string\ )?\ ("final"\ ("\#all"|"restriction" \\
& |"extension"))?\ ("nillable"\ ("true"|"false"))?\ ("id"\ id\ )? \\
& ("form"\ ("qualified"|"unqualified"))?\ ;
\end{aligned}
\tag{6}
$$

$$
\begin{aligned}
addelementref ::=\ & "element"\ "ref"\ qname\ ("minoccurs"\ int\ )? \\
& ("maxoccurs"\ int\ )?\ ("id"\ id\ )? < position >\ ;
\end{aligned}
\tag{7}
$$

Before going into detail, further components are necessary to localize or identify elements and node types in general. It is possible to localize node types in a content model under consideration of the node neighborhood with statement (8) and to identify a node type itself by using an absolute addressing (10).

$$
\begin{aligned}
position ::=\ & ("after"|"before"|\ ("as"("first"|"last")\ "into")\ |"in") \\
& < locator >\ ;
\end{aligned}
\tag{8}
$$

$$
locator ::=\ < xpathexpr >\ |\ emxid\ ;
\tag{9}
$$

$$
\begin{aligned}
xpathexpr ::=\ & ("/"("."\ |\ ("node()"\ |\ ("node()[@name =' "ncname"']")) \\
& ("["\ int\ "]")?\ )\ )+\ ;
\end{aligned}
\tag{10}
$$

An element reference statement (7) starts with "element ref", followed by the qualified name of the referenced element declaration (qname) and other, optional attributes for the frequency of occurrence ("minoccurs", "maxoccurs") or an XML Schema id ("id"). An element reference can be added "after", "before", "as first into", "as last into" or "in" the model-group node type with consideration of the node neighborhood and using statement (8). The identification of nodes is possible with a unique identifier of our conceptual model (emxid)[3]. Alternatively, a subset of XPath expressions can be used to create an absolute path (10). The given subset is sufficient for the simple localization or identification of every node type in the *Garden of Eden* style.

4.2   Deleting Elements

Compared to the add operation, deleting an element basically just requires some information of identification. The qualified name in general and in the case of references and wildcards also the position of an element is sufficient. The following statements realize the delete operation for elements:

$$
\begin{aligned}
delelement ::=\ & < delelementdef >\ |\ < delelementref > \\
& |\ < delelementwildcard >\ ;
\end{aligned}
\tag{11}
$$

$$
\begin{aligned}
delelementref ::=\ & "element"\ "ref"\ qname \\
& ("at"\ < locator >\ |\ < refposition >)\ ;
\end{aligned}
\tag{12}
$$

---

[3] Our conceptual model is EMX (Entity Model for XML Schema); see also section 6

$$refposition ::= ((''first''|''last''|''all''| (''at'' ''position'' \text{ int }))$$
$$''in'' < xpathexpr >) \mid \text{emxid} ;$$

(13)

The element reference statement (12) starts with "element ref", followed by the qualified name and information about the locator (introduced in section 4.1, statement (9)) or about the position of the reference (<refposition>). The reference position statement (13) enables the localization of one reference if more than one is given in the same group-model node type. With this statement the "first", the "last", "all" of them or a reference at a specific position ("at position") can be deleted. If the unique identifier of the conceptual model is known, the emxid can be used instead of the XPath expression.

4.3 Updating Elements

Updating elements is implemented by rule (4). Basically, all given information within an existing element can be updated. Also, adding new information to an existing element is considered, so the update specification for elements is similar to the one for adding new elements. The following statements realize the update operation for elements:

$$updelement ::= < updelementdef > \mid < updelementref >$$
$$\mid < updelementwildcard > ;$$

(14)

$$updelementref ::= ''element'' ''ref'' \text{ qname } ((''at'' < locator >)$$
$$\mid < refposition >) (''change'' (''ref'' \text{ qname })?$$
$$(''minoccurs'' \text{ int })? (''maxoccurs'' \text{ int })? (''id'' \text{ id })?)?$$
$$(''move'' ''to'' < position >)? ;$$

(15)

Element references are adapted with statement (15). Starting with "element ref", the qualified name and information about the position element references can be updated. The newly given or changed information are specified after "change". This information comprises a list of tuples of an identifier and the corresponding value, they are always optional (i.e. "?"). Furthermore, it is possible to move an element reference, that is why the "move to" component was inserted at the end of the statement (15). The move operation is a short form for deleting and inserting an element reference completely.

5 Example

In section 2 an XML Schema for exchange reports of the WSWC (Western States Water Council) was introduced. In general, three different steps are necessary to modify the old schema of figure 1. The result is represented in figure 2 (the steps are visualized in labelled rectangles).

1. Insert a new type, which contains up to ten reports

2. Insert a new element, which has the new introduced type of *step 1*

3. Update the "ReportDataType" type so that the report name is obligatory

Following, the necessary ELaX operations for the above mentioned steps are described. Furthermore, the replaced values of the data types are boldfaced in every following operation, e.g. the values of QNAME, NCNAME or XPath.

Step 1: First of all, a new complex type has to be inserted. The new complex type gets the NCNAME "ReportListDataType". After specifying the complex type, a group-model node type is inserted as a child of the new complex type (a sequence). The last operation within the first step is performed by inserting an element reference into the sequence. The necessary element declaration is "Report", this element can be referenced by the QNAME "WC:Report". Information about the occurrence is also given, up to ten reports can be collected in the report list. The following ELaX operations have to be executed, the sequence of applied rules are listed below every operation:

$$add\ complextype\ name\ \textbf{ReportListDataType}\ ;$$

Sequence of rules: (1), (2), addct$^{(see\ www.ls-dbis.de/elax)}$ $\qquad$ (16)

$$add\ group\ mode\ sequence\ in$$

$$\textbf{/node()/node()[@name='ReportListDataType']}\ ; \qquad (17)$$

Sequence of rules: (1), (2), addgroup$^{(see\ www.ls-dbis.de/elax)}$, (9), (10), (10)

$$add\ element\ ref\ \textbf{WC:Report}\ minoccurs\ \textbf{0}\ maxoccurs\ \textbf{10}\ in$$

$$\textbf{/node()/node()[@name='ReportListDataType']/node()}\ ; \qquad (18)$$

Sequence of rules: (1), (2), (5), (7), (8), (9), (10), (10), (10)

The correct order of the operations (16), (17) and (18) is specified in the *element information item* of XML Schema and represents the result of the implicitly given relationships of node types.

Step 2: After specifying a new complex type in *step 1*, a new element declaration with this complex type has to be defined. The NCNAME of this new element is "ReportList", the qualified name of the complex type is "WC:ReportList-DataType". Further information are not required.

$$add\ element\ name\ \textbf{ReportList}\ type\ \textbf{WC:ReportListDataType}\ ;$$

Sequence of rules: (1), (2), (5), (6) $\qquad$ (19)

Step 3: The last adaption of the XML Schema in figure 1 contains changing the occurrence of the report name. In our example it is changed from "minOccurs = '0'" to "minOccurs = '1'". Consequently, the name is no longer optional but obligatory. The corresponding operation is applied as follows:

$$update\ element\ ref\ \textbf{WC:ReportName}\ at\ /node()/node()$$

$$\textbf{[@name='ReportDataType']/node()[2]}\ change\ minoccurs\ \textbf{1}\ ; \qquad (20)$$

Sequence of rules: (1), (4), part I (15), (9), (10), (10), (10), part II (15)

6   Practical Use of ELaX

The transformation language ELaX was specified for dealing with XML Schema modifications. It is useful to describe and formulate different adaptions of an XML Schema. In general, this reflects all add, delete and update operations which are possible considering the underlying base model (XSD).

At the University of Rostock a research prototype named CodeX (Conceptual design and evolution for XML Schema) was developed for dealing with co-evolution. The idea behind CodeX is simple and straightforward at the same time: A given XML Schema is transformed to the specifically developed conceptual model (EMX - Entity Model for XML Schema [9]). With the help of this simplified model, the desired modifications are defined and logged (i.e. user interaction) and then used to automatically create transformation steps for adapting the XML documents (by using XSLT - Extensible Stylesheet Language Transformations). The mapping between EMX and XSD is unique, so it is possible to describe modifications not only on the EMX but also on the XSD.

ELaX is useful to unify the internal collected information (i.e. modifications) and additionally provides an interface for dealing directly with the underlying XML Schema.

7   State of the Art

Regarding other transformation languages, there are some which may possibly be used instead of ELaX. The most commonly to mention are XQuery and XSLT.

XSLT is a language for transforming XML documents into other XML documents so it can also deal with XML Schema. XSLT is very complex and difficult to understand, so the use and also the understanding of its results implies a huge overhead. This language is neither suitable for describing modifications nor for unifying the internal collected information within our context.

XQuery as a query language for different XML data sources and especially the extension of it through the update facility [2] "provide expressions that can be used to make persistent changes" to instances of the *abstract data model* (see section 3). By using all features of XQuery update, modifications can be produced that lead to non-regular XML Schema. This is why restrictions are required. However, by focussing on which parts of a complex update language (i.e. XQuery update) are suitable or which parts lead to non-valid XML Schema, again an unintentional overhead is produced.

In section 6 the integration of ELaX into our prototype was described in parts. Other prototypes for dealing with the evolution of XML Schema are e.g. X-Evolution [6], EXup [4], the GEA Framework [5] and XCase [7]. To our knowledge, XSchemaUpdate (used in X-Evolution and EXup) is the only XML Schema modification language which is comparable to ELaX. However, ELaX is closer to the base-model and considers wildcards, constraints and attributes, which are explicitly allowed in a node type considering the *element information item* of XML Schema. Moreover, it distinguishes between element declarations and references amongst others, so more fine-grained operations are possible, which simplifies the analyzing of modification steps.

8   Conclusion

In a heterogeneous and dynamic environment (e.g. the scenario of section 2), also "old" and longtime used XML Schema have to be modified to meet new requirements and to be up-to-date. Modifications of XML Schema documents urgently need a description and a formalism in order to be traceable. Therefore, we developed the new language ELaX (Evolution Language for XML-Schema).

ELaX is a base-model-oriented transformation language, which can be used to modify a given XML Schema. These modifications are in general add, delete and update operations on the node types of XML Schema which are specified in the *abstract data model* and implemented in the *element information item*. ELaX has a simple, intuitive and easy-understandable syntax, but nevertheless it is powerful enough to describe complex operations by combining the given operations. Moreover, it can be used to log modifications for the adaptions of XML documents associated with a given XML Schema, which represents an essential prerequisite for the co-evolution. One remaining step is the implementation of the language and the integration into our research prototype CodeX (Conceptual design and evolution for XML Schema). After it, a final evaluation is planned.

References

1. XQuery 1.0 and XPath 2.0 Data Model (XDM), 2nd edn. (December 2010), http://www.w3.org/TR/2010/REC-xpath-datamodel-20101214/ (Online; accessed May 24, 2013)
2. XQuery Update Facility 1.0 (March 2011), http://www.w3.org/TR/2011/REC-xquery-update-10-20110317/ (Online; accessed May 24, 2013)
3. W3C XML Schema Definition Language (XSD) 1.1 Part 1: Structures (April 2012), http://www.w3.org/TR/2012/REC-xmlschema11-1-20120405/ (Online; accessed May 24, 2013)
4. Cavalieri, F.: EXup: an engine for the evolution of XML schemas and associated documents. In: Proceedings of the 2010 EDBT/ICDT Workshops, EDBT 2010, pp. 21:1–21:10. ACM, New York (2010)
5. Domínguez, E., Lloret, J., Pérez, B., Rodríguez, Á., Rubio, A.L., Zapata, M.A.: Evolution of XML schemas and documents from stereotyped UML class models: A traceable approach. Information & Software Technology 53(1), 34–50 (2011)
6. Guerrini, G., Mesiti, M.: X-Evolution: A Comprehensive Approach for XML Schema Evolution. In: DEXA Workshops, pp. 251–255 (2008)
7. Klímek, J., Kopenec, L., Loupal, P., Malý, J.: XCase - A Tool for Conceptual XML Data Modeling. In: Grundspenkis, J., Kirikova, M., Manolopoulos, Y., Novickis, L. (eds.) ADBIS 2009. LNCS, vol. 5968, pp. 96–103. Springer, Heidelberg (2010)
8. Maler, E.: Schema design rules for ubl..and maybe for you. In: XML 2002 Proceedings by deepX (2002)
9. Nösinger, T., Klettke, M., Heuer, A.: A Conceptual Model for the XML Schema Evolution - Overview: Storing, Base-Model-Mapping and Visualization. In: Grundlagen von Datenbanken (2013)

StdTrip+K: Design Rationale in the RDB-to-RDF Process

Rita Berardi[1], Karin Breitman[1], Marco Antônio Casanova[1],
Giseli Rabello Lopes[1], and Adriana Pereira de Medeiros[2]

[1] Departamento de Informática,
Pontifícia Universidade Católica do Rio de Janeiro,
Rio de Janeiro, RJ – Brazil CEP 22451- 900
{rberardi,karin,casanova,grlopes}@inf.puc-rio.br
[2] Instituto de Ciência e Tecnologia,
Universidade Federal Fluminense,
Rio das Ostras, RJ – Brazil CEP 28890-000
adrianamedeiros@puro.uff.br

Abstract. The design rationale behind the triplification of a relational database is a valuable information source, especially for the process of interlinking published triplesets. Indeed, studies show that the arbitrary use of the *owl:sameAs* property, without carrying context information regarding the triplesets to be linked, has jeopardized the reuse of the triplesets. This article therefore proposes the StdTrip+K process that integrates a design rationale approach with a triplification strategy. The process supports the reuse of standard RDF vocabularies recommended by W3C for publishing datasets and automatically collects the entire rationale behind the ontology design, using a specific vocabulary called Kuaba+W.

Keywords: Triplification, mapping, matching, design rationale.

1 Introduction

Linked Data refers to a set of best practices for publishing and connecting structured data on the Web [3]. One of the most popular strategies to publish structured data on the Web is to convert relational databases to the Linked Data format, in a process known as *RDB-to-RDF* or *triplification* [11], [13].

One of the major challenges of publishing Linked Data is to investigate the value of information based on the trustworthiness of its sources, the time of validity, the certainty, or the vagueness asserted to specified or derived facts [6]. This challenge is associated with the lack of analytical information about the published Linked Data, i.e. information that answers questions such as: (1) Did the original relational database suffer *changes* when published as Linked Data that could impact its quality?; (2) Is the *translation* from the original relational database to Linked Data *correct*?; (3) Is the *chosen ontology* the most appropriate to represent the original relational database?; (4) Did the original relational database *lose* some relevant *information* when it was published as Linked Data? These details of the triplification process should answer the questions above mentioned which are reasoned in the decisions related to changes, correctness, choices and information losing during the triplification.

H. Decker et al. (Eds.): DEXA 2013, Part I, LNCS 8055, pp. 303–310, 2013.

In general, the decisions taken during a design process, the accepted and rejected options, and the criteria used are called *design rationale* (DR) [8], or *triplification rationale* by analogy. Besides helping the reuse of datasets, the triplification rationale has a potential value for supporting design of new ontologies because all the experience acquired during a design can be transmitted and augmented by the reuse of previous DRs in new designs. Although there are several triplification engines, we are unaware of any previous work that applies DR in the Linked Data domain, i.e. that captures the triplification rationale. The details intrinsically involved in the mapping activity should reflect all aspects related to how the concepts of the underlying conceptual schema are mapped to the RDF terms. Furthermore, these detailed information can explicit some problems in the mapping process. For instance, if an *entity element* of an ER is mapped to a *property element* in RDF, the *attribute elements* of this entity may not be represented due to the lack of the domain representation, since the domain is represented as a property.

The matching step involves domain expert decisions regarding the construction of the vocabulary. The details inherent in the matching step should reflect aspects related to the choice of each term of the vocabulary that will be used to publish the database. The decisions of the designer involved in this activity have to consider the database domain and context. For instance, considering a domain of an university publication database where the entity "Authors" has the attribute "name", the most adequate representation is *dc:creator* instead of *foaf:Person,* since *dc:creator* is more representative for the domain. Otherwise, if an entity "Students" has the same attribute "name", *dc:creator* is not the best choice although both entities are in the same domain of "University". The DR representation in the StdTrip+K process is executed through the Kuaba approach [12] that represents a more complete DR in respect to other DR approaches. So the major contribution of this paper is to address the incorporation of DR capture through the addition of Kuaba+W vocabulary in the StdTrip process [14], generating the StdTrip+K, that is, to the best of our knowledge, the first to address the capturing of the decisions behind the triplification task. The remainder of this article is organized as follows. Section 2 discusses related work. Section 3 details the StdTrip+K process along with a running example and describes the Kuaba+W vocabulary used to record the DR. Finally, Section 4 presents the conclusions and directions for future work.

2   Related Work

There are several approaches RDB-to-RDF with different mechanisms to tackle this translation process. The more relevant approaches for the RDB-to-RDF process are Triplify [1], D2RQ [2], Virtuoso RDF view [7] and RDBtoOnto [4]. Triplify motivates the need for a simple mapping solution using SQL (Structure Query Language) as a mapping language and transforms database query results into RDF triples and Linked Data. The mapping is done manually with no record of any rationale. D2RQ generates the mapping files automatically, using the table-to-class and column-to-predicate approach. It uses a declarative language, implemented as Jena graph, to define the mapping file, also with nothing about recording rationale. In the Virtuoso RDF view the mapping file, also called RDF view, is automatically generated with table-to-class approach. In this approach there is no reason to capture the rationale since it does not imply in options, arguments and decisions. RDBtoOnto brings a discussion on how to take advantage of database data in obtaining more accurate ontologies. This work also uses the table-to-class and column-to-predicate to create an initial ontology schema, which is then refined through identification of taxonomies guided by the tool. Although there is user interference, the decisions made are not recorded. There are other approaches like DB2OWL [5] and Ultrawrap [15], but still they do not cover the rationale issue. In the context of rationale models and tools, there are argumentation-based models such as IBIS [17], DRL [9], QOC [10] that allow the DR representation. However, they do not present a complete DR that includes accepted and rejected options and the reasons for that. Specifically in the Linked Data context, we have not found researches with this purpose. There are provenance models, like Open Provenance Model (OPM[1]), that records the history of creating a dataset in general terms. Despite been very important and essential for Linked Data quality, it lacks in terms of decisions during the creation of a mapping file. We can conclude that the approach followed by most tripliflying approaches has no concern with design rationale recording.

3   The StdTrip+K Process

The StdTrip+K process (Fig.1) is anchored in the principle of ensuring interoperability through the use of standards in schema design and through the DR recording. The process receives as input the RDB, the metamodels and the DR vocabulary Kuaba+W. At each stage, the respective DR is traced and recorded using Kuaba+W vocabulary that is incrementally recorded throughout the process execution. In the end, the process results in the RDB-to-RDF Mapping File, the OWL ontology and the final DR. The Kuaba+W vocabulary is described in Section 3.1 and the four steps (Mapping, Matching, Selection and Inclusion) of the StdTrip+K process are described in Section 3.2 using a motivation example.

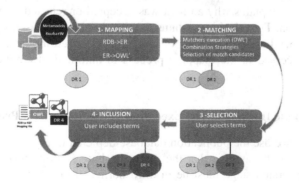

Fig. 1. StdTrip+K Process overview

---

[1] openprovenance.org/

3.1   Kuaba+W – A Design Rationale Vocabulary for RDB-to-RDF Process

Kuaba+W extends the Kuaba approach [12] in the sense that it eliminates elements not necessary in the RDB-to-RDF domain. Moreover, the Kuaba+W extension is related to the addition of the *Description* element, which is related to a *Justification* and carries information regarding the reasons for the domain expert to accept or reject an idea. A description is also related to a *Metamodel*, also new in the extension, since there is more than one metamodel involved in the RDB-to-RDF. A metamodel registers which formal artifact was involved in each step of design process, for instance ER and RDF metamodels. Fig. 2 shows the main elements of the ontology, using a UML-like graphical notation to help visualization. A **Reasoning element** represents the design issue that the ontology designer should deal with (question, ideas and arguments).

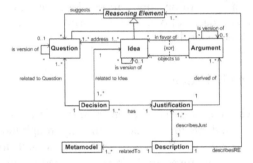

Fig. 2. The Kuaba+W ontology elements

An **Idea** represents a potential solution for the mapping or matching issue presented by the Reasoning Element Question. The **Argument** represents the criteria used to present an Idea for a Question. A **Decision** represents the acceptance or the rejection of an idea as a solution to a question. A **Justification** indicates the justification for each Decision that explains why an Idea was accepted or rejected as a solution for a particular Question. **Description** contains details about any Reasoning Element and justification, depending on the step of the process. **Metamodel** indicates which metamodel is accessed in the mapping process to automatically build the rationale RDF.

3.2   An Example Illustrating the Execution of the StdTrip+K Process

For the example we use the publication database depicted in Fig. 3. It is important to note that we implicitly assume that the input database is fully normalized, i.e., the input to the conversion stage must be in third normal form (3NF). Furthermore, we also assume that the user who follows this approach has some knowledge about the application domain of the databases. The result of the complete rationale captured can be seen in the illustration of the final stage (Fig. 4).

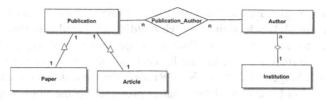

Fig. 3. Author-Publication ER diagram [14]

3.2.1 Stage 1 - Mapping

The general goal of this stage is to map the structure of the input relational database schema onto intermediate database ontology (we call OWL') and to trace the DR for the mapping (we call DR1). OWL' is not the final ontology because there is no execution of matching algorithms in this stage yet. To achieve the general goal, there are two sub stages: **(1.1)** *RDB-to-ER*, to transform the relational database schema into an Entity Relationship (ER) model and **(1.2)** *ER-to-OWL'*, to transform the ER model into an OWL ontology (OWL'). The rationale captured in this stage records the mapping rules used in the mapping since it is part of the domain expert decisions. The resulting (yet intermediate) OWL ontology is a model that simply mirrors the schema of the input relation database. To illustrate the rationale representation, we will consider only the part of the input database example regarding the mapping of the *Author* and *Institution* classes with their attributes and the relationship established between them, *ex:WorksFor*. We list the K-steps executed to capture the DR 1: **K1 - Identify reasoning elements from the ER model.** The reasoning elements *last_name*, *author*, *Author_Institution* and *Institution* were identified, because all of them are elements that will be mapped; **K2 - Identify the representation of the reasoning element in the ER model.** After having identified each reasoning element, the rationale representation records which element (Entity, Attribute, Relationship) it represents in the ER model, in order to keep the traceability of each element; **K3 - Record the corresponding mapping of the ER element onto the OWL element.** Having identified all the ER elements, the DR model records the correspondent OWL element mapped for each reasoning element; **K4 - Record the argument for the mapping.** For each reasoning element, the argument is the respective mapping rule used in the mapping. As the mapping rules are not rigid nor a consensus, this step records how each element was mapped as an argument form; **K5 - Record the corresponding OWL intermediate term**. Finally, this step records the intermediate term mapped for each element.

3.2.2 Stage 2 – Matching

The general goal of this stage is to find correspondences between the intermediate ontology obtained in the previous stage (Stage 1 - Mapping) and standard well-known RDF vocabularies. This stage comprises three sub stages: **(2.1) Matchers execution** – For each element in the intermediate ontology, there are partial candidates according to each matcher, with their respective similarity values. **(2.2) Combination strategies**

– aggregation strategies are applied to define an unified similarity value for each pair of ontology terms. **(2.3) Selection of match candidates** – until here there is still more than one match for each term, so the final sub stage aims at applying a selection strategy to choose one final match candidate for each ontology term. The steps for the DR representation of Stage 2, DR 2 are: **K6 - Record the candidates for each intermediate term.** It records each candidate that is presented to the domain expert; **K7 - Identify and record the arguments (*in favor of* and *objects to*).** For each candidate, there is a final similarity value that represents the reason for this candidate to be part of the list presented to the domain expert. As the Kuaba+W DR model defines arguments as "*in favor of*" and "*objects to*", they have to be identified and traced to keep all options the user currently have to make his or her decision. Due to space constraints, we illustrate only one case of different options with arguments *in favor of* and *objects to*, associated to *ex:last_name* example.

3.2.3 Selection Stage

The general goal of this stage is to select the terms resulting from the previous stages in order to build the final OWL ontology. In this stage, user interaction plays an essential role. Ideally, the user should know the application domain because he or she has to select the vocabulary elements that best represent each concept in the database. Similarly to the previous DR models, the DR of this stage (DR 3) is incrementally built from the preceding DR (DR 2) executing the following steps: **K8 – Record the user decision domain about each term.** The Kuaba+W model records all decisions involved in the acceptance (A) or rejection (R) of each term recommended by StdTrip+K. In the DR 3 model, these decisions are represented by the letters *A* and *R*, respectively; **K9 – Record the justification of the domain expert.** After each decision, the user expert justifies his or her choices. An example that represents the relevance of tracing the DR of this stage is related to the term *ex:last_name*, for which the expert domain decided to use the term with the lowest similarity value, and without the DR it would not be possible to know why.

3.2.4 Inclusion Stage

The general goal of this stage is to complete the final OWL ontology with terms that were not identified in the previous stages. This can happen when the Selection stage does not yield any result or when none of the suggestions in the list is considered adequate by the user. The DR 4 is recorded through the following step: **K10 – Record the new term and the justification.** The expert domain justifies the inclusion of a description which explains why this is an appropriate term in the input database context.

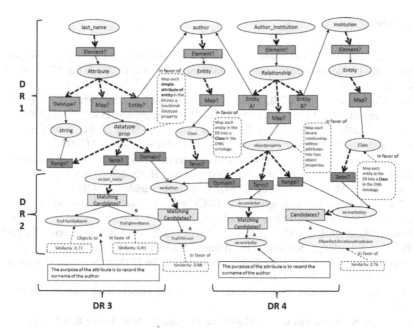

Fig. 4. Resulting design rationale captured for the example

4   Conclusions and Future Works

In this article, we introduced the StdTrip+K process. It allows the translation of a relational database to RDF triples reusing standard vocabularies and recording the DR from the translation. The StdTrip+K provides objective information about the RDB-to-RDF process and it is possible to answer the questions that still arise when using triple sets in the Linked Data cloud. (1) *Has the database suffered changes when published as RDF triples that could impact in its quality? May the original relational database have lost some relevant information when it is published as RDF triples?* As the DR shows the original form of the dataset (as ER model), it is possible to compare the database initial form and the mappings, and, consequently, evaluate the differences impact, if any. (2) *Is the chosen ontology the most appropriate to represent the database? Is the translation correct from the original relational database to RDF?* Once DR shows the options abandoned; accepted and the reasons for that, it is possible to evaluate the choices done. Also, the DR allows having access of one-to-one and one-to-many mappings despite not having been addressed in the running example of this article. We believe our work can be further improved as follows: Implementing the reuse of DR in the mapping process, adding recommendation functionality in the StdTrip+K making use of previous decisions regarding abandoned options in similar domains; Providing a more compact visualization of the captured DR allowing a detailed visualization just when required by the triple set consumer; and Incorporating the rationale model to other RDB-to-RDF strategies that presents different characteristic from StdTrip. The last further work emphasizes that the rationale model may be adapted to capture the triplification rationale in other RDB-to-RDF processes and it is not a specific solution for StdTrip approach.

References

1. Auer, S., Dietzold, S., Lehmann, J., Hellmann, S., Aumueller, D.: Triplify: light-weight linked data publication from relational databases. In: WWW 2009, pp. 621–630. ACM, New York (2009)
2. Bizer, C., Seaborne, A.: D2RQ-treating non-RDF databases as virtual RDF graphs. In: Proceedings of the 3rd International Semantic Web Conference, ISWC 2004 (2004)
3. Bizer, C., Heath, T., Berners-Lee, T.: Linked Data - The Story So Far. International Journal on Semantic Web and Information Systems (IJSWIS) 5(3), 1–22 (2009)
4. Cerbah, F.: Learning highly structured semantic repositories from relational databases: The RDBToOnto Tool. In: Bechhofer, S., Hauswirth, M., Hoffmann, J., Koubarakis, M. (eds.) ESWC 2008. LNCS, vol. 5021, pp. 777–781. Springer, Heidelberg (2008)
5. Cullot, N., Ghawi, R., Yétongnon, K.: DB2OWL: A Tool for Automatic Database-to-Ontology Mapping. In: SEBD, pp. 491–494 (2007)
6. Dividino, R., Schenk, S., Sizov, Staab, S.: Provenance, Trust, Explanations – and all that other Meta Knowledge. Künstliche Intelligenz KI 23(2), 24–30 (2009)
7. Erling, O., Mikhailov, I.: RDF support in the virtuoso DBMS. Networked Knowledge-Networked Media, 7–24 (2009)
8. Lee, J.: Design Rationale Systems: Understanding the Issues. IEEE Expert 12(13), 78–85 (1997)
9. Lee, J., Lai, K.: What's in Design Rationale. Human-Comput. Interaction 6(3-4), 251–280 (1991)
10. Maclean, A., Young, R., Bellotti, V., Moran, T.: Questions, Options, and Criteria: Elements of Design Space Analysis. Human-Comput. Interaction 6(3-4), 201–250 (1991)
11. McGuinness, D., Harmelen, F.: OWL web ontology language – W3C Recommendation (2004), http://www.w3.org/TR/owl-features/ (retrieved February 2013)
12. Medeiros, A.P., Schwabe, D.: Kuaba approach: Integrating formal semantics and design rationale representation to support design reuse. Artificial Intelligence for Engineering Design, Analysis and Manufacturing 22, 399–419 (2008)
13. Prud'hommeaux, E., Hausenblas, M.: Use cases and requirements for mapping relational databases to rdf (2010), http://www.w3.org/TR/rdb2rdf-ucr/ (retrieved November 27, 2012)
14. Salas, P., Viterbo, J., Breitman, K., Casanova, M.A.: StdTrip: Promoting the Reuse of Standard Vocabularies in Open Government Data. In: Wood, D. (ed.) Linking Government Data, pp. 113–134. Springer (2011)
15. Sequeda, J., Depena, R., Miranker: Ultrawrap: Using SQL views for RDB2RDF. In: Proceedings of International Semantic Web Conference. ISWC 2009 (2009)

Organizing Scientific Competitions on the Semantic Web

Sayoko Shimoyama, Robert Sidney Cox III, David Gifford,
and Tetsuro Toyoda

Integrated Database Unit, Advanced Center of Computing and Communication (ACCC)
RIKEN,
2-1 Hirosawa, Wako, Saitama, 351-0198, Japan
toyoda.tetsuro@gmail.com

Abstract. Semantic web techniques for Linked Open Data (LOD) are expected
to enhance the use of scientific data, and several data repositories for LOD have
been launched. Modifiable "Forkable Open-source programs" on code sharing
platforms make applications (Apps) utilizing data ready for reuse. In order to
organize a web-based scientific competition, platforms for both semantic data
resources and application programs need to be integrated so as to yield a
creative cycle between data publication and application development. We
developed the LinkData.org platform to integrate both data and application publishing platforms by recording dependency graphs, the utility of which we tested by organizing a scientific competition for synthetic biology on the platform.
It was found that participants to the competition generated many dependency
graphs by forking pre-existing applications or reusing schema of pre-existing
datasets. These creative activities could not be observed explicitly without being recorded such as by dependency graphs among the datasets and applications
on the platform. Hence we suggest a worldwide system needs to be established
to record and harvest such dependency graphs from distributed data platforms
and application-development platforms around the world, so that our intellectual and creative activities using open datasets for application development may
be recorded properly.

Keywords: Semantic Web, Linked Open Data, GenoCon, Synthetic Biology,
Open Science.

1 Introduction

Data repositories and directories for open data such as The Comprehensive
Knowledge Archive Network (CKAN) web-based system for data storage and distribution, supported by the Open Knowledge Foundation, help users register their data
resources and locate related data. The Resource Description Framework (RDF) graph
structure data format is the standard for sharing Linked Open Data (LOD) on the web
[1].

The LOD model with RDF and SPARQL endpoints [2] gives open data access to
external applications (Apps) worldwide, however, the act of separating data from
applications on the web hides synergic collaboration between data and applications;

H. Decker et al. (Eds.): DEXA 2013, Part I, LNCS 8055, pp. 311–318, 2013.

resulting in the situation that scientific competitions requiring development of both scientific linked data and applications using the data are not well organized without special attention to clearly displaying the dependency among these entities.

To overcome this situation, we developed its LinkData.org (http://linkdata.org) as a data publishing platform and LinkDataApp (http://app.linkdata.org) as an application publishing platform, and combined them by automatically recording dependency graphs that relate data and Apps using the data; thus LinkData acts as a repository of dependency graphs connecting Apps and datasets, as well as a repository of applications and datasets.

Here we introduce the dependency graph by using the online activities performed by participants for GenoCon2, the second international genomic design competition. We find that dependency graphs well represent the relationship between the GenoCon2 Data and Apps.

2   Methods

2.1   LinkData as an RDF Publishing Platform

Support Functions for Creating Table Data to Upload. As a support function that allows Users to easily define schema, LinkData provides a GUI by which anyone can create and download a template. When a User enters metadata for their data using this GUI, a table format Excel file using column names for RDF properties is generated, and this file can be downloaded. Users input their data to the template to create their own table data to upload.

Conversion to RDF Format and Publishing. The template data tables can be up-loaded, converted to RDF format, and published online. When a User uploads the table data file in Excel or TSV format, anyone accessing the page will be able to browse and download the table data, a template for table data, as well as in RDF for-mat. Please see the results section for an example.

Reuse Data Function. Schemas of all of published Data can be reused for publishing new datasets. Users can activate the reuse table data function at the published Data webpage and down-load a template with minimal modification.

Displaying Applications Relating to Particular Data. Applications that are using particular Data as an input are displayed on the page of each of published Data.

Creating Application by Editing Sample Program. A sample JavaScript program is automatically generated when a User selects a file from published Data as an input and creates a new App. Users can edit the sample program on a web browser to develop an original App. Anyone accessing to the published App page is able to execute, fork and download the App. Please see the result section for an example.

Forking Application to Publish as a New One. The Source code of published Apps can be reused by anyone. A forked program will be modified and published as a new App. The relationships between App and the Data used as input for the App are maintained so that users only need to make minimal changes to the program.

Changing Input Files to Create a New Application. When a User forks an App, input files can be changed to new Data provided by the User. With this function a user can create a new App without editing the program, and any User can create their own App even if they are not a JavaScript programmer.

Displaying Data relating to an App. Data used as an input to a particular App are displayed on the page of each published App.

2.2   Entities and Links of LinkData Concepts

The dependency graph uses three entity nodes: Data, App and User (Table 1). A user is a person registered to use the platforms. Data is a single data set which has been published by one or more Users of LinkData. It must have at least one file uploaded by the User. An App is a JavaScript application which has been published by a User in LinkDataApp, or an external web page registered by the user as an application using the Data. An App must load at least one file from Data. Relationships among entities are described by eight types of links shown in Table 2. A link relationship represents the graph association from one entity node to another. For the link new Data to old Data, a link termed "reuse," will be generated when a User creates new Data by reusing an existing schema from the old Data. The link Data to User, termed "contributed", will be generated when a User creates any Data. The link from a new App to an old App, termed "fork", will be generated when a User creates a new App by reusing an existing App's program code. The link from an App to Data, termed "load", will be generated when a User creates a new App by specifying some files as input from some particular Data or loads some files in their existing App. The link from User to User, termed "follow", will be generated when a User follows another User. For example, if User A follows User B, User A can receive updates about User B's Data or App and information about Data and Apps evaluated by User B. The links from User to Data and from User to App, termed "vote", will be generated when a user browses Data or an App and gives a Useful or Un-useful rating for that Data or App.

Table 1. Entities of LinkData Concepts

Entity	Definition
Data	A single data set which has been published by a User in LinkData
Application (App)	A single application which has been published by a User in LinkData
User	A user who had registered for a LinkData account

Table 2. Links of LinkData Concepts

Link	Term	Label	Definition
Data (new) → Data (old)	reuse	L_{dd}	Create new Data by reusing an existing Data.
Data → User	contributed	L_{du}	The relationship between Existing Data and the user who created the Data.
App (new) → App (old)	fork	L_{aa}	Create a new App by reusing an existing App's program code.
App → Data	load	L_{ad}	Create an App by specifying some files as input from some particular Data.
App → User	contributed	L_{au}	The relationship between an Existing App and the user who created the App.
User (A) → User (B)	follow	L_{uu}	User A follows user B to receive updates and information of evaluated Data and Apps by user B.
User → Data	vote	L_{ud}	A user gives a rating of Useful or Un-useful for considered Data
User → App	vote	L_{ua}	A user gives a rating of Useful or Un-useful for a considered App

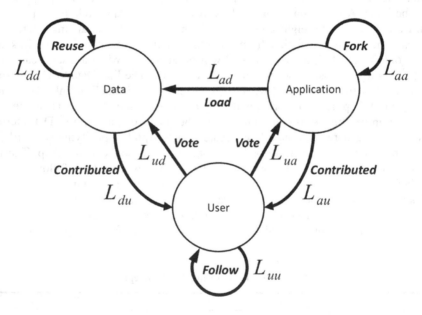

Fig. 1. Dependency graph for usability analysis. The graph comprises three types of nodes: data, application and user, and eight types of links: *Reuse - Ldd, Fork - Laa, Follow - Luu, Load - Lad, Vote for Data - Lud, Vote for App - Lua, Contributed Data - Ldu* and *Contributed App - Lau* are shown in the figure.

2.3   Dependency Graph for Analyzing the Creative Activities on the Platforms

The dependency graph represents the record of a user's creative activities on the platforms. The graph is composed of three types of nodes: data, application and user, and eight types of links between them: *Ldd, Laa, Luu, Lad, Lud, Lua, Ldu* and *Lau* are shown in Fig. 1, and the accumulated links represent the record of activities, such as loading data, data template reuse, and application forking (cloning and modification).

3   Results and Discussion

3.1   Count of Relationships among Three Entities Indicates Creative Synergy Cycle

LinkData hosts 557 datasets and 260 applications as of March, 2013. Datasets contain 350 public, 40 limited, and 162 private. Applications contain 160 public, 55 limited, and 45 private. There are a large number of Load (App to Data) relationships and many Fork (App to App) relationships (Table 3). In contrast, there are few Reuse (Data to Data) relationships. It is thus clear that there is a stronger synergy cycle between data resources and applications than "in data" (between Data and Data). In other words, this indicates that a platform which has both capabilities of publishing data resources and creating applications has higher creativity than one having only one capability of data resource creation.

Table 3. Count of relationships between data resources and applications hosted in LinkData

Kind of relationship		Count
Load	(App to Data)	166
Fork	(App to App)	137
Reuse	(Data to Data)	39
Follow	(User to User)	52
Vote	(User to Data)	244
Vote	(User to App)	89

All the dependency graphs are downloadable from the LinkData.org APIs (Table 4.).

Table 4. Downloadable Dependency Graph APIs on LinkData

APIs	TSV format	JSON format
reuse	http://linkdata.org/api/1/graph/reuse_tsv.txt	http://linkdata.org/api/1/graph/reuse_rdf.json
fork	http://linkdata.org/api/1/graph/fork_tsv.txt	http://linkdata.org/api/1/graph/fork_rdf.json
load	http://linkdata.org/api/1/graph/load_tsv.txt	http://linkdata.org/api/1/graph/load_rdf.json
follow	http://linkdata.org/api/1/graph/follow_tsv.txt	http://linkdata.org/api/1/graph/follow_rdf.json
vote to Data	http://linkdata.org/api/1/graph/vote_data_tsv.txt	http://linkdata.org/api/1/graph/vote_data_rdf.json
vote to App	http://linkdata.org/api/1/graph/vote_app_tsv.txt	http://linkdata.org/api/1/graph/vote_app_rdf.json

3.2   Organizing a Scientific Competition Using the Integrated Platform

For the synthetic biology competition GenoCon2 (http://genocon.org) [3], we challenged participants to design novel regulatory DNA for controlling gene expression in the thale cress plant Arabidopsis thaliana. Participant DNA designs will be synthesized and tested for tissue and time specificity in a real plant. Participants are allowed to create a *de novo* design tool as an App on the platform, or to modify by forking the App we provided as a computer aided design tool on the platform, called PromoterCAD [4].

Using PromoterCAD function modules, genes with the desired properties can be found and mined for regulatory motifs. These are introduced into the synthetic promoter by user choice of regulatory position. Repeating this process can create complex regulation at the promoter. Finally, the DNA design is exported for error and safety checking, DNA synthesis, and experimental characterization.

PromoterCAD rests on a rich set of high throughput microarray and DNA sequence data containing over one million measurements and annotations of 20,000 genes. These were uploaded to LinkData as a series of data mashup tables and data rank lists (Fig. 2). Where other DNA design tools act as sequence editors with DNA specific functions, PromoterCAD is able to pull sequence data directly from the data sources in the LinkData system, guided by the menu-driven interface. LinkData gives PromoterCAD the power to allow users to quickly perform advanced data queries, retrieve useful sequences, and organize them into their promoter sequence designs.

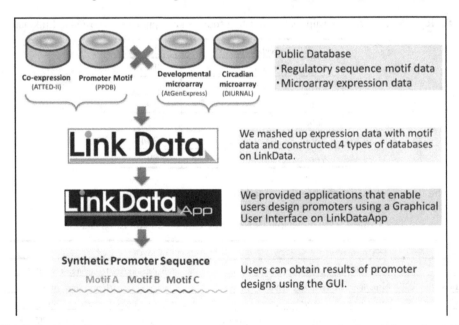

Fig. 2. PromoterCAD LinkData system for DNA design incorporates database information with user knowledge

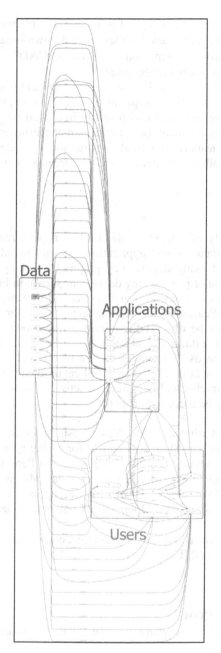

The LinkData system provides code extensibility to PromoterCAD. With the forking function, users can write their own JavaScript data mining modules to PromoterCAD, and draw upon the rich linked data in new ways. For example, one participant in GenoCon2 modified a PromoterCAD function to display the top 10 expressing genes in a specific plant tissue. Other GenoCon2 participants used this module, and the forked utility has since been incorporated as part of the main PromoterCAD functionality. Customizing applications with optional user created modules allows non-coding users to take advantage of the experience of more advanced coding users.

The architecture of PromoterCAD allows new LinkData sources to be added without any direct code modification. PromoterCAD provides downloadable template file of input data that is structured with spreadsheet (Excel) so that users only need to copy and paste gene expression values or regulatory sequence lists and upload it to LinkData. This function is intended to allow scientists who are not programmers to add their own databases to PromoterCAD. A user could adapt PromoterCAD to design regulatory DNA in other organisms such as mouse, human, or bacteria by replacing all of the data tables.

Fig. 3. Dependency **graph of the PromoterCAD application on LinkData.** This graph shows the interaction between the Data (Green color box), the Apps (Blue color box), and the Users (Grey color box). Red, Blue, Green and Grey lines indicate load, fork, contribute and vote, respectively.

The dependency graph in Fig. 3 shows an example of the cycle enhancing synergistic collaboration in this web-based scientific competition for synthetic biology promoter design. The source dataset http://linkdata.org/work/rdf1s339i "Speedup Lists of Developmental Coexpression" is a source for this graph. Data may be viewed and downloaded in mentioned formats. The App "GenoCon PromoterCAD" at http://app.linkdata.org/app/app1s94i is also shown in the graph.

The GenoCon2 promoter design contest generated active user groups and over 40 international submissions including from the USA, Egypt and Japan. Key users cooperated to create original designs that were modified and possibly improved by other users. Team collaboration was aided by the open nature of the design platform, and 13 high level promoter designs are being considered for final construction as transgenic plants. Application to further design challenge projects for other organisms is also planned.

4   Conclusion

A scientific competition was successfully organized on the LinkData platform that records the dependency graphs among datasets and applications. It was found that participants to the competition generated many dependency graphs by forking pre-existing applications or reusing the schema of pre-existing datasets as shown in Fig. 3, i.e., most of the participants did not develop a new *de novo* application by their own hands from scratch, but modified the applications or reused the datasets developed by others. These creative activities could not be observed explicitly without being recorded such as by dependency graphs among datasets and applications on the platform. Hence, we suggest a worldwide system needs to be established to record and harvest such dependency graphs from distributed data platforms and application-development platforms around the world, so that our intellectual and creative activities using open datasets for application development may be recorded properly.

Acknowledgement. We would like to acknowledge: Ms. Yuko Yoshida for development of converter and valuable discussion. Dr. Shuji Kawaguchi for giving advice on score calculation. Dr. Koro Nishikata for testing LinkData. Mr. Chanaka Perera, Mr. Uditha Punchihewa, Mr. Gayan Hewathanthri, Mr. Hiroaki Osada, Mr. Kazuro Fukuhara and Mr. Kiyoshi Mizumoto (Axiohelix Co., Ltd.) for web application and LinkData development. The Committee of Linked Open Data Challenge Japan for continuing interest and encouragement.

This work was supported by the National Bioscience Data-base Center (NBDC) of the Japan Science and Technology Agency (JST).

References

1. Manola, F., et al.: RDF Primer. W3C Rec. (2004)
2. Prud'hommeaux, E., et al.: SPARQL Query Language for RDF. W3C Candidate Rec. (2006)
3. Toyoda, T.: Methods for Open Innovation on a Genome – Design Platform Associating Scientific, Commercial, and Educational Communities in Synthetic Biology. Methods in Enzymology 498, 189–203 (2011)
4. Cox III, R.S., et al.: PromoterCAD: data-driven design of plant regulatory DNA. Nucleic Acids Research (in press)

An Inductive Logic Programming-Based Approach for Ontology Population from the Web

Rinaldo Lima[1], Bernard Espinasse[2], Hilário Oliveira[1], Rafael Ferreira[1], Luciano Cabral[1], Dimas Filho[1], Fred Freitas[1], and Renê Gadelha[1]

[1] Informatics Center, Federal University of Pernambuco, Recife, Brazil
{rjl4,htao,rflm,lsc4,dldmf,fred,rnsg}@cin.ufpe.br
[2] LSIS, Aix Marseille University, Marseille, France
bernard.espinasse@lsis.org

Abstract. Developing linguistically data-compliant rules for entity extraction is usually an intensive and time-consuming process for any ontology engineer. Thus, an automated mechanism to convert textual data into ontology instances (Ontology Population) may be crucial. In this context, this paper presents an inductive logic programming-based method that induces rules for extracting instances of various entity classes. This method uses two sources of evidence: domain-independent linguistic patterns for identifying candidates of class instances, and a WordNet semantic similarity measure. These two evidences are integrated as background knowledge to automatically generate extractions rules by a generic inductive logic programming system. Some experiments were conducted on the class instance classification problem with encouraging results.

Keywords: Ontology Population, Information Extraction, Pattern Learning, Inductive Logic Programming.

1 Introduction

Ontologies, from the computer science point of view, consist of logical theories that encode knowledge about a certain domain in a declarative way [2]. They also provide conceptual and terminological agreements among humans or computational agents that need to share information. On the other hand, the development of ontologies relies on domain experts that typically adopt a manual construction process, which turns out to be very time-consuming and error-prone. Hence, an automated or semi-automated mechanism able to extract the information contained in existing web pages into ontologies, *Ontology Population* (OP), is highly desired [2].

In this scenario, the main goal of this paper is to describe and evaluate a method to automatically induce, via an Inductive Logic Programming (ILP) framework, extraction rules for OP. The proposed method also exploits the semantic similarity between classes and candidate class instances. More precisely, this method relies on: (i) a natural language preprocessing which not only takes into account the typical lexical-syntactic aspects present in the English language, but also exploits semantic similarity between ontology classes and candidate class instances, and (ii) an ILP-based induction of symbolic extraction rules (expressed as Horn clauses) from examples.

H. Decker et al. (Eds.): DEXA 2013, Part I, LNCS 8055, pp. 319–326, 2013.

The rest of this paper is organized as follows: Section 2 is dedicated to related work. Section 3 presents some basic concepts about ILP. The ILP-based method for OP is described in Section 4. We present and discuss experimental results of our method for OP in Section 5. Finally, in Section 6 we conclude and outline future work.

2   Related Work

Several approaches have been developed for extracting class instances from textual data using machine learning techniques. KnowItAll [5] is a hybrid named-entity extraction system that combines Hearst's and some learned patterns for extracting class instances from the Web using a search engine. In order to assess the candidate instances, KnowItAll uses the PMI metric and a Naïve Bayes classifier for achieving a rough estimate of the probability that each candidate instance is correct. In [4], the authors proposed the idea that learned patterns could be used as both extractors (to generate new information) and discriminators (to assess the truth of extracted information). More recently, [9] reports some experimental results using ILP techniques to induce rules that extract instances of various named entities. Moreover, [9] also reported a substantial reduction in development time by a factor of 240 when ILP is used for inducing rules, instead of involving a domain specialist in the entire rule development process.

Although our approach has explored the same kind of surface patterns used by most of the approaches described above, our richer set of features (POS tagging, NER, and semantic similarity measure) seems to achieve promising results. Furthermore, our research hypothesis is that an ILP-based method would allow an easier and flexible integration of background knowledge (BK) provided by other levels of linguistic analysis, as potential future work.

3   Inductive Logic Programming

Inductive Logic Programming is a subfield of machine learning which uses first order clauses as a uniform representation for examples, background knowledge, and hypotheses [7]. According to [3], there are two main motivations for using ILP: (i) it overcomes the representational limitations of attribute-value (propositional) learning systems that employ a table-based example representation; (ii) it rather employs a declarative representation, which means that the hypotheses are understandable and interpretable by humans. Moreover, by using logic, ILP systems can exploit BK in its learning (induction) process. For instance, the BK can be expressed in the form of auxiliary predicate definitions provided by the user.

Informally, the ultimate goal of ILP is to explain all of the positive and none of the negative examples. More formally, given: (i) a set of examples $E = E^+ \cup E^-$, where E^+ (positive) and E^- (negative), and; (ii) background knowledge BK, the task of ILP is to find a hypothesis H such that: $\forall e \in E^+: BK \wedge H \models e$ (H is *complete*), and $\forall e \in E^-: BK \wedge H \not\models e$ (H is *consistent*).

Many existing ILP implementations, such as GILPS [10], are closely related to Prolog and, therefore, they impose the following typical restriction to the way of how the *BK* (in terms of predicates or rules) and examples are represented. In other words, the *BK* is restricted to Prolog clause in the form *head :- body₁, body₂, ...,* $body_n$. Thus, the head is implied by the body clauses, whereas E^+ and E^- are restricted to ground facts. We refer the reader to [3], [7], [10] for further details on ILP.

4   An ILP-Based Method for Populating Domain Ontologies

Our supervised method for OP takes profit of the high redundancy present in the Web content, considering it as a big corpus. Sharing the same idea, several authors pointed it out as an important feature because the amount of redundant information can represent a measure of its relevance [5], [4]. Moreover, we take into account the portability issue, i.e., the method should able to perform independently of the domain. We adopted the ILP framework as the core component for machine learning in our method because it can provide extraction rules in a symbolic form which can be fully interpreted by a knowledge engineer. Consequently, the user can either refine these rules or simply converting them to other rule formalisms.

One of the main advantages of ILP over other statistical machine learning algorithms is that not only the learned patterns are expressed in a symbolic form which is more easily interpreted by a knowledge engineer, but also allows the integration of considerable amount of prior knowledge as part of the solution to the problem under consideration. Moreover, according to [9], when compared with a handcrafting rule approach, an ILP-based method can provide a complete and consistent view of all significant patterns in the data at the level of abstraction specified by the knowledge engineer.

The proposed method is composed of four main steps as illustrated in Fig. 1.

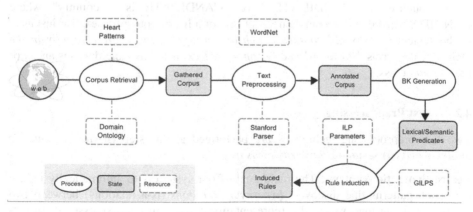

Fig. 1. Overview of the ILP-based ontology population system

In general terms, the method consists of a supervised approach to automatically generating extraction rules that subsumes lexico-syntactic patterns present in textual documents. As a result, the induced rules can be applied on an unseen set of preprocessed documents in order to extract instances that populate an ontology. In the remain part of this section, we present each system component in more detail.

4.1   Corpus Retrieval

The first step, the corpus retrieval process, starts retrieving sentences from web pages in order to constitute a corpus. We rely on a set of domain-independent linguistic patterns for this task. Fig. 2 presents the patterns used for gathering relevant documents containing candidate instances of concepts (classes) of a domain ontology.

After the user's choice of a class from this domain ontology, the system retrieves some documents based on both the *label* of the chosen class, and the patterns *P* (Fig. 2). For instance, selecting the *Country* class, the above patterns would match sentences in natural language such as: *"is a country"; "countries such as"; "such countries as"; "countries especially"; "countries including"; "and other countries"; "or other countries"*. These phrases likely include instances of the *Country* class in the CANDIDATE(S) part [2], [4], [5].

```
P1:        <CANDIDATE> is a/an <CLASS>
P2:        <CLASS>(s) such as <CANDIDATES>
P3:        such <CLASS>(s) as <CANDIDATES>
P4 and P5: <CLASS>(s) (especially/including) <CANDIDATES>
P6 and P7: <CANDIDATES> (and/or) other <CLASS>(s)
```

Fig. 2. Domain-independent Hearst patterns

Next, each query is submitted to a Web search engine, and the first N web documents are fetched for each pattern. We are interested in extracting sentences like, "such countries as CANDIDATES" or "CANDIDATE is a country" where CANDIDATE(S) denotes a single noun phrase or a list of noun phrases. For instance, in the sentence: *"Why did countries such as Portugal, France grow rapidly in the 1930's?"*, the terms *"Portugal"* and *"France"* are extracted as candidate instances of the *Country* class.

4.2   Text Preprocessing

Two text preprocessing techniques are performed at this step (ii) *lexico-syntactic analysis*, and (ii) *semantic similarity measuring*.

Lexico-Syntactic Analysis. The main goal of our system is to automatically induce extraction patterns that discovers *hypernymy relations* (is-a relations) between two terms. For doing that, we need a representation formalism that expresses these patterns in a simple and effective way. We defined a set of lexico-syntactic features produced by the preprocessing component in our architecture. These features are the building blocks that compose the background knowledge that will be used later by the component responsible for the induction of extraction rules. The prototype system developed in this work relies on the Stanford CoreNLP [11], a Natural Language Processing (NLP) tool. This NLP tool performs the following sequence of processing task: *sentence splitting, tokenization, Part-of-Speech* (POS) *tagging, lemmatization,* and *Named Entity Recognition* (NER) which labels sequences of words in a text into predefined categories such as *Person, Organization, Date,* etc.

Semantic Similarity Measuring. Semantic similarity measures based on WordNet [8] have been widely used in NLP applications, and they take into account the WordNet taxonomical structure to produce a numerical value for assessing the degree of the semantic similarity between two terms. We adopted the similarity measure proposed in [12] which provides the degree of similarity between the class C and a candidate class instance C_i. It relies on finding the most specific concept that subsumes both the concepts in WordNet.

4.3   Background Knowledge Generation

This step consist of identifying, extracting, and appropriately representing relevant BK for the task at hand. Previous research have shown that shallow semantic parsing can provide very useful features in several information extraction related tasks [6]. Accordingly, we explore the features listed in Tab. 1, which constitute the BK in our approach. These features provide a suitable feature space for the classification problem of candidate instances, as they describe each token in the corpus. Furthermore, we calculate the similarity degree between each token tagged as singular or plural noun by the POS tagger and a class in the domain ontology. We illustrate in Tab. 1 the BK that characterizes the candidate instance of the *Country* class, "*France*".

Table 1. ILP Predicates for the token "France"

Predicate Generated	Meaning
token (t_1)	t_1 is the token identifier
t_length (t_1, 6)	t_1 has length of 6
t_ner (t_1, location)	t_1 is a location entity according to the NER
t_orth (t_1, upperInitial)	t_1 has an initial uppercase letter
t_pos (t_1, nnp)	t_1 is a singular proper noun
t_next(t_1, t_2)	t_1 is followed by the token t_2
t_type (t_1, word)	t_1 is categorized as a word
t_wnsim(t_1, country, '09-10')	t_1 has a similarity score between 0.9 and 1.0 with the *Country* class

Given that the WordNet similarity values are in the [0,1] range, we perform a discretization of this numerical feature by creating 10 bins of equal sizes (0.1 each). Thus, for example, if the WordNet similarity value between the candidate class instance "*France*", and the class "*Country*" is 0.96, we put this value in the 10th bin which corresponds to the predicate *t_wnsim(t_id, country, '09-10')*. In other words, the predicate t_wnsim(A, country, '09-10') means that the token 'A' has a similarity score between 0.9 and 1.0 with the *Country* class.

4.4   Rule Induction

In this last step, we have to define the *language bias* which both delimits and biases the possibly huge hypothesis search space. In GILPS, this is achieved by providing appropriate *mode declarations*. Mode declarations characterize the format of a valid hypothesis (rule). They also inform both the *type*, and the *input/output modes* of the predicate arguments in a rule. Mode head declarations (*modeh*) state the target predicate, i.e., the head of a valid rule that the ILP system has to induce, whereas *mode body* declarations (*modeb*) determine the literals (or predicates), which may appear in a rule body. In addition, the *engine* parameter in GILPS permits the user to choose the way rules are specialized/generalized, i.e., how the hypothesis space are traversed, either in *top-down* or *bottom-up* manner. The top-down approach was selected because it enables the construction of shorter theories (in term of the number of clauses) [10]. Finally, GILPS induces a set of rules that can be applied on an unseen set of preprocessed documents in order to extract instances that populate an ontology.

5   Experimental Evaluation

In this section, we describe how the corpus was created and annotated. Next, we present and discuss the results of our experiments on the OP task.

5.1   Corpora Creation and Annotation of Examples

The corpus used in this evaluation was compiled using the 7 surface patterns listed in Section 4.1. We have performed an evaluation on 5 classes, namely *Country, Disease, Bird, Fish, and Mammal* classes. For each class, the system retrieved approximately 420 sentences equally distributed into sentences containing positive and negative candidate instances. We used the Bing Search Engine API [1] for collecting a total of 2100 sentences (420 sentences for 5 classes).

The task of inducing target predicates in GILPS requires that positive and negative examples be explicitly indicated before the generation of the classification model. Thus, two human annotators manually tagged the positive instances. There is no need to annotate the negative examples because they can be automatically identified as the complement of the positive ones.

5.2   Evaluation Measures and GILPS Parameters

The performance evaluation is based on the classical measures used in IR systems, i.e., *Precision* P, *Recall* R, and F1-*measure*. In all experiments, we used 10-fold cross-validation that provides unbiased performance estimates of the learning algorithm. GILPS was run with its default parameters, except for the following specific settings: *theory_construction = incremental, evalfn = compression, clause_length = 8, nodes = 1000*. In incremental theory construction, when the best hypothesis from an example is found, all the positive examples covered by this hypothesis are retracted from the training set, whereas the *nodes* parameter determines the maximum number of hypotheses that may be derived from a single positive example.

5.3   Results and Discussion

In order to estimate the classification performance of the learned rules for each class, we used two versions of the compiled corpus as described in Section 5.1. The first version of the corpus was only annotated with lexico-syntactic features (see Section 4.2). In the second version, we added a WordNet semantic similarity feature. Each class was assessed separately by building a binary classifier for each one.

Table 2. Classification performance of the induced rules

Class	No WordNet			With WordNet		
	P	R	F1	P	R	F1
Country	0.96	0.92	0.94	0.98	0.96	0.97
Disease	0.95	0.69	0.80	0.97	0.84	0.96
Bird	0.93	0.53	0.67	0.95	0.73	0.82
Fish	0.93	0.42	0.58	0.94	0.50	0.65
Mammal	0.93	0.39	0.55	0.93	0.49	0.64

The results shown in Tab. 2 are encouraging, since the proposed method seems to successfully extract a significant number of positive instances from the corpora. Considering the F1 score for the *Country* class, one can observe that, as being an entity type recognized by the parser (named entity = *location*), the *Country* class had a very tiny improvement on the sample with the additional WordNet predicate. On the other hand, for the other classes, the rules based only on lexico-syntactic predicates are highly precise, but its achieved recall score is lower than those ones when the WordNet predicate is used. This comparison shows that the semantic similarity measure provided by WordNet can be very useful. In fact, a statistical significance test (*paired Student t-test*) for the difference between the F1 scores of the two experiments above were performed. The test revealed that there is a significant difference at $\alpha = 0.05$ (95% confidence interval) between them. Thus, this assessment suggests that the additional WordNet similarity predicate in the BK actually contributed to achieve better performance results.

In the following, we list some induced rules expressed in terms of (*number of literals*), (*positive examples covered*), (*negative examples covered*), and the (*rule precision P*):

Rule 1: #Literals = 4, PosScore = 17, NegScore = 0, P = 100%
   is_a_mammal(A):- t_ner(A,misc), t_orth(A, upperinitial), t_pos(A, nn).
Rule 2: #Literals = 3, PosScore = 629, NegScore = 14, P = 97.8%
   is_a_country(A):- t_wnsim_country(A, '09-10'), t_ner(A, location).
Rule 3:   #Literals = 4, PosScore = 329, NegScore = 28, P = 92.0%
   is_a_disease(A):- t_length(A,8), t_type(A, word), wnsim(A, disease, '09-10').

The ILP-based system found a perfect extraction rule for the *Mammal* class (Rule 1), i.e., an instance beginning with an uppercase letter, tagged "miscellaneous" by the NER, and tagged as a singular noun. In Rule 2, the high precision score of the *Country* class is mainly due to the NER that has tagged the instance as a "*location*" combined with a high score similarity with the WordNet synset "*country*". Rule 3 classifies an instance of the *Disease* class if it is a term with 8 characters, and its similarity score with the WordNet synset "*disease*" is between 0.9 and 1.0.

6   Conclusion and Future Work

We have presented an ILP-based method for ontology population, which mainly relies on shallow syntactic parsing and a semantic similarity measure. Although we have achieved encouraging results so far, there are still some opportunities for improvement. Indeed, our method is currently based on a set of domain-independent extraction rules that usually fails to generalize on the most linguist variations. Thus, in order to improve its recall, we intend to use another sentence representation formalism based on *dependency grammar*, which was proven to be more robust to linguist variations. Finally, we intend to extract instances of relations as well.

References

1. Bing Search Engine API. API Basics,
   http://www.bing.com/developers/s/APIBasics.html
2. Cimiano, P.: Ontology Learning and Population from Text: Algorithms, Evaluation and Applications. Springer, New York (2006)
3. De Raedt, L.: Inductive Logic Programming. In: Encyclopedia of Machine Learning, pp. 529–537 (2010)
4. Downey, D., et al.: Learning Text Patterns for Web Information Extraction and Assessment. In: Proceedings of the 19th National Conference on Artificial Intelligence Workshop on Adaptive Text Extraction and Mining, San Jose, USA (2004)
5. Etzioni, O., et al.: Web-Scale Information Extraction in KnowItAll. In: Proc. of the 13th International World Wide Web Conference (WWW 2004), New York, USA, pp. 100–110 (2004)
6. Finn, A.: A Multi-Level Boundary Classification Approach to Information Extraction. Phd thesis, University College Dublin (2006)
7. Lavrac, N., Dzeroski, S.: Inductive Logic Programming: Techniques and Applications. Ellis Horwood, New York (1994)
8. Miller, G.A.: WordNet: a Lexical Database for English. Communications of the ACM 38(11), 39–41 (1995)
9. Patel, A., Ramakrishnan, G., Bhattacharya, P.: Incorporating Linguistic Expertise Using ILP for Named Entity Recognition in Data Hungry Indian Languages. In: De Raedt, L. (ed.) ILP 2009. LNCS, vol. 5989, pp. 178–185. Springer, Heidelberg (2010)
10. Santos, J.: Efficient Learning and Evaluation of Complex Concepts in Inductive Logic Programming. Ph.D. Thesis, Imperial College (2010)
11. Stanford CoreNLP Tools. The Stanford Natural Language Processing Group,
    http://nlp.stanford.edu/software/corenlp.shtml
12. Wu, Z., Palmer, M.: Verb Semantics and Lexical Selection. In: Proc. of the 32nd Annual Meeting of the Association for Comp. Linguistics, New Mexico, USA, pp. 133–138 (1994)

Incremental Algorithms
for Sampling Dynamic Graphs

Xuesong Lu, Tuan Quang Phan, and Stéphane Bressan

School of Computing, National University of Singapore
{xuesong,disptq,steph}@nus.edu.sg

Abstract. Among the many reasons that justify the need for efficient
and effective graph sampling algorithms is the ability to replace a graph
too large to be processed by a tractable yet representative subgraph. For
instance, some approximation algorithms start by looking for a solution
on a sample subgraph and then extrapolate it. The sample graph should
be of manageable size. The sample graph should preserve properties of
interest. There exist several efficient and effective algorithms for the sampling of graphs. However, the graphs encountered in modern applications
are dynamic: edges and vertices are added or removed. Existing graph
sampling algorithms are not incremental. They were designed for static
graphs. If the original graph changes, the sample must be entirely recom-
puted. Is it possible to design an algorithm that reuses whole or part of
the already computed sample?

We present two incremental graph sampling algorithms preserving
selected properties. The rationale of the algorithms is to replace a fraction of vertices in the former sample with newly updated vertices. We
analytically and empirically evaluate the performance of the proposed algorithms. We compare the performance of the proposed algorithms with
that of baseline algorithms. The experimental results on both synthetic
and real graphs show that our proposed algorithms realize a compromise between effectiveness and efficiency, and, therefore provide practical solutions to the problem of incrementally sampling the large dynamic
graphs.

1 Introduction

Graph sampling consists in the selection of a subgraph, the sample graph, from
an original graph. Graph sampling is useful, for instance, when the original graph
is too large to be processed. The original graph is then replaced by a representative sample graph of manageable size. A sample graph is representative if it
preserves selected properties of the original graph. The properties of interest are
usually the topological properties such as degree distribution, clustering coefficient distribution, etc.

In the pioneering paper [14], Leskovec et al. discuss several candidate graph
sampling algorithms. The authors empirically evaluate and quantify the ability
of these algorithms to preserve selected graph properties. Later, Hübler et al.
propose Metropolis algorithms to improve the quality of sampling general graph

H. Decker et al. (Eds.): DEXA 2013, Part I, LNCS 8055, pp. 327–341, 2013.
© Springer-Verlag Berlin Heidelberg 2013

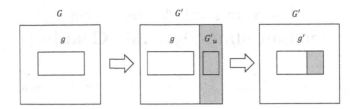

Fig. 1. Illustration of the rationale of incremental construction of a sample g' of G' from a sample g of G. G'_u is the subgraph induced by the updated vertices of G'. g' is constructed by replacing some vertices in g with selected vertices (in the smaller gray rectangle) of G'_u.

properties in [12]. Several authors propose graph sampling algorithms designed to preserve different specific properties (e.g. [10,16,17]). The above algorithms are sampling from static original graphs. Vertices and edges are neither added to nor removed from the original graph. However, the graphs encountered in modern applications are dynamic, that is, vertices or edges are added or removed. For example, Twitter was reported to have an average of 460,000 new users every day in February of 2011 [1]. Some of the properties of the Twitter user graph may have correspondingly changed over time.

For the sake of efficiency, at least, we want to avoid sampling the successive graphs from scratch. The challenge is to incrementally maintain a representative sample graph. We try and devise algorithms able to incrementally update the sample graph as the original graph is modified.

Formally, we consider two *undirected* graphs G and G'. We refer to G as the "old" graph and to G' as the "updated" graph. For the sake of simplicity, we consider that G' is obtained by the addition of vertices and edges to G. We call "updated" a vertex of G' that does not belong to G, or is the endpoint of an edge in G' that does not belong to G. We consider sample graphs of fixed size. The rationale of the proposed algorithms is to replace a fraction of the vertices in the sample graph g of the graph G with updated vertices to obtain a sample graph g' of the graph G'. We refer to g as the "old" sample graph and to g' as the "updated" sample graph. Figure 1 illustrates this rationale. We devise two variants of the idea above in which the vertices are replaced randomly or deterministically, respectively.

The rest of the paper is organized as follows. Section 2 discusses related work on graph sampling algorithms and incremental algorithms. Section 2.4 introduces the background knowledge of Markov Chain Monte Carlo and the Metropolis algorithm. Section 3 presents our algorithms. Section 4 analytically and empirically evaluates the performance of the proposed algorithms. We compare our algorithms with the state of the art algorithms proposed in [14] and [12], and show our incremental algorithms successfully realize the compromise between effectiveness and efficiency. Finally, we conclude in Section 5.

2 Related Work and Background Knowledge

2.1 Incremental Algorithms on Dynamic Graphs

In order to avoid the re-computation from scratch, a few incremental algorithms have been proposed for graph analysis problems on dynamic graphs[6,7,21]. In [6], Desikan et al. propose a method to incrementally compute PageRank for dynamic graphs. They argue that their approach can be generally used to incrementally compute any graph metric that satisfies the first order Markov property. In [7], Fan et al. propose incremental algorithms for graph pattern matching on dynamic graphs. They consider both unit update, i.e., a single-edge deletion or insertion, and batch updates, i.e., a list of edge deletions or insertions on the graph. In our work, we consider only batch updates. The reason is that unit updates change very little graph properties of large graphs, thereby the old samples are still representative. In [21], Roditty et al. study dynamic shortest-paths problems. They analyze the computational complexity of incremental single-source shortest-paths problems and propose a randomized fully-dynamic algorithm for the all-pairs shortest-paths problem.

2.2 Sampling from Static Graphs

Previous work concerned with graph sampling are mostly focusing on sampling from static graphs. Examples include sampling community structure of the original graph [17], sampling from disconnected or loosely connected graphs [20], sampling connected induced subgraphs [16], etc. Among these literatures, paper [14] and [12] aim to sample general graph properties from an original graph.

In the pioneering paper [14], Leskovec et al. discuss a series of candidate algorithms for sampling from large graphs. The sample graphs are supposed to preserve properties of interest in the original graphs. Among the discussed algorithms, *Random Walk (RW)* sampling and *Forest Fire (FF)* sampling perform best overall. RW selects uniformly at random a starting vertex and simulates a random walk on the graph. In each step, RW jumps back to the starting vertex with a certain probability and restarts the random walk. Similarly to RW, FF selects a starting vertex v uniformly at random from the graph. Then a random number x is drawn from a geometric distribution with mean $\frac{p_f}{1-p_f}$, where p_f is called forward burning probability. FF hereafter selects x neighbor vertices of v that have not yet been selected. Recursively, FF applies the same process to these newly selected vertices, until enough vertices are sampled. Both RW and FF have the problem of getting stuck. The solution is to select a new starting vertex uniformly at random.

RW and FF are efficient but do not guarantee to generate the most or nearly the most representative sample graphs with respect to specific graph properties. Hübler et al. improve this issue in [12]. They adopt the idea of random walk on Markov chain and propose the *Metropolis Graph Sampling (MGS)* algorithm for sampling representative sample graphs. The algorithm belongs to *Markov Chain Monte Carlo (MCMC)* methods [5,11]. In particular, MGS selects an initial subgraph $g_{current}$ of size n uniformly at random from G, where n is the sample graph size. MGS sets $g_{best} = g_{current}$ and starts a random walk from $g_{current}$. At each step, MGS randomly replaces a vertex v in $g_{current}$ with a vertex w randomly selected from $(G \setminus g_{current}) \cup \{v\}$. MGS forms a new subgraph $g_{new} = (g_{current} \setminus \{v\}) \cup \{w\}$. Then MGS transfers from $g_{current}$ to g_{new} with probability $\min(1, \frac{\Delta_{G,\sigma}(g_{current})}{\Delta_{G,\sigma}(g_{new})})$, where $\Delta_{G,\sigma}(g)$ measures the distance between the original graph G and the sample graph g with respect to property σ. In case of a success, if $\Delta_{G,\sigma}(g_{current}) < \Delta_{G,\sigma}(g_{best})$, MGS sets $g_{best} = g_{current}$. The random walk terminates after a sufficiently large number of steps. MGS then outputs g_{best} as the sample graph. Overall, MGS generates subgraphs with probability inversely proportional to their distance measure Δ to the original graph, and stores the subgraph with the smallest distance measure as the sample graph.

MGS outperforms RW and FF in preserving the properties of original graphs. However, MGS is not efficient because of two reasons. First, it requires a large number of random walk steps in order to achieve the convergence of the Markov chain. The number of random walk steps usually scales with the size of the original graph. Second, it computes the graph property σ of the subgraph and the distance measure at each random walk step, which requires additional cost.

2.3 Sampling from Graph Streams

In a recent work [3], Ahmed et al. considers the situation such that one cannot access the entire original graph from main memory. This situation happens when the original graph is too large or dynamic over time. They propose to solve the problem by considering the original graph as a stream of edge lists and generating a sample graph from the graph stream. The main proposed algorithm is called *Partially-Induced Edge Sampling (PIES)*. PIES maintains a reservoir of size n, where n is the size of the sample graph. It inserts into the reservoir the first n vertices of the graph stream and randomly replaces existing vertices afterwards by newly arrival ones. At each step, an edge is added to the sample graph if its two incident vertices are in the reservoir. Once a vertex is replaced, all the incident edges are removed accordingly from the sample graph. The sampling process continues until the end of the graph stream.

Our work is different from the work in [3] in two aspects. First, we consider batch updates on the original graph, that is, we update the old sample graph after a number of vertices and edges are added to the original graph, whereas in [3] the authors consider unit updates, that is, they update the old sample graph whenever a new edge is added to the original graph. As we discussed in Section 2.1, unit updates are cumbersome as they change very little graph properties of original graphs. Second, we focus on generating most representative sample graphs, whereas PIES focus on sampling from a graph stream using a single pass, and, therefore has no theoretical guarantee on sample quality.

2.4   Markov Chain Monte Carlo

The Metropolis algorithm belongs to Markov Chain Monte Carlo methods [5,11]. Markov Chain Monte Carlo methods are a category of algorithms that sample from a Markov chain via random walk. The Markov chain is constructed by all the candidate samples or states. The current state depends only on its adjacent states. If the constructed Markov chain is *ergodic*, that is, the Markov chain is aperiodic and irreducible[1], it has a *stationary distribution* over all the states, which is achieved by a sufficient large number of random walk steps on the Markov chain.

The Metropolis algorithm [18] can draw samples from an ergodic Markov chain with any desired probability distribution $\mathbf{P}$ by satisfying the *detailed balance* condition $p(s)T(s, s\prime) = p(s\prime)T(s\prime, s)$, where $p(s)$ is the stationary probability of state s and $T(s, s\prime)$ is the transition probability from s to $s\prime$, for any state s and $s\prime$. The transition probability $T(s, s\prime) = t(s, s\prime) \times a(s, s\prime)$, where $t(s, s\prime)$ is the probability of transferring from s to $s\prime$ and $a(s, s\prime)$ is the probability of accepting the transition. To ensure the detailed balance, the acceptance probability has to be set as $a(s, s\prime) = \min(1, \frac{p(s\prime)}{p(s)} \times \frac{t(s, s\prime)}{t(s\prime, s)})$. If the constructed Markov chain is regular, that is, all the states have the same number of adjacent states, we have $t(s, s\prime) = t(s\prime, s)$. In such kind of scenario, the acceptance probability is simplified as $a(s, s\prime) = \min(1, \frac{p(s\prime)}{p(s)})$.

3   Proposed Algorithms

We now turn to the problem of incremental construction of a updated sample graph g' of G' from an old sample graph g of G. The two shortcomings discussed in Section 2.2 make MGS not appropriate for sampling from very large graphs. Therefore, we incrementally use the idea of MGS to avoid sampling from scratch, while keeping the sample graph quality still competitive. The incremental algorithms should either reduce the number of random walk steps or the cost of computing σ and the distance measure.

3.1   Incremental Metropolis Sampling

One factor that determines the efficiency of the MGS algorithm is the number of random walk steps required to achieve the stationary distribution. If we sample g' from G' using the MGS algorithm, the number of random walk steps is relevant to the size of G'. The smaller the size of G' is, the fewer subgraphs of G' are in the constructed Markov chain, thereby the fewer random walk steps are required to achieve the stationary distribution. Therefore the efficiency of the MGS algorithm can be improved if we can apply it to sample g' from a subgraph of G'. On the other hand, as we discussed in Section 1, the rationale of incremental sampling is to replace some vertices in g with the same number of updated

---

[1] Any state is reachable from any other state in a finite number of random walk steps.

vertices in G'. The key challenge is how to select the replaced vertices and the updated vertices. We denote by G'_{temp} the subgraph induced by the vertices in g and the updated vertices in G'. We find that the Metropolis algorithm can spontaneously solve this problem via random walk on the Markov chain whose states are the subgraphs induced by the vertices in G'_{temp}.

In particular, we first construct the subgraph G'_{temp} and set $g_{best} = g_{current} = g$. We start the random walk from the old sample graph g. At each step, we randomly select a pair of vertices (v, w), where $v \in g_{current}$ and $w \in (G'_{temp} \setminus g_{current}) \cup \{v\}$. We construct a new subgraph $g_{new} = (g_{current} \setminus \{v\}) \cup \{w\}$. We accept the transition from $g_{current}$ to g_{new} with probability $\min(1, \frac{\Delta_{G',\sigma}(g_{current})}{\Delta_{G',\sigma}(g_{new})})$. Reader notice that, differently from the MGS algorithm, we construct the subgraphs $g_{current}$ and g_{new} from G'_{temp} but compute the distance measure Δ_σ between the subgraphs and the updated graph G'. In case of a success, we set $g_{current} = g_{new}$ and, if $\Delta_{G',\sigma}(g_{current}) < \Delta_{G',\sigma}(g_{best})$, we store $g_{current}$ as g_{best}. We output g_{best} as the updated sample graph g' after a sufficient large number of random walk steps. In summary, we sample g' from the subgraph G'_{temp} of G'. We construct a Markov chain whose states are the induced subgraphs of size n of G'_{temp}. We perform a random walk starting from g on this Markov chain using the Metropolis algorithm. The replacement of the vertices is spontaneously realized during the random walk. We output the subgraph as the new sample g' that has the smallest distance measure to G' with respect to property σ. We call this algorithm *Incremental Metropolis Sampling (IMS)*. The pseudo code is described in Algorithm 1. The parameter p affects the convergence of the Markov chain. We set $p = 10 \times \frac{E}{V} \lg V$ in the experiment according to the parameter setting in [12], where E and V are the number of edges and the number of vertices of G'_{temp}, respectively.

We next prove the *Ergodicity* of the constructed Markov chain in Algorithm 1.

Lemma 1. *The Markov chain constructed by the IMS algorithm is ergodic on the space of subgraphs of size n induced by the vertices of G'_{temp}.*

Proof. (i) Aperiodicity. It is straightforward to prove the Markov chain is aperiodic because there is self loop by applying the Metropolis algorithm.

(ii) Irreducibility. We prove that there is always a path from subgraphs g_1 of size n to subgraph g_2 of size n, for any g_1 and g_2 of G'_{temp}. Suppose $g_1 \setminus g_2 = \{v_1, v_2, \ldots, v_k\}$ and $g_2 \setminus g_1 = \{w_1, w_2, \ldots, w_k\}$, where $1 \le k \le n$. Denote by a pair (v_i, w_i) replacing v_i by w_i in g_1. The sequence of replacement pairs $(v_1, w_1), (v_2, w_2), \ldots, (v_k, w_k)$ forms a path from g_1 to g_2. Since g_1 and g_2 are any subgraphs of size n of G'_{temp}, we proved the irreducibility.

By (i) and (ii), we proved the ergodicity of the Markov chain.

The proof of the detailed balance of the Markov chain is similar to that in [12] for proving the detailed balance of the Markov chain whose states are the connected graphs. Since the Markov chain is ergodic and satisfies the detailed balance condition (see Section 2.4), we can sample the subgraphs of size n using IMS with probability inversely proportional to their distance measure Δ_σ to the updated graph G'.

Algorithm 1. Incremental Metropolis Sampling

Input: g : a sample graph of G, n : the sample size, G' : the updated graph,
$\quad$ $\#it$: the number of random walk steps.
Output: g' : a sample graph of G'.

1 G'_{temp} = the subgraph of G' induced by the vertices of g and the updated
$\quad$ vertices of G';
2 $g_{best} = g_{current} = g$;
3 **for** $i = 1$ to $\#it$ **do**
4 $\quad\quad$ $v \leftarrow$ a vertex selected uniformly at random from $g_{current}$;
5 $\quad\quad$ $w \leftarrow$ a vertex selected uniformly at random from $(G'_{temp} \setminus g_{current}) \cup \{v\}$;
6 $\quad\quad$ $g_{new} = (g_{current} \setminus \{v\}) \cup \{w\}$;
7 $\quad\quad$ $\alpha \leftarrow$ random number from interval $[0, 1]$;
8 $\quad\quad$ **if** $\alpha < (\frac{\Delta_{G',\sigma}(g_{current})}{\Delta_{G',\sigma}(g_{new})})^p$ **then**
9 $\quad\quad\quad\quad$ $g_{current} = g_{new}$;
10 $\quad\quad\quad\quad$ **if** $\Delta_{G',\sigma}(g_{current}) < \Delta_{G',\sigma}(g_{best})$ **then**
11 $\quad\quad\quad\quad\quad\quad$ $g_{best} = g_{current}$;
12 $\quad\quad\quad\quad$ **end**
13 $\quad\quad$ **end**
14 **end**
15 $g' = g_{best}$;
16 Return g';

3.2 Sample-Merging Sampling

Another factor that affects the efficiency of the MGS algorithm is the computational cost at each random walk step. As we need to compute the property σ of the current subgraph and the distance measure of the current subgraph to the original graph, this cost is in turn determined by the size of the subgraph. Therefore the second method to speed up the MGS algorithm is to reduce the size of the sampled subgraphs in the Markov chain.

To achieve this goal, we propose an algorithm based on the following conjecture. If g_x is a representative sample graph of G_x and g_y is a representative sample graph of G_y, $g_x \cup g_y$ is a representative sample graph of $G_x \cup G_y$. We assume that $G_x \cap G_y = \emptyset$. Note that $g_x \cup g_y$ is an induced subgraph of $G_x \cup G_y$. The intuition is that, $g_x \cup g_y$ fully preserves the properties of $G_x \cup G_y$ if $g_x = G_x$ and $g_y = G_y$, and $g_x \cup g_y$ do not preserve any properties of $G_x \cup G_y$ if both g_x and g_y have only one vertex. Therefore, if g_x is a normal sample graph of G_x and g_y is a normal sample graph of G_y, $g_x \cup g_y$ should preserve the properties of $G_x \cup G_y$ to some extent.

The algorithm then works as follows. We construct a subgraph g_{non} that is induced by the non-updated vertices of g. We denote by G'_{non} the subgraph induced by the non-updated vertices of G'. We consider g_{non} as a representative sample graph of G'_{non}. Suppose the size of g_{non} is s_{non}. We denote by G'_u the subgraph induced by the updated vertices of G'. We then sample a subgraph g'_u of size $n - s_{non}$ from G'_u using the MGS algorithm. According to the conjecture, $g_{non} \cup g'_u$ is a representative sample graph of $G' = G'_{non} \cup G'_u$. We merge g_{non}

and g'_u and form the sample graph g', that is, g' is the subgraph induced by the vertices of g_{non} and g'_u. In summary, we deterministically keep in g' the non-updated vertices of g. We select the rest of the vertices of g' from the updated vertices of G' using the MGS algorithm. By this method, we achieve the goal of reducing the size of subgraphs in the Markov chain.

We call this algorithm *Sample-Merging Sampling (SMS)*. The pseudo code is described in Algorithm 2.

Algorithm 2. Sample-Merging Sampling

Input: g : a sample subgraph of G, n : the sample size, G' : the updated graph,
  $\#it$: the number of random walk steps.
Output: g' : a sample graph of G'.

1 g_{non} = the subgraph induced by the non-updated vertices of g;
2 s_{non} = the size of g_{non};
3 G'_u = the subgraph induced by the updated vertices of G';
4 Sample g'_u of size $n - s_{non}$ from G'_u using the MGS algorithm;
5 $g' = g_{non} \cup g'_u$;
6 Return g';

The problem of SMS is that it may generate sample graphs of lower quality with respect to property σ, that is, $\Delta_{G',\sigma}(g') > \Delta_{G',\sigma}(g)$. The possible explanations are as follows. First, g is already a representative sample graph of G' with respect to property σ; second, g_{non} is not a representative sample graph of G'_{non} as we expected; lastly, the merging phase introduces extra error to the distance measure between G' and g'. In this case, we restore g as a representative sample graph of G', as we are concerned with preserving graph properties rather than updating the old sample graphs.

4   Performance Evaluation

4.1   Analytical Evaluation

Complexity Analysis. We present the time complexity of the algorithms in this section. In the MGS algorithm, we compute at each step the graph property σ of a subgraph of size n and compute the distance measure Δ between the subgraph and the original graph. We denote by $O(C_{\sigma,n})$ the time complexity of computing property σ of a subgraph of size n. We select the Kolmogorov-Smirnov D-statistic to compute Δ. Below we see that the time complexity scales linearly with the size of the distribution (the property) of the sample graph plus that of the original graph. For the sake of simplicity, we denote by $O(n + N)$ the corresponding time complexity, where N is the size of the original graph. Therefore the time complexity of MGS at each step is $O(C_{\sigma,n} + n + N)$. The number of random walk steps required is at least equal to the mixing time of the Markov chain. The mixing time of a Markov chain is theoretically studied in [22]. There are also empirical techniques for evaluating the mixing time, for instance, the *Geweke* diagnostics [8]. However, it is estimated that the number of random walk steps in the order of size of the original graph is sufficient for the convergence of our Markov chain [9,12]. Therefore the total time complexity of MGS is $O(C_{\sigma,n}N + nN + N^2)$. Accordingly, the time complexity of IMS is $O(C_{\sigma,n}N_{temp} + nN_{temp} + N_{temp}^2)$, where N_{temp} is the size of subgraph G'_{temp}. The time complexity of SMS is $O(C_{\sigma,n_u}N_u + nN_u + N_u^2)$, where n_u is the size of subgraph g'_u and N_u is the size of subgraph G'_u. We use the number of random walk steps discussed in this section in the experimental study below.

4.2 Empirical Evaluation

We conduct the experimental study in this section. We empirically evaluate the algorithms on their abilities of preserving a selected set of graph properties. All the properties are considered as distributions. We choose the Kolmogorov-Smirnov D-statistic to compute distance measure Δ_σ between the sample graph and the original graph on property σ, following the convention in [12,14]. We compare the efficiency of the algorithms. We run experiments on both synthetic datasets and real datasets.

The Graph Properties. We select four graph properties for evaluation, which are also used in [14].

Degree distribution: the degree distribution is defined as the probability distribution of all the degrees over a graph.

The distribution of clustering coefficient: the clustering coefficient of a vertex is defined as the ratio of existing edges between its neighbor vertices. Suppose a vertex v connects to k other vertices, where $k > 1$. The clustering coefficient of v is computed as $\frac{m}{k(k-1)/2}$, where m is the number of existing edges between these k vertices. The distribution of clustering coefficient [14] is defined as the distribution of average clustering coefficient over all vertices of the same degree.

The distribution of component size: a graph may consist of many disjoint components. The distribution of component size is defined as the distribution of the numbers of components over all the sizes. A relevant property is the *graphlet distribution* or the distribution of network motifs [13,19]. A graphlet is a connected induced subgraph usually having 3, 4 or 5 vertices. The graphlet distribution is defined as the probability distribution of all the graphlets over a graph.

Hop-plot: $P(h)$ is defined as the number of reachable pairs of vertices at distance less than or equal to h. The hop-plot is the distribution of the numbers of vertex pairs over all the distances less than or equal to h [14].

All the properties are computed using the *Snap* [2] library.

Kolmogorov-Smirnov D-statistic. We select the Kolmogorov-Smirnov D-statistic to compute the distance measure $\Delta_{G,\sigma}(g)$ between the sample graph g and the original graph G on property σ, following the conventions in [12,14]. The distribution (normalized) of a property of the sample graph and that of the original graph usually have different sizes. The KS D-statistic is appropriate for this distance measure since it can measure the difference between two distributions of different sizes.

Given distribution $\mathbf{D} = [D_1, D_2, \ldots, D_{n_D}]$ and $\mathbf{d} = [d_1, d_2, \ldots, d_{n_d}]$, the KS D-statistic is computed as follows. We first define a range of random variables $\mathbf{x} = [x_1, x_2, \ldots, x_k]^2$. Then we compute the cumulative distribution function F_D and F_d for $\mathbf{D}$ and $\mathbf{d}$ over $\mathbf{x}$, respectively, as Equation 1,

$$F_D(i) = \frac{1}{n_D} \sum_{j=1}^{n_D} I_j, \quad i = 1, 2, \ldots, k, \tag{1}$$

where $I_j = 1$ if $D_j \leq x_i$ and $I_j = 0$ otherwise.

Then the KS D- statistic is computed as,

$$D = \sup_i |F_D(i) - F_d(i)|, \quad i = 1, 2, \ldots, k. \tag{2}$$

A smaller value of KS D-statistic means the two distributions in comparison are more similar to each other. Therefore, in our scenario smaller values of KS D-statistic mean higher quality of sample graphs.

Datasets. We generate Barabási-Albert random graphs [4] and Forest Fire random graphs [15] for the synthetic datasets. Both of the graph models simulate the evolution of real graphs. Barabási-Albert random graphs simulate the preferential attachment phenomenon in real graphs. Forest Fire random graphs reproduce the behavior of densification power laws and shrinking diameters in real graph evolution.

We generate 10 random graphs with different densities for each model. In the Barabási-Albert model, the density of the graph is determined by the parameter d, which is the number of edges each new vertex generates. Higher values of d lead to the generation of denser graphs. We set $d = 1, 2, \ldots, 10$ for the ten graphs, respectively. In the Forest Fire model, the density is relevant to both the forward burning probability p and the backward burning probability p_b. We set $p_b = 0.3$ and vary the value of p from 0.31 to 0.4 with increments of 0.01. Higher values of p lead to the generation of denser graphs. In order to generate multiple disjoint components in the Barabási-Albert graphs, we randomly form a new small component with a fixed probability at the arrival of each new vertex. In the Forest Fire graphs, we generate multiple components by setting the probability of a new vertex being an orphan to be 0.1. However, we discard the orphan vertices

---

[2] In our experimental study, we select the random variables varying in the range $\{0, 1\}$ with increments 0.01.

and keep only the vertices having degree at least 1. Each Barabási-Albert graph has 10, 000 vertices and Each Forest Fire graph has 100, 000 vertices.

The real dataset is a Facebook friendship graph. The graph contains a five-year friendship list within 15 schools in USA, with complete time-stamp for the confirmation of each pair of friends. The graph has 331, 667 vertices (users) and 1, 391, 866 edges (pairs of friends).

Experimental Setup. We implement RW, FF, MGS, PIES, IMS and SMS using C++. We run all the algorithms on each of the graphs described above. We evaluate their ability of sampling each of the selected properties separately. For each Barabási-Albert graph, we generate a sample graph of size 200 whenever the graph size increases by 1, 000. For each Forest Fire graph, we generate a sample graph of size 1, 000 whenever the graph size increases by 10, 000. For the Facebook friendship graph, we generate ten sample graphs of size 1, 000 whenever the graph size increases by 30, 000. FF, RW and MGS always generate the sample graphs from scratch, whereas IMS and SMS incrementally generate the updated sample graphs based on the old sample graphs in previous sampling stage. PIES updates sample graphs whenever a new edge is added to the original graph. However, we evaluate the sample graphs only in the stages described above. We compute the average distance measure between the sample graphs and the current original graphs for each graph property. We compute the average execution time for generating the sample graphs of all four properties for each graph model. All the experiments are run on a cluster of 54 nodes, each of which has a 2.4GHz 16-core CPU and 24 GB memory.

Effectiveness. Figure 2, 3 and 4 show the average results of preserving degree distribution of the Barabási-Albert graphs, the Forest Fire graphs and the Facebook friendship graph, respectively.

We observe that the IMS and SMS algorithms perform very similar to the MGS algorithm and perform much better than the RW and FF algorithms. RW and FF start from a random vertex and tend to sample subgraphs in a local area around the vertex. Therefore they are not able to capture the global degree distribution of the original graph. On the contrary, IMS can find the optimal sample graph induced by the selected vertices of G'_{temp} that preserves the degree distribution of the original graph G', and SMS can find the optimal subgraph induced by the updated vertices that preserves the degree distribution of the updated graph. PIES performs better than RW and FF as it samples globally from the entire original graph. However, PIES does not perform as good as IMS and SMS since it has no theoretical guarantee on sample quality. IMS and SMS perform slightly worse than MGS since they do not search the entire space of the subgraphs of G'. On the other hand, the IMS performs slightly better than SMS. This is because SMS samples local optimal subgraphs from the updated graph and introduces extra error in the merging phase.

Figure 5, 6, and 7 show the average results of preserving the distribution of clustering coefficient of the Barabási-Albert graphs, the Forest Fire graphs and the Facebook friendship graph, respectively.

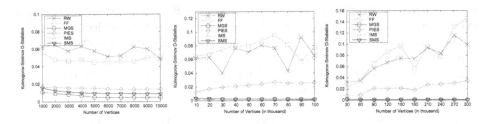

Fig. 2. Degree distribution: Barabási-Albert graphs

Fig. 3. Degree distribution: Forest Fire graphs

Fig. 4. Degree distribution: Facebook friendship graph

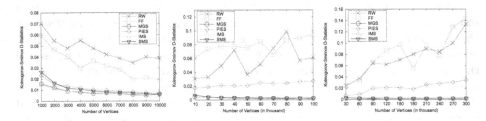

Fig. 5. Clustering co-efficient distribution: Barabási-Albert graphs

Fig. 6. Clustering coefficient distribution: Forest Fire graphs

Fig. 7. Clustering co-efficient distribution: Facebook friendship graph

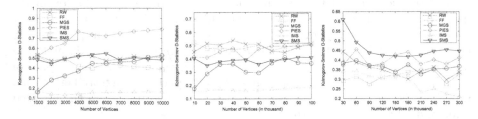

Fig. 8. Component size distribution: Barabási-Albert graphs

Fig. 9. Component size distribution: Forest Fire graphs

Fig. 10. Component size distribution: Facebook friendship graph

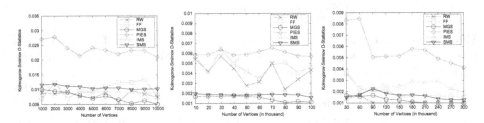

Fig. 11. Hop-plot: Barabási-Albert graphs

Fig. 12. Hop-plot: Forest Fire graphs

Fig. 13. Hop-plot: Facebook friendship graph

We observe the similar results to those of preserving degree distribution. MGS, IMS and SMS perform better than RW, FF and PIES. IMS performs slightly better than SMS. Both IMS and SMS perform very close to MGS.

Figure 8, 9, and 10 show the average results of preserving the distribution of component size of the Barabási-Albert graphs, the Forest Fire graphs and the Facebook friendship graph, respectively.

We observe that RW, FF, MGS, IMS and SMS perform close to each other on the Barabási-Albert graphs and the Facebook friendship graph. This is because there is a giant component with 99% vertices in each Barabási-Albert graph and in the Facebook friendship graph. The distribution of component size is dominated by this giant component. Since RW and FF tend to generate a connected subgraph, the corresponding distribution of component size is similar to the distribution of graphs with a giant component. On the other hand, the Forest Fire graphs have several large components. The distributions of component size of the Forest Fire graphs are much flatter. Therefore we observe that IMS and SMS perform much better than RW and FF on the Forest Fire graphs. PIES always performs the worst.

Figure 11, 12, and 13 show the average results of preserving hop-plot of the Barabási-Albert graphs, the Forest Fire graphs and the Facebook friendship graph, respectively.

We observe that IMS outperforms the other five algorithms. By checking the result of each graph, we find that MGS and SMS do not perform well on sparse graphs. MGS may get stuck in local optimal because for a sparse original graph new subgraphs after replacement usually do not have change on the paths. For SMS, in addition to the former reason, the merging phase may introduce more errors on hop-plot for sparser graphs, especially when there form connections between the two parts of the sample graphs.

Overall, we show that MGS, IMS and SMS perform better than RW, FF and PIES on preserving the four properties. IMS and SMS are as effective as MGS and IMS performs better than SMS in general. Among the four properties, we observe that SMS performs not so well on preserving the hop-plot property. The possible explanation is that hop-plot is a global property of the original graph, whereas SMS generates the local optimal sample graph from the updated graph. The merging phase also introduces extra error.

Efficiency. Figure 14, 15 and 16 show the average execution time of sampling the four graph properties for the Barabási-Albert graphs, the Forest Fire graphs and the Facebook friendship graph, respectively.

We observe that RW, FF and PIES are faster than the other three algorithms as expected. Our IMS and SMS algorithms are much faster than the MGS algorithm, and SMS is faster than IMS. This is as expected because IMS and SMS require fewer random walk steps than MGS does, and SMS computes the graph properties and distance measure of subgraphs of smaller sizes than IMS does.

The difference between the execution time of MGS and IMS and that between MGS and SMS increases very quickly as the original graph grows. This is because the total number of vertices of the original graph increases much faster than the

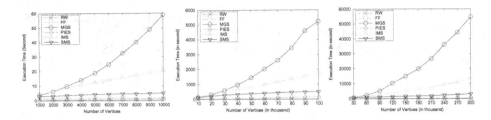

Fig. 14. Execution Time: Barabási-Albert graphs

Fig. 15. Execution Time: Forest Fire graphs

Fig. 16. Execution Time: Facebook friendship graphs

number of updated vertices does as the original graph grows. Therefore the number of random walk steps required by MGS increases much faster than that required by IMS and SMS.

5 Conclusion

We have presented two incremental graph sampling algorithms preserving the distributions of degree, of clustering coefficient, of component size and of hop-plot. The rationale of the algorithms is to replace a fraction of vertices in the old sample with newly updated vertices. The two algorithms differ in their way of selecting the vertices to be replaced and updated in the sample graph.

We analytically and empirically evaluated the performance of the proposed algorithms: Incremental Metropolis Sampling and Sample-Merging Sampling. We compared their performance with that of baseline algorithms: Random Walk, Forest Fire, Metropolis Graph Sampling. The experiment results on both synthetic and real graphs showed that our proposed algorithms realize a compromise between effectiveness and efficiency.

Metropolis Graph Sampling is the most effective yet slowest of the three baseline algorithms. Our algorithms are as effective as Metropolis Graph Sampling and significantly more efficient than Metropolis Graph Sampling. Although very efficient, Random Walk and Forest Fire prove not effective. Our algorithms are less efficient that than Random Walk and Forest Fire but significantly more effective. Of the two algorithms, Incremental Metropolis Sampling is slightly more effective than Sample-Merging Sampling, while Sample-Merging Sampling is more efficient than Incremental Metropolis Sampling.

We have provided practical algorithms for the incremental sampling of the large dynamic graphs being created in real life such as coauthor network, social network, World Wide Web and so on.

Acknowledgements. We thank Prof. Edoardo Airoldi (airoldi@fas.harvard.edu) for providing the Facebook friendship dataset.

References

1. http://blog.twitter.com/2011/03/numbers.html
2. http://snap.stanford.edu

3. Ahmed, N.K., Neville, J., Kompella, R.: Space-efficient sampling from social activity streams. In: BigMine, pp. 53–60 (2012)
4. Barabási, A.-L., Albert, R.: Emergence of Scaling in Random Networks. Science 286(5439), 509–512 (1999)
5. Berg, B.A.: Markov Chain Monte Carlo Simulations and Their Statistical Analysis: With Web-based Fortran Code. World Scientific Publishing Company (2004)
6. Desikan, P.K., Pathak, N., Srivastava, J., Kumar, V.: Incremental page rank computation on evolving graphs. In: WWW (Special Interest Tracks and Posters), pp. 1094–1095 (2005)
7. Fan, W., Li, J., Luo, J., Tan, Z., Wang, X., Wu, Y.: Incremental graph pattern matching. In: SIGMOD Conference, pp. 925–936 (2011)
8. Geweke, J.: Evaluating the accuracy of sampling-based approaches to the calculation of posterior moments. In: Bayesian Statistics, pp. 169–193 (1992)
9. Gionis, A., Mannila, H., Mielikäinen, T., Tsaparas, P.: Assessing data mining results via swap randomization. In: KDD, pp. 167–176 (2006)
10. Gjoka, M., Kurant, M., Butts, C.T., Markopoulou, A.: Walking in facebook: A case study of unbiased sampling of osns. In: INFOCOM, pp. 2498–2506 (2010)
11. Hastings, W.K.: Monte Carlo sampling methods using Markov chains and their applications. Biometrika 57(1), 97–109 (1970)
12. Hübler, C., Kriegel, H.-P., Borgwardt, K.M., Ghahramani, Z.: Metropolis algorithms for representative subgraph sampling. In: ICDM, pp. 283–292 (2008)
13. Kashtan, N., Itzkovitz, S., Milo, R., Alon, U.: Efficient sampling algorithm for estimating subgraph concentrations and detecting network motifs. Bioinformatics 20(11), 1746–1758 (2004)
14. Leskovec, J., Faloutsos, C.: Sampling from large graphs. In: KDD, pp. 631–636 (2006)
15. Leskovec, J., Kleinberg, J.M., Faloutsos, C.: Graph evolution: Densification and shrinking diameters. TKDD 1(1) (2007)
16. Lu, X., Bressan, S.: Sampling connected induced subgraphs uniformly at random. In: Ailamaki, A., Bowers, S. (eds.) SSDBM 2012. LNCS, vol. 7338, pp. 195–212. Springer, Heidelberg (2012)
17. Maiya, A.S., Berger-Wolf, T.Y.: Sampling community structure. In: WWW, pp. 701–710 (2010)
18. Metropolis, N., Rosenbluth, A.W., Rosenbluth, M.N., Teller, A.H., Teller, E.: Equation of State Calculations by Fast Computing Machines. The Journal of Chemical Physics 21(6), 1087–1092 (1953)
19. Milo, R., Shen-Orr, S., Itzkovitz, S., Kashtan, N., Chklovskii, D., Alon, U.: Network Motifs: Simple Building Blocks of Complex Networks. Science 298(5594), 824–827 (2002)
20. Ribeiro, B.F., Towsley, D.F.: Estimating and sampling graphs with multidimensional random walks. In: Internet Measurement Conference, pp. 390–403 (2010)
21. Roditty, L., Zwick, U.: On dynamic shortest paths problems. Algorithmica 61(2), 389–401 (2011)
22. Sinclair, A.: Algorithms for Random Generation and Counting: A Markov Chain Approach (Progress in Theoretical Computer Science). Birkhäuser, Boston (1993)

CoDS: A Representative Sampling Method for Relational Databases

Teodora Sandra Buda[1], Thomas Cerqueus[1], John Murphy[1],
and Morten Kristiansen[2]

[1] Lero, Performance Engineering Lab
School of Computer Science and Informatics, University College Dublin
`teodora.buda@ucdconnect.ie`, {`thomas.cerqueus,j.murphy`}`@ucd.ie`
[2] IBM Collaboration Solutions, IBM Software Group, Dublin, Ireland
`morten_kristiansen@ie.ibm.com`

Abstract. Database sampling has become a popular approach to handle large amounts of data in a wide range of application areas such as data mining or approximate query evaluation. Using database samples is a potential solution when using the entire database is not cost-effective, and a balance between the accuracy of the results and the computational cost of the process applied on the large data set is preferred. Existing sampling approaches are either limited to specific application areas, to single table databases, or to random sampling. In this paper, we propose CoDS: a novel sampling approach targeting relational databases that ensures that the sample database follows the same distribution for specific fields as the original database. In particular it aims to maintain the distribution between tables. We evaluate the performance of our algorithm by measuring the representativeness of the sample with respect to the original database. We compare our approach with two existing solutions, and we show that our method performs faster and produces better results in terms of representativeness.

Keywords: Relational database, Representative database sampling.

1 Introduction

Nowadays applications are generally faced with the challenge of handling large number of users that produce very large amounts of data up to terabytes in size. The storage space, administration overhead of managing large datasets and the analysis of this data is a real challenge in different fields. For instance, in data mining, a balance between the accuracy of the results and the computational cost of the analysis is generally preferred to overcome this challenge. Moreover, in software validation, the operational data available for a system under development could serve as a realistic testing environment. However these databases consist of large amount of data, which is computationally costly to analyze.

Database sampling is a potential solution to this problem: a smaller database can be used instead of the original one. Olken's major contribution to random

H. Decker et al. (Eds.): DEXA 2013, Part I, LNCS 8055, pp. 342–356, 2013.

sampling from large databases proves sampling to be a powerful technique [14]. Database sampling methods aim to provide databases that (i) are smaller in size, (ii) are consistent with the original database (e.g. conformance of the schema), (iii) contain data from the original database, (iv) are representative of the original database. The last criteria is crucial because the accuracy of the results of the following analysis to be performed on the sample is expected to be significantly higher if the sample is representative of the original database. For instance, a representative sample of the production environment would determine the sample contain realistic test data, encompassing a variety of scenarios the user created. In particular, in functional testing, a small realistic sample of the production environment would suffice to test the core functionality of the system under development, while maintaining the accuracy of the results. The problem raised in this work is to define a method that produces a representative database sample targeting relational databases.

Existing databases sampling methods involve random sampling [6], target single-table databases [11], or they are specific to an application area [10,4]. For instance, in [13], the reader is presented a representative sampling approach that aims to handle scalability issues of processing large graphs. However, most of today's structured data is stored in relational databases, consisting of multiple tables linked through various constraints. Single table sampling methods applied on relational databases produce an inconsistent sample database with regards to the referential integrity. Moreover, we expect that random samples provide poor accuracy in the results of the analysis to be performed on the large data set (e.g. testing purposes, data mining methods, approximate query evaluation). For instance a random sample of the production environment could sample only one test case and not detect high priority errors of the system.

In this paper, we propose the CoDS system: a novel approach for database sampling, targeting relational databases, with the purpose of creating a smaller representative sample, that respects the referential integrity constraints. We consider that a sample is representative if it follows the same distribution for specific fields. The fields considered by CoDS are the foreign key constraints. A foreign key constraint in a database is used to create and enforce a link between the data in two tables. Thus, these constraints represent invaluable inputs for our system to depict the relationships between data and produce a representative sample. If the sample database follows the same distribution as the original database for these fields, it is feasible to expect that the results of the following analysis to be performed on the sample will produce the same results as the ones performed on the original database. The sampling mechanism proposed is independent of the application area and will result in producing a consistent representative sample.

The remainder of this paper is organized as follows: Section 2 discusses related work and describes various application areas in which representative sample database may be of interest. Section 3 describes the main contribution of this work: the representative database sampling system. Sections 4 describe the experimental evaluation of CoDS, and its comparison with previous approaches. Section 5 concludes the paper.

2   Related Work

Several database sampling methods have been proposed in specific application areas proving sampling to be a useful and powerful technique. However, most of them are designed for specific application areas: software testing, data mining, query approximation. Before presenting methods built for these different areas, we present general approaches.

General approaches. The database sampling approach presented in [3] is oriented towards relational databases focusing on the advantage of using prototype databases populated with operational data. Data items that follow a set of integrity constraints (e.g. foreign-key constraints, functional dependencies, domain constraints) are randomly selected from the original database, so that the resulting sample database is consistent with the original database. Furthermore, there are a few commercial applications that support sampling from databases. For instance, IBM *Optim*[1] is used for managing data within many database instances. Its component, *Move*, can be used for sampling by using the option to select every n^{th} row of each table from the original database. *Optim* ensures that the referential integrity is respected by the sample database. As a recognized value of database sampling, Oracle DBMS supports the possibility to query a sample of a given table instead of the whole table by using the *Sample* statement[2].

Software testing. Analyzing the production environment, its constraints, and generating relevant testing data are just a few of the challenges encountered during the testing process. Existing methods for populating the testing environment commonly generate synthetic data values or use some type of random distribution to select the data that must be included in the resulting database [18,15]. In [20] the reader is presented with a privacy-preserving approach that uses the operational data available for testing purposes focusing on the importance of the representativeness of the data as it can increase the probability of detecting crucial faults of the system. Moreover, the testing environment would encompass scenarios created by the user, useful for testing the core functionality of the system. However, the production environment generally consists of large amounts of data. Database sampling is a potential solution to overcome this challenge.

Data mining. Various sampling approaches have been proposed in the data mining community, proving that sampling is a powerful technique for achieving a balance between the computational cost of performing data mining on a very large population and the accuracy of the results [17,12]. However, the approaches devised in this community are generally oriented towards the data mining algorithm used on the sample [4,19,16] and most of the standard methods for data mining are built on the assumption that data is stored in single-table databases.

---

[1] http://www-01.ibm.com/software/data/data-management/optim-solutions/
[2] http://docs.oracle.com/cd/B19306_01/server.102/b14200/statements_10002.htm

In [11], the authors propose a static sampling approach which uses the distribution of the sample data as an evaluation criterion to decide whether the sample reflects the large dataset. However, it is limited to single-table datasets and to univariate analysis. Some recent work in the data mining community [8,21] avoided this shortcoming and target relational databases. In [21], authors present a sampling algorithm for relational databases that focuses on improving the scalability and accuracy of multi-relational classification methods.

Approximate Query Evaluation. Numerous papers proposed random sampling for approximate query answering [2,9], and statistics estimation for query size result [5], allowing approximate but faster answers to queries. A more recent approach that extended the table-level sampling to relational database sampling is presented in [7]. The authors propose a sampling mechanism called *Linked Bernoulli Synopses* based on *Join Synopses* [1] aiming to provide fast approximate query answers for join queries over multiple tables. Their solution imply maintaining the foreign key integrity of the synopses. Both approaches are probabilistic and require the processing of each tuple in a database. In the case of JS, each tuple from the set of tables is sampled with a probability equal to the sampling rate. After this insertion of tuples in the sample database, JS ensures the referential integrity of the sample database remains intact by visiting all the tables, starting with the root, and adding the missing referenced tuples in the sample database. LBS is run only one time over the entire database. The decision of whether or not to include the row in the sample is different in LBS. LBS requires the retrieval of every tuple from each table and calculates the probability of a tuple t, being inserted in the sample database based on the probabilities of the tuples referencing tuple t to be inserted in the sample. The computation of this probability is described in detail in [7]. In the case that one of referencing tuples has already been included in the sample, the tuple under analysis is also included in the sample, thus avoiding the referential integrity to be broken.

3   CoDS: A Representative Sampling System

The CoDS[3] system proposes a method to produce a representative α-sample of an original database, where α represents the sampling rate for the original database and is given as an input by the user of the system. The objective is to maintain the distribution between the tables of a database to ensure the representativeness property, while maintaining the referential integrity of the data. The system targets relational databases, in third normal form. We assume that the schema of the original database forms a connected graph. CoDS aims to analyze and preserve the distributions between a starting table and the rest of the tables of the database, through various joins when needed. CoDs computes a set of identifiers that need to be sampled from the starting table to preserve these distributions, along with a representative error measure when a perfect representative sample could not be created. CoDS is composed of four phases:

---

[3] Chains of Dependencies-based Sampling.

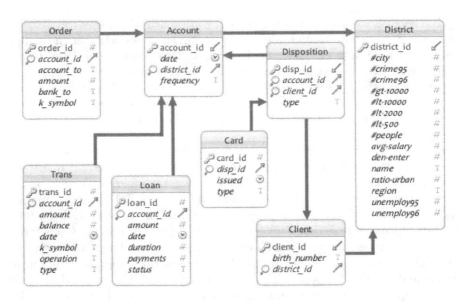

Fig. 1. The *Financial* database schema

- The system identifies the starting table (section 3.2).
- The system detects the relationships of the starting table with the rest of the tables of the database by following the foreign key constraints from the metadata. Then it generates the scatter plots associated to these relationships (section 3.3).
- The system analyzes the generated scatter plots between the starting table and the rest of the tables in order to compute a set of identifiers of the starting table that need to be sampled to preserve the distribution of these relations (section 3.4).
- Finally, the system proceeds in sampling the tuples associated with the set of identifiers of the starting table computed in the previous step, and to sampling all the related tuples from the rest of the tables (section 3.5).

Before presenting each phase in detail, we introduce the formal model and definitions used in the remaining of the document.

3.1   Model and Definitions

Relational database. A relational database is a set of n tables $T = \{t_1, ..., t_n\}$. Each table t_i of the database is composed of a set of attributes $A_i = \{a_1, ..., a_m\}$. The set of attributes that allow to uniquely identify a tuple in table t_i is called the primary key noted PK_i. A foreign key is a set of attributes that refers to another table's attributes. For instance, in Fig. 1, an example of such foreign key is *client_id*, declared in table *Disposition*. A table may contains several foreign keys. We denote by FK_i^j the set of attributes of table t_i that reference table t_j. When a

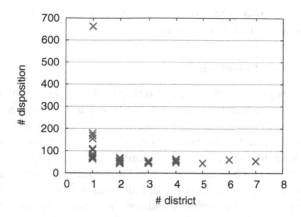

Fig. 2. Scatter plot associated to chain $\langle District, Client, Disposition \rangle$

table t_i defines a foreign key constraint to another table t_j, we say that t_i directly references t_j and we denote it by: $t_i \rightarrow t_j$. Symmetrically, t_j is directly referenced by t_i, and denote it by $t_j \leftarrow t_i$. We refer by children of table t by $children(t)$ to the set of tables that t references: $children(t) = \{t_i \in T : t \rightarrow t_i\}$. We refer by $desc(t)$ to the set of all the descendants of t:

$$desc(t) = children(t) \cup \bigcup_{t_i \in children(t)} desc(t_i)$$

Similarly, we refer to parents of t by: $parents(t) = \{t_i \in T : t \leftarrow t_i\}$. We define the set of related tables of t_i as follows: $RT(t) = parents(t) \cup children(t)$. We denote by $O(t)$ and $S(t)$ the tuples of t in the original and the sample database.

Chain of dependencies. A chain of dependencies is a sequence of tables $\langle t_1, \ldots, t_k \rangle$ such that $\forall i \in [1, k-1]$, $t_i \in RT(t_{i+1}) \land \forall i, j \in [1, k], i \neq j \Rightarrow t_i \neq t_j$. An example of a chain of dependencies in Fig. 1 is $\langle District, Client, Disposition \rangle$. The set of all chains of dependencies of t (i.e. $t_1 = t$) is denoted by $Ch(t)$.

Scatter plot and data point. Given a chain $\langle t_1, \ldots, t_k \rangle$ we consider the scatter plot associated to this chain between table t_1 and table t_k. We denote by $Sp(t)$ the set of all scatter plots associated to $Ch(t)$. A scatter plot is composed of a set of points corresponding to that plot. Each point of a scatter plot is called a *data point*. A data point situated at the coordinate (x, y) means that: (i) if $t_1 \leftarrow t_2$: x tuples of table t_1 are indirectly referenced by y tuples of table t_k, (ii) if $t_1 \rightarrow t_2$: x tuples of table t_k are indirectly referenced by y tuples of table t_1. For instance, for the scatter plot of chain $\langle District, Client, Disposition \rangle$ presented in figure 2, we can see that only one district is indirectly referenced by 663 dispositions, or that 7 districts are indirectly referenced by 54 dispositions. Each data point is uniquely identified by its y value, and contains identifiers of table t_1 (i.e. contains a set of values of PK_1) from the original database. For instance, in Fig. 2, the data point with $y = 663$ contains the identifier of the single district that is indirectly referenced by 663 dispositions.

3.2   Starting Table Selection

The objective of this phase is to select a starting table for the sampling, which we denote by t_*. In CoDS, a leaf table (i.e. a table that has no children) is chosen as starting table. If the database has more than one leaf tables the system chooses the one with the maximum number of tuples. The reason for this is to avoid choosing a leaf with few tuples, as this would critically impact the sampling method by having very little influence on the tuples selected from the related tables, and thus on the representativeness of the sample database. CoDS selects a leaf table as a starting table in order to reduce the computational cost of analyzing the chains generated by using a bottom-up approach. We show in section 3.3 that the computational cost of analyzing a chain using a bottom-up approach is lower in contrast with a top-down approach. Moreover, we expect that using a leaf table produces less errors related to the sample size and representativeness.

3.3   Generation of Chains

In the second phase, we aim to discover the relationships between the starting table and the rest of the tables of the database, generate the set of chains of dependencies of the starting table and construct their associated scatter plots. These scatter plots will be used for the selection of identifiers of the starting table. The system generates $Ch(t_*)$ by following all the possible paths through the arrows between the tables, starting with t_*. Note that each table is visited only once in this representation and the shortest path is preferred. If two chains with equal t_k have the same length but are composed of different tables, both chains are considered. Let us consider the relational database presented in Fig. 1. CoDS generates the following chains of dependencies for $t_* = District$: $Ch(District) = \{\langle District, Client\rangle, \langle District, Client, Disposition\rangle, \langle District, Client, Disposition, Card\rangle, \langle District, Account\rangle, \langle District, Account, Disposition\rangle, \langle District, Account, Order\rangle, \langle District, Account, Trans\rangle, \langle District, Account, Loan\rangle, \langle District, Account, Disposition, Card\rangle\}$.

For each chain of dependency discovered, $ch = \langle t_1, t_2, \ldots, t_k\rangle$, a scatter plot is generated with the following properties:

If $t_1 \rightarrow t_2$: The scatter plot is interpreted as x tuples of table t_k are indirectly referenced by y tuples of table t_1 (i.e. the x-axis corresponds to the t_k, while the y-axis corresponds to t_1). The reason for this is that t_1 is directly or indirectly referencing each table in the chain. Thus, table t_1 is in relation 1:1 or 1:N with table t_2 and indirectly with all tables from ch. In this case, if the x-axis corresponds to t_1, the scatter plot would be formed of a single point, corresponding to all of the identifiers of t_1. Each scatter plot is composed of a set of data points, which are composed of the set of identifiers of table t_1. In order to compute these identifiers in this case, the following query is run by the system: SELECT $t_k.PK_k$, $t_1.PK_1$ FROM $t_1 \bowtie \ldots \bowtie t_k$. For each value of $t_k.PK_k$, CoDS will count the number of distinct values for $t_1.PK_1$ from the previous query and this will determine the values for y. For each value of y, CoDS will count how many identifiers of $t_k.PK_k$ (i.e. x value) are associated with y distinct tuples of t_1. A nested SQL query in this case would result in losing information about the identifiers of each data point, or would require an extra query for each value on the y-axis. In order to avoid multiple queries, CoDS constructs the set of data points from the above query. The method constructs the data point with the values of $t_1.PK_1$. The data point will appear at coordinates (x, y).

If $t_1 \leftarrow t_2$: The scatter plot is interpreted as x tuples of table t_1 are indirectly referenced by y tuples of table t_k (i.e. the x-axis will correspond to the t_1, while the y-axis will correspond to t_k). The reason for this is that each table in the chain is directly or indirectly referencing t_1. Symmetrically with the previous case, if we considered the axes inverted, the scatter plot would consist of a single point. The data points associated to this scatter plot are computed using the following query:

SELECT $t_1.PK_1$, COUNT(DISTINCT $t_k.PK_k$) AS y FROM $t_1 \bowtie \ldots \bowtie t_k$
GROUP BY $t_1.PK_1$

The query is distinct in this case as the grouping of values of the y-axis is performed by the identifiers of t_1. Thus, after this query is performed, CoDS constructs for each value of y the set of identifiers of t_1 associated with y number of tuples of table t_k and a data point dp composed of the identifiers discovered with coordinates $(\|dp\|, y)$. In this case, we also do not use a nested query in order to be able to instantiate each data point with the associated values for $t_1.PK_1$ without using additional queries. For instance for $ch = \langle District, Client, Disposition \rangle$ the following query is constructed:

SELECT $District.district\_id$, COUNT(DISTINCT $Disposition.disp\_id$) AS y
FROM $District \bowtie Client \bowtie Disposition$ GROUP BY $District.district\_id$

The system proceeds in counting how many districts (i.e. x value) have the same y number of dispositions associated. For each value of y it then constructs the data point with the associated values of $district\_id$. Each unique data point will appear at the computed coordinates (x, y) on the scatter plot associated. The scatter plot is presented in Fig. 2. Finally, we observe that a bottom-up approach will determine the processing of smaller results for the queries used and will require less internal processing of data by CoDS, delegating this task to the database management system.

3.4   Identification of Tuples to Sample

The third phase of the system consists of the selection of identifiers from the starting table to sample so that the size of the starting table will be $\alpha \cdot \|O(t_\star)\|$. The output of this phase is a set of identifiers from the starting table that are required to be included in the sample for preserving the distribution along the discovered chains. We refer to this set of identifiers to sample with Id_S. The input of this phase is a set of chains of dependencies generated previously by the system, with their associated scatter plots and data points. A key point is identifying data which has the same characteristics across all the scatter plots, as they represent the same scenario. As data points consist of a set of identifiers with the same characteristic on the y and x axis, CoDS considers each data point as a group of identifiers from the starting table with the same characteristics. However, as data points are distributed across multiple scatter plots, a set of identifiers grouped in one scatter plot might be distributed in another. The objective is to produce an α-sample of each of these data points. The *current* number of identifiers of a data point dp represents the number of identifiers of t_* that have been included in Id_S. It is calculated using the following formula:

$$CurrentNo(dp) = \|Id_S \cap dp\|$$

The *expected* number of tuples of data point dp represents the number of identifiers dp should contain in the sample database. It is defined as:

$$ExpectedNo(dp) = \alpha \cdot \|dp\|$$

The objective of CoDS is to meet the following condition:

$$\forall dp \in \cup_{sp \in Sp(t_*)} sp : CurrentNo(dp) = ExpectedNo(dp)$$

It is not always feasible that all data points in all scatter plots verify this condition. The system proceeds in checking for each data point dp of all scatter plots whether this condition is met or not. While the latter is true the system calls balance(dp). In order to avoid an infinite loop, the maximum number of iterations for calling balance(dp) is: $\lceil ExpectedNo(dp) - CurrentNo(dp) \rceil$. The function **balance** represents the core functionality of the sampling algorithm. The function is presented in detail in algorithm 1, where $\|dp\|$ represents the number of identifiers that the data point dp contains. In order to decide which identifier should be added to a data point, the system computes for each data point dp the set of related data points, RDP(dp) by intersecting dp with all the data points from all the rest of the generated scatter plots:

$$RDP(dp) = \{dp' \in \cup_{sp' \in Sp(t_*) \backslash sp} sp' : dp' \cap dp \neq \emptyset\} \tag{1}$$

This information is used to calculate the impact factor of an identifier $id \in dp$:

$$IF(id) = \sum_{dp' \in RDP(dp)} \frac{CurrentNo(dp')}{ExpectedNo(dp')}$$

The impact factor suggests how much impact adding an identifier will have. Adding an identifier with low impact factor will not trigger major differences between the current number and the expected number of any of the related data points, facilitating a balanced insertion. Situations when no identifier is found to insert in Id_S (i.e. as this would disrupt the distribution with the current number of a related data point higher than the expected number) are best avoided using this strategy.

After all data points are balanced by CoDS, if $\|S(t_\star)\| \neq \alpha \cdot \|O(t_\star)\|$, the system finally checks for each value of $PK_\star$ in $O(t_\star)$ whether it can be added to Id_S. The reason for this is to try to fill data points that have the current number 1, and expected number 0 or 1 as these are hardly influenced by balance(dp).

Algorithm 1. balance(dp)

```
1  if CurrentNo(dp) < ExpectedNo(dp) then
2  |    c ← 0;
3  |    RDP(dp) ← computeRDP(dp) ;                    // see equation (1).
4  |    while c < ‖dp‖ do
5  |    |    // Retrieve identifier with the c-th smallest impact factor
6  |    |    id ← dp.getIdNthSmallestIF(c);
7  |    |    Id_S ← Id_S ∪ {id};
8  |    |    // Checking whether adding id disrupts any scatter plot
9  |    |    if ∃dp' ∈ RDP(dp): CurrentNo(dp') > ⌈ExpectedNo(dp')⌉ then
10 |    |    |    Id_S ← Id_S \ {id};
11 |    |    else break;
12 |    |    c++ ;
```

3.5 Creation of the Database Sample

The final phase consists in creating and populating the tables in the sample database. For each table of the original database, we create a new table in the sample database following the same specifications (attributes, types, primary key, foreign keys, etc.). After the insertion of the tuples corresponding to the Id_S of the $t_\star$, fillRT$(t_\star)$ is called. This method ensures that the related tables of $t_\star$ will be filled with referencing or referenced tuples of S$(t_\star)$. The algorithm is presented in detail in algorithm 2 and it represents a bottom-up breadth-first recursive approach. In this algorithm, $isFilled(t)$ determines whether a table t has already been filled, and $filled(T)$ defines the set of tables of T that have been filled: $filled(T) = \{t_i \in T : isFilled(t_i)\}$. Note that a table t with multiple children is not filled until either all its children or all its children reachable by the already filled tables have been filled. The reason for this is to avoid the space overhead that might be triggered between the children of t. For instance, in Fig. 1, $children(Disposition) = \{Account, Client\}$. Considering $t_\star = District$, filling table $Account$ will trigger inserting tuples in table $Disposition$. Filling table $Disposition$ with tuples referencing existing tuples in $Account$ might trigger inserting tuples in $Client$ to avoid missing references. This would trigger inserting tuples in $District$ to avoid missing references, and results in a cyclic insertion flow that should be avoided. The function buildAndExecuteQuery(t_1, T') (algorithm 2, lines 3 and 8) is used to insert tuples in table t_1 based on already the inserted tuples in tables $T' = \{t_2, t_3, \ldots, t_j\}$. The function executes one of the following queries:

If $t_1 \leftarrow t_2$: INSERT INTO $S.t_1$ (SELECT $*$ FROM $O.t_1$ WHERE PK_1 IN
(SELECT FK_2^1 FROM $S.t_2$) AND ... AND PK_1 IN (SELECT FK_j^1 FROM $S.t_j$))
If $t_1 \rightarrow t_2$: INSERT INTO $S.t_1$ (SELECT $*$ FROM $O.t_1$ WHERE FK_1^2 IN
(SELECT PK_2 FROM $S.t_2$) AND ... AND FK_1^j IN (SELECT PK_j FROM $S.t_j$))

Algorithm 2. fillRT(t)

---

1 $Crt \leftarrow \emptyset$;
2 **for** $t_i \in children(t)$: NOT $isFilled(t_i)$ **do**
3 | buildAndExecuteQuery$(t_i, filled(parents(t_i)))$;
4 | $Crt \leftarrow Crt \cup \{t_i\}$;

5 **for** $t_i \in Crt$ **do** fillRT(t_i);
6 **for** $t_i \in parents(t)$ AND NOT $isFilled(t_i)$ **do**
7 | **if** $(\forall t_j \in children(t_i) : isFilled(t_j)$ OR $\nexists d \in desc(t_j): isFilled(d))$ **then**
8 | | buildAndExecuteQuery$(t_i, filled(children(t_i)))$;

9 **for** $t_i \in parents(t)$: $isFilled(t_i)$ **do** fillRT(t_i);

---

4 Evaluation

In this section we evaluate our method and compare it to the Join Synopses approach (JS) [1], and Linked Bernoulli Synopsis approach (LBS) [7]. Both methods aim to construct a consistent database sample of a relational database and are described in detail in section 2.

4.1 Environment and Dataset

JS, LBS, and CoDS were implemented against MySQL databases, using Java 1.6. CoDS was deployed on a machine with quad-core 2.5GHz processor, 16GB RAM, and 750GB Serial ATA Drive with 7200 rpm. Each experiment was run with 12GB maximum size of the memory allocation pool. We consider the *Financial* database[4] from PKDD'99 Challenge Discovery (see Fig. 1). It contains typical bank data, such as its clients information, their accounts, transactions, loans, and credit cards. The database contains 8 tables, and a total of $1,079,680$ tuples. The sizes of the tables range from 77 (table *District*) to $1,056,320$ tuples (table *Trans*). The average number of tuples per table is $134,960$.

4.2 Measures

Representativeness. In this work, we aim to produce a representative sample of a relational database. In order to measure the accuracy of our approach, we propose to measure the representativeness of a sample as follows. We evaluate the sample database by comparing the distributions between consecutive linked

---

[4] http://lisp.vse.cz/pkdd99/Challenge/berka.htm

tables in the graph representation of the database (e.g. in Fig. 1: *District* and *Account, Account* and *Order, Disposition* and *Card, ...*) with their associated distributions in the original database. The representativeness error of the relationship between two tables:

$$\delta(t,t') = \frac{1}{\|sp_t^{t'}\|} \sum_{dp_t^{t'} \in sp_t^{t'}} \min\left(\frac{|S - \lfloor E \rfloor|}{\lfloor E \rfloor}, \frac{|S - \lceil E \rceil|}{\lceil E \rceil}\right)$$

where $\|sp_t^{t'}\|$ represents the number of data points in the scatter plot between table t and table t', $S = CurrentNo(dp_t^{t'})$, and $E = ExpectedNo(dp_t^{t'})$. The average representativeness error for a sample database:

$$\delta(T) = \frac{1}{\|T\|} \sum_{t \in T} \left(\frac{1}{\|RT(t)\|} \sum_{t' \in RT(t)} \delta(t,t')\right)$$

Sample size error. When sampling a database $O(T)$ with a sampling rate α, we expect that each table will be reduced in size by α. As a consequence, we expect the database size to be reduced by a factor α. An accurate sampling method should produce a sample database $S(T)$ with size $\alpha \cdot \|O(T)\|$. We measure the *global sample size error* of a sample with respect to a database as:

$$global\_sample\_size\_error(T) = \frac{S_T - \alpha \cdot O_T}{\alpha \cdot O_T}$$

where $O_T = \sum_{t \in T} \|O(t)\|$ and $S_T = \sum_{t \in T} \|S(t)\|$.

Time. We measure the time needed to sample a database in seconds.

4.3   Results and Observations

In this section, we present the results of running CoDS, JS, and LBS with regards to the metrics described in the previous section. The starting table identified by CoDS is the leaf table *District*. A diamond pattern described in detail in [7] is contained in the *Financial* database between the following tables: *District, Client, Account, Disposition* (see Fig. 1). The proposed solutions for applying LBS in this situation is to store the *District* table completely, or switch to JS method. For comparison purposes and due to the small number of tuples of the *District* table, we have chosen to store it completely when applying LBS.

Representativeness error. Figure 3 shows the results for the average representativeness error for the sample database. We observe that CoDS method performs best for all values of α. The error varies between 23.2% and 8.9% for the LBS method with an average of 29.9%, resulting in LBS being the less accurate method for this measure. JS method is quite close to LBS, with the representative error varying between 23.5% and 7.2% and an average of 29.2%. CoDS method is less sensible to the variation of α, with the error varying between 7% and 5%. We observe that the CoDS method produces the most representative sample, with an average error of 6.3%.

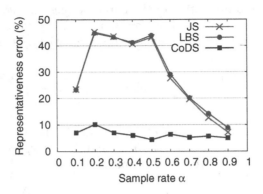

Fig. 3. Representativeness error for JS, LBS, and CoDs

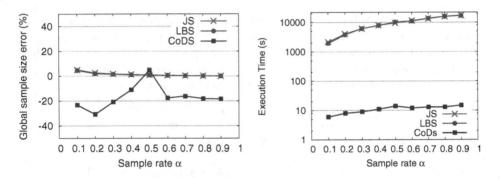

Fig. 4. Global sample size error   **Fig. 5.** Execution time

Sample size error. Figure 4 shows the results for the global sample size error. The error can be negative in the case that not enough tuples have been inserted in the sample database. We observe that the LBS technique produces the best sample database with a global sample size error varying between 4.55% and 0.1%. We notice that JS performs quite close to LBS with regards to this metric, with the error varying between 4.96% and 0.1% for the JS method. CoDS method generally samples less data than expected, with the global sample size error varying between −23.3% and −18.3%. The worst case for all methods occurs when $\alpha = 0.1$ due to the fact that the sample size error is relative to the expected sample size. Thus, for small values of alpha, a small variation between the sample and its expected size determines higher values for the error. Moreover, the reason why CoDS generally produces a sample database with less tuples than desired is because the method is cautious and does not insert tuples that might disturb the representativeness of the sample.

Execution time. Figure 5 shows the execution time for CoDS, JS, and LBS methods. We observe that CoDS outperforms JS, and LBS producing a sample database 300-1000 times faster, for α ranging from 0.1 to 0.9. The execution time in the case of JS and LBS is dependent on the processing of each tuple of each table in the original database.

In conclusion, we observe that CoDS produces the best results in terms of representativeness except for $\alpha = 0.9$. We observe that CoDS is very close to the best solution in terms of global sample size error and outperforms JS and LBS method with regards to the execution time for all values of α by producing a sample database between 300 and 1000 times faster.

5   Conclusion

In this paper, we proposed CoDS, a novel approach for relational database sampling. CoDS aims to produce a representative consistent sample by taking into consideration the dependencies between the data in a relational database. To do so, CoDS analyses the distribution between a certain table (called the starting table) and all the other tables. We conducted experiments on the *Financial* database. Results show that CoDS outperforms the previous existing consistent sampling approaches in terms of representativeness and also in terms of execution time. The sampling algorithm aims to significantly decrease the storage space needed for the original database, while achieving a balance between the computational cost of running the analysis on the original database and the accuracy of the results by preserving the properties of the original database.

As future work, we plan to extend our method to take into account other characteristics of the database. In particular we aim to consider the distribution of attributes values in order to produce a sample that is realistic not only at the table-level, but also at the attribute-level. We plan to study how to improve our method's accuracy in terms of sample size error, while maintaining the representativeness of the sample. Last but not least, we plan to apply our approach to populate testing environments. This work will be done in collaboration with IBM. The objective is to significantly decrease the time it takes to populate the testing environment, and demonstrate in a real situation that the representativeness of a sample allows to find more anomalies in the code in comparison with random-based samples.

Acknowledgments. This work was supported, in part, by Science Foundation Ireland grant 10/CE/I1855 to Lero - the Irish Software Engineering Research Centre (www.lero.ie).

References

1. Acharya, S., Gibbons, P.B., Poosala, V., Ramaswamy, S.: Join synopses for approximate query answering. In: International Conference on Management of Data (SIGMOD), pp. 275–286 (1999)
2. Agarwal, S., Iyer, A.P., Panda, A., Madden, S., Mozafari, B., Stoica, I.: Blink and it's done: interactive queries on very large data. VLDB Endowment 5(12), 1902–1905 (2012)

3. Bisbal, J., Grimson, J., Bell, D.: A formal framework for database sampling. Information and Software Technology 47(12), 819–828 (2005)
4. Chakaravarthy, V.T., Pandit, V., Sabharwal, Y.: Analysis of sampling techniques for association rule mining. In: 12th ACM International Conference on Database Theory (ICST), pp. 276–283 (2009)
5. Chaudhuri, S., Das, G., Srivastava, U.: Effective use of block-level sampling in statistics estimation. In: ACM International Conference on Management of Data (SIGMOD), pp. 287–298 (2004)
6. Ferragut, E., Laska, J.: Randomized sampling for large data applications of SVM. In: 11th IEEE International Conference on Machine Learning and Applications (ICMLA), vol. 1, pp. 350–355 (2012)
7. Gemulla, R., Rösch, P., Lehner, W.: Linked bernoulli synopses: Sampling along foreign keys. In: Ludäscher, B., Mamoulis, N. (eds.) SSDBM 2008. LNCS, vol. 5069, pp. 6–23. Springer, Heidelberg (2008)
8. Goethals, B., Le Page, W., Mampaey, M.: Mining interesting sets and rules in relational databases. In: 25th ACM Symposium on Applied Computing (SAC), pp. 997–1001 (2010)
9. Haas, P.J., König, C.: A bi-level bernoulli scheme for database sampling. In: ACM International Conference on Management of Data (SIGMOD), pp. 275–286 (2004)
10. Ioannidis, Y.E., Poosala, V.: Histogram-based approximation of set-valued query-answers. In: 25th International Conference on Very Large Data Bases (VLDB), pp. 174–185 (1999)
11. John, G., Langley, P.: Static versus dynamic sampling for data mining. In: 2nd International Conference on Knowledge Discovery and Data Mining (KDD), pp. 367–370 (1996)
12. Köhler, H., Zhou, X., Sadiq, S., Shu, Y., Taylor, K.: Sampling dirty data for matching attributes. In: ACM International Conference on Management of Data (SIGMOD), pp. 63–74 (2010)
13. Lu, X., Bressan, S.: Sampling connected induced subgraphs uniformly at random. In: Ailamaki, A., Bowers, S. (eds.) SSDBM 2012. LNCS, vol. 7338, pp. 195–212. Springer, Heidelberg (2012)
14. Olken, F.: Random Sampling from Databases. PhD thesis, University of California at Berkeley (1993)
15. Olston, C., Chopra, S., Srivastava, U.: Generating example data for dataflow programs. In: International Conference on Management of Data, pp. 245–256 (2009)
16. Palmer, C.R., Faloutsos, C.: Density biased sampling: an improved method for data mining and clustering. In: ACM International Conference on Management of Data (SIGMOD), pp. 82–92 (2000)
17. Provost, F., Jensen, D., Oates, T.: Efficient progressive sampling. In: 5th ACM International Conference on Knowledge Discovery and Data Mining (KDD), pp. 23–32 (1999)
18. Taneja, K., Zhang, Y., Xie, T.: MODA: Automated test generation for database applications via mock objects. In: IEEE/ACM International Conference on Automated Software Engineering (2010)
19. Toivonen, H.: Sampling large databases for association rules. In: 22nd International Conference on Very Large Data Bases, VLDB (1996)
20. Wu, X., Wang, Y., Guo, S., Zheng, Y.: Privacy preserving database generation for database application testing. Fundamenta Informaticae 78(4), 595–612 (2007)
21. Yin, X., Han, J., Yang, J., Yu, P.: Efficient classification across multiple database relations: a crossmine approach. IEEE Transactions on Knowledge and Data Engineering (TKDE) 18(6), 770–783 (2006)

Publishing Trajectory with Differential Privacy: A Priori vs. A Posteriori Sampling Mechanisms

Dongxu Shao[1], Kaifeng Jiang[2], Thomas Kister[1], Stéphane Bressan[2], and Kian-Lee Tan[1]

[1] School of Computing, National University of Singapore
{shaodx,thomas.kister,tankl}@comp.nus.edu.sg
[2] Center for Maritime Studies, National University of Singapore
{cmsjk,steph}@nus.edu.sg

Abstract. It is now possible to collect and share trajectory data for any ship in the world by various means such as satellite and VHF systems. However, the publication of such data also creates new risks for privacy breach with consequences on the security and liability of the stakeholders. Thus, there is an urgent need to develop methods for preserving the privacy of published trajectory data. In this paper, we propose and comparatively investigate two mechanisms for the publication of the trajectory of individual ships under differential privacy guarantees. Traditionally, privacy and differential privacy is achieved by perturbation of the result or the data according to the sensitivity of the query. Our approach, instead, combines sampling and interpolation. We present and compare two techniques in which we sample and interpolate (a priori) and interpolate and sample (a posteriori), respectively. We show that both techniques achieve a $(0, \delta)$ form of differential privacy. We analytically and empirically, with real ship trajectories, study the privacy guarantee and utility of the methods.

1 Introduction

With the increasing pervasiveness of high quality location-acquisition technologies, geolocation becomes the bread and butter of many applications. In those applications traditionally concerned with navigation such as shipping, new analytical and operational opportunities are created. However, the possibility to collect and share trajectory data for any ship in the world by various means such as satellite and VHF systems creates new risks for privacy breach with detrimental consequences on the security and liability of the stakeholders. Moreover, trajectory data generally contain sensitive information [1]. Any improper publication of such sensitive data can lead to privacy breach. In fact, as every position is potentially sensitive, it is critical to protect the privacy of each individual position in the trajectory. This motivates us to investigate the problem of publishing trajectory data with differential privacy.

ε-differential privacy was first introduced by Dwork in 2006 [2], and it is now a widely accepted privacy standard. It requires that the output answer by the

H. Decker et al. (Eds.): DEXA 2013, Part I, LNCS 8055, pp. 357–365, 2013.

randomized mechanism to a query function be insensitive to any change of a single element in the underlying database. The insensitivity is controlled by the parameter ε. Hence it is very difficult for attackers to obtain truthful information about any position by analyzing the published results.

There are mainly two methods for achieving differential privacy: Laplacian mechanism [3] and Exponential mechanism [4]. The first one adds random noise following the Laplace distribution to the true answers. The second one returns a sampled result from the collection of all possible outputs. Both of the two mechanisms perturb the results according to the sensitivity of the query. The query in our consideration here is the trajectory itself. Its sensitivity is very high because the velocity of a ship can be very fast. Hence it is very hard to get good utility by using these two common methods. In order to obtain reasonable utility, we adopt a relaxed version of ε-differential privacy.

Dwork et al. [5] proposed (ε, δ)-differential privacy, where δ bounds the probability that ε-differential privacy does not happen. In this paper, we propose two mechanisms using combination of sampling and interpolation to preserve $(0, \delta)$-differential privacy. Therefore, our proposal guarantees that the strongest version of ε-differential privacy happens except for a little probability δ. This privacy preserving is obtained by the sampling stage. The interpolation stage is designed to deliver trajectories with good utility. These two stages can be in any order. We also compare a priori sampling mechanism and a posteriori sampling mechanism.

Contribution: In this paper, we consider the problem of publishing trajectories via the differential privacy model. The key challenge is to improve the utility of the mechanism while preserving privacy level. Our proposed mechanisms are able to achieve the strongest differential privacy except a small probability. We comparatively evaluate the performance of this mechanism both qualitatively by illustrating the publication of real ship trajectories and quantitatively by measuring the error between the published and original trajectory.

- We propose a priori sampling mechanism (**SFI**[1]) and a posteriori sampling mechanism (**IFS**[2]) to publish trajectories with $(0, \delta)$-differential privacy.
- We compare these two mechanisms analytically and empirically.
- We conduct numerical experiments to evaluate the utilities of **SFI** and **IFS**. The numerical results show that the **SFI** mechanism has better performance.

The rest of this paper is organized as follows. Section 2 reviews related work. Formal definitions and problem statement are introduced in Section 3. We present our two sampling-based differentially private mechanisms in Section 4. We report the numerical results in Section 5 and give the final conclusion in Section 6.

---

[1] SFI stands for Sampling First and Interpolation.
[2] IFS stands for Interpolation First and Sampling.

2   Related Work

Generally there are mainly two different types of trajectory publishing. One type aims to publish a group of trajectories and considers each trajectory as one individual record. The other type considers one trajectory as a database and each position in the trajectory as one individual record.

Recent privacy-preserving technology for the first type starts with (k, δ)-anonymity proposed by Abul et al.[6]. The intuition is to disturb the trajectory so that at least k many different trajectories co-exist in a cylinder with radius δ. Chen et al.[7] were among the first to connect trajectory publishing and differential privacy. They proposed a data-dependent sanitization mechanism by building a noisy prefix tree according to the underlying data.

To our knowledge, not much work has been done on the second type. This is the focus of our paper. The privacy of each position can be preserved by sampling. Besides, we use interpolation to retrieve the information of sampled-out positions. The interpolation method in our proposal is widely used in robot route planning. To find smooth enough paths passing through a sequence of given waypoints, with requirements on velocity and direction, a classic technology is using Bézier spline. To obtain a continuous curve matching given direction and velocity at waypoints, cubic Bézier was applied in [8].

The random sampling method would allow others to study the statistical patterns about the entire population based upon the collected sample data. Intuitively, a simple random sampling already provides certain amount of privacy guarantees for the underlying population. Chaudhuri and Mishra [9] have shown that a simple random sampling mechanism only does not preserve ε-differential privacy, and under certain conditions it may guarantee that the ε-differential privacy is preserved with probability at least $1 - \delta$. Gehrke et al. [10] introduced a new definition of privacy called crowd-blending privacy, which is a relaxation of differential privacy. The authors show that the crowd-blending mechanism, with a pre-sampling from the underlying population, can both guarantee differential privacy and the stronger notion of zero-knowledge privacy.

In the above mentioned random sampling results, only the sampled data would be released to public for statistical studies. However, in order to monitor the ship's navigation, we still would like to estimate the ship's possible positions in the time interval between any two sampled positions. Significant events may happen in some time interval. Hence it is of great importance to infer the ship's positions during the navigation. This could also be achieved by the interpolation. Our proposal in this paper can achieve $(0, \delta)$-differential privacy for small δ. Moreover, a large number of experiments conducted on real ship trajectories demonstrate good utility of our mechanisms.

3   Preliminaries

Differential privacy has been widely used to protect the privacy of the individual participants while providing useful statistical information about the underlying population.

Definition 1 ((ε, δ)-Differential privacy[5]). *A randomized mechanism $\mathcal{K}$ gives the (ε, δ)-differential privacy if for every two databases D and D' differing in at most one row, and for every $S \subseteq \mathrm{Range}(\mathcal{K})$,*

$$\Pr[\mathcal{K}(D) \in S] \leq \exp(\varepsilon) \times \Pr[\mathcal{K}(D') \in S] + \delta.$$

Note that if $\delta = 0$, $(\varepsilon, 0)$-differential privacy is ε-differential privacy [2, 3]. (ε, δ)-differential privacy is a relaxed version of ε-differential privacy that allows privacy breaches to occur with a very small probability controlled by δ.

A trajectory is a sequence of positions on the 2-dimensional plane representing the moving path of a vehicle with additional information such as the direction, velocity and timestamp at every position. For simplicity, we assume that the information of the starting position and the terminal position is known to public.

Definition 2. *A trajectory T is a sequence $\langle (P_0, \theta_0, v_0, t_0), \ldots, (P_n, \theta_n, v_n, t_n) \rangle$, where P_i is the coordinate of the i-th position, θ_i is the direction, v_i is the velocity and t_i is the timestamp.*

After we deliver an output for an input trajectory, we have to measure the utility of our delivery. There are many ways to measure the distance of two trajectories, based on different intuitions. Two measures are adopted here for two purposes.

Definition 3. *Given two trajectories T and $\widetilde{T}$, the MAX distance between them is*

$$MAX(T, \widetilde{T}) = \max\{||P_i - \widetilde{P}_i||_2 : \ 0 \leq i \leq n\}.$$

The MAX distance measures the maximum of the distance between positions with the same timestamp. To calculate the MAX distance, the two trajectories must have the same timestamps. Since the goal of this paper is to publish a perturbed trajectory $\widetilde{T}$ while preserving $(0, \delta)$-differential privacy, we have to define neighboring trajectories to be with the same timestamp sequence.

Definition 4. *Two trajectories T and T' are neighboring if they have the same timestamp sequence and differ at exactly one tuple.*

Alternatively, one may be interested in the similarity between T and $\widetilde{T}$. The Dynamic Time Warping distance (DTW) [11] is an ideal to fulfill this task. The Dynamic Time Warping (DTW) algorithm defined recursively as:

$$\mathrm{DTW}(i,j) = \begin{cases} 0 & \text{if } i = -1 \text{ and } j = -1, \\ +\infty & \text{else if } i = -1 \text{ or } j = -1, \\ dist(P_i, \widetilde{P}_j) + min(\ \mathrm{DTW}(i-1, j), & \text{otherwise} \\ \qquad\qquad\qquad\quad \mathrm{DTW}(i, j-1), \\ \qquad\qquad\qquad\quad \mathrm{DTW}(i-1, j-1)) \end{cases}$$

where $dist(P_i, \widetilde{P}_j)$ is the cost function between the two points. We choose to define that function as the Euclidean distance between P_i and $\widetilde{P}_j$. The algorithm consists in walking along both trajectories, pairing points between the both of them, but allowing that for the next step only one of the trajectories is walked to its next point. Therefore a point can be paired with one or more consecutive points on the other trajectory. This allows us to measure the similarity between both trajectories' pattern.

4 Sampling-Based Differentially Private Schemes

In this section, we shall present two differentially private schemes to protect an individual trajectory. Both schemes are based on sampling and interpolation: Apriori Sampling Scheme (**SFI**) and Aposteriori Sampling Scheme (**IFS**).

To analyze the privacy of our mechanisms, we first recall the composition lemma proved by Dwork et al. [12], which implies the adaptive combination of a (ε, δ)-differentially private algorithm and a deterministic algorithm is also (ε, δ)-differentially private. Hence, the privacy in our proposal is fully taken care of by the sampling stage. Moreover, the interpolation method employed is cubic Bézier interpolation. It is a classic method of route planning.

An extended version of this paper with proofs, details of Bézier interpolation and pseudocodes is available in the technical report [13].

4.1 A Priori Sampling

Given a trajectory T and a privacy parameter δ, we first compute an integer $k = \lceil \frac{1}{\delta} \rceil$. Then we partition T into groups with k positions. By sampling an integer l from $\{1, ..., k\}$ uniformly, we keep the l-th position in each group and remove all other positions. Then we interpolate positions removed by using cubic Bézier interpolation.

Theorem 1. *The mechanism* **SFI** *is* $(0, \delta)$-*differentially private.*

In fact, the behaviour of the privacy parameter δ is dependent on the number of positions in the underlying trajectory. To achieve the $(0, 0)$-differential privacy where $\delta = 0$, no intermediate positions can be sampled, and the output is based on the interpolation of the two endpoints only. The next strong $(0, \delta)$-differential privacy happens for $\delta = \frac{1}{n}$, where only one intermediate position is sampled. In other words, if the input δ is between 0 and $\frac{1}{n}$, then the mechanism is the same as that for $\delta = 0$. Generally, a non-trivial δ is one element of the discrete set $\{\frac{k}{n} : k = 0, \ldots, n\}$.

4.2 A Posteriori Sampling

In the **SFI** mechanism for small δ, few intermediate positions are sampled for the interpolation. In other words, much information between two consecutive waypoints is lost. Hence, the interpolation may not reflect the real trajectory very well. An alternative way to avoid this is to do interpolation first and then sample a sub-trajectory. Since all information is kept in the interpolation stage, the sampled sub-trajectory will be more similar to the real one.

Let T be an input trajectory. The first step is to interpolate the curve in each time interval by using the cubic Bézier interpolation. Let $B(t)$ be the resulted Bézier-spline. Then we sample m timestamps uniformly from these n many time intervals, say $t'_1, \ldots, t'_m$. So the intermediate trajectory T_{mid} is $\langle (B(t'_i), B'(t'_i)) \rangle_{i=1,\ldots,m}$ with two endpoints.

So far, T_{mid} is an alternative version of T. It can be proved the process to output T_{mid} is $(0, \delta)$-differentially private by setting $m \leq \frac{\ln(1-\delta)}{\ln(1-\frac{2}{n})}$. However, T_{mid} and T may not have the same timestamps. To do the comparison, we have to interpolate the positions at the timestamps of T.

Theorem 2. *The mechanism **IFS** is* $(0, \delta)$-*differentially private.*

Generally, a non-trivial δ is one element of the discrete set $\{1-(1-\frac{2}{n})^m : m \in \mathbb{N}\}$.

5   Experimental Results

In order to compare our two algorithms, we use real trajectories of ships captured in the Singapore Straits during one hour (2012-09-09 from 08:00 to 09:00 UTC time). Because of space constraint, we have selected two representative trajectories with different shapes, one from a tug boat (Ship 1) and one from a cargo ship (Ship 2). As a summary, we present the average error for these two mechanisms on all real trajectories we have.

For each value of δ we generate 100 trajectories and choose to present one randomly. We then compute for both mechanisms the average distance between the original trajectory and the published trajectories, according to the *MAX* and *DTW* distances.

For the two selected representative ships, Figure 1 reports the average *MAX* error of the **SFI** and **IFS** mechanisms, and Figure 2 reports the average *DTW* error of the **SFI** and **IFS** mechanisms. Figure 3 reports the average errors for two mechanisms on all real trajectories we have, where each mechanism generates 100 trajectories for each real trajectory.

All these figures show a similar trend. The utility of **SFI** mechanism is better when δ is small. Smaller value of δ implies worse accuracy of the second interpolation. When δ reaches some value, the utility of **IFS** mechanism becomes no

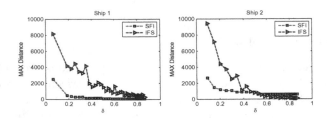

Fig. 1. MAX

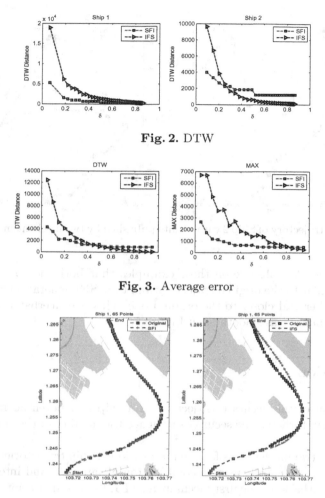

Fig. 2. DTW

Fig. 3. Average error

Fig. 4. Original trajectory of Ship 1 with results published by two mechanisms ($\delta = 0.1$)

worse than that of **SFI** mechanism. This is because the number of sampled way-points m goes to infinity while δ increases to 1, which will provide more accurate information. Hence, the **IFS** mechanism would be chosen for high toleration of privacy breach.

Another consequence of Figure 1 and Figure 2 is that the **SFI** mechanism works better for the trajectory of ship 1 with almost all δ. This behavior is common in our experiments and the trajectory of ship 1 is representative. Hence, it is reasonable to conclude that the utility of **SFI** for not-so-smooth trajectories is better than that of **IFS** mechanism.

We can now illustrate the end result with the trajectories of two selected ships in Figure 4 and Figure 5. They illustrate the original trajectories together with their published trajectories with **IFS** and their published trajectories with

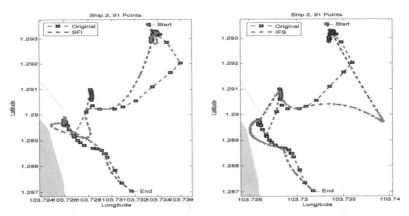

Fig. 5. Original trajectory of Ship 2 with results published by two mechanisms ($\delta = 0.1$)

SFI, respectively. We observe on these examples that both methods generate trajectories similar to the original trajectory and that **SFI** generates trajectories that are smoother and closer to the original one. This is exacerbated when the original trajectory is less smooth as in Figure 5.

6   Conclusion

The publication of the accurate trajectory of a ship is a potential menace to privacy that may threaten the security or engage the liability of the ship and its stakeholders.

We proposed two mechanisms for the publication of ship trajectories with differential privacy guarantees, using a combination of sampling and interpolation to create a perturbation. The first mechanism, **SFI**, follows an a priori approach in which a trajectory is sampled and interpolated. The second mechanism, **IFS**, follows an a posteriori approach in which a trajectory is interpolated, sampled (and possibly interpolated and sampled again).

We showed that both **SFI** and **IFS** are (ϵ, δ)-differentially private with $\epsilon = 0$. We analytically and empirically compared the two mechanisms and showed that both of them are effective in publishing realistic trajectories similar to the original trajectory. We showed that the utility of **SFI** is better than that of **IFS** for smaller values of δ and not-so-smooth trajectories.

We are currently fine tuning the general approaches discussed in this paper to take care of special cases. We are also studying the extension of our techniques to take into account prescribed constraints such as further speed, acceleration and other maneuvering limits and forbidden areas.

Acknowledgements. This research was funded by the A*Star SERC project "Hippocratic Data Stream Cloud for Secure, Privacy-preserving Data Analytics Services" 102 158 0037, NUS Ref: R-702-000-005-305 and R-252-000-433-305.

References

[1] Agard, B., Morency, C., Trépanier, M.: Mining public transport user behaviour from smart card data. In: The 12th IFAC Symposium on Information Control Problems in Manufacturing, INCOM (2006)

[2] Dwork, C.: Differential privacy. In: Bugliesi, M., Preneel, B., Sassone, V., Wegener, I. (eds.) ICALP 2006. LNCS, vol. 4052, pp. 1–12. Springer, Heidelberg (2006)

[3] Dwork, C., McSherry, F., Nissim, K., Smith, A.: Calibrating noise to sensitivity in private data analysis. In: Halevi, S., Rabin, T. (eds.) TCC 2006. LNCS, vol. 3876, pp. 265–284. Springer, Heidelberg (2006)

[4] McSherry, F., Talwar, K.: Mechanism design via differential privacy. In: 48th Annual IEEE Symposium on Foundations of Computer Science, FOCS 2007, pp. 94–103. IEEE (2007)

[5] Dwork, C., Kenthapadi, K., McSherry, F., Mironov, I., Naor, M.: Our data, ourselves: Privacy via distributed noise generation. In: Vaudenay, S. (ed.) EUROCRYPT 2006. LNCS, vol. 4004, pp. 486–503. Springer, Heidelberg (2006)

[6] Abul, O., Bonchi, F., Nanni, M.: Never walk alone: Uncertainty for anonymity in moving objects databases. In: Proceedings of the 2008 IEEE 24th International Conference on Data Engineering, ICDE 2008, pp. 376–385. IEEE Computer Society, Washington, DC (2008)

[7] Chen, R., Fung, B.C.M., Desai, B.C.: Differentially private trajectory data publication. CoRR abs/1112.2020 (2011)

[8] Mandel, C., Frese, U.: Comparison of wheelchair user interfaces for the paralysed: Head-joystick vs. verbal path selection from an offered route-set. In: Proceedings of the 3rd European Conference on Mobile Robots, ECMR 2007 (2007)

[9] Chaudhuri, K., Mishra, N.: When random sampling preserves privacy. In: Dwork, C. (ed.) CRYPTO 2006. LNCS, vol. 4117, pp. 198–213. Springer, Heidelberg (2006)

[10] Gehrke, J., Hay, M., Lui, E., Pass, R.: Crowd-blending privacy. Cryptology ePrint Archive, Report 2012/456 (2012), http://eprint.iacr.org/

[11] Sakoe, H., Chiba, S.: Dynamic programming algorithm optimization for spoken word recognition. IEEE Transactions on Acoustics, Speech and Signal Processing 26(1), 43–49 (1978)

[12] Dwork, C., Rothblum, G., Vadhan, S.: Boosting and differential privacy. In: 2010 51st Annual IEEE Symposium on Foundations of Computer Science (FOCS), pp. 51–60. IEEE (2010)

[13] Shao, D., Jiang, K., Kister, T., Bressan, S., TAN, K.L.: Publishing trajectory with differential privacy: A priori vs a posteriori sampling mechanisms. Technical Report: TRA4/13 (2013),
https://dl.comp.nus.edu.sg/dspace/handle/1900.100/3932

Towards Automated Compliance Checking in the Construction Industry

Thomas H. Beach[1,*], Tala Kasim[1], Haijiang Li[1], Nicholas Nisbet[2], and Yacine Rezgui[1]

[1] School of Engineering, Cardiff University, 5 The Parade, Roath, Cardiff, UK
[2] AEC3 UK Ltd

Abstract. The Construction industry has a complex structure of regulatory compliance, consisting of statutory requirements and performance based regulations. The increasing importance of sustainability has further intensified this, with a new building's compliance against sustainability assessment methodologies now often an important contractual requirement.

Automatic compliance checking against these requirements has been long sought after within this industry and several approaches have attempted to achieve this goal. The key improvement that can be made to many existing approaches is enabling the development and maintenance of the regulations by those who are most qualified to do this the domain experts. This is illustrated by the fact that in many cases regulatory compliance systems are closed and when modifications are needed they must be made by software's developers. This process is simply not viable in this industries rapidly changing environment.

In this paper we describe our framework for compliance checking, showing the potential for utilising an integrated process to enable domain experts to create and maintain their own regulations that can then be executed by an open source rule engine. We will describe our process, the methodology and software developed to support it. We will present our initial results in the form of two case studies illustrating progress towards automation of commonly used regulations. Finally, we will also discuss how our approach could be generalised to other related sectors to enable the adoption of a similar approach towards automatic regulatory compliance.

1 Introduction

One of the major concerns for professionals in any industry is ensuring compliance of their work against the plethora of statutory, contractual and performance based requirements their industry or clients may impose upon them. To help ease these problems the use of computer systems to support regulatory compliance has become increasingly common, but, in many cases, there is one key problem; the conversion of regulations designed to be readable by humans into computer executable code is a difficult challenge. Performing this task often requires close

H. Decker et al. (Eds.): DEXA 2013, Part I, LNCS 8055, pp. 366–380, 2013.

co-operation of domain experts with expertise relating to the regulations themselves and computing experts with experience in the systems with which these compliance checking systems are built.

The problem is particularly acute within the AEC (Architecture Engineering and Construction) sector. This particular sector has a complex structure of regulatory and contractual requirements [14]. These contractual requirements are often specified based on the client's desires as to how the building should perform, i.e. a "green" charity will often want a building with a low carbon footprint. Additionally, these requirements vary between countries and even sometimes between local authorities. This, coupled with the fact that a building is built by a large multi-organisational team, leads to a challenging environment in which regulatory compliance must be measured with a high level of certainty right from the building's conception.

In addition to this, working within the AEC sector presents additional challenges. Firstly, AEC regulations are highly complex and, especially in the case of energy performance regulations, constantly evolving. Secondly, the AEC sector is only beginning to tackle coordination issues between partners within a construction project [15] because of this building data within the AEC sector is still often stored in a series of incompatible proprietary data formats for application such as AutoDesk Revit [5] and Bentley Systems' Micro-station [6], currently only one open data standard exists - The IFCs (Information For Construction) [1].

This paper outlines our work in developing an integrated approach for the development of a regulatory compliance system. There were several key goals behind the development of this methodology:

- To produce a methodology to allow domain experts to specify the regulations within the system.
- To produce a methodology that will allow the freedom to move between different data standards.
- To separate, as much as possible, the domain expertise from the computing expertise.

While our initial work has been focused on regulatory compliance checking within the AEC sector we also believe that our approach is generalisable to other regulations in related fields of industry. Within the AEC sector itself the adoption of our three key goals has proved especially important. As regulations are rapidly evolving, the ability of the experts that truly understand the regulations to update them vastly improves the maintainability of the system. This also gives increased understand of what the regulatory compliance system is actually checking allowing validation of the system to be conducted with a far higher level of certainty.

In this paper, Section 2 will outline the background of the regulations that are being considered in this work, Section 4 will describe existing work in the field of regulatory compliance, focusing specifically on other efforts within the AEC sector. The architecture of our system will then be described and, finally, two case studies that have been used for validation will be discussed in Section 8.

2 Background

Prescriptive national building standards were first introduced in the UK in 1965. Since then, and with increasing focus on low carbon initiatives and sustainability, additional regulations have been added. Two of the most common performance based regulations are the Code For Sustainable Homes (CSH) [9] and BREEAM (BRE Environmental Assessment Method) [11]. These two regulations are optional, but their use is often stipulated by clients when purchasing buildings. Both of these regulations are termed "balanced scorecard methodologies", meaning that each section of the regulations award a set number of points (also called credits) and the credit total is used to provide an overall rating for each building.

2.1 Code for Sustainable Homes

The Code for Sustainable Homes (CSH) [9] is the national standard for assessing, rating and certifying the sustainability performance homes. It aims to encourage continuous improvement in sustainable home design and to promote higher standards over the current statutory requirements. The code provides nine measures for sustainable design; namely: energy, water, materials, surface water runoff, waste, pollution, health and well-being, management and ecology. Each of these sections awards credits and according to the performance in these sections an overall rating is given of between 1 to 6 stars. Code for Sustainable Homes is a voluntary scheme that applies in England, Wales and Northern Ireland.

2.2 BREEAM

BREEAM (Building Research Establishment Environmental Assessment Method) [11], established in the UK in 1990, is the first comprehensive building performance assessment method. The main aim of introducing BREEAM was to mitigate the impact of buildings on the environment and to increase recognition of buildings according to their environmental benefits. The basis of the scheme is to grade the individual building according to environmental performance. There are nine different dimensions assessed in BREEAM, namely: management, materials, health and well-being, energy, transport, water, land use and pollution. Each issue is then divided into sub categories which are required to meet certain criteria to achieve a BREEAM rating benchmark [11]. BREEAM certification is awarded on a scale ranging from unclassified, pass, good, very good, excellent and outstanding.

3 Related Work

There have been considerable efforts in the past towards performing automated regulation checking, with various approaches being adopted, Giblin et al [10] describe their developments in the use of regulations being expressed as logical models, in their work they have specifically targeted regulations designed for the regulation business activities. Similar work has been done by Cheng et al [7] in development of an XML framework for expressing regulations.

Within the AEC sector itself there have been several efforts to automate compliance of building regulations; Liebich et al [13] is one of the earliest successful examples within the industry of the implementation of a compliance checking system. This particular work was targeted at Singapore's Building Regulations. However, in their work the authors focus mainly on the processing of rules in relation to industry standard data formats and not on the critical aspect of the of extracting the rules from the regulation documents. This work has, however, been one of the most successful to date in terms of actual use within the AEC industry.

Yang et al [16] expand on the idea of compliance checking by utilising an object orientated approach to modelling requirements. Their approach allows the extraction of entities from within the building regulations, however from the descriptions presented by the authors, their approach seems a largely manual process of extracting rules in the form of computerised code directly from the regulations document. More recently, Eastman et al [8] outlined the architecture that a building compliance checking must take and summarised existing efforts within the industry, including various approaches of extracting regulatory information from human readable documents. In their work the authors raised concerns about the different types of data formats used in the AEC sector and they also described initial work in embedding meta-data relating to IFC object based format directly into building regulations.

In terms of commercial products, the main compliance checking product currently on the market for the AEC sector is Solibri Model Checker [3]. The Solibri model checker enables users to perform several common pieces of regulatory checking, i.e. the distance a fire escape and evacuation distance, as users are designing the building. However, one major problem with the Solibri system is that the regulations within the application are closed, users cannot edit them or add new regulations. The only way in which new regulations are added is when Solibri updates the software. This is a critical problem with the vast variety and inter-national differences between regulations int the AEC sector.

4   System Architecture

One of the key goals in developing our architecture is that the process of developing the rules used for compliance checking in the AEC sector is kept as close as possible to the domain specialists; the people that understand the regulations and the people that understand the industry specific data formats. To achieve this the architecture shown in Figure 1 has been developed.

Figure 1 shows how our architecture is divided into three distinct domains with the main software components supporting the process of the regulatory compliance checking in the centre. Each domain encompasses a set of users within the AEC sector:

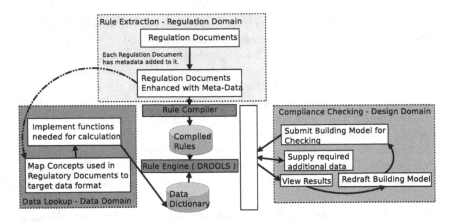

Fig. 1. System Architecture

Regulation Domain - The experts that create and/or maintain the regulations. Currently within the AEC sector companies operating in this domain will include BRE (Building Research Establishment), who maintain and operate BREEAM and UK central government that manage CfSH.

Data Domain - The experts that define and implement the data formats used by the AEC sector. Commonly used data formats include the IFCs [1] which are the only open standard currently used within the AEC industry. The IFCs are maintained by BuildingSMART [2]. Other data formats commonly used include proprietary formats from Bentley Systems [6] and AutoDesk [5].

Design Domain - The users of the system; i.e. architects, engineers. These individuals will utilise an automatic compliance checking system as a design tool when designing their building.

Our architecture also consists of three main software components:

- A Rule Compiler that converts the information extracted from the Rule documents into a machine readable form. This will be described further in Section 5.
- A Data dictionary containing mappings between the data needed for the execution of the rules and the data available in one of more industry specific data formats. This will be described further in Section 6.
- A web services interface allowing communication with add-ons developed for industry standard design packages.

The following sections will describe how our architecture is used in each of the domains in Figure 1, firstly describing how the rules are extracted from the regulation documents, then how the data required by these rules is mapped into an industry standard data model, and finally how the rule engine uses the output from these two domains to perform regulatory compliance checking on submitted building models.

5 Extracting Rules from Regulatory Documents

The process of extracting computable rules from the regulation documents has been undertaken by the addition of meta data to the regulation documents. In order to do this the RASE methodology [12], which has been used previously in similar efforts, has adopted and expanded. The RASE methodology allows the addition meta-data to the regulation document at the block level (i.e paragraph level) and inline (i.e. individual words or groups of words).

At the block level, series of nested boxes are used to surround paragraphs (or groups of paragraphs) enabling the expression of complex and/or nested groupings. Each box normally represents one decision, which contains one or more inline which define what this decision is.

RASE provides users with four "tags" each of which has a well defined logical meaning:

- Application: Restricts the Scope
- Selection: Increases the Scope
- Exception: Allows the specification of exceptions to the rule being specified.
- Requirement: Specifies the definitive requirements that must be met.

In short, Application, Selection and Exception define the scope of the decision and the requirements define the decision itself. This is shown in the Venn Diagram in Figure 2. In practice, these block and inline tags are added as HTML tags to XHTML versions of the regulations using an specifically created application, an example of the output of this application is shown in Figure 3.

Additionally, when adding metadata to the documents the domain specialist must also specify extra data describing each tag that they add:

- The topic i.e *Building, Window, Door.*
- The property i.e. *type, width, height.*

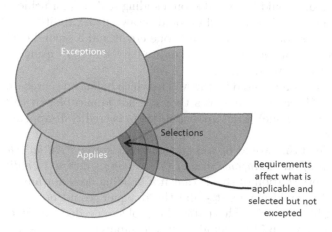

Fig. 2. Defining the Scope

Where the principal contractor achieves compliance with the criteria of a compliant scheme, CCS score between 24 and 31.5

Fig. 3. Adding MetaData to Regulation Documents

- The comparison i.e. $=$, $>$, $<$
- The value.
- The Unit i.e *m, cm, litres.*

This information is specified in plain English based on the text of the regulation itself. Within the application which is used to add the metadata, users have the ability the pick from items that have been previously used in each of the fields to promote the re-use of terminology where appropriate.

5.1   Extending RASE

One important consideration that had to be taken into account is that BREEAM and the Code for Sustainable Homes do not just produce pass/fail results. They give numeric scores (credits) which are then computed in a post-process to give an overall rating. In order to deal with this "Balanced Scorecard" approach the RASE terminology needed to be expanded to include the concepts of *total* and *output.*

Totals are needed as each part of the regulations (known as an issue) has a varying maximum number of credits that can be awarded depending on factors such as the buildings type, whether it has laboratory features and, in some cases, more complex decisions need to be made to determine the total number of credits available for an issue.

Output is needed to allow us to model the amount of credits awarded i.e. in some issues credits could be awarded on a sliding scale i.e. for achieving a certain performance between 1 and 5 credits could be awarded. Credits may also need to be broken down so one decision awards one credit and a second decision awards a second credit (giving a total of two credits for the issue) - using an output tag carefully allows this to modelled.

This is further complicated by the way in which balanced scorecard regulations within the AEC sector treat issues that are not applicable. In this case, if an issue (or part of an issue) is not applicable then its credits disappear (their totals are set to zero).

An example of this would be a regulation that stated *"All bicycle racks must be within 10m of the main entrance of the building - award 1 credit"*. However, the correct interpretation of this is that if a building does not have bicycle racks then the credit would not exist - and the designer would not be penalised for not including bicycle racks. This particular problem needed careful consideration when converting the RASE metadata into computer executable rules, which is discussed further in Section 7.

5.2   Practical Implementation of the Approach

In order to implement our approach individuals with the knowledge of the various regulations were initially instructed in a workshop style environment, this was followed up by individual ad-hoc meetings to enable the various individuals to utilise the software correctly. All together, two days of training were undertaken by the users. Following this training, it was found that, given the pre-existing knowledge of the complex logical structure of these regulations, the process of actually applying the RASE methodology was relatively fast, with each issue within a regulation document taking between two hours and a full day for the longest complex issues.

There was, however, one modification that was made to our approach as the process was being undertaken. This was to to utilise decision flowcharts (in spreadsheet format) that had been created for many BREEAM issues. These flowcharts contain sufficiently rich information regarding the logical structure of BREEAM to enable us to generate a large amount of the markup code automatically. This was adopted as it greatly speeded the translation of BREEAM (the most complex of our regulations) into a machine executable form, and enabled additional, more detailed, validation to take place within the BREEAM regulations.

6   Integration of Rules with an Industry Standard Data Format

Once the task of adding metadata to a regulation has been completed, the next challenge is to map the data into an industry standard data format. For our work the IFCs [1] have been chosen, the reason for this selection is that the IFCs are currently the only open standard within the AEC sector.

The first step is to determine the data requirements of the regulations. This is done using a tool developed to read the meta-data previously added to regulations and produce list of all the data items it requires. Also at this stage the data-type (double, integer, boolean, enumerated type or string) of the data needs to be worked out, this is done by automatically applying a series of heuristics on the comparison and value properties set on the inline tags within the document, additionally the name of the data items will also be converted into more conventional variable names by removing spaces and any invalid characters.

The use of the metadata that has been embedded into regulation documents to produce this data listing leads to the data structures generated naturally using the terms that occur within the regulation documents. To enable to translation between this "language" and the "language" of the IFCs the concept of a dictionary that performs the mapping between these two languages is used.

Figure 4 shows an example a few such mappings, in this Figure the solid arrows represent mappings intra-context and the dotted arrows show mappings between the terminology used in the regulations and the IFCs (inter-context). Figure 4 shows object *Building* and it's properties *UsedForFlammableStorage*,

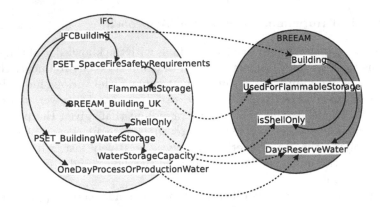

Fig. 4. Example Mappings to the IFCs

DaysReserveWater and *isShellOnly* (extracted from the meta-data added to the regulations). These are mapped by our dictionary to their counterparts in the IFC model. "Building" is mapped to *IFCBuilding*, *UsedForFlammableStorage* is mapping to a *FlammableStorage* data item within the *SpaceFireSafetyRequire-ments* property-set of the *IFCBuilding* object and *isShellOnly* does not map directly to data within IFCs so it is mapped to a newly added data item within a newly created IFC extension. *DaysReserveWater* however, needs data from two data items within the IFCs, which can be seen as there are two dotted arrows leading to it. In this example three mappings have been described, however in reality a single issue will have many mappings (often between 10 and 25) - there is, however, significant re-use of mappings across a complete set of regulations.

From this process that there emerges three distinct types of mappings:

1. A 1-1 mapping to an IFC data item.
2. A 1-1 mapping to some new data item that are being added to the IFCs
3. A 1-many mapping to several existing data items within the IFC model.

The first and second of these are relatively simple to map, with the first occurring when a data item exists within an IFC model, and the second occurring when a data item does not exist, and it cannot be calculated in some way. The third type is more complex and this can be subdivided into several sub-types; where the mapping is done by some standard calculation (i.e. area, volume), where the mapping is done by some procedure or more complex calculation that must be defined and finally when the calculation may need to use the results of an external application.

To undertake this process of mapping, the individual performing the mapping will undertake the two tasks shown on Figure 1. They will firstly, by entering data in a spreadsheet that is automatically generated from the meta-data produced previously. In completing this spreadsheet they will identify the target data items within the IFCs and also specify (in the case of 1 to many mappings) if it will be computed as a calculation, a procedure or using an external application.

Once the data entry is completed the spreadsheet will be processed to generate an XML data dictionary and a set of java method stubs for each calculation, procedure and call to external program. An example method stub for the calculation relating to "DaysReserveWater" is shown below:

```
public static IFCReal DaysReserveWater_Calculation(
    IFCProject project, IFCBuilding building ) { .... }
```

In this example the method is given two pieces of input data the IFCProject (which is the root element of an IFC Model) and the Building being considered.The returned type is defined based on the type of data the dictionary is expecting (in this case an IFCReal). The domain data specialist needs only to implement the contents of this method to calculate the value needed.

7   Execution of Rules Using the DROOLS Rule Engine

Once the addition of meta-data to regulation documents, and the data mapping has been completed, the regulations are now ready for processing.

At the centre of our processing is the DROOLS [4] rule engine. This rule engine was selected as the execution engine for our compliance checking systems because its open source and widely used.

In order for the DROOLS rule engine to process the meta-data that has been added to the regulation documents the meta-data tags must be converted into a format understandable by the rule engine, namely DRL (DROOLS Rule Language). This is done by using our rule compiler which converts RASE into DRL. The DRL is then converted by DROOLS into executable code bytecode.

This process of converting the RASE into DRL is done by utilising a series of logical formulas based on the RASE tags within each block level metadata tag. Generally speaking each block level tag is treated as a rule. When executing a rule it is done in two steps: firstly determine if the rule is in scope and secondly determine if the rule has been passed or failed.

The logical formula used to determine this is shown in Figure 5.

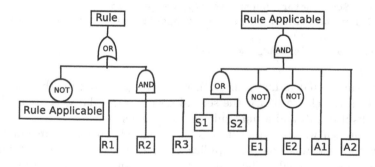

Fig. 5. Rule Logic Expansion - adapted from [12]

The first step is to determine if a rule is within scope. Within figure 5 *S1,S2* represents the Select Tags - *E1,E2* represent Exception tags and *A1,A2* represent the applicable tags. This means in order for a rule to be applicable: All the exceptions must not be met (i.e. are false), all the applicabilities must be met and at least one selection must be met.

Once it has been determined if a rule is in scope it must then be determined if it has passed/failed. This is done using the second formula shown in Figure 5. This figure shows that in order for a rule to be passed as true it must either be out of scope or must pass all of its requirements.

Using these formulas our rule compiler converts tagged documents into DRL rules. In our previous example (Figure 3) a simple sample clause was shown. This clause is verifying if the principal contractor has a suitable score from the compliant construction scheme (CCS) and awards 1 credit if successful. This particular clause has one application: that the rule only applies to contractors that are the "principal contractor and two requirements: that the CCS score is greater than 24 and that it is less than 31.5. The DRL produced by running this clause through our rule compiler is shown below:

```
rule BREEAM_MAN2_1-1SF salience 100 when
    not exists Contractor(type=='principal')
then result.isNA('BREEAM_MAN_2_1-1')
end


rule BREEAM_MAN2_1-1ST salience 100 when
    not exists Contractor(type=='principal')
then result.total('BREEAM_MAN_2_1',1)
end


rule BREEAM_MAN_2_1-1F when
    not exists Result(id='BREEAM_MAN2_1-1',na==true)
    exists Contractor(type=='principal', ccs_score <24 , ccs_score > 31.5)
then results.fail('BREEAM_MAN2_1-1')
end


rule BREEAM_MAN_2_1-1T when
    not exists Result(id='BREEAM_MAN2_1-1',na==true)
    forall ($contractor: Contractor(type=='principal')
        Contractor(this=$contractor, ccs_score >=24 , ccs_score <=31.5)
    )
then results.award('BREEAM_MAN2_1',1)
end
```

This DRL shows that in fact four DRL rules are generated for this clause. The first checks if the clause is NA, the second checks if the clause is applicable, the third checks if it fails and a final rule checks if it passed.

This approach was taken as it was not desirable to adopt an approach of assuming the rule was true (or applicable) just because it did not fail. The reason for this is to enable the system to handle situations where information may be missing from data files and enable us to give a third outcome "unknown".

Within the DRL code shown above; the Result object enables the rule engine to look up the results of previous rule executions. Additionally, it can be seen that testing to see if a rule has failed only requires testing to see if there exists one object that violates it, whereas testing for truth requires testing of each object in turn using *forall*. Finally, it should be noted that the scope rules are given a salience value, the reason for this value being specified is to ensure that the rule engine executes the scope rule prior to the rules determining if it is true or false.

It is worth examining the implementation of output and total. Processing an output is relatively simple as it is just a matter of giving the credits ID (BREEAM_MAN2_1) and the number of credits scored to the award method. However, the total is only also awarded is the rule determining if the regulation is in score is found to be true. This enables correct modelling of the behaviour of these regulations.

This example has shown how the DRL will be generated for a single clause, but (as described in Section 5) it is possible to group together clauses using block level RASE tags. This is generally much simpler to generate DRL for as it is simply applying and/or semantics to the results of rules such as those described in this section.

8 Case Study

Initial trials of the system have taken place and the results have been successful. Two case study issues have been tested; ECO5 from Code for Sustainable Homes and MAN2 from BREEAM. In order to test these issues several sets of sample building data were used. In each case, this building data was assessed using the standard manual process and then automatically by our system, with the two sets of results then compared to ensure the accuracy the automated system. In all cases so far the automated system has proven to be accurate. In this section we will describe one test from each of the two case studies in detail, to show how the developed system has functioned.

In our first test, the ECO (Ecology) 5 was examined. This issue tests the protection of ecological features within the construction site. Figure 6 shows the visualised and raw results from one of the Eco5 tests. It can be seen from this figure that rule is structured as a single OR with 6 possible options.

In this particular test, two of these options have been passed, meaning that the result of the overall regulation is true. The raw result output in the figure also shows the passes, the awarding of a total and the credit that have been awarded. It is interesting to note that the credits are awarded twice for Eco5, this is because two of the options have passed when only one was required, the post processing, however, understands this possibility and only considers the highest value awarded.

For our second case study we tested the Man (Management)2 issue. This issue tests whether construction sites are managed in an environmentally, socially responsible and accountable manner. Figure 7 shows an equivalent set of results

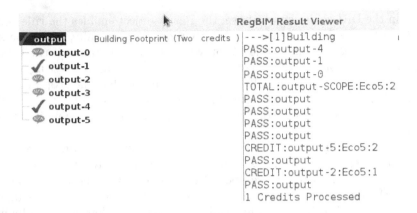

Fig. 6. CfSH Eco5 Results

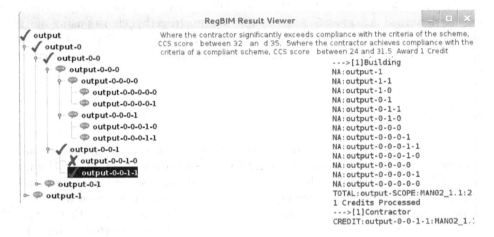

Fig. 7. BREEAM Man2 Results

for the more complex Man2 issue. This issue has a more deeply nested structure of requirements, a large number of which are not applicable in this example.

In this particular example the only branch that is applicable is output-0-0-1 this is an OR choice between two options. This structure is due to the fact that within the Man2 issue there are several sections that apply only for buildings with self contained dwellings, i.e. flats and sheltered housing. Equally the section is applicable in Figure 7 is relevant only to buildings without self contained dwellings. The results of the output-0-0-1 branch shows that one of its sub-regulations fails and the other passes, meaning the regulation itself is true. Once again the raw output shows that total/credits have been awarded for the pass result.

9 Conclusion

Automating Regulatory compliance within the AEC sector is key within the industry due to its complex regulatory structure and the many differences inter and intra nationally in the regulations that apply to buildings [14].

Our integrated approach described in this paper allows various domain experts with experience in the regulations and domain specific data files to create and maintain the regulatory compliance system. Using our approach those who maintain the regulations can alter the metadata that has been added to the regulations in order to change their functionality without needing to understand the industry data file formats or even how the underlying rule engine will work.

The decoupling of the data formats from the rule execution also allows for the development of multiple data file back-ends for the system. This will easily allow, although our initial work has focused on the IFC format, for additional data formats. Our work in mapping into the IFC format has also enabled us to identify additional data items that need to be added to the IFC specification and this will be contributed back to BuildingSmart (the standardisation body for the IFCs) in the form of an extension proposal covering regulatory compliance.

Even though only a few issues have been developed so far to a fully working state, the remainder of the issues within BREEAM and CSH have been analysed to ensure our system is able to cope with all the types of requirements present within these regulations. In the future we will expand our system to integrate it closely with an industry standard design package to enable dynamic requirement checking as a designer designs their building.

We believe that our system provides significant advantages over existing approaches. Currently the only widely used regulation compliance system in the AEC sector is Solibri[3]. However, their system currently only provides a small subset of the regulations used by the construction industry. More importantly, however, the rule system implemented by Solibri is closed, meaning that rule modifications must be made the company itself. The key advantage of our system over their product is our integrated approach, allowing the domain experts, who truly understand the regulations, to work together using software tools to produce the computer executable rules. To the best of our knowledge this approach, utilising experts that are familiar with each aspect of the process (regulations and industry specific data formats) is unique within the construction industry.

While our initial work has focused on the development of a regulatory compliance system for the AEC sector it is anticipated that our approach is generalisable to many other related industries. However, when adapting the approach modifications may need to be made to the meta-data used to support specific ways in which a particular industry operates, this will be similar to the modifications made to adapt RASE in Section 5.1 to support the balanced-scorecard regulations common in the AEC sector.

References

1. Industry Foundation Classes ISO/PAS 16739:2005
2. BuildingSmart, http://www.buildingsmart.org/ (access March 2013)
3. Solibri Model Checker, http://www.solibri.com/ (access March 2013)
4. DROOLS Expert - Rule Engine, http://www.jboss.org/drools/ (accessed March 2013)
5. Autodesk Revit Architecture – available at: http://usa.autodesk.com/revit-architecture/ (last accessed: January 15, 2013)
6. Bentley Systems – available at: http://www.bentley.com (last accessed: January 15, 2013)
7. Cheng, C.P., Lau, G.T., Law, K.H.: Mapping regulations to industry-specific taxonomies. In: Proceedings of the 11th International Conference on Artificial Intelligence and Law, ICAIL 2007, pp. 59–63. ACM, New York (2007)
8. Eastman, C., Lee, J.M., Jeong, Y.S., Lee, J.K.: Automatic rule-based checking of building designs. Automation in Construction 18(8), 1011–1033 (2009)
9. UK Government Department for Communities and Local Government. Code for Sustainable Homes Technical Guide. Technical report (2010)
10. Giblin, C., Liu, A.Y., Müller, S., Pfitzmann, B., Zhou, X.: Regulations expressed as logical models (realm). In: Proceedings of the 2005 Conference on Legal Knowledge and Information Systems: JURIX 2005: The Eighteenth Annual Conference, Amsterdam, The Netherlands, pp. 37–48. IOS Press (2005)
11. BRE Global. BREEAM Technical Guide V2.0. Technical report, BRE Global (2011)
12. Hjelseth, E., Nisbet, N.: Exploring semantic based model checking. In: Proceedings of the 2010 27th CIB W78 International Conference, vol. (54) (2010)
13. Liebich, T., Wix, J., Forester, J.: Speeding-up Building Plan Approvals:The Singapore e-Plan Checking project offers automatic plan checking based on IFC. In: European Conferences on Product and Process Modelling (2002)
14. Rezgui, Y., Miles, J.: Harvesting and Managing Knowledge in Construction: From theoretical foundations to business applications. Taylor & Francis (2011)
15. Rezgui, Y., Beach, T., Rana, O.F.: A Governance Approach for BIM Management across Lifecycle and Supply Chains Using Mixed-Modes of Information Delivery. Journal of Civil Engineering and Management 2 (2013)
16. Yang, Q.Z., Xu, X.: Design knowledge modeling and software implementation for building code compliance checking. Building and Environment 39(6), 689–698 (2004)

Quantifying Reviewer Credibility
in Online Tourism

Yuanyuan Wang, Stephen Chi Fai Chan, Grace Ngai, and Hong-Va Leong

Department of Computing, The Hong Kong Polytechnic Univesity,
Hung Hom, Kowloon, Hong Kong, China
{csyywang,csschan,csgngai,cshleong}@comp.polyu.edu.hk

Abstract. With the growing interconnectedness of the world and advances in transportation and communication, more and more people are travelling as independent tourists, putting together their own itineraries and activities from information researched from social media. However, many reviewers post reviews without validation, leading to the explosive growth of reviews and the proliferation of uninformative, biased or even false information. This makes it very challenging for travellers to find credible reviews. Previous work has shown that credibility assessment of sources and messages are fundamentally interlinked. Hence, there has been much work on measuring the credibility of reviewers. However, most current work investigates the factors impacting the perception of reviewer credibility without quantitative evaluation. This paper presents a method that quantifies the credibility of reviewers in TripAdvisor. An Impact Index is proposed to measure reviewer credibility by evaluating the expertise and trustworthiness based on the number of reviews posted by the reviewer and the number of helpful votes received by the reviews. Furthermore, the Impact Index is improved into the Exposure-Impact Index by considering in addition the number of destinations on which the reviewer posted reviews. Our experimental results show that both Impact Index and Exposure-Impact Index outperform the state-of-the-art method in measuring the credibility of reviewers to help travellers search for credible reviews.

Keywords: reviewer credibility, credible review, tourism.

1 Introduction

The growing interconnectedness of the world and the advances in transportation and communication have swelled the number of travellers travelling as independent tourists, putting together their own itineraries and activities from information researched from social media. Moreover, an increasing number of travellers post reviews online to share their experiences and opinions, which has become one important source of information [1–3]. However, tourism websites usually do little to verify the identity of reviewers and the content of reviews. Reviewers are usually allowed to register at the website using nickname and email address,

H. Decker et al. (Eds.): DEXA 2013, Part I, LNCS 8055, pp. 381–395, 2013.
© Springer-Verlag Berlin Heidelberg 2013

without other identifying information, such as real name, photo and occupation. Even worse, they can freely post reviews without going through a rigorous editorial process for factual verification [3]. This leads to the explosive growth of reviews and the presence of uninformative, biased or even false information, which make it very time consuming and challenging for the travellers to find credible reviews [4, 5].

Some investigations into the cues that influence the perception of the credibility of reviews in tourism [3, 9, 10] provide travellers with some guidelines for judging credible reviews. However, these studies did not develop a method to search for credible reviews automatically. Credibility assessment has been studied from three perspectives: source credibility, message credibility and medium credibility [6–8]. In tourism, the review can be considered the message, the reviewer, the source and the tourism website, the medium. Rieh et al. [8] and Fragale et al. [11] have pointed out that credibility assessment of sources and messages are fundamentally interlinked. In light of this insight, this paper focuses on measuring the credibility of reviewers to help travellers search for credible reviews.

There has been much work on measuring the credibility of reviewers. Some researchers have applied surveys to explore the factors affecting the evaluation of the expertise and trustworthiness of the reviewer in online tourism [10, 12–14], which are also two key dimensions of source credibility [7, 15, 16]. However, these studies did not make a quantitative evaluation of reviewer credibility. Lee et al. [4] was the first team, so far we can determine, to use the average Review Helpful rating (RHR), which is the total number of helpful votes (or total RHR in their term) that a reviewer received from travellers, divided by the total number of reviews posted by the reviewer, to quantitatively represent reviewer credibility in TripAdvisor[1]. This method tends to favor the reviewers who have posted few reviews (possibly implying a narrow range of expertise) but nevertheless obtained high average RHR (which implies high trustworthiness).

In this paper, we present a method that quantitatively measures the credibility of reviewers in TripAdvisor, which is the most popular travel community in the world. An Impact Index is proposed to compute the reviewer credibility by evaluating the expertise and trustworthiness jointly, based on the number of reviews posted by the reviewer and the number of helpful votes received by the reviews. Reviewers who have a high Impact Index are those who have posted more reviews, which manifests their expertise, and each of the reviews having obtained more helpful votes, implying their trustworthiness. Compared to the previous method of measuring average RHR, the Impact Index considers expertise and trustworthiness simultaneously, and does not emphasize one dimension only. To better represent the multi-faceted nature of credibility, the Impact Index is further improved into the Exposure-Impact Index by considering in addition the number of destinations on which a reviewer has posted reviews. Then, we examine the effectiveness of the Impact Index and the Exposure-Impact Index by comparing them to the previous method of measuring average RHR.

---

[1] http://www.tripadvisor.com/

The rest of this paper is organized as follows. Related work is presented in Second 2. Section 3 presents the Impact Index and the Exposure-Impact Index. In Section 4, comparison experiments are presented to demonstrate the effectiveness of the Impact Index and the Exposure-Impact Index measurement. Finally, we conclude this paper by giving some directions for future work.

2   Related Work

In tourism, the products are the intangible and experiential service purchases, which can not be evaluated before their consumption. Therefore, travellers tend to search for information before travelling to reduce the degree of uncertainty and risk, and facilitate decision making [1, 2]. The development of social media technologies and travellers' willingness to post reviews online sharing experience and opinion have created a market for tourism websites, such as TripAdvisor. However, Kusumasondjaja et al. [3] pointed out, tourism websites usually lack the mechanism to rigorously verify reviewers and review contents, leading to the explosive growth of reviews and the presence of uninformative, biased or even false information, which make it very time consuming and challenging for the travellers to find credible reviews [4, 5].

To address this problem, some researchers have investigated the cues that affect the perception of the credibility of reviews [3, 9, 10, 12]. For instance, Kusumasondjaja et al. [3] investigated the impact of the review valence and the reviewer's identity on the perception of credibility, and found that a negative review with the reviewer identity disclosed could enhance the perceived credibility of reviews. The work of Xie et al. [9] indicated that hotel reviews with the presence of personal profile information were perceived as being more credible by travellers. Sidali et al. [12] found that a review must be perceived as expert so as to be trusted. The study conducted by Gretzel et al. [10] indicated that the detailed description, the type of website, and the date the review was posted were very important for the evaluation of a travel review. However, most of the previous work developed qualitative guidelines based on surveys to help the travellers distinguish credible reviews, without developing a method to search credible reviews automatically.

Previous literatures [8, 11] pointed out that credibility assessment of sources and messages are fundamentally and positively interlinked. Source credibility has been widely investigated since the "Yale Group", led by Carl Hovland, defined it as expertise and trustworthiness [15]. Although many studies on source credibility have explored several different dimensions, the focus is still the initial two dimensions [7, 15, 16]. In terms of these two dimensions of source credibility, there has been much work on investigating the credibility of reviewers. Gretzel et al. [10] found that reviewer credibility is most frequently judged based on the reviewer's travel experience. The result of the survey conducted by Sidali et al. [12] showed that the number of posted reviews and travelling a lot are important to judge the expertness. Vermeulen et al.[14] have applied experience as proxy of expertise. However, these studies did not make a quantitative evaluation of the reviewer credibility. Additionally, Lee et al. [4] used the average RHR to represent the credibility of reviewers. This approach can evaluate the reviewer credibility quantitatively, but it tends to favor the trustworthiness.

3   Quantifying the Credibility of Reviewer

In this section, we present a method and its improved approach to measure the credibility of reviewers.

3.1   Reviewer Credibility

This paper focuses on measuring reviewer credibility by considering two key dimensions: expertise and trustworthiness. In previous literature [15, 20, 21], the expertise of source refers to the source's knowledge, ability or skill to know the truth and provide valid information, and it is usually described by the terms "experienced", "knowledgeable", and "competent". The trustworthiness of the source refers to the source's willingness, moral inclination or motivation to tell the truth, and it is commonly described by the terms "well-intentioned", "truthful", and "unbiased".

Based on previous studies [10, 12–14], the experience of reviewers in tourism can be used to represent expertise because expertise increases as experience increases. It can be extracted from reviewers' contribution history [12]. TripAdvisor records reviewers' contribution factors, as shown in Figure 1, and the descriptions of them are presented in Table 1. In TripAdvisor, helpful vote is a feedback from a traveller who considers the review helpful. Previous work [4] has stated that the number of helpful votes can signal the quality of an online review, and

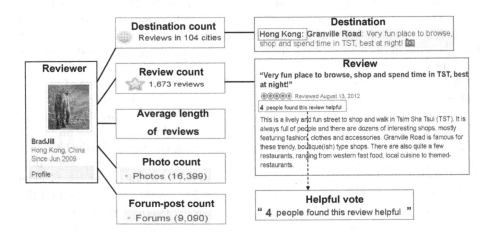

Fig. 1. A reviewer in TripAdvisor and her/his contribution factors and helpful votes

Table 1. Contribution factors of reviewers in TripAdvisor

Contribution Factor	Description
Destination Count	No. of cities that a reviewer has visited
Review Count	No. of reviews that a reviewer has posted
Average Length of Reviews (ALR)	Total no. of words of all reviews posted by a reviewer divided by Review Count
Photo Count	No. of photos that a reviewer has uploaded
Forum-post Count	No. of posts that a reviewer has posted in forum

serve as a reputation proxy of the reviewer. Therefore, this paper used the contribution factors and the number of helpful votes as the indicators to evaluate the expertise and trustworthiness of reviewer credibility in TripAdvisor.

3.2 Impact Index

Inspired by the H-Index [22], which is an index measuring both the productivity and impact of the published work of a scientist or scholar, we develop a method to measure both the expertise and trustworthiness of the reviewer. H-Index is computed based on the set of the most cited papers and the number of citations the papers have received. Among different contribution factors of reviewers in TripAdvisor, the number of reviews can directly reflect the quantity of reviewer's contribution. Moreover, the number of helpful votes cast on reviews can represent the impact of the reviews. So we propose an Impact Index to measure reviewer credibility based on the number of reviews and the number of helpful votes. It is defined as:

A reviewer has an Impact Index of L if the reviewer has posted at least L reviews, each of which has received at least L helpful votes, and the other reviews have less than L helpful votes.

Reviewers with higher Impact Index have following characteristics: first, they have posted more reviews on many things about their travels, such as different attractions, restaurants or hotels. It manifests that they have more experience and knowledge, and become relatively more competent to write helpful reviews, which indicates their higher level of expertise. Second, they have sufficient number of reviews that have received more helpful votes. It manifests that more travellers believe their reviews are helpful and reliable, which implies their high level of trustworthiness. The algorithm for computing the Impact Index of a reviewer is as follows:

Algorithm for computing Impact Index

1. Input N reviews posted by the reviewer and the number of helpful votes received by each review;
2. Initialize the Impact Index: L=0;
3. Rank the reviews based on the number of helpful votes in descending order and get their ranking indexes. The number

of helpful votes of the i'th review (i=1,2, ...,N) is denoted
as H(i);
4. i=1;
5. If H(i)>=i, i=i+1 and goto step 5; Otherwise L=i-1;
6. Output the Impact Index L.

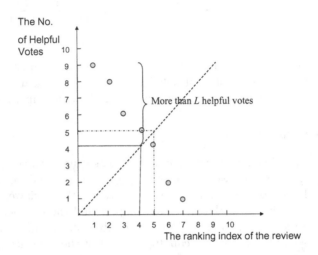

Fig. 2. Geometrical representation of the Impact Index of a reviewer

For instance, as shown in Figure 2, a reviewer has posted 7 reviews, which are
ranked according to their number of helpful votes, from the most to the least,
with the ranking indexes as {1,2,...,7}. For each of the reviews, from the first to
the fourth, the number of helpful votes is larger than its ranking index. But for
the fifth review, its number of helpful votes is smaller than the ranking index.
Therefore, the Impact Index L of the reviewer is 4. A reviewer cannot have a
high Impact Index without posting a substantial number of reviews. Meanwhile,
these reviews need to receive more helpful votes from travellers in order to count
for the Impact Index.

Although the Impact Index measures the credibility of a reviewer by considering both expertise and trustworthiness, only using the number of reviews is not
enough to represent the multi-faceted nature of credibility. For instance, a reviewer may have posted a lot of reviews on only one destination, which indicates
that this reviewer is only knowledgable about this particular destination, with
limited rather than diverse and broad experience. Therefore, we need to consider
more contribution factors as indicators to evaluate the reviewer credibility.

3.3   Exposure-Impact Index

With the Exposure-Impact Index, we further consider the number of destinations on which a reviewer has posted reviews as another dimension of reviewer

credibility. If a reviewer has posted reviews on many destinations, she or he tends to have high exposure and has experienced many attractions, hotels or restaurants at different destinations, implying her/his diverse and broad experience and knowledge. So the reviewer is possibly more competent and is better able to provide comprehensive and reliable information. Therefore, the Impact Index is further improved into Exposure-Impact Index, which is defined as:

A reviewer has an Exposure-Impact Index of E if there are at least E destinations on which the reviewer has posted reviews, and the reviews on each of the E destinations have received at least E helpful votes, and the reviews on other destinations have received less than E helpful votes.

The Exposure-Impact Index makes use of the number of destinations and the number of helpful votes as two direct elements and also indirectly considers the impact of the number of reviews. If the Exposure-Impact Index of a reviewer is higher, two conditions should be satisfied: on one hand, the reviewer has posted many reviews on more destinations, and has higher exposure, which indicates the diversity and the breadth of the experience and knowledge of the reviewer, and further implies her/his wider range of expertise; on the other hand, the reviews on each destination have received more helpful votes, which indicates that the reviewer has posted more helpful and reliable reviews, implying the trustworthiness. Therefore, the reviewer's expertise and trustworthiness can be evaluated simultaneously by the Exposure-Impact Index.

The algorithm for computing the Exposure-Impact Index is similar to that for computing Impact Index. Figure 3 shows an example to calculate the Exposure-Impact Index of a reviewer, who has been to 6 destinations. The destinations are ranked based on the total number of helpful votes, which is the sum of the number of helpful votes received by all the reviews on each destination. The

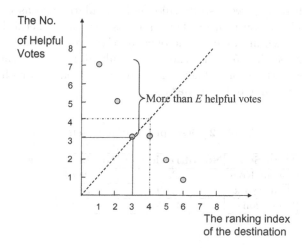

Fig. 3. Geometrical representation of the Exposure-Impact Index of a reviewer

ranking index of the 6 destinations are {1,2,...,6}. For each destination, from the first to the third, the number of helpful votes is larger than or equal to its ranking index. However, for the fourth destination, its number of helpful votes is smaller than the ranking index. So the Exposure-Impact Index E of the reviewer is 3. Reviewers cannot have high Exposure-Impact Index if they have not posted multiple reviews on a substantial number of destinations. Meanwhile, the reviews on each of the destinations need to receive enough helpful votes in order to contribute to the Exposure-Impact Index.

4   Evaluation

We evaluated the Impact Index and the Exposure-Impact Index as follows. Firstly, experiments were carried out on three data sets collected from TripAdvisor to compare the effectiveness of our methods against the average RHR method in assessing the credibility of reviews. In the experiment, we invited a team of raters to assess the quality of the reviews posted by the reviewers using each of the three methods. Secondly, linear regression analysis was applied to examine the relationship between the contribution factors and the credibility of the reviewer as assessed by different methods.

4.1   Data Collection

We developed a web crawler to collect data from TripAdvisor. As evaluation data, we downloaded reviews written in English that were posted between Aug 10, 2012 and Oct 3, 2012 on three destinations: Hong Kong, New York, and London. These destinations were chosen because they are popular destinations located in different continents (thus ensuring a disparate enough group of reviewers) and also contain a significant percentage of English-language reviews. Along with each review, we also downloaded metadata such as the number of helpful votes and the contribution factors of its author. General description of the data sets is shown in Table 2. The overlap of the reviewers between any two destinations was less than 5%, and there were only 20 reviewers who had posted reviews on three destinations. Therefore, our evaluation is, for all intents and purposes, based on three different data sets.

Table 2. Descriptions of data sets

Data Sets	No. of reviewers	No. of reviews
D-HongKong	4205	86612
D-NewYork	21879	384901
D-London	21375	362791

4.2   Evaluation of Effectiveness by Human Raters

This paper focuses on quantifying reviewer credibility to help travellers find credible reviews. Therefore, to assess the effectiveness of our methods, we conducted a survey and invited human raters to evaluate the credibility of reviews posted by the reviewers with high value of Impact Index, Exposure-Impact Index and average RHR. Based on previous results of investigations [6–8, 17–19] into criteria in evaluating the credibility of messages or reviews, we focus on three dimensions: organization, information and reliability. The organization dimension judges if the review is written in a well-organized structure, with a clear topic and in fluent language that make it easy to read and understand [6, 7, 17]. The information dimension decides if the review contains diverse, detailed and sufficient relevant information about what it reviews on [6, 7, 18, 19]. And the reliability dimension assesses if the review is telling the truth, and expressing unbiased opinion based on reviewer's personal experience [6, 18, 19]. The 5-point scale was used to evaluate each dimension, which is described in detail as Table 3, 4 and 5.

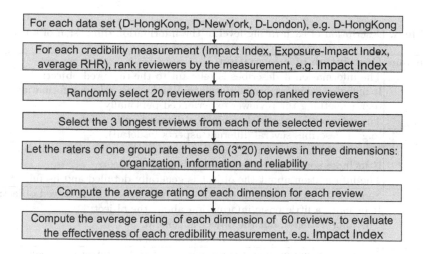

Fig. 4. Implementation flow of survey

We recruited 15 raters who are graduate students, and divided them into three groups equally. For each data set, one group of raters were invited to check the credibility of the reviews posted by the reviewers with high value of one measurement. Figure 4 shows the survey procedures, using the data set D-HongKong as an example.

Survey Results. For each data set, the evaluation results of the reviews chosen from the reviewers returned by each of the three measurement methods were collected from the corresponding group of raters. The average rating of the reviews

Table 3. Description of each rating level in the organization dimension of a review

Ratings	Description of organization
5	Its structure is clear and easily recognized; Most of the paragraphs have clear topic with detailed description; Its language is fluent and easy to read and understand.
4	It is paragraphed clearly, but not organized in a clear structure; Some of the paragraphs have clear topic with detailed description; Its language is generally fluent and easy to read and understand.
3	It is not paragraphed clearly or just has one long paragraph; Most of the paragraphs do not have clear topic. If it has one paragraph, the topic is not clear but with detailed description; Its language has no obvious problem for reading and understanding.
2	It has only one paragraph; It has no clear topic or detailed description; Its language is not easy to read and understand.
1	It has only one paragraph with few sentences; It has no topic or detailed description; Its language is very poor, and difficult to read and understand.

Table 4. Description of each rating level in the information dimension of a review

Ratings	Description of information
5	The information it describes is relevant to the reviewed object; It includes sufficient basic information and some unique information (e.g. something the reviewer experienced personally); The information about the object is detailed and useful (e.g. it describes several different aspects in detail).
4	Most of the information it describes is relevant to the reviewed object; It includes some basic information about the object; The information about the object is generally detailed and useful.
3	Some of the information it describes is relevant to the reviewed object; It includes a little basic information about the object; The information about the object is general. (e.g. it describes only one or no aspect in detail)
2	A little information it describes is relevant to the reviewed object; It includes just one piece of basic information about the object; Or it includes some basic information about the object, which is too common and easily obtained.
1	The information it describes are not relevant to the reviewed object.

in each dimension posted by the reviewers ranked top by three methods for each data set are shown in Figure 5. From the figure, we can observe that the reviews posted by the reviewers returned by Impact Index and Exposure-Impact Index obtain much higher rating in each dimension than those returned by average RHR. For instance, the rating of reviews posted by the reviewers returned by Impact Index and Exposure-Impact Index is higher in the organization dimension than those returned by average RHR, by 7%-22% and 17%-27%, respectively.

Table 5. Description of each rating level in the reliability dimension of a review

Ratings	Description of reliability
5	The information of the object includes comprehensive and convincing specifics, examples, or data, and can be accepted as truth; The opinion is fair and unbiased, with detailed personal experience (e.g. including date, time, person, or what happened) as evidence, which can support the opinion.
4	The information of the object includes some convincing specifics, examples, or data, and can be generally accepted as truth; The opinion is generally fair, with some detailed personal experience as evidence, which can generally support the opinion, though not sufficiently. And there may be a few biased opinions.
3	The information of the object includes a few convincing specific, example, or data, and can be generally accepted as truth; Some of the opinions are generally fair, with a few personal experiences as evidence. And there are some biased and unfair opinions.
2	The information of the object is very general without any detail. And it is difficult to accept the information as truth; The opinion is biased and unfair, or the personal experience can not support the opinion.
1	There is little or no basic information of the object. The information is seems to be false; The opinion is expressed in a very emotional and extreme way, and is unfair without any evidence. Its purpose is to boast of or attack the reviewed object.

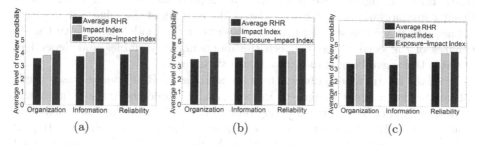

Fig. 5. The average rating of reviews in each dimension posted by reviewers ranked top by three methods on three data sets (a) D-HongKong (b) D-NewYork (c)D-Longdon

The survey results suggest that the reviews posted by the reviewers returned by Impact Index and Exposure-Impact Index are more credible than those returned by average RHR. Therefore, our methods work more effectively than the average RHR to help find credible reviews. We believe that this is because both the Impact Index and Exposure-Impact Index methods evaluate reviewer credibility by considering two dimensions, including expertise and trustworthiness, while the average RHR method tends to favor the reviewers who have posted few reviews, which possibly implies a narrow range of expertise, but nevertheless obtained high average RHR.

Moreover, the rating of reviews posted by the reviewers returned by the Exposure-Impact Index is higher in each dimension than that returned by the Impact Index. Therefore, we suggest that the Exposure-Impact Index method performs better than the Impact Index method, because it assesses the expertise of the reviewer by directly considering the number of destinations.

4.3   Regression Analysis on Contribution Factors versus Credibility Formulation

To give us further insight into the differences between various methods, we performed a linear regression analysis experiment to investigate the relationship between the contribution factors and the three evaluation methods: namely, Impact Index, Exposure-Impact Index and average RHR, respectively. The independent variables were the contribution factors, and the dependent variable was the value returned by each measurement method. For each data set, the linear regression analysis investigated the relationship between the dependent and independent variables, in order to find the contribution factors which are strongly related to reviewers' Impact Index, Exposure-Impact Index and average RHR.

Linear Regression Analysis Results. The results obtained by linear regression analysis are presented in Figure 6 which shows the most relevant contribution factors to reviewers' Impact Index, Exposure-Impact Index and average RHR. Each sub-figure, such as Figure 6(a), represents the model generated by linear regression which best fits the data. The weights of the edges are the Beta values, which are the standardized coefficients of the model. They give the estimates of the correlations between the independent variables and dependent variable that have been standardized with variance 1.

From the figure, we can observe that the average RHR of reviewers is strongly related to the Average Length of Review (ALR), while the Impact Index and Exposure-Impact Index are positively related to the Destination Count, Review Count and ALR. This shows that the Impact Index and the Exposure-Impact Index behave similarly when compared to the average RHR. This suggests that considering only the average RHR will give us some credible reviewers but it

Table 6. Three reviewers of D-HongKong who are ranked lower by average RHR

Reviewer name	Average RHR ranking	Impact Index ranking	Exposure -Impact Index ranking	Destination Count	Review Count	No. of helpful votes
ct-cruisers	908	15	5	125	416	460
Fiver75	3266	28	6	87	983	391
bongkeh	441	29	17	56	165	258

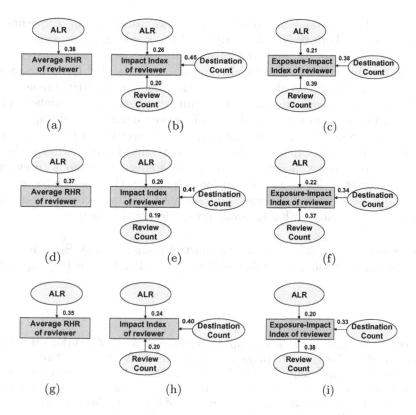

Fig. 6. Results of linear regression analysis between contribution factors (ALR denotes Average Length of Reviews) and reviewer's average RHR, Impact Index, and Exposure-Impact Index. (a)-(c) D-HongKong; (d)-(f) D-NewYork; (g-i) D-London.

misses some reviewers who have a high level of exposure and expertise, but the average RHR is not so high. For instance, Table 6 shows three reviewers in D-HongKong who have posted a lot of reviews on many destinations, and received a fair number of helpful votes, and who also have much exposure and expertise. They would not have been discovered by the average RHR method, but are ranked much higher by both Impact Index and Exposure-Impact Index methods.

5   Conclusions

To help travellers search credible reviews online, this paper proposes one measurement and a variant to quantify the credibility of reviewers in TripAdvisor. The Impact Index is proposed to evaluate the reviewer credibility by considering expertise and trustworthiness based on the number of reviews and the number of helpful votes. To represent the the multi-faceted nature of credibility, the Impact Index is further improved into Exposure-Impact Index by considering in addition the number of destinations on which a reviewer has posted reviews. Experimental results show that both the Impact Index and the Exposure-Impact Index work more effectively than average RHR to quantify the credibility of reviewers to help find credible reviews. Additionally, the Impact Index and Exposure-Impact Index can discover some credible reviewers that the average RHR missed.

So far, in this research we have put in equal emphasis on the number of destinations and the number of helpful votes. However, different weighting schemes may be more appropriate for different purposes. Therefore, for the future work, we will investigate the impact of adjusting the weight of two dimensions for the Exposure-Impact Index, and then develop more effective methods to evaluate the credibility of reviewers in tourism for helping travellers search for credible reviews. It is believed that this method to quantify reviewer credibility is applicable to other domains with a reviewer-review-feedback structure.

Acknowledgement. This project was partially supported by Hong Kong Research Grants Council. The number of the grant is PolyU 5116/08(B-Q13F).

References

1. Litvin, S.W., Goldsmith, R.E., Pan, B.: Electronic Word-of-mouth in Hospitality and Tourism Management. Tourism Management 29(3), 458–468 (2008)
2. Sparks, B.A., Browning, V.: The Impact of Online Reviews on Hotel Booking Intentions and Perception of Trust. Tourism Management 32(6), 1310–1323 (2011)
3. Kusumasondjaja, S., Shanka, T.: Credibility of Online Reviews and Initial Trust: The Roles of Reviewer's Identity and Review Valence. Journal of Vacation Marketing 18(3), 185–195 (2012)
4. Lee, H., Law, R., Murphy, J.: Helpful Reviewers in TripAdvisor: An Online Travel Community. Journal of Travel & Tourism Marketing 28(7), 675–688 (2011)
5. Metzger, M.J., Flanagin, A.J., Medders, R.: Social and Heuristic Approaches to Credibility Evaluation Online. Journal of Communication 60(3), 413–439 (2010)
6. Metzger, M.J., Flangin, A.J., Eyal, K., Lemus, D.R., McCann, R.M.: Credibility for the 21st Century: Integrating Perspectives on Source, Message, and Media Credibility in the Contemporary Media Environment. Communication Yearbook 27, 293–335 (2003)
7. Flanagin, A.J., Metzger, M.J.: Digital Media and Youth: Unparalleled Opportunity and Unprecedented Responsibility. The John D. and Catherine T. MacArthur Foundation Series on Digital media and Learning, pp. 5–27. MIT Press, Cambridge (2008)
8. Rieh, S.Y., Danielson, D.R.: Credibility: A Multidisciplinary Framework. Annual Review of Information Science and Technology 41(1), 307–364 (2007)
9. Xie, H., Miao, L., Kuo, P.J., Lee, B.Y.: Consumers' Responses to Ambivalent Online Hotel Reviews: The Role of Perceived Source Credibility and Pre-decisional Disposition. International Journal of Hospitality Management 30(1), 178–183 (2011)
10. Gretzel, U., Yoo, K.H., Purifoy, M.: Online Travel Review Study: Role and Impact of Online Travel Reviews. Laboratory for Intelligent Systems in Tourism, Texas A&M University (2007)

11. Fragale, A.R., Heath, C.: Evolving Information Credentials: The (mis) Attribution of Believable Facts to Credible Sources. Personality and Social Psychology Bulletin 30(2), 225–236 (2004)
12. Sidali, K.L., Schulze, H., Spiller, A.: The Impact of Online Reviews on the Choice of Holiday Accommodations. In: Information and Communication Technologies, pp. 87–98. Springer Wien, New York (2009)
13. Yoo, K.H., Lee, Y., Gretzel, U., Fesenmaier, D.R.: Trust in Travel-related Consumer Generated Media. In: Information and Communication Technologies in Tourism, pp. 49–60. Springer, New York (2009)
14. Vermeulen, I.E., Seegers, D.: Tried and Tested: The Impact of Online Hotel Reviews on Consumer Consideration. Tourism Management 30(1), 123–127 (2009)
15. Hovland, C.I., Janis, I.L., Kelley, H.H.: Communication and Persuasion: Psychological Studies of Opinion Change. Yale University Press, New Haven (1953)
16. Pornpitakpan, C.: The Persuasiveness of Source Credibility: A Critical Review of Five Decades. Journal of Applied Social Psychology 34(2), 243–281 (2004)
17. Fogg, B.J., Soohoo, C., Danielson, D.R., Marable, L., Stanford, J., Tauber, E.R.: How do Users Evaluate the Credibility of Web Sites? A Study with Over 2,500 Participants. In: Proceedings of the 2003 Conference on Designing for User Experiences, pp. 1–15 (2003)
18. Rieh, S.Y.: Judgment of Information Quality and Cognitive Authority in the Web. Journal of the American Society for Information Science and Technology 53(2), 145–161 (2002)
19. Metzger, M.J.: Making Sense of Credibility on the Web: Models For Evaluating Online Information and Recommendations for Future Research. Journal of the American Society for Information Science and Technology 58(13), 2078–2091 (2007)
20. Fogg, B.J., Tseng, H.: The Elements of Computer Credibility. In: Proceedings of the SIGCHI Conference on Human Factors in Computing Systems: the CHI is the Limit, pp. 80–87 (1999)
21. Cho, J., Kwon, K., Park, Y.: Q-rater: A Collaborative Reputation System Based on Source Credibility Theory. Expert Systems with Applications 36(2), 3751–3760 (2009)
22. Hirsch, J.E.: An Index to Quantify an Individual's Scientific Research Output. Proceedings of the National academy of Sciences of the United States of America 102(46), 16569 (2005)

Classifying Twitter Users Based on User Profile and Followers Distribution

Liang Yan[1], Qiang Ma[2], and Masatoshi Yoshikawa[2]

[1] Corporate Software Engineering Center, Toshiba Corporation,
1, Komukai-Toshiba-cho, Saiwai, Kawasaki, 212-8582, Japan
an@swc.toshiba.co.jp
[2] Graduate School of Informatics, Kyoto University,
Yoshida Honmachi, Sakyo, Kyoto, 606-8501, Japan
{qiang,yoshikawa}@i.kyoto-u.ac.jp

Abstract. We propose methods to classify Twitter users into open accounts and closed accounts. Open accounts (shop accounts, etc.) are the accounts who publish information to general public and their intentions is to promotion products, services or themselves. On the other hand, closed accounts tweet information on their daily lives or use Twitter as a communication tool with their friends. To distinguish these two different kinds of Twitter users can help us to search for local and daily information on Twitter. We classify Twitter accounts based on user profiles and followers distributions. The features of profile of open accounts include clue keywords, telephone number, detailed address, and so on. Follower distribution is another notable feature: most open accounts have followers from variety community. The experimental results validate our methods.

1 Introduction

To search for some real time information, such as "where can I buy iPhone5 now in Kyoto", Twitter is a valuable source since there are million users publish local and daily (shopping, etc.) information. To search for the real time information on Twitter, we need analyze the tweets with using some clue phrases[1][2]. We notice that there are two groups of users on Twitter, from which we can get different clue phrases. For example, in the query that searching for the shops where can buy iPhne5, the two group can be consumers and sellers. Consumers are more like to use "buy", while sellers use some words about "sell" in their tweets. Not only the different expression in tweets contents, but also the conflicts of the results need to be considered. For example, a shop might say that "There are still some inventories in our shop", while a consumer said "iPhone5 is sold out in that shop" at the same time. The message from consumer has high probability of being right in this case.

It is to say that, to distinguish different type of users is helpful to discovery right and real information from Twitter. In this paper, we propose novel notions of *open account* and *closed account* to distinguish Twitter users. An open

H. Decker et al. (Eds.): DEXA 2013, Part I, LNCS 8055, pp. 396–403, 2013.

account is the account with a purpose for promotion and most of their tweets are advertisement, such as a shop, a singer, a news agency and so on. A closed account is the account with a purpose for making friends or communication and the tweets are almost about daily log or feeling show.

We classify the two groups by user profiles and followers distributions. There are some features in an open account' profile, such as some clue words, telephone and the detailed address. We apply machine learning technologies, such as SVM (support vector machine) by using these features to user classification. We also notice the followers distributions of open accounts more scattered than those of close accounts. The followers of an open account is more likely from anywhere with no relationship with each other, while the closed account's followers are more likely to be the acquaintances in the real world. We calculate the entropy/standard deviation to estimate the diversity of its followers. We also propose an integrated user classification method by using features of both user profiles and followers distributions. From the experiments, we found both user profile and followers distribution are instrumental for classifying Twitter users.

2   Related Work

Marco Pennacchiotti et al.[3] employ a machine learning approach to classify Tweet users into many fields, such as political orientation or ethnicity, by analyzing user profile features, tweeting behavior and social network.

A. Java et al.[4] propose method to classify the Twitter users into information seeking users, information sharing users and friend making users by comparing the quantities of a user's friends and followers. Their purpose is different from our work. Usually, open account is an information sharing user. But sometime, they also follow a lot of accounts for get the information from the customers or competitors, and in that case, open account can also be an information seeking user at the same time.

What's more, Z. Chu et al.[5] also study on classifying Twitter user into human, bot and cyborg by comparing the tweets content they wrote and the frequency they post tweets. But the goal is totally different from us because open account, such as shops and celebrities, are also human accounts.

3   Classification of Twitter Users

3.1   Open Account and Closed Account

Open accounts are the accounts with a purpose for advertising or spreading information, tweeting to general public. Closed accounts are the accounts with a purpose for making friends or communication, tweeting to a certain range of accounts.

Usually, a closed account is created by one person. They publish tweets about their daily lives, share experience, contact friends, or just for fun. However, open accounts have more need to be done that they are intending to be broadcast centers or information sources. They can be an organization, a company, an agent or an institution and so on. They active in the world of Twitter to better advertise or publicize benefit with the huge amount of Twitter users.

3.2   User Classification by Using User Profile

Usually, open accounts have some clue words in profile, such as "shop" "bot", which help us to know them. Also, an open account is more likely to have telephone numbers or a postal address. For example, a travel agent gives a telephone number for clients to contact. However, closed accounts, almost nobody will give a telephone number or a postal address. Therefore, we propose SVM based classification method by using the features of clue words, telephone number and address.

The feature of clue words is the frequency of clue words appearing in a profile. For example, we find "shop" and "sell" in a shop's profile and both word only appear once, the feature of clue words would be "2". The features of telephone number and address are "1" or "0". "1" means there is telephone number or address in the profile, while "0" is opposite. Currently, we use libsvm-3.14[1] to realize the SVM based classification.

We propose the method to collect clue Words of Open Accounts' Profiles as follow.

1. First, we give a seed set of clue words R
2. Then, classify the given accounts by using R. As a result, we have D_o and D_c. D_o is the profiles set of open accounts, while D_c is the profiles set of closed accounts
3. Extract nouns set K_i from $i \in D_o$. Let K be the union of all the K_i.
4. For each k $\in$ K compute $s(k) = df(k, D_o)/(1 + df(k, D_c))$. Here, $df(k, D_o)$ and $df(k, D_c)$ denote the frequency of k in D_o and D_c, respectively. This formula means that a noun with high frequency in profiles of open accounts and low frequency in profiles of closed accounts will have high probability of being clue word to distinguish open and closed accounts.
5. If $s(k) > \theta$, add k into R. Here, *theta* is a pre-specified threshold.
6. Repeat 2-6 until there is no new clue words can be added into R.
7. Output R as the clue words set.

3.3   User Classification Based on Followers' Network

The followersf distributions are different between open account and closed account. For a closed account, his/her followers are from a small number of communities, such as his/her friends, classmates, colleagues and relatives and so on. However, in the case of an open account, the followers can from anywhere and they never know each other usually. For example, the followers of a news media's account are more likely the persons who are interested in reading news from various communities. Figure 1 and Figure 2 shows the examples followers distributions of an open account and a closed account.

---

[1] http://www.csie.ntu.edu.tw/~cjlin/libsvm/

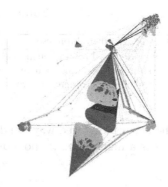

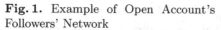

Fig. 1. Example of Open Account's Followers' Network

Fig. 2. Example of Closed Account's Followers' Network

We suppose the followers distribution of an open account is more scattered and have a higher diversity. The method to classify users by compare their followers distribution is in three steps.

1. Creating followers' network. To a given account, we construct its followers' network with its followers, and the followers and friends of its followers.
2. Detecting the communities in the network. We utilize Gephi[6] to detect communities from the followers' network of each given account.
3. Calculating the entropy/Standard deviation of the followers' network by only considering the followers, but the followers' followers or friends.

The followers diversity based on entropy is computed as follows. In the formula, n is the number of detected communities.

$$E = -\sum_{i=1}^{n}(p_i \log p_i) \tag{1}$$

where $p_i = \frac{x_i}{n_f}$ is the possibility of a node belong to community i.

The follower diversity based on standard deviation is computed as follows.

$$S_s = \sqrt{\frac{1}{n-1}\sum_{i=1}^{n}(\frac{x_i - u}{n_f})^2} \tag{2}$$

where u is the average nodes number of a community. x_i is the nodes number of community i. n_f is the whole nodes number of the network.

Higher Entropy and Lower standard deviation are expected in open accounts.

3.4   Integrated Classification Methods

As one of the integrated ways, we use the follower diversity (entropy) as a new feature and apply the machine learning method.

Table 1. Clue Words Set

Original Word in Japanese	Meaning in English
shoppu,eigyo,jyoho,shinbun,koushiki,	shop,business,information,newspaper,official,
nyusu,hasshin,kachudo,unnei,kaishya,	news,transmission,acivity,management,company,
kigyo,tenpo,chiiki,oen,ibento,kaisai	enterprise,store,region,support,event,hold

As the other integrated way, we define an integrated formula as follows. If an account has a high *score*, its possibility of being open account is high.

$$score = \alpha * E + \beta * score_p \tag{3}$$

$$score_p = log(w_1 k + w_2 t + w_3 a) \tag{4}$$

where, α , β, w_1, w_2 and w_3 are pre-specified parameters. k is clue words, t is telephone and a is address.

4   Experiments

4.1   Data Set

We used three data sets in our experiments. We collected a list of 85 Twitter accounts whose tweets had been highly ranked by using the Twitter search engine[2] with a query "iPhone5 near:'Kyoto' within:50mi" (in Japanese). We classified these 85 accounts into open and closed accounts based on analysis of their profiles and tweets manually. As the result, there are 16 open and 69 closed accounts. We collected the 85 accounts' profiles by using Twitter API[3] and the profiles of the 85 accounts constitute data set A.

We also collect data need to construct the followers' network of the 85 accounts. The followers' network of the 85 accounts is data set B. We collected the profiles from the followers' network of an open account from data set B. We get 488 accounts' profiles as data set C, and there are 149 open accounts and 339 closed accounts.

4.2   Experiment on Classifying Method Users Based on SVM

We constructed an initial clue words set by using 20 open and 20 closed accounts. Then, with the method described in Section 3.2, we have collected the clue words set shown in Table 1.

Then, for data sets A and C, we classify them by our SVM based method. We tested three kinds of combinations of features in this experiment. k means keyword, t means telephone number and a means address. Table 2 shows the results of data set A, while Table 3 shows that of data set C. We calculate Precision (p), Recall (r) and F-measure to evaluate the results. From the results, we conclude as follow.

---

[2] http://twitter.com/#!/search
[3] https://dev.twitter.com

Table 2. Result of Data Set A by SVM Based User Classification

Feartures	Precision	Recall	F-measure
k&t&a	1	0.437500	0.608696
k	1	0.18750	0.315789
t&a	1	0.312500	0.476190

Table 3. Result of Data Set C by SVM Based User Classification

Feartures	Precision	Recall	F-measure
k&t&a	0.745098	0.510067	0.605578
k	0.551724	0.322148	0.406780
t&a	1	0.288591	0.447917

- Feature words substantial increase recall ratio of identifying open accounts, with an undesirable precision ratio.
- Telephone number and address achieve a good precision of identifying open accounts, but the recall need to be improved.
- Considering feature words, telephone number and address achieved best performance to identify open accounts.

We found about 70% of open accounts' profiles have clue words (the words in our clue words set), while there are some (about 20%) closed accounts' profiles also have some clue words. The closed accounts which have clue words are easily classified into open accounts wrongly. That is why clue words give us a high recall but low precision. We also found that, the open accounts' profiles, which have telephone number and address are only about 30%, but almost all the closed accounts' profile don't have telephone number or address. This leads a high precision and low recall when we only consider telephone number and address.

4.3   Experiment on Classifying Based on Followers Distribution

In this experiment, at first, we compared the diversity scores of open and closed accounts. The results are shown in Table 4.

From the results, we can see that, entropy based diversity of open account is higher than that of closed account, while standard deviation based diversity of open account is lower. It is to say, the diversity of an open account's followers is higher.

We chose receiver operating characteristic (ROC) to evaluate our method. We set open accounts as positives, while closed accounts as negatives. ROC curve of

Table 4. Average of Entropy and Standard Deviation based Diversity Scores

Average	Entropy based Diversity	Standard Deviation based Diversity
Open Account	2.7811	0.0463
Closed Account	2.4282	0.0750

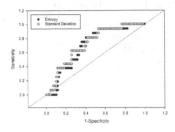

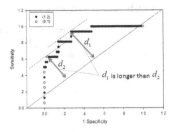

Fig. 3. ROC Curves of Diversities of Entropy and Standard Deviation

Fig. 4. ROC curve for Comparing case (1,2) and (0,1)

Table 5. Results on Data Set A (integrated classification method based svm)

Feartures	Size of Training Data Set	Precision	Recall	F-measure
e&k&t&a	10	0.875000	0.437500	0.583333
e&k&t&a	20	0.888889	0.500000	0.640000
k&t&a	10	1	0.375000	0.545455
k&t&a	20	1	0.437500	0.608696

diversities based on entropy and standard deviation are shown in Figure 3. Y-axis is sensitivity and X-axis is 1-specificity. Sensitivity measures the proportion of actual positives which are correctly identified, while specificity measures the proportion of negatives which are correctly identified. Usually, the result is good when the result curve is in the left side of the middle line.

Both of the ROC curves of entropy and standard deviation are on the left side. It is to say, followers distribution is helpful in classification.

4.4 Experiment on Integrated Classification Methods

We carried out the experiments on our integrated classification methods by using data set A and data set B. The results of integrated method based on SVM are shown in Table 5. In the table, e means the feature of entropy. From the result, we can say, entropy based follower diversity is a useful feature to improve the recall ratio to identify open accounts.

We also carried out experiment on integration calculation to integrate user profile and followers' network for user classification. We use data set C to decide the parameters of w_1, w_2 and w_3. We calculate the value of precision (p), recall (r) and F-measure (F) and set the value of w_1, w_2 and w_3 as 1, 2 and 2 because of the best average score of F measure. Then, we set a and b as (1,1), (2,1), (1,2) and (0,1) and we found the value of (1, 2) achieve the best performance. Also, we compare the results when a and b are (0,1) and (1,2) by using ROC curves as Figure 4 shows. We can see, the best point is the farthest node from the middle line and the best point of case (1, 2) is farther from the middle line.

From the experimental results, we found that both user profile and followers' network are useful for user classification. As to integrated classification methods, we found that the result of integration calculation is better than machine learning technologies by comparing with the best score of F measure (0.714286 vs. 0.640000). What's more, we also found that result comes better when we consider followers' network rather than only use user profile.

5   Conclusion

In this paper, we proposed novel concepts of open and closed accounts. We also proposed methods for classifying Twitter users into open accounts and closed accounts based on user profiles and followers' distribution. The experimental results reveal that both user profile and followers distribution can help classification and we will achieve better results when we use both of them.

Further experiments using large data set are necessary to improve the classification methods. Applications for Twitter search based on the user classification are another direction of our future work.

Acknowledgment. This work was partly supported by JSPS KAKENHI Grant Number 25700033.

References

1. Yan, L., Ma, Q., Yoshikawa, M.: Where can I Buy iPhone4S Now?: Spatio-Temporal Entity Retrieval on Twitter. DEIM Forum 2012 (2012)
2. Yan, L., Ma, Q., Yoshikawa, M.: Classifying Twitter Users for Spatio-temporal Entity Retrieval. IPSJ Technical Reports 2012-DBS-156(15), 1–6 (2012)
3. Pennacchiotti, M., Popescu, A.: A Machine Learning Approach to Twitter User Classification. In: ICWSM 2011, pp. 281–288 (2011)
4. Java, A., Song, X., Finin, T., Tseng, B.: Why We Twitter: Understanding Microblogging Usage and Communities. In: SNA-KDD 2007, pp. 56–65 (2007)
5. Chu, Z., Gianvecchio, S., Wang, H., Jajodia, S.: Who is Tweeting on Twitter: Human, Bot, or Cyborg? In: ACSAC 2010, pp. 21–30 (2010)
6. Bastian, M., Heymann, S., Jacomy, M.: Gephi: An Open Source Software for Exploring and Manipulating Networks. In: ICWSM 2009, pp. 361–362 (2009)

Fast Community Detection

Yi Song and Stéphane Bressan

School of Computing,
National University of Singapore
{songyi,steph}@nus.edu.sg

Abstract. We propose an algorithm for the detection of communities in networks. The algorithm exploits degree and clustering coefficient of vertices as these metrics characterize dense connections, which, we hypothesize, are indicative of communities. Each vertex, independently, seeks the community to which it belongs by visiting its neighbour vertices and choosing its peers on the basis of their degrees and clustering coefficients. The algorithm is intrinsically data parallel. We devise a version for *Graphics Processing Unit* (GPU). We empirically evaluate the performance of our method. We measure and compare its efficiency and effectiveness to several state of the art community detection algorithms. Effectiveness is quantified by five metrics, namely, modularity, conductance, internal density, cut ratio and weighted community clustering. Efficiency is measured by the running time. Clearly the opportunity to parallelize our algorithm yields an efficient solution to the community detection problem.

1 Introduction

A Community forms when a group of vertices in a network is more interconnected than its vertices are connected to other vertices in the network. The knowledge of such groups or communities helps finding efficient ways to distribute and gather information in online social networks for example. Community detection is a useful tool in fields such as sociology, biology and marketing. In this paper, we propose an efficient yet effective algorithm for the detection of communities in networks.

We model a network as a simple graph $G(V, E)$, where V is a set of vertices and E is a set of edges. G is undirected, un-weighted, and has no self-loop. The idea of our method is, for each vertex, to seek the community to which it belongs by visiting its neighbour vertices. Decisions are made based on the degrees, clustering coefficients of the neighbours and the number of common neighbours. Our method starts from a micro perspective, which is different from that of previous works such as GN (see [13] and [25]). Considering the size of networks in modern applications, we try and design a scalable method in order to be able to deal with the large networks in a reasonable time. Therefore we try and minimize the number of pairs-wise computation among vertices. Instead of comparing all pairs of vertices in a graph, we only explore each vertex'

H. Decker et al. (Eds.): DEXA 2013, Part I, LNCS 8055, pp. 404–418, 2013.

immediate neighbourhood. Indeed vertices in the same community are more likely to be neighbours. This significantly reduces the complexity except in the case of dense graphs. In our algorithm, as vertices can independently explore their neighbourhood and join a community by following an immediate neighbour, the algorithm is intrinsically data parallel. We devise a parallel algorithm and implement it on a *Graphics Processing Unit* (GPU).

We empirically evaluate the performance of our algorithm with both real world networks and synthetic networks. We evaluate the quality of communities using metrics from different classes (see [36]), as well as with one metric recently proposed in [28]. The metrics include modularity, conductance, internal density, cut ratio, and weighted community clustering (*WCC*). Those metrics indicate the community quality from different perspectives. We measure the running time. We compare our algorithm with several the state-of-the-art algorithms.

The rest of the paper is organized as follows. Section 2 briefly reviews the related works on graph clustering and community detection. Section 3 presents the algorithm we propose. Section 4 shows the experiment setting, experiment results and results analysis. Finally we conclude in Section 5.

2   Related Work

Graph clustering and community detection methods can be categorized into several classes. Several authors ([27,18,37,17,33]) use random walks. For example, Pons and Latapy in [27] use random walk to calculate the similarities, which they call distance between each pair of adjacent vertices, and then use Ward's agglomerative hierarchical clustering approach to find communities. Jin et al. in [18] propose an algorithm based on Markov random walk to unfold the communities, and extract them with a cutoff criterion in terms of conductance. Dongen in [33] uses Markov Clustering, which simulates the random walks.

Several authors ([25,4,26,15,16]) focus on *modularity* which is first proposed by Girvan and Newman in [13]. Modularity is defined as the number of edges inside groups minus the expected number in an equivalent graph with edges placed at random. An equivalent graph here means that the graph has the same number of edges and the same degree distribution. For example, Clauset in [4] defines a local measurement of community structure called *locally modularity* and proposes an agglomerative algorithm to maximize the *local modularity* of the communities detected. Girvan and Newman in [25] propose a divisive method to identify community. The edges with highest betweenness are removed iteratively, thus disconnecting the graph and creating communities. The best partition has the highest *modularity*.

Some authors, e.g., in [9] and [12], use cliques. For example, Du et al. in [9] use maximal cliques for community detection. An algorithm called *ComTector* is proposed. It enumerates all maximal cliques for finding clustering kernel, assigns the rest vertices to closest kernels, and merges fractional communities. Palla et al. in [12] design the clique percolation method (CMP) which finds all cliques of size k. Communities are connected union of k-cliques.

The authors of [30],[5] and [1] detect community in an agglomerative way. Ahn et al. in [1] define clusters as sets of edges. Their method group edges with an agglomerative hierarchical clustering technique. Clauset et al. in [5] propose a greedy hierarchical agglomerative algorithm. It starts from each vertex being a community and then joins two communities at each iteration. The two communities are selected based on the idea of maximizing modularity increment. They use dendrogram to represent the whole process.

Besides those ideas, Jierui and Boleslaw in [34] propose speaker-listener label propagation algorithm for overlapping community detection. Zhang et al. in [38] propose a method that combines spectral mapping, fuzzy clustering and the optimization of a quality function. Yan and Gregory in [35] propose an optimization for existing community detection algorithms. Pairwise vertex similarities are measured beforehand, and existing algorithms are applied on the graph with the vertex similarities as edge weights. Rosvall and Bergstrom in [29] use an information theoretic approach to detect community in weighted and directed network.

Some methods, such as those presented in [2], [3], [14], [19] and [6], detect community locally. For example, Baumes et al. in [2] and [3] propose two heuristics to detect locally dense subgraphs as communities. Two subgraphs with significant overlap can be locally optimal and thus are overlapping communities. The first heuristic finds disjoined clusters by deleting high-ranking vertices and then adds the deleted vertices to one or more clusters. The second heuristic starts from randomly chosen seeds and then adds or deletes one vertex at a time untill the density metric cannot be further improved. Goldberg et al. in [14] propose an additional requirement based on the work in [2] and[3], which requires the community to be a connected sub-graph. so that the algorithm is able to examine the connectivity of the cluster found.

3   Algorithm

We propose an algorithm that delegates the job of finding communities to individual vertices. Each vertex seeks its community independently. The decisions of which community to join are made based on the degrees and clustering coefficients of neighbours, as well as on the number of common immediate neighbours. We hypothesize that vertices tend to join groups with more connections. In other words, the vertices try to attach themselves to dense structures, i.e. structures with more connections among vertices in this structure.

The algorithm starts by calculating degrees and local clustering coefficient for each vertex (line 1). The local clustering coefficient is defined as

$$cc[i] = \frac{e_{jk} : j, k \in V, e_{jk} \in E}{degree[i] * (degree[i] - 1)}$$

It is the ratio between the number of edges between vertices within its neighbourhood and the number of edges that could possibly exist between them. It quantifies how close the vertex connects with its neighbours.

Algorithm 1. Fast Community Detection

Input: graph $G(V, E)$ with $|V|$ vertices, $|E|$ edges;
Result: Clusters C_i, $i \in (1, 2, ..., k')$

```
1  Compute degree[v] and cc[v], v ∈ V;
2  for each v do
3      if degree[v]<degree[vⱼ] then                    /* vⱼ ∈v_neighbour */
4          g[v] ← vᵢ, where degree[vᵢ]=max(degree[vⱼ]) ;
5      else
6          g[v] = v;
7  for each v do
8      if g[v] = v and degree[v]=degree[vᵢ] then
9          if v and vᵢ has more than half common vertices;
10         then
11             g[v]← vᵢ, if vᵢ has smaller id;
12     else
13         v_g ← g[v];
14         c1 ← number of common neighbours between v and j;
15         c2 ← number of common neighbours between v and (v_neighbour \ v_g);
16         if c1<c2 then
17             g[v] ← vᵢ, where degree[vᵢ]=max(degree[vⱼ]), vⱼ ∈(v_neighbour \ v_g)
18 for each v do
19     if g[v]≠ v then
20         i ← g[v];
21         repeat
22             i ← g[i] ;
23         until g[i]= i, find standalone vertex;
24         g[v]←i;
25 k ← different numbers in g[v];
26 for i from 1 to k do
27     for v ∈ Cᵢ do
28         find the cluster Cⱼ where v has the maximum number of immediate
           neighbours;
29         if i ≠ j then
30             Cluster v into Cⱼ;
31 Return Cᵢ,i ∈ (1, 2, ..., k');
```

In the second step, each vertex look around its immediate neighbours. If the degree of the vertex, for example vertex v, is the largest among its immediate neighbours, vertex v stands alone and does not follow other vertices. If the degree of vertex v is not the largest among its immediate neighbours and itself, vertex v follows the neighbour with the largest degree among v's immediate neighbours (line 2-6). If more than one vertex among the immediate neighbours have the largest degree, then vertex v follows the one with the largest clustering coefficient compared to other neighbours.

In the second round, each vertex adjusts their decisions (line 7-17). If the standing-alone vertex v has neighbours with the same degree, check the number of common neighbours of vertex v and v's neighbour that has the same degree. If there are enough common neighbours, these two vertices are suggested to be in the same community. If the vertex v does not stand alone but follows some neighbour, we check the number of common neighbours vertex v has with the vertex that it follows, and the number of common neighbours it has with the other neighbours. If vertex v has more common neighbours with its other neighbours than the one it follows, then vertex v turn to the vertex with the second largest degree in the neighbourhood or stands alone if itself has the second largest degree.

In the third round, each vertex finalizes the community which it desired to join (line 18-24). If the vertex that vertex v follows is also following vertex v_i, than vertex v also turn to vertex v_i. In the end, each vertex follows a vertex that stands alone. With all the other vertices that follow this vertex, they form a community.

After each vertex chooses its community (line 25), we post-process the memberships to refine the communities (line 26-30). If any vertex has more connections outside the community than inside the community, it changes its membership. This refine process may change the number of communities from the last step.

The only input of the algorithm is the graph itself. No pre-defined number of communities is needed. In experiments the graph is given as an edge list. The output is the communities.

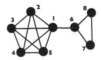

Fig. 1. Example

Figure 1 shows a graph with 8 vertices and 14 edges. After the first round, vertex 2, 3, 4, 5, 6 all follow vertex 1 (g[1]=1, g[2]=1, g[3]=1, g[4]=1, g[5]=1, g[6]=1), while vertex 7 and 8 follow vertex 6 (g[7]=6, g[8]=6). In the second round for each vertex, the status of vertex 1 is unchanged. The status of vertex 2, 3, 4, 5 is also unchanged, because they have more common neighbours with vertex 1 that they follow than with other vertices ({vertex 2, 3, 4, 5}\themselves), vertex 7 and 8 still follow 6, while vertex 6 changes to be standing alone instead of following vertex 1 because vertex 6 has more common neighbours with 7 and 8 than with vertex 1. No more changes happen in the third round and the refinement, and thus final result is that we find two communities: one community is labelled by vertex 1, and has vertex 1, 2, 3, 4, 5; the other community is labelled by vertex 6, and has vertex 6, 7, 8.

We devise a parallel version. Both the first and second rounds are parallelized. In the first round the vertices look for the vertex with the largest degree in the neighbourhood at the same time. In the second round, each vertex make decisions concurrently. The rest of the algorithm is sequential.

The time complexity for calculating clustering coefficient is $\mathcal{O}(n*d^2)$, where n is the number of vertices and d is the maximum degree of vertices in graph. The complexity for the first round is $\mathcal{O}(n*d)$. The complexity for the second round is $\mathcal{O}(n*d^2)$. The complexity for the third round is $\mathcal{O}(n^2)$. The complexity for the refinement is $\mathcal{O}(n*d^2)$. Therefore the time complexity for the whole algorithm is $\mathcal{O}(n*d^2 + n^2)$. For the parallel version, The complexity for the first round is $\mathcal{O}(d)$. The complexity for the second round is $\mathcal{O}(d^2)$. The rest is the same as that of the sequential version. Thus the time complexity for the whole parallel algorithm is $\mathcal{O}(d^2 + n^2)$.

4 Experiment

We conduct experiments on both synthetic and real world graphs including three benchmarks for community detection. We ran the sequential algorithms on an 2.83GHz Inter Core, 2 Quad CPU machine with 2GB of main memory under Windows 8 OS. The parallel algorithm ran on the same machine with a GeForce GTX 560 Ti graphics card having 2048 MB of global memory, 8 multiprocessor and 48 CUDA cores per multiprocessor. All algorithms were implemented in Visual C++ 10.0. The parallel algorithm is implemented using the application programming interface CUDA for the C language. CUDA [7], the C language Compute Unified Device Architecture, is provided by NVIDIA and works on NVIDIA graphic cards. The CUDA programming model consists of a sequential host code combined with a parallel kernel code.

We compare our algorithm with three state-of-the-art algorithms: *InfoMap* [29], *WalkTrap* [27] and Girvan and Newman (*GN*)[13][25]. *InfoMap* is based on information theory. *Walktrap* is based on random walk. *InfoMap* has been empirically shown to have better performance compared with other algorithms for community detection [11].

4.1 Dataset

We generate a batch of benchmark graphs [20] with known community structure, number of vertices, the average degree, maximum degree, minimum and maximum size of micro and macro community due to the hierarchical structure, and fraction of edges between vertices belonging to same or different communities. In our experiments, we generate graphs with 2000 vertices and different average degrees while the other parameter are the same. They have no overlapping communities.

The real-world benchmark graphs we use are listed as followings. Among them Zachary's Karate Club data, American College Football data and Dolphin network are widely used for evaluating community detection algorithms.

Karate Club data is a social network of karate club members studied by the sociologist Wayne Zachary. The network has 34 members (vertices) and they separated into two different groups due to a controversy between one of the instructors and administrator of the club.

American College Football data is a network with 115 teams (vertices) which are separated into 12 conferences. An edge exists between two vertices if there is match between two teams. More games happen among teams within the same conference than teams from different conferences.

Dolphin Network is collected by David Lusseaua [23]. The network represents frequent associations between 62 dolphins (vertices) in a community living off Doubtful Sound, New Zealand.

Email-URV data is collected by Guimer et al. [8]. The network contains user-to-user (address- to-address) links from the network of e-mail interchanges among faculty and graduate students at Rovira i Virgili University of Tarragona, Spain. It's available on Alex Arenas Website [10].

Arxiv HEP-PH collected by Leskovec et al. [22], is a collaboration network contains scientific collaborations between authors who submitted papers to High Energy Physics. It's available on SNAP website [31].

Wiki-Vote, collected by Leskove et al. [21], contains user-to-user (who-vote-whom) links from Wikipedia network. It's available on SNAP website [31]. Each vertex represents a user. An edge is created from a user to a candidate if a user votes for Wikipedia admin candidates.

Email-Enron data set contains user-to-user (address-to-address) links. It was made public by the Federal Energy Regulatory Commission during its investigations. We obtained it from [31]. Each vertex represents an email address. An edge exists between vertex i and vertex j if address i sends at least one email message to address j.

Epinions data set contains user-to-user (who-trust-whom) links from Epinions network. It was collected by Epinions staff P. Massa. We obtained it from *trustlet* website [32][24]. Each vertex represents a user. An edge corresponds to a trust or distrust statement from one user to another user.

We extract the largest component of the networks that have more than one component. The number of vertices and the number of edges of each data are listed in Table 1

Table 1. Statistics of datasets

	Number of Vertices	Number of edges
Karate Club	34	78
Dolphin	62	159
American College Football	115	610
Email-URV	1,133	5451
Wiki-Vote	7,066	100,736
Arxiv HEP-PH	11,204	117,649
Email-Enron	33,696	180,811
Epinions	119,130	704,276

4.2 Metrics

We use five metrics to qualify the communities: modularity, conductance, internal density, cut ratio and weighted community clustering. Modularity, conductance, internal density and cut ratio are selected from four classes of metrics for community [36] so that we can eliminate the bias of having only one kind of metric. Weighted community clustering is a recently proposed metric [28].

The **Modularity** [25] is defined as

$$modularity = \frac{1}{2m} \Sigma_{i,j \in V} (A_{ij} - \frac{k_i k_j}{2m}) \delta(c_i, c_j)$$

where $A_{ij} = 1$ if i and j are connected, otherwise $A_{ij} = 0$, and $\delta(c_i, c_j) = 1$ if i and j belong to the same cluster, otherwise $\delta(c_i, c_j) = 0$.

The **Conductance** for a set of vertices S is defined as

$$conductance(S) = \frac{c_s}{2m_s + c_s}$$

where $c_s = |(u, v) \in E : u \in S, v \notin S|$. It is the number of edges with one end in the set and the other end outside the set. $m_s = |(u, v) \in E : u \in S, v \in S|$. It is the number of edges in S.

The **Internal Density** for a set of vertices S is defined as

$$InternalDensity(S) = \frac{m_s}{n_s(n_s - 1)/2}$$

where m_s is the same as above. n_s is the number of vertices in S. Internal Density is the internal edge density of S.

The **Cut Ratio** for a set of vertices S is defined as

$$CutRatio(S) = \frac{c_s}{n_s(n - n_s)}$$

Cut Ratio is the fraction of existing edges out of all possible edges having one end outside the cluster.

The Weighted Community Clustering for a is defined as

$$WCC(S) = \frac{1}{|S|} \sum_{x \in S} f(x, S)$$

where $f(x, S) = \frac{t(x,S)}{t(x,V)} * \frac{vt(x,V)}{|S \setminus x| + vt(x, V \setminus s)}$ if $t(x, V) \neq 0$; $f(x, S) = 0$ if $t(x, V) = 0$. $t(x, S)$ is the number of triangles that vertex x closes with vertices in S and $vt(x, S)$ is the number of vertices of S that form at least one triangle with x.

In our experiments, we take the average of the conductances of communities found for the conductance of the whole network, and same for the other metrics except modularity.

4.3   Experimental Assessment and Analysis

Figure 2 shows the communities found in the Karate Club network by each algorithm. Figure 3 shows the communities found in the Dolphin new network by each algorithm. Vertices in the same color are in the same community.

Figure 4 shows the measurement results on the four real data sets. X-axis is labelled by the names of data sets. Y-axis is the value of metrics. For each data set, the metric values for the communities detected by each algorithm are compared. Figure 4 (a) shows that the communities that FCD and ParallelFCD found have

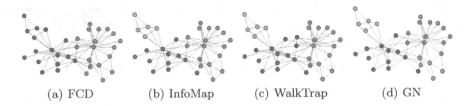

(a) FCD   (b) InfoMap   (c) WalkTrap   (d) GN

Fig. 2. Communities for Karate Club data by different algorithms

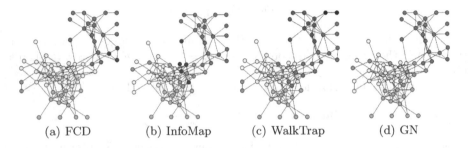

(a) FCD   (b) InfoMap   (c) WalkTrap   (d) GN

Fig. 3. Communities for Dolphin data by different algorithms

lower modularity on these four datasets. However, this does not indicate that our algorithm is not better than the other three algorithms. Figure 2 shows that our algorithm identifies two communities that coincide with the truth that the members of the Karate Club separated into two different groups due to the controversy, and thus the result of our algorithm is actually more reasonable than the other three algorithms even though the modularity values are lower. Figure 4 (b) shows the conductance results. The lower the conductance, the better the communities found. In this case, our algorithm has the lowest conductance on two data sets and highest conductance on the other two data sets. Figure 4 (c) shows the internal density results. The higher the internal density, the better the communities found. In this case, our algorithm has highest internal density in three of the four data sets, and lowest in one data set. Figure 4 (d) shows the cut ratio results. The lower the cut ratio the better the communities found. In this case, our algorithm has the lowest cut ratio in one of the four data set, and the highest in the other three data sets. Figure 4 (e) shows the weighted community clustering results. The higher the WCC, the better the communities found [28]. In this case our algorithm has lower WCC in three of the four data sets. Figure 4 (f) shows the running time. For all the four data sets here, FCD performs fastest among the algorithms. $ParallelFCD$ performs faster than $InfoMap$, $WalkTrap$ and GN on the Email-URV data.

To sum up the results on these four real data sets, our algorithm, FCD and its parallel version, find communities with better values in terms of internal density and conductance, but not with other metrics. However, as we see from the results for Karate Club, the communities detected by our algorithm stay more truthful

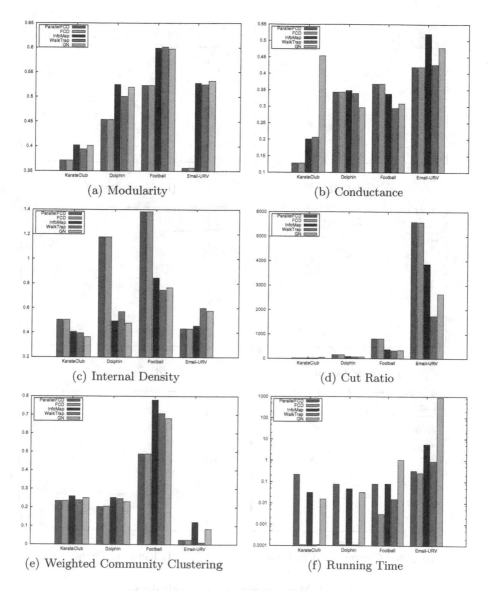

(a) Modularity

(b) Conductance

(c) Internal Density

(d) Cut Ratio

(e) Weighted Community Clustering

(f) Running Time

Fig. 4. Measurements on real world data

than those by the other algorithms. In this sense, our algorithm is effective. From the comparison of the running time, *FCD* is obviously more efficient than the others.

Figure 5 shows the results on the benchmark graphs. It shows the value changing as the graphs increase average degrees. X-axis is the average degree of the graphs. Y-axis is the value of metrics. Each dot represents one metric value for the communities detected by one algorithm. Figure 5 (a) shows the modularity

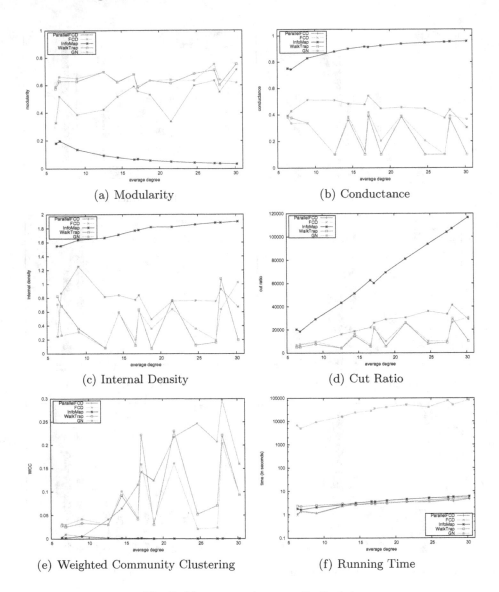

(a) Modularity

(b) Conductance

(c) Internal Density

(d) Cut Ratio

(e) Weighted Community Clustering

(f) Running Time

Fig. 5. Measurements on synthetic data

results. It shows that *WalkTrap* has the highest modularity in general though in some cases *GN* and *FCD* has the highest modularity, and *FCD* has higher modularity than *InfoMap*. Figure 5 (b) shows the conductance results. It shows that *InfoMap* has the highest conductance and *GN* has the the lowest. Figure 5 (c) shows the internal density results. It shows that *InfoMap* has the highest internal density, and *GN* has the lowest density. Figure 5 (d) shows the cut ratio results. It shows that *InfoMap* has the highest cut ratio, and *GN* has the lowest.

Figure 5 (e) shows the *WCC* results. It shows that *FCD* and *WalkTrap* have higher *WCC*, and *InfoMap* and *GN* has lower *WCC*. Figure 5 (f) shows the running time. *FCD* and *ParallelFCD* are shown to be faster in most cases. *GN* is much slower than *InfoMap*, *WalkTrap* and *FCD*. *ParallelFCD* is not obviously faster than *FCD* due to the data communication between the host *CPU* and device *GPU*.

To sum up the results on these synthetic graphs, *FCD* (*ParallelFCD*) performs more stable than *InfoMap* and *GN* in terms of effectiveness. *InfoMap* is the best in terms of internal density but the other three algorithms are better in terms of conductance, cut ratio and *WCC*. *GN* is the best in terms of conductance and cut ratio but the other three algorithms are better in terms of internal density and modularity. Compare with *WalkTrap*, *FCD* is better in most cases in terms of modality and internal density.

FCD is faster than the other three in general. In other words, *FCD* is more efficient.

Another set of experiment tests running time on large networks: Wiki-Vote, Arxiv HEP-PH, Email-Enron, and Epinion network. We sample subgraphs from the networks. Every subgraph contains k percentage vertices of the original networks, where $k = 10, 20, ..., 90$. We run *FCD* and *InfoMap* algorithm on these subgraphs and the original graphs. The running time is recorded. Figure 6 shows

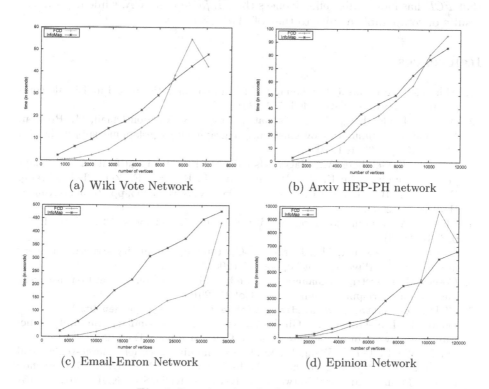

(a) Wiki Vote Network   (b) Arxiv HEP-PH network

(c) Email-Enron Network   (d) Epinion Network

Fig. 6. Running time for large networks

the running time changing as the number of vertices of networks increases. Each figure shows the results for one data set. X-axis is the number of vertices. Y-axis is the time measured in seconds. Due to *WalkTrap* and *GN* algorithms' scalability on large graphs, we only compare *InfoMap* and *FCD* algorithm here. The results show that both algorithms are able to work with graphs with more than 10,0000 vertices. For graphs such as Email-Enron with 33,696 vertices, the algorithms are able to finish the task in a few minutes. In most cases *FCD* is faster than *InfoMap*.

5 Conclusions

In this paper we propose a fast community detection algorithm. It initiates each vertex to seek for the community in its neighbourhood independently. Each vertex chooses its community and peers based on the knowledge of degrees and clustering coefficients of neighbours and the number of common neighbours. The algorithm is parallelizable and thus we devise a GPU version of the algorithm for parallel computation. We empirically evaluate the performance of *FCD*, and compare it with *InfoMap*, *WalkTrap* and *GN* algorithms. We find that *FCD* is the fastest. We assess effectiveness based on the values of modularity, conductance, internal density, cut ratio and weighted community clustering. We find that *FCD* has more stable effectiveness than *InfoMap* and *GN*, while it produces results of comparable quality to that of *WalkTrap*.

References

1. Ahn, Y.-Y., Bagrow, J.P., Lehmann, S.: Link communities reveal multiscale complexity in networks. Nature 466, 761 (2010)
2. Baumes, J., Goldberg, M.K., Krishnamoorthy, M.S., Magdon-Ismail, M., Preston, N.: Finding communities by clustering a graph into overlapping subgraphs. In: IADIS AC, pp. 97–104 (2005)
3. Baumes, J., Goldberg, M., Magdon-Ismail, M.: Efficient identification of overlapping communities. In: Kantor, P., Muresan, G., Roberts, F., Zeng, D.D., Wang, F.-Y., Chen, H., Merkle, R.C. (eds.) ISI 2005. LNCS, vol. 3495, pp. 27–36. Springer, Heidelberg (2005)
4. Clauset, A.: Finding local community structure in networks. Phys. Rev. E 72, 026132 (2005)
5. Clauset, A., Newman, M.E.J., Moore, C.: Finding community structure in very large networks. Phys. Rev. E 70, 066111 (2004)
6. Coscia, M., Rossetti, G., Giannotti, F., Pedreschi, D.: Demon: a local-first discovery method for overlapping communities. CoRR (2012)
7. CUDA-Zone, http://www.nvidia.com/object/what_is_cuda_new.html
8. Danon, L., Diaz-Guilera, A., Giralt, F., Arenas, A.: Self-similar community structure in a network of human interactions. Physical Review E 68 (2003)
9. Du, N., Wu, B., Pei, X., Wang, B., Xu, L.: Community detection in large-scale social networks. In: Proceedings of the 9th WebKDD and 1st SNA-KDD 2007 Workshop on Web Mining and Social Network Analysis, WebKDD/SNA-KDD 2007, pp. 16–25. ACM (2007)

10. Email-URV, `http://deim.urv.cat/~aarenas/data/welcome.htm`
11. Fortunato, S., Lancichinetti, A.: Community detection algorithms: a comparative analysis: invited presentation, extended abstract. In: VALUETOOLS 2009. ICST, Brussels (2009)
12. Gergely Palla, I.F., Derenyi, I., Vicsek, T.: Uncovering the overlapping community structure of complex networks in nature and society. Nature 435, 814–818 (2005)
13. Girvan, M., Newman, M.E.J.: Community structure in social and biological networks. Proceedings of the National Academy of Sciences 99(12), 7821–7826 (2002)
14. Goldberg, M.K., Kelley, S., Magdon-Ismail, M., Mertsalov, K., Wallace, A.: Finding overlapping communities in social networks. In: SocialCom/PASSAT, pp. 104–113 (2010)
15. Gregory, S.: An algorithm to find overlapping community structure in networks. In: Kok, J.N., Koronacki, J., Lopez de Mantaras, R., Matwin, S., Mladenič, D., Skowron, A. (eds.) PKDD 2007. LNCS (LNAI), vol. 4702, pp. 91–102. Springer, Heidelberg (2007)
16. Gregory, S.: A fast algorithm to find overlapping communities in networks. In: Daelemans, W., Goethals, B., Morik, K. (eds.) ECML PKDD 2008, Part I. LNCS (LNAI), vol. 5211, pp. 408–423. Springer, Heidelberg (2008)
17. Harel, D., Koren, Y.: On clustering using random walks. In: Hariharan, R., Mukund, M., Vinay, V. (eds.) FSTTCS 2001. LNCS, vol. 2245, p. 18. Springer, Heidelberg (2001)
18. Jin, D., Yang, B., Baquero, C., Liu, D., He, D., Liu, J.: A Markov random walk under constraint for discovering overlapping communities in complex networks. Journal of Statistical Mechanics: Theory and Experiment (2011)
19. Lancichinetti, A., Fortunato, S., Kertész, J.: Detecting the overlapping and hierarchical community structure in complex networks. New Journal of Physics 11 (2009)
20. Lancichinetti, A., Fortunato, S., Radicchi, F.: Benchmark graphs for testing community detection algorithms. Physical Review E (Statistical, Nonlinear, and Soft Matter Physics) 78(4) (2008)
21. Leskovec, J., Huttenlocher, D., Kleinberg, J.: Predicting positive and negative links in online social networks. In: Proceedings of the 19th International Conference on World Wide Web. ACM (2010)
22. Leskovec, J., Kleinberg, J.M., Faloutsos, C.: Graph evolution: Densification and shrinking diameters. TKDD 1(1) (2007)
23. Lusseau, D., Schneider, K., Boisseau, O.J., Haase, P., Slooten, E., Dawson, S.M.: The bottlenose dolphin community of doubtful sound features a large proportion of long-lasting associations. Behavioral Ecology and Sociobiology 54(4), 396–405 (2003)
24. Massa, P., Avesani, P.: Trust metrics in recommender systems. In: Computing with Social Trust. Springer, London (2009)
25. Newman, M., Girvan, M.: Finding and evaluating community structure in networks. Phys. Rev. E 69, 026113 (2004)
26. Nicosia, V., Mangioni, G., Carchiolo, V., Malgeri, M.: Extending the definition of modularity to directed graphs with overlapping communities. Journal of statistical Mechanics: Theory and Experiment (2009)
27. Pons, P., Latapy, M.: Computing communities in large networks using random walks. In: Yolum, p., Güngör, T., Gürgen, F., Özturan, C. (eds.) ISCIS 2005. LNCS, vol. 3733, pp. 284–293. Springer, Heidelberg (2005)

28. Prat-Pérez, A., Dominguez-Sal, D., Brunat, J.M., Larriba-Pey, J.-L.: Shaping communities out of triangles. In: Proceedings of the 21st ACM International Conference on Information and Knowledge Management, CIKM 2012. ACM (2012)
29. Rosvall, M., Bergstrom, C.T.: Maps of random walks on complex networks reveal community structure. Proceedings of the National Academy of Sciences of the United States of America 105 (2008)
30. Schaeffer, S.E.: Graph clustering. Computer Science Review 1(1), 27–64 (2007)
31. SNAP, http://snap.stanford.edu/data
32. TrustLet, http://www.trustlet.org/
33. van Dongen, S.M.: Graph clustering by flow simulation. PhD thesis, University of Utrecht (2000)
34. Xie, J., Szymanski, B.K.: Towards linear time overlapping community detection in social networks. In: Tan, P.-N., Chawla, S., Ho, C.K., Bailey, J. (eds.) PAKDD 2012, Part II. LNCS, vol. 7302, pp. 25–36. Springer, Heidelberg (2012)
35. Yan, B., Gregory, S.: Detecting communities in networks by merging cliques. CoRR (2012)
36. Yang, J., Leskovec, J.: Defining and evaluating network communities based on ground-truth. In: Proceedings of the ACM SIGKDD Workshop on Mining Data Semantics, MDS 2012. ACM (2012)
37. Yen, L., Vanvyve, L., Wouters, D., Fouss, F., Verleysen, F., Saerens, M.: Clustering using a random-walk based distance measure. In: Proceedings of ESANN 2005 (2005)
38. Zhang, S., Wang, R.S., Zhang, X.S.: Identification of overlapping community structure in complex networks using fuzzy c-means clustering. Physica A: Statistical Mechanics and its Applications 374(1), 483–490 (2007)

Force-Directed Layout Community Detection

Yi Song and Stéphane Bressan

School of Computing,
National University of Singapore
{songyi,steph}@nus.edu.sg

Abstract. We propose a graph-layout based method for detecting communities in networks. We first project the graph onto a Euclidean space using Fruchterman-Reingold algorithm, a force-based graph drawing algorithm. We then cluster the vertices according to Euclidean distance. The idea is a form of dimension reduction. The graph drawing in two or more dimensions provides a heuristic decision as whether vertices are connected by a short path approximated by their Euclidean distance. We study community detection for both disjoint and overlapping communities. For the case of disjoint communities, we use k-means clustering. For the case of overlapping communities, we use fuzzy-c means algorithm. We evaluate the performance of our different algorithms for varying parameters and number of iterations. We compare the results to several state of the art community detection algorithms, each of which clusters the graph directly or indirectly according to geodesic distance. We show that, for non-trivially small graphs, our method is both effective and efficient. We measure effectiveness using modularity when the communities are not known in advance and precision when the communities are known in advance. We measure efficiency with running time. The running time of our algorithms can be controlled by the number of iterations of the Fruchterman-Reingold algorithm.

1 Introduction

Communities detection is instrumental in fields of study such as sociology [9], biology [8], and marketing [21]. Communities exist when nodes in the network form a group in which they are better connected to each other than to the rest of the network. In this paper, we propose methods for finding communities.

We model a network as a simple graph $G(V, E)$, which is undirected, unweighted and without self-loop. V is a set of vertices. E is a set of edges. Our main idea is to obtain a representation of the graph in a Euclidean space and then cluster the vertices based on the Euclidean distance. This is different from what common graph clustering algorithms in that most of them cluster the graph and detect communities directly or indirectly according to geodesic distance. We use Fruchterman-Reingold's force-directed algorithm (FR) (see [11].) This graph layout approach transforms the connections among vertices based on attractive forces and repulsive forces pulling vertices together and pushing them apart, respectively, into proximity in a Euclidean space. In this way, vertices with more

H. Decker et al. (Eds.): DEXA 2013, Part I, LNCS 8055, pp. 419–427, 2013.

connections are closer while vertices without or with less connections are relatively further from each in the Euclidean space of possibly lower dimension that is intrinsic of the graph. Such a dimension reduction is a good opportunity enables the use of the graph layout and Euclidean distance to heuristically detect communities. While the original FR algorithm is presented for two dimensions, we consider versions for one, two, three and more dimensions and three dimensions. We evaluate the role of the number of dimensions as a parameter of our methods in terms of its impact on effectiveness and efficiency. For disjoint community detection, the data clustering technique we use is k-means clustering (KM), while for overlapping community, we use Fuzzy C-mean clustering (FCM), which can indicate the strength between each vertex and communities, and thus does not restrict each vertex to belonging to one group only. All these algorithms' complexities are not high and neither is FR's. Our method building on these techniques is thus efficient for large social networks.

We evaluate effectiveness by measuring modularity. Modularity is defined based on this idea that edges between nodes in the same community are dense, and are sparse between different communities. To find communities with natural division, modularity is defined as the number of edges falling within groups minus the expected number in an equivalent (same number of edges and the same degree distribution) graph with edges placed at random [19]. For graphs with known community structures, we measure the precision as well by comparing memberships to communities that our approach discovers with those to the known communities.

The rest of the paper is organized as follows. Section 2 introduces the related works on graph clustering and community detection. Section 3 briefly reviews the background and presents our approach to discovery community structures in networks. Section 4 describes the data sets that we use, and present and analyze the results of our experiments. Finally we conclude in Section 5.

2   Related Work

Community detection is a form of graph clustering. Graph clustering methods can be categorized into partition clustering, hierarchical clustering, divisive global clustering, and agglomerative global clustering [23]. Large amount of specific methods are proposed such as Star clustering [3], Repeated Random Walks [17], and Markov Clustering [24].

Some methods are specifically proposed for disjoint community detection. Girvan and Newman, in their pioneering works on community detection ([13,18]), propose a divisive method to identify community. The edges with highest betweenness are removed iteratively, which splits the graph into communities. Clauset et al. in [6] propose a greedy hierarchical agglomerative algorithm which starts from each vertex being a community and then joins two communities at each iteration. Rosvall and Bergstrom in [22] use information theoretic approach to detect community in weighted and directed network.

For overlapping community detection, Du et al. in [7] use maximal cliques for community detecting. They propose an algorithm that enumerates all maximal cliques, finds clustering kernel in each group of the overlapping maximal cliques. Palla et al. in [12] propose the clique percolation method which finds all cliques of size k and communities are detected by finding connected union of k-cliques. It 's based on an assumption that the network must have a large quantity of similar cliques. Chen et al. in [5] detect overlapping communities utilizing concept from game theory. Lancichinetti et al. in [15], considering various networks features, present a method called Order Statistical Local Optimization Method. It can be applied to weighted, directed graphs besides simple graphs. Jierui and Boleslaw in [25] propose the Speaker-listener Label Propagation Algorithm for overlapping community detection in large-scale networks.

3   Algorithm

3.1   Background

Force-Directed Algorithms. The idea of force-directed algorithms is to achieve a "aesthetically pleasing" graph layout by simulating the whole graph as a physical system. Edges in the graph are seen as springs binding vertices. Vertices are virtually pulled closer together or pushed further apart according to physical forces. The positions of the vertices are adjusted and this procedure continues until the the system comes to an equilibrium. In addition, Fruchterman and Reingold's force-directed algorithm [11] aims at even vertex distribution. The authors define the attractive force and the repulsive force as $f_a(d) = d^2/k$ and $f_r(d) = -k^2/d$, where $k = C\sqrt{\frac{area}{number\_of\_vertices}}$, and d is the distance between every pair of vertices. $area$ is the windows size for display the graph.

k-Clustering. K-means clustering [16] partitions objects to k clustering, assigns each object the cluster with the nearest mean and adjusts their membership untill an optimum is reached. As a soft version of k-means, Fuzzy C-means clustering (FCM) [4] assigns each object a fuzzy degree of membership to each cluster. Instead of belonging to only one cluster, objects classified via this algorithm can belong to several clusters with different strengths. As a general version of k-means, Expectation-maximization algorithm (EM) [2] models clusters using statistic distributions. The reason we adopt k-means, rather than EM is that k-means is effective enough for this problem and k-means is more efficient. We experimentally show this in Section 4.

3.2   Algorithm

We propose an algorithm that can systematically enumerate all possible number of clusters and find the configuration with the highest modularity. Therefore, the algorithm iterates by changing the value of k from 1 to $|V|$ which is the number of vertices in the network. We show the changes of modularity with the change of k value. If the number of clusters is prior knowledge, we can set the number of iterations to be 1 to this numeber.

Our method starts from the *FR* algorithm. The inputs for the algorithm are the edges of the graph only. Output is the coordinates of vertices in Euclidean Space. Then we sort the degrees of the vertices and initialize the centers of the clusters for the clustering by the vertices with highest degrees. The idea is that the vertices with high degree have higher chance of being the community centers. The centers may change during the clustering. We refine the clusters after the data clustering in Euclidean space. If there's any vertex that doesn't have any connection with other vertices in the same cluster, or it has less connections inside its cluster than outside its cluster, then it will be grouped to the cluster where it has the maximum number of connections. In other words, this vertex will be grouped to the cluster that has most immediate neighbors. The refinement process may change the number of clusters, which is actually good for those who only roughly know the number of clusters. They can input the maximum number of clusters they believe and let our method find out the exact number of clusters in the network without trying all the value of k from 1 to $|V|$.

Algorithm 1. Force-directed Layout Community Detection Algorithm

---

Input: graph G with n vertices, the number of trials t, $t \leq n$.;
Result: Clusters C_i, $i \in (1, 2, ..., k')$

1 $v = Fruchterman\_Reingold(G)$, $v \in R^{n*2}$, $v = [v_1; v_2; ...; v_n]$;
2 Sort_degree(G);
3 $k \leftarrow 1$;
4 **for** *each $k \leq t$* **do**
5   $\quad$ $C'_i = K$-means(v);
6   $\quad$ $C_i = Refinement(C'_i)$;
7   $\quad$ Calculate modularity and record the maximum;
8 **Return** $C_i, i \in (1, 2, ..., k')$ with the maximum modularity;

Algorithm 2. Refinement

Input: Clusters C_i, $i \in (1, 2, ..., k)$;
Result: Clusters C'_i, $i \in (1, 2, ..., k')$;

1 **for** *i from 1 to k* **do**
2   $\quad$ **for** $v \in C_i$ **do**
3   $\quad\quad$ find the cluster C_j where v has the maximum number of immediate neighbors;
4   $\quad\quad$ **if** $i \neq j$ **then**
5   $\quad\quad\quad$ Cluster v into C_j;
6 **Return** $C'_i, i \in (1, 2, ..., k')$;

We call the above algorithm *FR-KM* for the experiments. The other two versions of the algorithm are similar to *FR-KM* but depend on different clustering methods. We name the one using expectation-maximization algorithm *FR-EM* and the one using fuzzy *c*-means algorithm *FR-FCM*. For *FR-FCM*, there's no refinement of the memberships for the vertices, since we intend to deal with overlapping community.

The modularity we use is the same as [18], defined as $\frac{1}{2m}\Sigma_{i,j\in V}(A_{ij} - \frac{k_ik_j}{2m})\delta$ (c_i, c_j), where $A_{ij} = 1$ if i and j are connected, otherwise $A_{ij} = 0$, and $\delta(c_i, c_j) = 1$ if i and j belong to the same cluster, otherwise $\delta(c_i, c_j) = 0$.

4 Experiment

We conduct experiments on both synthetic and real world graphs including two benchmark graphs for community detection algorithm. The experiments ran on an Inter Core, 2 Quad CPU, 2.83GHz, 2GB machine running Windows 8 OS. The algorithms are implemented in C.

We use a batch of benchmark graphs [14] to evaluate the effectiveness of our method. The real-world benchmark graphs we use are Zachary's Karate Club data and American College Football data. We also test on the Email-URV data set, Wikipedia data set, and Facebook data set. They represent large online social network data. See [1] for detailed description of the data sets ,and results and analysis of overlapping community detection.

4.1 Analysis of Non-overlapping Community Detection

We compare our method to the algorithms of Girvan and Newman(GN) ([13,18]), one of the state-of-the-art algorithms in community detection. Modularity is first proposed in this algorithm. We also compare our method with $Walktrap$ algorithm ([20]) and $InfoMap$ algorithm ([22]), which has been shown to perform quite well for community detection (see [10]).

Table 1. Performance Comparison between FR-EM,FR-KM and GN

	KarateClub		AmericanFootball		EmailURV	
	modularity	running time	modularity	running time	modularity	running time
GN	0.4013	0.016	0.5976	1.014	0.5323	3193.532
$Walktrap$	0.3944	0.0000001	0.6015	0.015	0.5250	0.92
$InfoMap$	0.402038	0.015	0.599176	0.047000	0.521420	5.912000
FR-KM	0.417406	0.020000	0.601731	2.179000	0.542659	15.388000

Table 1 shows the performance of the algorithms. In this comparison, we use the normal two dimension FR algorithm with its iteration equal 400 for Karate-Club and AmericanFootball data and 1000 for EmailURV data. The number of trials is set to 30. For all three graphs, our method produces partitions with highest modularity among the four algorithms. Although $Walktrap$ and $InfoMap$ are faster than our method and GN is faster than our method for smaller graphs, the running time of our method is still tolerable. As the size of graph becomes larger, our method becomes faster. If the number of clusters is known in advance, then the number of trial is 1 instead of 30 that we set. If so, our method takes much less time. GN is much slower for larger graphs. For the other two real-world data sets, Wiki-Vote and Facebook, we are unable to make the comparison due to *GN*'s scalability, but we will show the running time of clustering these two graphs by our method.

Figure 1 shows the performance comparison between multiple dimension *FR-KM* and *GN*. We extend the normal two dimension FR algorithm to one dimension and three, four, five dimensions. We set the number of trials 30. For karate club data, the number of trial is equal to its number of vertices. We run each *FR-KM* with the number of iterations of *FR* changing from 100 to 2000 with interval 100. We find that the larger the number of iterations of *FR-KM*, the longer time it takes. However, the number of iterations of *FR* doesn't have decisive influence on the modularity. This suggests that there is no need to increase the number of iterations to get higher modularity. In terms of dimension, we find that for small graphs, projecting them to one dimension or three dimension may get higher modularity sometimes, but for large graphs, the two dimension *FR-KM* performs best. It's faster and clusters graph with higher modularity. That is why we shall adopt the normal two dimension *FR* in our algorithm when it comes to large graphs. *FR-KM* outperforms in both effectiveness and efficiency with large graphs compared with *GN*.

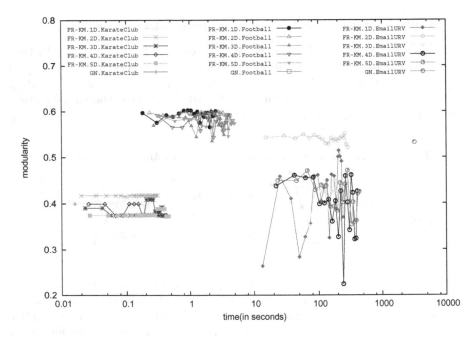

Fig. 1. Performance Comparison between multiple dimension *FR-KM* and *GN*

Figure 2 shows the modularity when the initial input number of clusters, k varies. The final number of clusters may be different from the values of k on X-axis here. Our method changes the number of clusters during cluster refinements, which produces local optimum number of clusters. Therefore, we can see from the result that the trend of the line is horizontal in general. This suggests that even without knowing the number of clusters beforehand, we can find a local optimum around initial k value. This local maximum is probably the global optimum or close to the global optimum.

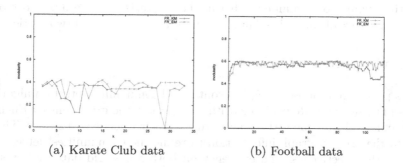

| (a) Karate Club data | (b) Football data |

Fig. 2. Modularity for varying number of clusters

Figure 3 shows the running time for varying number of clusters for Email-URV data, Wiki-Vote data and Facebook data. For each data set, the time for projecting the graph onto Euclidean space is the same, but the clustering time differs. *KM* running time keeps the same in general as the initial number of clusters increases while *EM*'s running time linearly increases as the initial number of clusters increases. Compared with *KM*, *EM* takes much more time. The trends are similar among results for the three data sets.

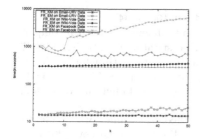

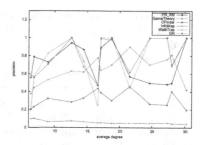

Fig. 3. Running time for varying number of clusters for Email-URV, Wiki-Vote, and Facebook data set

Fig. 4. Precision for varying average degree of synthetic graphs

We compare our method with *GN*, *InfoMap* and *WalkTrap* algorithm, and two other community detection algorithms, CFinder ([12]) and GameTheory algorithm ([5].) Figure 4 shows the precision achieved by the algorithms on the generated graphs with different average degrees. Since the community structures are known, precision is obtained by counting the number of correctly clustered vertices. The results show that our method outperforms the CFinder, *GN* and *InfoMap*, and produces results comparable with GameTheory algorithm and *WalkTrap*. The reason for CFinder having the low precision may be that not every vertex in the graph are clustered. The clusters consists of 3-cliques only in our experiment. The reason for *InfoMap* having the low precision may be that the number of community this method detects is large and most of the communities are of small size. Many communities are of size of two vertices only.

5   Conclusions

In this paper, we propose a graph-layout based community detection algorithm. We use Fruchterman-Reingold algorithm to project the graph onto a Euclidean space and we cluster the vertices according to their Euclidean distance. Then we refer to the original graph information to refine the communities detected. We evaluate the effectiveness and efficiency on both real-world data and synthetic data. For disjoint community detection, the results show that *FR-KM* is more effective on both small graphs and large graphs than *GN*, and is much more efficient than *GN* on large networks. *FR-KM* is also more effective than *Walktrap* and *InfoMap* algorithms in terms of modularity. Compared with *GN*, *CFinder*, *InfoMap*, *WalkTrap* and Gametheory algorithms on the synthetic graphs with known communities in advance, our method is more effective than *GN* and *CFinder* and has good performance comparable with *WalkTrap* and Game Theory algorithm according to the results of precision. For overlapping detection, the result for Karate Club data shows that *FR-FCM* is reasonably effective.

References

1. https://dl.comp.nus.edu.sg/dspace/handle/1900.100/4144
2. Dempster, A.P., Laird, N.M., Rubin, D.B.: Maximum likelihood from incomplete data via the em algorithm. Journal of the Royal Statistical society, Series B (1977)
3. Aslam, J.A., Pelekhov, E., Rus, D.: The star clustering algorithm for static and dynamic information organization. J. Graph Algorithms Appl. 8, 95–129 (2004)
4. Bezdek, J.C.: Pattern Recognition with Fuzzy Objective Function Algorithms. Kluwer Academic Publishers, Norwell (1981)
5. Chen, W., Liu, Z., Sun, X., Wang, Y.: A game-theoretic framework to identify overlapping communities in social networks. Data Min. Knowl. Discov. (2010)
6. Clauset, A., Newman, M.E.J., Moore, C.: Finding community structure in very large networks. Phys. Rev. E 70, 066111 (2004)
7. Du, N., Wu, B., Pei, X., Wang, B., Xu, L.: Community detection in large-scale social networks. In: WebKDD/SNA-KDD, ACM (2007)
8. Eisen, M.B., Spellman, P.T., Brown, P.O., Botstein, D.: Cluster analysis and display of genome-wide expression patterns. Proc. Natl. Acad. Sci. U.S.A. (1998)
9. Etling, B., Kelly, J., Faris, R., John, P.: Mapping the arabic blogosphere: Politics, culture, and dissent. Berkman Center Research Publication (2006-06) (2009)
10. Fortunato, S., Castellano, C.: Community structure in graphs. In: Encyclopedia of Complexity and Systems Science, pp. 1141–1163 (2009)

11. Fruchterman, T.M.J., Reingold, E.M.: Graph drawing by force-directed placement. Softw. Pract. Exper. 21(11), 1129–1164 (1991)
12. Gergely Palla, I.F., Derenyi, I., Vicsek, T.: Uncovering the overlapping community structure of complex networks in nature and society. Nature (2005)
13. Girvan, M., Newman, M.E.J.: Community structure in social and biological networks. Proceedings of the National Academy of Sciences (2002)
14. Lancichinetti, A., Fortunato, S., Radicchi, F.: Benchmark graphs for testing community detection algorithms. Physical Review E 78 (2008)
15. Lancichinetti, A., Radicchi, F., Ramasco, J.J., Fortunato, S.: Finding statistically significant communities in networks. CoRR (2010)
16. Lloyd, S.P.: Least squares quantization in pcm. IEEE Transactions on Information Theory 28, 129–137 (1982)
17. Macropol, K., Can, T., Singh, A.K.: Rrw: repeated random walks on genome-scale protein networks for local cluster discovery. BMC Bioinformatics (2009)
18. Newman, M., Girvan, M.: Finding and evaluating community structure in networks. Phys. Rev. E 69, 026113 (2004)
19. Newman, M.E.J.: Modularity and community structure in networks. Proceedings of the National Academy of Sciences 103(23), 8577–8582 (2006)
20. Pons, P., Latapy, M.: Computing communities in large networks using random walks. In: Yolum, p., Güngör, T., Gürgen, F., Özturan, C. (eds.) ISCIS 2005. LNCS, vol. 3733, pp. 284–293. Springer, Heidelberg (2005)
21. Reddy, P.K., Kitsuregawa, M., Sreekanth, P., Rao, S.S.: A graph based approach to extract a neighborhood customer community for collaborative filtering. In: Bhalla, S. (ed.) DNIS 2002. LNCS, vol. 2544, pp. 188–200. Springer, Heidelberg (2002)
22. Rosvall, M., Bergstrom, C.T.: Maps of random walks on complex networks reveal community structure. Proceedings of the National Academy of Sciences of the United States of America 105 (2008)
23. Schaeffer, S.E.: Graph clustering. Computer Science Review 1(1), 27–64 (2007)
24. van Dongen, S.: Graph Clustering by Flow Simulation. PhD thesis (2000)
25. Xie, J., Szymanski, B.K.: Towards linear time overlapping community detection in social networks. In: Tan, P.-N., Chawla, S., Ho, C.K., Bailey, J. (eds.) PAKDD 2012, Part II. LNCS, vol. 7302, pp. 25–36. Springer, Heidelberg (2012)

On the Composition of Digital Licenses
in Collaborative Environments

Marco Mesiti, Paolo Perlasca, and Stefano Valtolina

DI, Department of Computer Science, University of Milano, Italy
{mesiti,perlasca,valtolina}@di.unimi.it

Abstract. In the era of Web 2.0, users are not any longer just consumers of resources but they can actively produce, share and modify content, by composing and enhancing digital resources and services. In this context, the intellectual property of the users collaborating in authoring activities should be preserved. Starting from a model for digital licences generation and management useful in collaborative environments like the Web 2.0, in this paper we propose the algorithms of a DRM component responsible for the composition and modification of digital resources and the generation of the related licenses. Then, the paper presents a compliant architecture based on a composition of web services.

1 Introduction

The complexity and expanding scale of most collaborative projects being carried out nowadays in the context of Web 2.0 require more cooperation among users in the production of digital contents and services. We are also observing the generation of communities of users (belonging to the same company or group of interest) working together for the creation of new resources (like wikies, social networks, and mashups). Users are not any longer just consumers of resources but they can actively produce, share and modify content/services eventually created by other users. In this context, the intellectual property of the users collaborating in the authoring activities should be preserved.

Different Rights Expression Languages (like ccREL [6], ODRL [8], MPEG-21 REL [10]) have been develop for the specification of licenses to preserve the intellectual property. These proposals differ from the scope and the granularity according to which it is possible to specify and manage each aspect directly or indirectly related to the license specification and management processes. Digital Rights Management (DRM) systems enable the creation, adaptation, distribution and consumption of multimedia contents and services according to the permissions and constraints specified by the content creators or rights issuers [1, 13]. MPEG-21 REL [10] and ODRL [8] natively support the specification of rights for the modification and composition of resources. Few approaches for the specification of licenses of composed resources have been proposed in the context of the creative common licences expressed by ODRL [3–5]. However, these licences have a different purpose than those created by MPEG-21 REL

H. Decker et al. (Eds.): DEXA 2013, Part I, LNCS 8055, pp. 428–442, 2013.

and current DRM systems are mainly tailored for the protection of resources rather than for supporting users in the composition process.

This paper provides the building-blocks for the realization of a tool helping an user to visualize his/her own resources (or the resources created by the communities she/he belongs to), to compose resources, and to generate related licenses. We first introduce a formal model that is compliant with the MPEG-21 REL for the representation of licenses as collection of grants. By abstracting many verbose details of MPEG-21, we create the basis for reasoning on the composition issues. Moreover, the model points out different approaches for specifying the principal of a grant (either by specifying a single user identifier, a group of user identifiers or a predicate to be evaluated by considering user certificates) that can be very useful in our collaborative environments to reduce the number of licenses to be generated and, at the same time, to make a broader usage of digital resources. Then, we provide an approach for evaluating the weak and strong compatibility of two grants that is the basis for the composition and update of resources. In the evaluation of weak compatibility, the user profile and the conditions of grants are not considered. This is useful in the early stage of composition design to quickly checking the basic conditions of composition without loosing time in the evaluation of the user profile and also when external services needed for their evaluation are unavailable. Afterward, to actually enable the user to compose resources, and to generate the final license, user profile and conditions of grants are taken into account in the evaluation of strong compatibility. We finally propose an architecture supporting the composition of resources and the generation of a new license based on the components' licenses. Key features of this architecture are the weak and strong compatibility service for the two-steps evaluation of licenses compatibility and the process of resources aggregation and generation of the corresponding license.

The paper is organized as follows. Section 2 presents the license data model and how a license is evaluated. Section 3 deals with the issue of checking whether two grants are compatible for composition and can be exercised at the current time. Section 4 provides the basic algorithms for the generation of a new license when resources are composed or updated, whereas Section 5 deals with the enabling architecture. Related work and concluding remarks are finally presented.

2   License Data Model

In this section we provide a formal model for the representation of licenses that supports the specification of communities to whom a grant is released. Finally, we discuss the mechanism for the evaluation of an access request.

Principals and Issuers. The principals to whom rights are granted can be specified through the user identifiers or relying on the possession of a given certificate. For the sake of simplicity, in our model the fact that users hold certificates are represented through predicates. Such predicates are verified on a given user, if and only if she/he holds the corresponding certificate released

$$u \downarrow p \text{ iif } \begin{cases} p = u & p \in \mathcal{U} \\ u \in p & p \text{ is a set} \\ p(u) & isPred(p) \end{cases} \qquad p_1 \sqcap p_2 \text{ iif } \begin{cases} p_1 = p_2 & p_1, p_2 \in \mathcal{U} \\ p_1 \cap p_2 \neq \emptyset & p_1, p_2 \text{ are sets} \\ p_1 \in p_2 & p_1 \in \mathcal{U} \text{ and } p_2 \text{ is a set} \\ p_2 \in p_1 & p_2 \in \mathcal{U} \text{ and } p_1 \text{ is a set} \\ isPred(p_1) \vee isPred(p_2) \end{cases}$$

(a)   (b)

Fig. 1. Predicates on Principal specifications

by a Credential Authority. A predicate can also be evaluated on a set of users and, in this case, it is verified when each user in the set holds the corresponding certificate. The use of certificates in our collaborative environment is particularly useful to specify with a single license the community of users that can exercise a given right. In the remainder, $\mathcal{U}$ denotes the set of user identifiers.

Definition 1 (Principal). *The principal of a license can be: a unique identifier associated with a user, a set of user identifiers, or a predicate* m. *A user* i *satisfies the predicate* m, *if and only if the user* i *holds the corresponding certificate* m.

Example 1. Let $\{i_1, ..., i_{10}\} \subseteq \mathcal{U}$ be a set of user identifiers. The principal of a license can be: i_1, the set $\{i_3, i_5, i_7\}$, or uniMi, where uniMi is a predicate assessing the employment at the University of Milan. In the paper we will use the user identifiers Alice, Bob, and Tom. Alice and Tom are uniMi employees. □

The user issuing a license is named *issuer*. The issuer of a license is always a single user identifier. With $\mathcal{I}$ we denote the set of issuers, $\mathcal{I} \subseteq \mathcal{U}$. Given a principal specification p, we might need to establish whether a user u is a principal. The predicate $\downarrow$ is satisfied in the cases reported in Figure 1(a) (function *isPred* is true when the argument is a predicate). Moreover, the predicate $\sqcap$ in Figure 1(b) is used to weakly identify when two principal specifications p_1 and p_2 can have non empty intersection. Since the evaluation of predicates requires to access the user profile, when p_1 or p_2 is a predicate, $\sqcap$ is considered verified, and its actual evaluation deferred when the strong conditions are taken into account.

Resources. A Resource can be a digital work (such as an e-book, an audio file, or an image), a service (such as an email service or a transaction service), or even a fragment of information characterizing a principal (such as a name or an email address). Resources can be the aggregation of different ones. In order to allow the specification of digital licenses at different granularities (from the entire resource to one of its component), resources can be represented through a tree. Internal nodes are labeled by resource identifiers, whereas leaves are the resources. Resource identifiers are exploited as references in digital licenses. We denote with $\mathcal{O}$ the set of resources, and with $\hat{\mathcal{O}}$ the corresponding identifiers.

Example 2. Suppose that o_{txt} and o_{img} are resources representing a text and a picture. The resource o_c obtained by their concatenation is denoted, by adopting a JSON notation [12], as $\{\hat{o}_c : \{\hat{o}_{\text{txt}} : o_{\text{txt}}, \hat{o}_{\text{img}} : o_{\text{img}}\}\}$. □

Rights. A principal can be granted to exercise a right against a resource. Typically, a right specifies an action (or activity) or a class of actions that a principal may perform on or using the associated resource. Rights can be classified in 3 types: *use*, through which the principal can `play`, `print`, `execute` a resource; *manage*, through which the principal can `install`, `uninstall`, `move` or `delete` a resource; *transformation*, through which the principal is authorized to manipulate the resource. The MPEG-21 transformation rights we deal with are: `reduce`, `enlarge`, `modify`, `diminish`, `enhance`, `adapt` and `embed`. These rights present a subsumption relationship existing among them: the `modify` right subsumes the `reduce` and `enlarge` rights. Another subsumption exists between the transformation rights and the `play` right because, in order to transform a resource, the user should be able to access/play the resource itself. The `modify` right allows one to apply any modification to a given resource, whereas `reduce` and `enlarge` allow a specific kind of modification. The rights `adapt`, `diminish`, and `enhance` present the same semantics and subsumption relationships of the previous three rights but their application produces a new modified resource, and leave the original ones unaltered. The `embed` right allows one to attach or include another resourse in a given resource. This right is important in order to correctly operate a composition among resources. With $\mathcal{R}$ we denote the set of rights.

Conditions. MPEG-21 provides a set of conditions for the verification of terms and obligations under which rights can be exercised by a given principal. Let NC be the set of name of MPEG-21 conditions we consider.

Example 3. `ExerciseLimit`, `ValidityInterval`, and `FeeFlat` are samples of names of conditions available in MPEG-21, representing: the number of times a given right can be exercised, temporal interval of right validity, and the obligation of the payment of a fee, respectively. □

Each condition name can be associated with a set of basic constraints representing limitations (like temporal and spatial constraints) that need to be verified to consider the condition satisfied. Let NP_c be the set of property names associated with the condition identified by the condition name $c \in NC$, a constraint is $np_c \ op \ v$, where: $np_c \in NP_c$ is the name of a property, $op \in \mathcal{OP} = \{<, >, \leq, \geq, =, \neq, \in, \notin\}$ is a comparison operator, and v is a valid value for the property name np_c (the set of legal values for a property name np_c is denoted $\mathcal{V}_{np_c}$).

Definition 2 (Condition). *Let* $c \in NC$ *be a condition name,* $\{np_{c_1}, \ldots, np_{c_m}\} \subseteq NP_c$. *A condition* $\bar{c}$ *is a pair* $(c, \{\langle np_{c_1} op_1 v_1 \rangle, \ldots \langle np_{c_m} op_m v_m \rangle\})$, *where* $v_i \in \mathcal{V}_{np_{c_i}}$, $op_i \in \mathcal{OP}$, $1 \leq i \leq m$

Example 4. Consider the condition names presented in Example 3. A property, named `count`, can be specified for `ExerciseLimit`, representing the number of times a right can be exercised. Therefore, the following condition $c_1 = (\texttt{ExerciseLimit}, \{\langle \texttt{count} = 5 \rangle\})$ states that the associated right can be exercised up to five times. The condition $c_2 = (\texttt{ValidityInterval}, \{\langle \texttt{time} > "2012\text{-}01\text{-}01T00{:}00{:}00", \texttt{time} < "2013\text{-}01\text{-}01T00{:}00{:}00"\rangle\})$ represents a validity time interval between midnight of January 1, 2012 and, midnight of January 1 2013. The condition $c_3 = (\texttt{FeeFlat}, \{\langle \texttt{amount} = 5 \rangle\})$ is satisfied whenever the principal has paid the amount of 5.   □

Grants and Licenses. A grant describes the terms of a license. In the following to identify the ith component of a tuple t, we use the notation $t[i]$.

Definition 3 (Grant). *A grant g is a 4-tuple $\langle p, r, \hat{o}, C \rangle$ where p is a principal, $r \in \mathcal{R}$ is a right, $\hat{o} \in \hat{\mathcal{O}}$ is a resource identifier, and C is a set of conditions.*

Example 5. Suppose that $o_{\texttt{txt}}$, $o_{\texttt{img}}$ and $o_{\texttt{mpg}}$ are resources representing a text, a picture and a piece of music. Moreover, c_1, c_2, c_3 are the conditions specified in Example 4. The following grants specify that user $\texttt{Alice}$ can play $o_{\texttt{img}}$ without restrictions, she can adapt, play and print $o_{\texttt{txt}}$ under the satisfaction of condition c_1 and c_2. Moreover $\texttt{Tom}$, $\texttt{Bob}$ and users belonging to $\texttt{uniMi}$ can embed $o_{\texttt{mpg}}$ under the satisfaction of condition c_3. $\texttt{Tom}$ can reduce and print, like $\texttt{Alice}$, $o_{\texttt{txt}}$ under the satisfaction of condition c_1 and c_2.

$$g_1 = \texttt{< Alice, play, } \hat{o}_{\texttt{img}}, \{\} \texttt{ >} \qquad g_2 = \texttt{< Alice, adapt, } \hat{o}_{\texttt{txt}}, \{c_1, c_2\} \texttt{ >}$$
$$g_3 = \texttt{< Alice, play, } \hat{o}_{\texttt{txt}}, \{c_1, c_2\} \texttt{ >} \qquad g_4 = \texttt{< Alice, print, } \hat{o}_{\texttt{txt}}, \{c_1, c_2\} \texttt{ >}$$
$$g_5 = \texttt{< uniMi, embed, } \hat{o}_{\texttt{mpg}}, \{c_3\} \texttt{ >} \qquad g_6 = \texttt{< \{Tom, Bob\}, embed, } \hat{o}_{\texttt{mpg}}, \{c_3\} \texttt{ >}$$
$$g_7 = \texttt{< Tom, print, } \hat{o}_{\texttt{txt}}, \{c_1, c_2\} \texttt{ >} \qquad g_8 = \texttt{< Tom, reduce, } \hat{o}_{\texttt{txt}}, \{c_1, c_2\} \texttt{ >}\ \square$$

With $\mathcal{G}$ we denote the set of grants. A License is conceptually a container of grants expressing the rights that can be exercised on the identified resources.

Definition 4 (License). *A license l is a pair $< i, G >$, where: $l[1] = i \in \mathcal{I}$ is the party issuing the license, and $l[2] = G = \{g_1, \ldots, g_n\} \subset \mathcal{G}$ is a set of grants.*

In the following, $\mathcal{L}$ denotes the set of licenses, $\mathcal{L}_u$ is the set of licenses of user u, $G_u = \{g \in \bigcup_{l[2] \in \mathcal{L}_u} \mid u \downarrow g[1]\}$ is the set of grants issued to the user u, $G_u^o = \{g \in G_u \mid g[3] = \hat{o}\}$ denotes the set of grants of user u that refers to resource o, and $G_u^{r,o} = \{g \in G_u^o \mid g[2] = r\}$ is the subset of G_u^o referring to the right r. These sets contain all the grants referring the resource o and the resources that o aggregates.

Example 6. Consider the grants of Example 5. $\texttt{Bob}$ can issue the following licenses to $\texttt{Alice}$, $\texttt{Tom}$ and to himself respectively.

- $l_1 = \texttt{< Bob}, \{g_1, g_2, g_3, g_4, g_5\} \texttt{ >} \in \mathcal{L}_{\texttt{Alice}}$
- $l_2 = \texttt{< Bob}, \{g_5, g_6, g_7, g_8\} \texttt{ >} \in \mathcal{L}_{\texttt{Tom}}$
- $l_3 = \texttt{< Bob}, \{g_6\} \texttt{ >} \in \mathcal{L}_{\texttt{Bob}}$

Relying on these licenses, we can compute the following sets.

- $\mathcal{L} = \{l_1, l_2, l_3\}, \mathcal{L}_{\texttt{Alice}} = \{l_1\}, \mathcal{L}_{\texttt{Tom}} = \{l_2\}, \mathcal{L}_{\texttt{Bob}} = \{l_3\}$
- $G_{\texttt{Alice}} = \{g_1, g_2, g_3, g_4, g_5\}, G_{\texttt{Tom}} = \{g_5, g_6, g_7, g_8\}, G_{\texttt{Bob}} = \{g_6\}$
- $G_{\texttt{Alice}}^{\hat{o}_{\texttt{mpg}}} = \{g_5\}, G_{\texttt{Tom}}^{\hat{o}_{\texttt{mpg}}} = \{g_5, g_6\}, G_{\texttt{Bob}}^{\hat{o}_{\texttt{mpg}}} = \{g_6\}$ □

License Evaluation. The *user profile* is used to store the information of context (for example, her/his location and time of execution), the state of given constraints, and the certificates he/she holds. This information is used when granting the access to resources The profile management will be presented when discussing the system architecture.

Definition 5 (User Profile). *Let* $u \in \mathcal{U}$ *be an user. The profile of user* u, *denoted* $\mathcal{P}ro(u)$, *is a set of tuples* $(c, np_{c_j} w_j)$, *where* $c \in NC$, $np_c \in NP_c$ *and* $w_j \in \mathcal{V}_{np_{c_j}}$, *and a set of certificates asserting the partecipation to a community whose validity are certified by a Credential Authority.*

In the following, $[\![np_{c_j}]\!]_{\mathcal{P}ro(u)} = w_j$ denotes that the current value for the property np_{c_j} in the profile of u is w_j, whereas $eval_u$ is a predicate for the evaluation of a constraint/grant for a user u.

Definition 6 (Condition and Grant Evaluation). *Consider* $u \in \mathcal{U}$ *and a condition* $(c, \{\langle np_{c_1} op_1 v_1 \rangle, \ldots \langle np_{c_m} op_m v_m \rangle\})$. c *is satisfied wrt* u *if and only if:*

$$eval_u(c) = \bigwedge_{i=1}^{m} eval_u([\![np_{c_i}]\!]_{\mathcal{P}ro(u)} op_i v_i) = \text{true}$$

A grant $g = \langle p, r, \hat{o}, C \rangle$ *is satisfied w.r.t.* u *if and only if:*

$$eval_u(g) = u \downarrow p \wedge \bigwedge_{c \in C} eval_u(c) = \text{true}$$

Given an authorization request $<u, r, \hat{o}>$, representing the request of user $u \in \mathcal{U}$ to exercise the right $r \in \mathcal{R}$ on the resource $o \in \mathcal{O}$, we now present its evaluation.

Definition 7 (Authorization Request Evaluation). *Let* $ar = < u, r, \hat{o} >$ *be an authorization request and* G_u *the set of grants issued for* u. ar *is granted to* u, *denoted by* $[\![ar]\!]$, *if and only if* $\exists g = \langle p, r, \hat{o}, C \rangle \in G_u$ *such that* $eval_u(g) = \text{true}$.

3 Grant Compatibility

In this section, we first discuss whether two rights are compatible, that is, whether they can be applied at the same time either on the same resource or on a pair of resources. Then, we present the notion of grant compatibility. We differentiate between weak and strong grant compatibility because of performance issues, the latter notion is more complex to be evaluated (and it is useless to compute if the licenses are not weak compatible). Without loss of generality, we restrict ourselves to pairs of licenses.

The rights compatibility notion depends on the context (same resource or different resources) on which the rights have been specified. In order to introduce this notion, we first need to establish when two rights are in *conflict*, that is, when they cannot be exercised at the same time by the same user. The pairs of privileges (`enlarge, reduce`) and (`enhance, diminish`) are in conflict because it does not make sense to apply at the same time, or within the same transaction, two opposite privileges on the same resource. Moreover, since these pairs of privileges are subsumed by the corresponding modify/adapt privilege, the grant issuer, wishing to grant a wider rights to the principal, can exploit this privilege instead of the conflicting privileges. The pairs of privileges (embed, reduce) and (embed, diminish) are in conflict when specified on different resources, because it is not possible to embed the first resource into the second one if we are only allowed to reduce/diminish the second resource.

We also need to introduce the notion of *non comparable rights*. Two rights are non comparable on different resources, when the possibility to exercise the first right does not influence the possibility to exercise the second one. The rights $\mathcal{R}_T = \{$reduce, enlarge, modify, diminish, enhance, adapt$\}$, that is, the *transformation* rights with the exception of embed, are non comparable among each others. Formally, each pair in $\mathcal{R}_T \times \mathcal{R}_T$ is not comparable.

Definition 8 (Rights Compatibility). *Let $(r_i, r_j) \in \mathcal{R} \times \mathcal{R}$ be a pair of rights. r_i is compatible with r_j, if and only if: when specified on the same resource, they are not in conflict, i.e., $(r_i, r_j) \notin \{$(enlarge, reduce), (enhance, diminish)$\}$; when specified on different resource, they are not in conflict and are comparable, i.e. $(r_i, r_j) \notin \{$(embed, reduce), (embed, diminish)$\} \cup \mathcal{R}_T \times \mathcal{R}_T$.*

Relying on the notion of rights compatibility, we can introduce the concept of weak grants compatibility.

Definition 9 (Weak Grants Compatibility). *Let $g_i = \langle p_i, r_i, \hat{o}_i, C_i \rangle$ and $g_j = \langle p_j, r_j, \hat{o}_j, C_j \rangle \in \mathcal{G}$ be two grants. g_i is weak compatible with g_j ($g_i \simeq g_j$) if and only if: $p_i \sqcap p_j$ and r_i is compatible with r_j.*

The fact that an user holds grants that are weak compatible does not mean that she/he can exercise them on the resources. The user profile should be considered and the grants that are weak compatible evaluated by taking the user profile into account.

Definition 10 (Strong Grants Compatibility). *Let g_i and g_j be two weak compatible grants ($g_i \simeq g_j$), $\mathcal{P}ro(u)$ the profile of user u. g_i and g_j are strong compatible w.r.t. u ($g_i \approx_u g_j$) if and only if $eval_u(g_i) = eval_u(g_j) = $ true*

Example 7. Consider the license l_2 of Example 6. It follows that
$$g_5 \simeq g_6, \; g_5 \simeq g_7, \; g_6 \simeq g_7, \; g_7 \simeq g_8 \qquad \text{and} \qquad g_5 \not\simeq g_8, \; g_6 \not\simeq g_8$$
Suppose now that $eval_{\text{Tom}}(c_1) = eval_{\text{Tom}}(c_3) = $ true whereas $eval_{\text{Tom}}(c_2) = $ false. Consequently, only $g_5 \approx_{\text{Tom}} g_6$ holds. □

4   License Composition

The creation of a new license for the composition of two resources or for a modified resource can be realized by exploiting the License Generation Service of our architecture (details in the next section). Therefore, a request of license

---

Algorithm 1. The `compose` licenses request

Input: o_a a resource, o_b a resource, $r \in \{\texttt{enlarge}, \texttt{modify}, \texttt{enhance}, \texttt{adapt}, \texttt{embed}\}$ a right, o_c the composed resourse, u a user

1: **if** $\exists g_a \in G_u^{\texttt{embed}, o_a}$ and $\exists g_b \in G_u^{r, o_b}$ s.t. $g_a \approx_u g_b$ **then**
2:      **if** $r \in \{\texttt{enhance}, \texttt{adapt}\}$ **then**
3:          Generate the license $l_c = <u, G_u^{o_a} \cup G_u^{o_b}[{}^{o_c}/{}_{o_b}]>$
4:      **else**
5:          **if** $r = \texttt{embed}$ **then**
6:              Generate the license $l_c = <u, G_u^{o_a} \cup G_u^{o_b} \cup \{(u, \texttt{play}, \hat{o}_c, \{\})\}>$
7:          **else**
8:              Generate the license $l_c = <u, G_u^{o_a} \cup G_u^{o_b}>$
9:          **end if**
10:     **end if**
11: **end if**

Output: The generated license (or denial to generate a license)

---

generation should be sent to this service. We consider the following kinds of basic license generation requests: `compose`, `update`, `add`, `remove`. The first two are used when the user wishes to automatically generate a license for the composition of two resources, or the modification of a single resource. By contrast, the last two are used to specify extra grants for new added resources, or to remove grants that do not apply any longer to the modified resource.

For the sake of understandability we present the algorithms for the evaluation of the basic licenses generation requests. However, the approach can be easily extended to consider more sophisticated sequences of combinations and modifications of resources. In these cases, the actual generation of the license can be postponed at the end of the sequence of basic modification operations. We also remark that analogous algorithms have been developed for simply checking whether there are the conditions for the generation of a new license. These checking algorithms simply exploit the weak compatibility notions instead of the strong compatibility notions discussed in the previous section. In the remainder with the notation $G[{}^{o'}/{}_o]$ we mean that in the licenses contained in G, all the references to the resource o are substituted with the reference o'.

Algorithm 1 generates the new license when there is a `compose` licenses request. Whenever an user u wishes to compose a resource o_a with a resource o_b, she/he should at least hold an `embed` right on o_a and a right r that allows to update (or append information to) the resource o_b. If $r \in \{\texttt{enhance}, \texttt{adapt}\}$, the resource o_b should be left unchanged and a new resource, named o_c, should be generated. Otherwise, o_c is o_b itself. Whenever user u holds the grants for performing the composition operation, a new license is generated containing the union of the grants user u holds on the original resources. In case a new resource is generated, the references to o_b occurring in the grants should be substituted with the references to o_c (line 3 in Algorithm 1). If $r = \texttt{embed}$, the resource o_c contains the concatenation of o_b and o_a. In this case we need to introduce a grant that allows the user u to access the container o_c, otherwise the access to the components would be forbidden.

Algorithm 2 is used to generate a new license when a resource is modified through a `enhance`/`adapt`/`diminish` right. For the other transformation rights

---

Algorithm 2. The update licenses request

Input: o a resource, $r \in \{\text{enhance}, \text{adapt}, \text{diminish}\}$ a right, o_c the modified resourse, u a user
1: **if** $\exists g \in G_u^{r,o}$ s.t. $[\![<u, r, o>]\!]$ **then**
2:     Generate the license $l_c = <u, G_u^o[{}^o\!/\!_o]>$
3: **end if**
Output: The generated license (or denial to generate a license)

---

there is no need to generate a new license (the original license is still valid). Since a new resource o_c is generated, we need to substitute the references to o_b occurring in the grants with the references to o_c.

Example 8. Let $o_{txt} = \{\hat{o}_{txt} : \{\hat{o}_{txt_1} : o_{txt_1}, \hat{o}_{txt_2} : o_{txt_2}\}\}$ and $o_{txt_3} = \{\hat{o}_{txt_3} : \{\hat{o}_{txt_4} : o_{txt_4}, \hat{o}_{txt_5} : o_{txt_5}\}\}$ be two resources, $l_4 = <\text{Bob}, \{g_9, g_{10}, g_{11}, g_{12}\} > \in \mathcal{L}_{\text{Tom}}$ be another license issued from Bob to Tom consisting of the following grants:

$$g_9 = <\text{Tom}, \text{enlarge}, \hat{o}_{txt}, \{c_1, c_2\} > \qquad g_{10} = <\text{Tom}, \text{embed}, \hat{o}_{txt_3}, \{c_1, c_2\} >$$
$$g_{11} = <\text{Tom}, \text{embed}, \hat{o}_{txt}, \{\} > \qquad g_{12} = <\text{Tom}, \text{enhance}, \hat{o}_{txt}, \{\} >$$

Suppose that Tom wishes to compose o_{txt_3} into o_{txt} by enhancing, enlarging or embedding the latter. At the time of the request, we have that: $\mathcal{L}_{\text{Tom}} = \{l_2, l_4\}$, $G_{\text{Tom}}^{\text{embed}, \hat{o}_{txt_3}} = \{g_{10}\}$, $G_{\text{Tom}}^{\text{enlarge}, \hat{o}_{txt}} = \{g_9\}$, $G_{\text{Tom}}^{\text{modify}, \hat{o}_{txt}} = G_{\text{Tom}}^{\text{adapt}, \hat{o}_{txt}} = \emptyset$, $G_{\text{Tom}}^{\text{enhance}, \hat{o}_{txt}} = \{g_{12}\}$, $G_{\text{Tom}}^{\text{embed}, \hat{o}_{txt}} = \{g_{11}\}$. Finally, suppose that $eval_{\text{Tom}}(c_1) = eval_{\text{Tom}}(c_2) = \textbf{true}$. Consequently, $g_{10} \cong_{\text{Tom}} g_9$, $g_{10} \cong_{\text{Tom}} g_{11}$ and $g_{10} \cong_{\text{Tom}} g_{12}$ are valid and thus, in all cases, the Tom's requests can be satisfied. Referring to Algorithm 1, if Tom requires to compose o_{txt_3} into o_{txt}

by enlarging o_{txt}. The structure of the resulting updated resource o_{txt} is $o_{txt} = \{\hat{o}_{txt} : \{\hat{o}_{txt_1} : o_{txt_1}, \hat{o}_{txt_2} : o_{txt_2}, \hat{o}_{txt_3} : \{\hat{o}_{txt_4} : o_{txt_4}, \hat{o}_{txt_5} : o_{txt_5}\}\}\}$ whereas the generated license is $l_5 = <\text{Tom}, \{g_7, g_8, g_9, g_{10}, g_{11}, g_{12}\} > \in \mathcal{L}_{\text{Tom}}$.

by embedding o_{txt}. A new resource o_{txt_6} is generated and its structure is $o_{txt_6} = \{\hat{o}_{txt_6} : \{\hat{o}_{txt} : \{\hat{o}_{txt_1} : o_{txt_1}, \hat{o}_{txt_2} : o_{txt_2}\}, \hat{o}_{txt_3} : \{\hat{o}_{txt_4} : o_{txt_4}, \hat{o}_{txt_5} : o_{txt_5}\}\}\}$. A new license is generated for o_{txt_6}: $l_6 = <\text{Tom}, \{g_7, g_8, g_9, g_{10}, g_{11}, g_{12}, g_{13}\} > \in \mathcal{L}_{\text{Tom}}$, where $g_{13} = <\text{Tom}, \text{play}, \hat{o}_{txt_6}, \{\} >$ is a new grant that allow Tom to access to the concatenated resources.

by enhancing o_{txt}. A new resource o_{txt_7} is generated and its structure is $o_{txt_7} = \{\hat{o}_{txt_7} : \{\hat{o}_{txt_1} : o_{txt_1}, \hat{o}_{txt_2} : o_{txt_2}, \hat{o}_{txt_3} : \{\hat{o}_{txt_4} : o_{txt_4}, \hat{o}_{txt_5} : o_{txt_5}\}\}\}$. A new license is generated for o_{txt_7}: $l_7 = <\text{Tom}, \{g_7', g_8', g_9', g_{10}, g_{11}', g_{12}'\} > \in \mathcal{L}_{\text{Tom}}$ where the grants are updates as follows in order to refer to the new resource identifier:

$$g_7' = <\text{Tom}, \text{print}, \hat{o}_{txt_7}, \{c_1, c_2\} > \qquad g_8' = <\text{Tom}, \text{reduce}, \hat{o}_{txt_7}, \{c_1, c_2\} >$$
$$g_9' = <\text{Tom}, \text{enlarge}, \hat{o}_{txt_7}, \{c_1, c_2\} > \qquad g_{11}' = <\text{Tom}, \text{embed}, \hat{o}_{txt_7}, \{\} >$$
$$g_{12}' = <\text{Tom}, \text{enhance}, \hat{o}_{txt_7}, \{\} > \qquad \qquad \square$$

The two presented algorithms allow the automatic generation of a new license when resources are composed and modified. However, the user might decide to introduce further grants (e.g. on new introduced resources) or to remove useless grants (e.g. grants that refer resource components that have been removed). These operations are performed through the add and remove algorithms that we do not report for space constraints. The add algorithm checks the consistency of

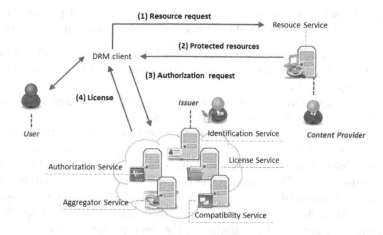

Fig. 2. A simplified DRM scenario

the new inserted grants with respect to the other grants occurring in the license, whereas the `remove` algorithm compacts the license by removing grants that cannot any longer be evaluated on the modified resource.

5   Enhanced DRM Architecture

This section presents a DRM architecture enabling a set of services and technologies to govern the authorized use of digital resources and to manage any consequences of that use throughout their entire life cycle. The architecture is approached by identifying the main involved entities: the resource provider, the users and the issuer. A DRM system is a set of DRM related services offered by the issuer to users and resource providers to enable the consumption of resources. In a typical simplified scenario, illustrated in Figure 2, a user exploits its DRM client to contact a Resource Service (1). This service enables the access to the protected resources (2). After an authorization has been requested (3) and evaluated according to the user profile (4) through a set of services, the resources can be consumed. For evaluating the user-request and for issuing the corresponding license, the architecture relies on a workflow of service-requests. In the reminder of the section, we first report a set of high level services that are needed for enabling such composition, and then we describe the correct sequence of service-requests used to compose a new resource and to generate its proper license. For the lack of space we concentrate only on the issue of composition, the update of a single resource is handled analogously.

Service-Oriented Architecture. The architecture in [14] is extended for supporting the license generation when a new resource is created by a user (henceforward the producer) through the composition of different resources. For the sake of clarity, we avoid to present the external tracking and payment services focusing only on services devoted to check the composition compatibility.

1. *Resource Service.* This service is in charge of the management of the resources and answers to the access/composition requests posed by an user through a DRM compliant client.
2. *License Service.* This service is in charge of the management of the licenses and of issuing new licenses upon users request.
3. *Authorization Service.* The Authorization Service is responsible for the evaluation of licenses upon an access/composition request is received by the Resource Service. For its activity the Authorization Service gets in touch with the License Service for obtaining the licenses, with the Compatibility Service for checking compatibility issues in case of composition/update of resources, with the Identification Service for the evaluation of principal specification on licenses' grants, and for acquiring the user profile for the evaluation of grants, and with External Services for checking specific constraints (like for example the external Payment Service in case the principal has to pay for consuming the resources) occurring in licenses.
4. *Identification Service.* This service is responsible for the authentication of the users and for checking the users' certificates. Moreover, it gets in tough with the DRM client to obtain temporal, geographical and context of use information to be included in the user profile. Finally, the Identification Service is in charge of storing and keep updated the users profiles.
5. *Compatibility Service.* This service is responsible to check the weak and strong compatibility conditions discussed in Section 3. The Compatibility Service, like the Authorization Service, can get in tough with External Services for the strong verification of conditions of grants.
6. *License Generation Service.* This service is in charge of executing the algorithms discussed in Section 4, and of creating the new licence of the composed resource by interacting with the License Service to retrieve licenses and to store the generated ones.

DRM Services Composition in Collaborative Environment. In a collaborative environment, we should define a workflow of service-requests able to support the generation of a license of a new resource created by composing different ones. Specifically, our proposal relies on a workflow consisting of two main activities: the composition of different resources and the definition of the license on the resulting resource. To support these activities, our architecture should adopt specific composition and integration management services (see Figure 3).

For composing different resources, the DRM client forwards the composition request obtained by the producer to the Resource Service. In order to authorize the operation, the Resource Service gets in touch with the Authorization Service. This service collects the licenses of the involved resources from the License Service. By considering the grants contained in the obtained licenses, the Authorization Service evaluates the composition request by means of the Compatibility Service. The first check concerns the validity of the weak compatibility

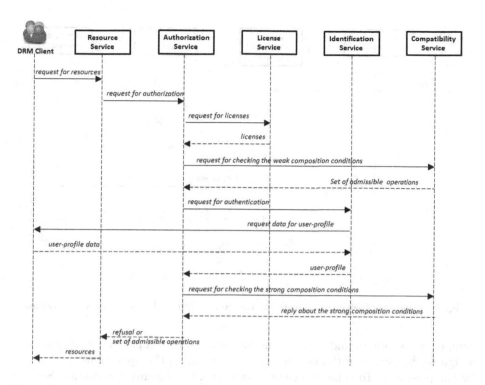

Fig. 3. Sequence of service-requests for composing a set of resources in the DRM system

conditions of the resources to compose and leads to the identification of a set of operations that the producer can carry out on the original resources according to her/his licenses. Once the Compatibility Service has checked the weak compatibility conditions, the Authorization Service has to authenticate the producer and control the grant conditions according to the producer profile. Authentication is realized in the following way. The Authorization Services asks to the Identification Service to authenticate the producer by providing the producer profile. The Identification Service, in turn, needs to query the DRM client for gathering information about the producer's context of use and for the evaluation of predicates relying on the producer certificates. The evaluation of the grant conditions is realized through the Compatibility Service by considering the producer profile and External Services. After carrying out these two activities, the Authorization Service can authorize or refuse the access to resources. In case the compatibility check comes to a satisfactory reply, it will provide the set of operations that the producer can carry out for composing the new resource.

Once the resources have been composed, the producer requests the generation of a new license. First of all, the DRM client has to contact the License Generation Service that takes care of all steps needed to create the new license and to send it to the License Service for its storage (see Figure 4). The License Generation Service requires to the License Service the producer's licenses of the

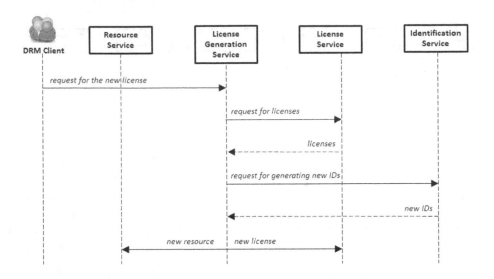

Fig. 4. Sequence of service-requests for the generation and storage of a new license

component resources and exploits the information to compose a set of grants to assign to the license of the new resource according to the operations carried out by the producer. To complete these operations, the License Generation Service has to ask to the Identification Service to generate new identifiers for resources enhanced and adapted during the composition process. By exploiting the identifiers generated by the Identification Service, the License Generation Service generates the license of the new resource, if it is possible. If the process is positively concluded, the License Generation Service sends the new resource to the Resource Service for its storage and sends the new license to the License Service.

6   Related Work

Different Rights Expression Languages (like ccREL [6], ODRL [8], MPEG-21 REL [10]) have been proposed for expressing digital licenses, terms and conditions. These proposals differ from the scope and the granularity according to which it is possible to specify and manage each aspect of the license specification and management processes. Although some of them support the necessary rights to compose resources, none of them explicitly defines how to create and assign a new license to the result of the composition.

The issue of license composition and compatibility analysis is difficult to address in a systematic way since license terms and conditions can be expressed in different ways, including the natural language, by using terms whose meaning would be ambiguous or not interpretable uniquely. This aspect makes obviously more difficult to determine which correct result should be returned from the composition process. In [3–5] the authors address the problem of service license[1] composition and compatibility analysis by using the ODRL-s language [2], an extension of ODRL [7, 8] to implement the clauses of service licensing. A similar approach is used in [19] where the authors present a framework to associate licensing terms to web data in order to combine licenses for the data resulting from queries; the resulting composite license is still expressed as licensing terms. Finally, in [1] the authors address the issue of identifying the kind of licenses, expressed through MPEG-21 REL [10, 11], a user should hold in order to compose or transform resources and present use-cases for the main operations.

Our approach differs from them from several points of view. First, our application scenario is different from that of service and web data licensing. In their case, the focus is on the management of the business and legal contractual information expressed as CC terms, with respect to the ODRL CC profile [9] and the CC license schema [6] respectively. This is used to establish if different services are compatible and composable among them and to verify wether users can access to web data provided by different data sets and released under different licensed terms. Conversely, our formal model, compliant with MPEG-21 REL [10], is mainly focused on the issue of determining the conditions under which the transformation/composition of resources can be performed in a systematic way and to provide a license for the transformed/composed resource. Finally, the weak/strong evaluation strategy we propose allows the evaluation of the authorization conditions only when strictly required.

With the proliferation of Internet-based applications and the ready availability of powerful file sharing and distribution tools, DRM has become a critical concern in the Internet domain. The literature is rich of proposals exploiting DRM platforms based on the distribution of digital content via Internet, like: the Open and Secure Digital Rights Management Solution (OSDRM) [18]; the Microsoft Windows Media DRM platform (MSDRM) [17]; the IBM Electronic Media Management System (EMMS) [16]; and the Real Networks HelixDRM [15]. These solutions aim at creating a secure framework for delivering multimedia on the Internet, and caters for the creation of secure contents, payment collection, distribution and rendering of multimedia. Most of this literature describes a DRM system as a platform containing several functionality, including content registration and protection; it offers publication, content search, purchase and licensing, authorization and access control. Nevertheless, these systems are not able to support users in the composition process by taking into account the users' licenses on the component resources.

7   Conclusions and Future Work

Starting from a formal model for the representation of licenses, in this paper we provided algorithms for the evaluation of compatibility of license's grants and for the generation of a new license of composed/updated resources. Moreover, we discussed an architecture that allows the composition and modification of

---

[1] A service license describes the terms and conditions for the use and access of the service in a machine readable way that services could be able to understand.

resources. These results are the building blocks for the realization of a tool for helping a user to compose resources in collaborative environments that we are currently implementing. We remark that our licenses can be easily translated into corresponding MPEG-21 REL licenses [10].

As future work we plan to enhance our approach by allowing the composition of resources belonging to different users. In this scenario, a negotiation of users' grants should take place in order to determine the grants to be assigned to the generated resource. Moreover, we are considering the issue of distribution of licenses to the users belonging to the user's communities.

References

1. Delgado, J., et al.: Definition of mechanisms that enable the exploitation of governed content. In: AXMEDIS, pp. 136–142 (2006)
2. Gangadharan, G., et al.: ODRL service licensing profile (ODRL-s). In: Proc. 5th Int'l Workshop for Technical, Economic and Legal Aspects of Business Models for Virtual Goods, Germany (2007)
3. Gangadharan, G., D'Andrea, V.: Service licensing: conceptualization, formalization, and expression. Service Oriented Computing and Applications 5(1), 37–59 (2011)
4. Gangadharan, G.R., et al.: Consumer-specified service license selection and composition. In: ICCBSS, pp. 194–203 (2008)
5. Gangadharan, G.R., Weiss, M., D'Andrea, V., Iannella, R.: Service license composition and compatibility analysis. In: Krämer, B.J., Lin, K.-J., Narasimhan, P. (eds.) ICSOC 2007. LNCS, vol. 4749, pp. 257–269. Springer, Heidelberg (2007)
6. Abelson, H., et al.: ccREL: The Creative Commons Rights Expression Language. Creative Commons Wiki (2008)
7. Iannella, R.: Open digital rights management. In: Workshop on Digital Rights Management for the Web, France (2001)
8. Iannella, R.: Odrl specification 1.1 (2002)
9. Iannella, R.: Odrl creative commons profile specification (2005)
10. Information Technology-Multimedia Framework. Part 5: Rights expression language, iso/iec 21000-5 (2004)
11. Information Technology-Multimedia Framework. Part 6: Rights data dictionary, iso/iec 21000-6 (2004)
12. json. Javascript object notation
13. Ku, W., Chi, C.-H.: Survey on the technological aspects of digital rights management. In: Zhang, K., Zheng, Y. (eds.) ISC 2004. LNCS, vol. 3225, pp. 391–403. Springer, Heidelberg (2004)
14. Michiels, S., et al.: Towards a software architecture for DRM. In: DRM, pp. 65–74 (2005)
15. RealNetworks. Helixcommunity-the foundation of great multimedia applications (2013)
16. RealNetworks. The IBM electronic media management system (2013)
17. RealNetworks. Windows media digital rights management (2013)
18. Serrão, C., et al.: Open SDRM - An open and secure digital rights management solution. In: Proc. Int. Association for Development of the Information Society, Portugal (2003)
19. Villata, S., Gandon, F.: Licenses compatibility and composition in the web of data. In: COLD (2012)

The *Hints from the Crowd* Project

Paolo Fosci[1], Giuseppe Psaila[1], and Marcello Di Stefano[2]

[1] University of Bergamo, Dept. of Engineering,
Viale Marconi 5, I-24044 Dalmine, BG, Italy
{paolo.fosci,psaila}@unibg.it
[2] University of Palermo, Dept. of Computer Science Engineering,
Viale delle Scienze Ed. 6, I-90100 Palermo, Italy
marcello.distefano@unipa.it

Abstract. Can the crowd be a source of information? Is it possible to receive useful hints from comments, blogs and product reviews? In the era of Web 2.0, people are allowed to give their opinion about everything such as movies, hotels, etc.. These reviews are *social knowledge*, that can be exploited to suggest *possibly interesting items* to other people.

The goal of the *Hints From the Crowd* (HFC) project is to build a *NoSQL* database system for large collections of product reviews; the database is queried by expressing a natural language sentence; the result is a list of products ranked based on the relevance of reviews w.r.t. the natural language sentence. The best ranked products in the result list can be seen as the best hints for the user based on crowd opinions (the reviews).

The HFC prototype has been developed to be independent of the particular application domain of the collected product reviews. Queries are performed by evaluating a text-based ranking metric for sets of reviews, specifically devised for this system; the metric evaluates the relevance of product reviews w.r.t. a natural language sentence (the query).

We present the architecture of the system, the ranking metric and analyze execution times.

1   Introduction

Users of modern Web 2.0 applications consider a matter of fact being allowed to write their own comments or reviews about any kind of product. Similarly, they expect to find reviews posted by other users, so that they can exploit them to make decisions about, e.g., a phone, a hotel room, a movie.

All these reviews constitute an incredible source of information. They are previous *social knowledge* about products, by means of which users would like to get useful *hints*. But how could a user obtain them? Typically, the user has a wish and would like to find out products that match those wishes, based on opinions of other users. This approach must be necessarily supported by a system. In fact, looking at the problem by a database technology point of view, product reviews constitute a text, yet moderately structured, database; user's wishes can be seen as natural language queries over the set of reviews and the

H. Decker et al. (Eds.): DEXA 2013, Part I, LNCS 8055, pp. 443–453, 2013.

user wants to obtain the products whose set of reviews matches the query at the highest degree. In other words, such a system is a *NoSQL* database system, where queries are natural language sentences.

This is the premise that motivated the *Hints From the Crowd* (HFC) project: we wanted to build a *NoSQL* database system for large collections of product reviews; the database is queried by expressing a natural language sentence; the result is a list of products ranked based on the relevance of reviews w.r.t. the natural language sentence. The best ranked products in the result list can be seen as the best hints for the user based on crowd opinions (the reviews).

We started the development of the HFC prototype from our previous work [2], where we studied several ranking metrics for retrieving blogs, but we introduced semantic tagging and semantic expansion based on the WordNet ontology.

The main goal of the project was to demonstrate that it is possible to obtain an answer to a query in acceptable time on a large set of reviews. So, the paper is focused on the architecture of the prototype, on the precise definition of the ranking metric, and on the study of execution time performance at query time.

Three main categories of *NoSQL* databases are usually considered [7] : *Key-Value stores*, *Document stores* and *Column Family stores*. Furthermore, even *Graph databases* can be considered as well belonging to the world of *NoSQL* databases [6]. However, the work in [1] clearly highlights that a key aspect of *NoSQL* databases is performance in terms of execution times; therefore, this motivates our focus on proposing efficient solutions for our proposal.

As far as the analysis of customer reviews is concerned, usually data mining techniques are adopted. In particular, [5] presents a system to compare opinions about products is presented, where association rule mining is exploited to assign a positive or negative polarity to words (namely, adjectives) in product reviews, and use this polarity to rank the opinion about products; [4] extracts, by means of an association rule mining technique, relevant features that summarize product reviews. Notice that we do not propose a data mining technique. We simply take inspiration from the concept of frequent itemset.

The paper is organized as follows. Section 2 shows and describes the architecture of the system. Section 3 describes how the query engine works and the ranking measure. Section 4 presents the results of the performance study we conducted by means of the IMDb dataset. Finally, Section 5 draws the conclusions.

2   The HFC System

As stated in Section 1, the goal of the *Hints From the Crowd* project is to build a NoSQL database system that deals with collections of product reviews, that can be queried by expressing a natural language sentence. The system extracts those products whose reviews better match the natural language query (in the rest of the paper, we use the term *query* in place of *natural language query*).

Due to lack of space, in this Section we briefly present the architecture of the overall system and data structures defined to foster query performance.

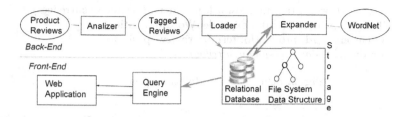

Fig. 1. Architecture of the HFC System

Architecture. The HFC system is composed by several components, each one devoted to perform a specific task. We distinguish between the *back-end*, responsible for the initial general indexing, and the *front-end*, where the *Query Engine* and its user interface actually resides. Figure 1 shows the architecture.

Back-end. In this side of the system, we find the components (denoted as rectangles) that prepares the data structure on which queries are executed. These components operate on source data and intermediate results (denoted as ovals) and upload data structures in the *Storage Box*.

- *Analyzer.* This component analyzes product reviews, identifying words and their grammar categories by a *pos-tagger*[1], obtaining tagged reviews.
- *Loader.* The goal of this component is to actually load tagged reviews into the data structures on which queries are performed (see paragraph *Data Structure*).
- *Expander.* After completion of loading of tagged reviews, *tagged words* are expanded on the base of their grammar category by means of an ontology[2], in order to afterwards allow the *Query Engine* to capture a wider set of results related with a query.

Front-end. The *Query Engine* is the key component of the front-end. It exploits the preliminary work performed by back-end components, to answer the query provided by users through the *Web Application*.

Data Structure. The HFC system must provide reasonable performance at query time, but at the same time all the components must easily get all the information they need. For these reasons, both a relational database and a file system data structure have been adopted.

Figure 2 depicts the schema of the data structure. The key table of the schema is table **Terms**, which describes each single term managed by the system. In effect, we distinguish between simple words and *tagged* words (i.e., terms), that is, words with a tag that describes the grammatical category (adjective, noun,

---

[1] For pos-tagging we used the *Stanford Parser*:
http://nlp.stanford.edu/index.shtml
[2] We used *WordNet 3.1*: http://wordnet.princeton.edu/

Terms(**id**, word, tag, taggedword, products, reviews, occurrences)
Term2Expansion(termId, expandedWordId, relation)
Occurrences(**id**, productId, termId, review, position)
Product(**id**, *domain specific attributes*)

Fig. 2. Schema of the Relational Database

verb, or adverb), based on the classification used by *WordNet*; therefore, the same simple word can appear several times in the table, one for each different possible grammatical category for that word. Attributes `products`, `reviews` and `occurrences` counts the number of products and the number of reviews in which a term occurs, and the total number of occurrences, respectively.

If a term is obtained by expanding another term, based on the WordNet ontology, it is described in the same table `Terms`. The expansion relation is described by table `Term2Expansion`, that associates a term (attribute `termId`) to the expanded term (attribute `expandedTermId`). Attribute `relation` denotes the typology of expansion, i.e., synonym, hypernym, meronym to name a few.

Table `Occurrences` describes all occurrences of terms in product reviews; in particular, notice attribute `position`, that indicates the position of the occurrence in the review. This table is actually represented as a pool of files where each file is the inverted index for a term.

Finally, table `Products` describes each single product, and its schema is context-dependent since attributes are defined based on the application domain. For example, since we use *IMDb* movie data set to test the prototype, we defined attributes concerning movies, such as title, director, year, and so on.

3   Query Engine

We now describe the key component, i.e., the query engine. Based on a natural language sentence (the query) it extracts those products whose reviews are mostly relevant for the query. Relevance is evaluated by means of a *ranking metric*; retrieved products are returned as a list sorted in reverse order of relevance. Hereafter, we describe how the ranking metric is defined.

3.1   Termsets

In this paper we consider a query q as a *set of terms* (or briefly, a *termset*). Thus, we describe a query containing a number n of terms as $q = \{t_1, \ldots, t_n\}$, and we investigate only those queries where $n > 1$ or, in other words, $|q| > 1$.

With I, we denote a generic termset that is a subset of q for which applies $|I| > 1$. With D_q, we denote the set of termsets I derived from q. Notice that the cardinality of D_q is $|D_q| = 2^n - (n+1)$, i.e. D_q is the power set of q without the empty set and the n single terms that compose q.

With I_l we denote an l-termset of q, that is a termset composed by l terms, i.e. $|I_l| = l$. With $D_{q,l}$ we denote the set of l-termsets I_l. Notice that the cardinality of $D_{g,l}$ is $|D_{q,l}| = \binom{n}{l}$.

3.2   Termset Weight

We now define the concept of *weight* for a termset.

Definition 1: The weight of a l−termset is a function of its length and the length of the query q ($|q| = n$); it is denoted as $w_q(l)$.

For $n = 2$, there is only one 2-termset and its weight is $w_q(2) = 1$ by definition. For $n > 2$, the weight of the single n−termset q is, by definition, $w_q(n) = 0.5$, while for $2 < l < n$, it is $w_q(l) = w_q(l + 1)/((\binom{n}{l}) + 1)$ and for $l = 2$ it is $w_q(2) = w_q(3)/(\binom{n}{2})$.   □

The rationale behind Definition 1 is the following. The topmost termset, corresponding to the whole query, is the most important one, and its weight is equal to the overall weight of all the shorter termsets. The same principle is valid for any generic termset I_l (with $2 < l < n$), whose weight is equal to the overall weight of all lower level termsets (even those that are not subset of I_l). This way, reducing the size of termsets, the contribution of each level quickly decreases.

Notice, that the overall weight of all termsets is exactly 1 ($\sum_{I \in D_q} w_q(|I|) = 1$).

3.3   Query Expansion and Semantic Coefficient

As stated in Section 2, reviews are processed performing several operations. Similar operations are performed on a user-query in natural language as well.

Pos-tagging. By means of *Stanford Parser*, each word of a user query is tagged with an attribute that denotes its grammar role (*verb, noun, adjective* to name a few) in the query.

Stopwords filtering. Stopwords are those words that are too common in reviews (such as *articles, conjunctions* or common verbal forms like *is* or *have*). These words hold a small semantic meaning, so after pos-tagging operation *stopwords* are discarded from the query.

Thus, denoting with SW the set of possible stopwords, frm now on the notation:

$$q = \{t_1, \ldots, t_n\}$$

includes only those terms $t_i \notin SW$, and, as stated in Section 3.1, we consider only those queries q such that $|q| > 1$ (actual length without stopwords).

Term expansion. By means of *WordNet* ontology, each tagged term $t_i \in q$ is expanded with all those terms directly associated to t_i based on its grammar tag. For example, a *noun* is expanded with all its *synonyms, hypernyms* or *hyponyms* and so on, while a *verb* is expanded with all its *synonyms* or *meronyms*, to name a few. There are actually a total of 15 possible different relationships between a tagged term and its expanded word.

We denote with t_i^* the generic expanded term of t_i, and with $ET(t_i)$ the set of all expanded terms of t_i. By definition, $t_i \in ET(t_i)$ with an *identity* relation, thus, $|ET(t_i)| \geq 1$.

Notice that, given a generic expanded term t^*, it can happen that $t^* \in ET(t_i)$ and $t^* \in ET(t_j)$, with $i \neq j$. In other words, we cannot state a-priori that $ET(t_i) \cap ET(t_j) = \oslash$ with $i \neq j$. As an example, the term *colour* can be an *hypernym* expansion for both terms *red* and *black*.

Query expansion. An *expanded query* q^* is each combination of $\{t_1^*, \ldots, t_n^*\}$. We consider *valid* a combination $q^* = \{t_1^*, \ldots, t_n^*\}$ only if $t_i^* \neq t_j^*$ $\forall i \neq j$ Notice that the original query q is a particular q^* itself, and it is *valid* by definition.

Expanded Termsets. Previous considerations about query q and its expansions, are applicable to each termset I_l. With I_l^* we denote an *expanded termset* $I_l^* = \{t_1^*, \ldots, t_l^*\}$, and similarly I_l^* is *valid* only if $t_i^* \neq t_j^*$ $\forall i \neq j$.

With $EI(I)$ we denote the set of all possible expanded termset I^* that can be derived from I. The cardinality of $EI(I) = \prod_{t \in I} |ET(t)|$, that is the number of all possible combinations of the expanded terms of those terms that compose I.

Finally, with D_q^*, we denote the set of all valid expanded termsets that are included in q and all its valid expansions q^*.

Semantic coefficient. Each $t^* \in ET(t)$ has a *semantic coefficient* $sc_t(t^*)$, with $0 < sc_t(t^*) \leq 1$, that depends on the cardinality of $ET(t_i)$.

Definition 2: For each $t^* \in ET(t)$ except t, $sc_t(t^*) = 0.5/|ET(t)|$, and $sc_t(t) = 0.5 + 0.5/|ET(t)|$.   □

The rationale of semantic coefficient, is the following. A term describes a semantic concept that is mostly expressed by the term itself, but receives a small contribution from expanded terms: the greater the number of expansions, the smaller the semantic contribution of a single expanded term. Notice that $\sum_{t^* \in ET(t)} sc_t(t^*) = 1$.

With $sc_I(I^*)$ we denote the semantic coefficient for an expanded termset I^* derived from I.

Definition 3: Given an expanded termset $I^* = \{t_1^* \ldots t_l^*\}$ derived for a termset $I = \{t_1 \ldots t_l\}$, it is $sc_I(I^*) = \prod_{t_i^* \in I_l^*} sc_{t_i}(t_i^*)$.   □

This way, a termset that contains only original terms gives the highest semantic contribution, while augmenting the number of expanded terms in the termset, the semantic contribution decreases.

Notice that, according to the above definition, $\sum_{I^* \in EI(I)} sc_I(I^*) = 1$.

3.4   Product Reviews and Termsets

Consider a product p (a movie, a camera, etc.); its set of reviews is denoted by $R(p) = \{r_1, \ldots, r_k\}$. Each review is a text, i.e., a sequence of term occurrences $r_i = <t_1, \ldots, t_s>$.

With $T(R(p))$ we denote the set of terms appearing in reviews for product p, and with $T(r_i)$ the set of terms appearing in review $r_i \in R(p)$.

Definition 4: A termset I is said *relevant* for product p if $\exists r_i | I \subseteq T(r_i)$.   □

The set of relevant termsets for product p is denoted as $RD_{p,q}$. In an analogous way, $RD^*_{p,q}$ is the set of all relevant expanded termsets for product p. Notice that $RD_{p,q} \subseteq D_q$, and also $RD^*_{p,q} \subseteq D^*_q$.

3.5 Termset Average Density

In a preliminary work [3], we assumed that every termset occurrence in product reviews contribute to the *support* of the termset with the same weight, i.e. 1, since the support, by definition, is the number of reviews containing the termset on the total amount of reviews.

Given a termset I, in a single review, terms in I can be very dense or very sparse. We had the intuition that a review in which the occurrences of terms in I are dense is more relevant for the query than a review where occurrences are sparse. Thus, we introduce the concept of *Termset Density* of an termset I for a single review.

Definition 5: Consider a product p, a review $r \in R(p)$, and a termset I_l. The *Termset Review Density* $d_r(I_l)$ is defined as

$$d_r(I_l) = l / minWin_r(I_l)$$

where $minWin_r(I_l)$ is the size of the minimal window in review r that includes all the terms of termset I_l.   □

Notice that for *Termset Review Density*, it holds that $0 < d(I_l, r) \leq 1$

The next step is to define a *Termset Average Density* for a generic termset I (we omit the subscript l so as not to burden notation) w.r.t. a product p.

Definition 6: Consider a product p and its set of reviews $R(p)$. With $R_I(p)$ we call the subset of $R(p)$ of those reviews containing termset I. The *Termset Average Density* for product p, denoted as $ad_p(I)$, is defined as:

$$ad_p(I) = (\textstyle\sum_{r \in R_I(p)} d_r(I))/|R(p)|$$

□

The Termset Average Density is analogous to termset support, with the difference that the contribution of the occurrence of a termset I in a review r is not 1 but its density $d_r(I)$. Notice that $ad_p(I) \leq s_p(I) \leq 1$, where with $s_p(I)$ we denote the support of a termset I for a product p.

3.6 Product Ranking Metric

Finally, we can now define the *Product Ranking Metric PRM*.

Definition 7: Consider a query q, the set of termsets D^*_q derived from q, the weights $w_q(|I^*|)$ and semantic coefficients $sc_q(I^*)$ for each expanded termset $I^* \in D^*_q$.

Consider a product p, the set of reviews $R(p)$ and the set of relevant expanded termsets $RD^*_{p,q}$ that can be actually extracted from $R(p)$. Given, for each $I^* \in$

Table 1. Data set general informations

Movies	Reviews	Reviews per Movie			Data Size
		Max	Min	Avg	
109,221	2,207,678	4,876	1	20	3,091Mb

Table 2. Indexed schemes

Schema	A	B	Diff %
Pos-Tagger	active	inactive	
Distinct tagged terms	1,151,827	776,852	-32.55%
Occurrences	216,345,522	216,345,522	0.00%
Analyzer Time (A = Ps+Pt)	2226.80h	3.82h	-99.83%
Parsing Time (Ps)	2.11h	2.42h	+14.74%
Pos-tagging Time (Pt)	2224.69h	1.40h	-99.94%
Loader Time (D)	56.05h	49.76h	-11.23%
Expander Time (E)	3.73h	2.67h	-28.49%
Total Time (T = A+D+E)	2286.58h	56.25h	-97.54%

$RD^*_{p,q}$, the average termset density $ad_p(I^*)$, the *Product Relevance Value* for product p is defined as

$$PRM_q(p) = \sum_{I^* \in RD^*_{p,q}} (w_q(|I^*|) \times ad_p(I^*) \times sc_q(I^*))$$

$\square$

The rationale of the above definition is the following. For each termset I^* included in the query q and actually relevant in the reviews, its contribution to the overall relevance value is given by its weight $w_q(|I^*|)$ (that depends on its size) multiplied by its *average density* $ad_{p,q}(I^*)$ and its semantic coefficient $sc_q(I^*)$.

The system of weights and semantic coefficients has been designed to obtain a $PRM_q(p) = 1$ for an *ideal* set of reviews for product p, where each review contains every expanded termset I^* that can be derived from q with a density $d_r(I^*) = 1$, and every expanded termset I^* is *valid*.

4   Evaluation

Our dataset (downloaded from IMDb.com web site) is described in Table 1.

Experiments have been run on a PC with two Intel Xeon Quad-core 2.0GHz/ L3-4MB processors, 12GB RAM, four 1-Tbyte disks and Linux operating system.

Indexing. While indexing data set, as described in the *back-end* of HFC in Section 2, we figured out how *pos-tagging* affects the HFC system.

Disabling pos-tagging means tagging each term with a unique trivial tag, and considering for each term every possible expansion regardless of its role inside the query; in other words, disabling pos-tagging implies a significant reduction of the number of managed distinct terms because words are not distinguished anymore on the basis of their grammar category; however, the counter effect is that the query engine has to likely retrieve a greater number of term occurrences and calculate a greater number of termsets (briefly see Table 4).

Table 2 reports data collected during dataset indexing. Column *A* shows data regarding indexing with *pos-tagging activated*, while Column *B* shows data regarding indexing with *pos-tagging deactivated* (respectively *Schema A* and *Schema B* in the rest of the paper). For each *Schema*, the number of identified tagged terms and the total number of indexed term occurrences are shown. As stated before, disabling pos-tagging reduces the number of tagged terms ($\sim 33\%$) while the number of occurrences remains unchanged. Table 2 reports also data relative to execution time during the tasks described in the back-end of the system architecture (see Figure 1) in the indexing phase. In particular, the *Analyzer Time* has been splitted into a *Parsing Time*, that is basically the share of time due to reading data from data set and identify terms occurrences, and a *Pos-tagging Time*, that considers only the execution time of *Stanford Parser* when pos-tagging is active, or the simple operation of labeling each term with the same tag when the pos-tagging is inactive. It is clearly evident how much pos-tagging affects the Analyzer Time: 2224.69 hours equivalent to more than 92 days! In order to reduce this waiting time, we exploited all the 8 cores of our machine, parallelizing the Analysis phase in 8 independent processes, splitting data set into 8 different independant sub-data sets, and reducing the actual waiting time to about 13 days.

To conclude, it is interesting to notice how the difference of tagged terms recognized in *Schema A* w.r.t. *Schema B* affects more (in percentage) the execution time of the *Expander* than the *Loader* (that include also occurrences loading).

Query Performance. For our query performance tests, we prepared a set of 25 standard user queries in natural language like *I want to know more about the history of Greece and Persian wars*, or *All those moments will be lost in time, like tears in rain*[3]. Due to the lack of space, we don't report the other queries.

Basically, the query engine evaluates a query performing 4 different steps: (1) *query expansion*, (2) *occurrences loading*, (3) *product ranking*, (4) *result sorting*. While Steps 1 and 4 must be performed by a single unique thread, Steps 2 and 3 can be parallelized (and performed in different threads).

On the base of this consideration, we realized also a multi-thread version of the query engine, with 5 different running threads for Steps 2 and 3. In order to do that, it was necessary to horizontally split the pool of files that represent Table Occurrences described in section 2, into 5 different and independant file systems.

The first test we made, on *Schema A*, shows the benefits on average performance per query (Table 3) from the *single-searching-thread* version of the query engine (SST-QE) to the *5-searching-threads* verion of query engine (5ST-QE).

The analysis shows that most of of execution times is spent during *occurrences loading*: this is mostly due to our storage system based on classical hard disks.

---

[3] From the movie *Blade Runner*.

Table 3. SST-QE vs 5ST-QE

	SST-QE	5ST-QE	Diff %
Average Time (T=QE+TG+TE+TM+S)	2,501.12 ms	1,994.66 ms	-20.25%
1. Query Expansion (QE	286.44 ms	286.40 ms	-0.01%
Thread generation (TG)	0.40 ms	1.88 ms	370.00%
Thread execution (TE ≤ O+R)	2,199.64 ms	1,691.60 ms	-23.10%
2. Occurrences Loading (O)	1,962.52 ms	1,639.84 ms	-16.44%
3. Ranking.(R)	237.12 ms	75.12 ms	-68.32%
Thread merging (TM)	1.64 ms	1.80 ms	9.76%
4. Sorting (S)	13.00 ms	12.98 ms	-0.17%

Table 4. Pos tagging vs No-pos tagging

Schema	A	B	Diff %
Pos Tagging	active	inactive	
Total time	1,995 ms	3,480 ms	74.47%
Movies	2,067	2,994	44.85%
Occurrences Retrieved	107,200	226,994	111.75%
Termsets analyzed	5,414	13,795	154.80%

With more modern solid-state storage devices, that are at least one order of magnitude faster, we are confident to dramatically improve performance.

Another issue is about threads parallelization. From the compared analysis, at first glance could seems that the 5ST-QE has not significantly improved performance, since there is only a 16.44% of gain in *occurrences loading*. However, this is due to the fact that using a single machine data are transferred to main memory through a single *system bus*. We are confident that by parallelizing the process on different machines, performance should significantly increase. As a matter of fact, the compared analysis of *ranking* execution times, that do not involve disk use, tells that 5ST-QE is 68.26% faster than SST-QE.

The second test, is a comparison between average performance of the 5ST-QE on *Schema A* and *Schema B* (as described in the *Indexing* paragraph). Table 4 provides average *data-per-query*. It can be noticed that when pos-tagger is inactive, there is a growing of average execution time, mostly due to the larger number of occurrences to load, and also to the larger number of termsets to analyze. On the other hands, the number of retrieved movies increases, since deactivating pos-tagging has the effect of increasing the number of expanded terms to search (causing generation of false positive movies, i.e., movies whose reviews are not actually relevant w.r.t. the original query).

5   Conclusions

The scope of this paper was to present the architecture and the query engine of *HFC NoSQL* database system. Although performance of the system can be further be improved, the considerations in Section 4 show that the approach is feasible in terms of query response time.

We are aware we did not discuss about system effectiveness, but it was beyond the scope of the paper. However the web-interface we developed is designed to collect users opinions about the system, and by means of that, in the future work we intend to deeper investigate effectiveness of the system. Moreover, as far as effectiveness is concerned, in the future work we intend to integrate term expansion with *linked-data* as a source for semantic ontology about terms.

References

1. Cattell, R.: Scalable sql and nosql data stores. SIGMOD Record 39(4), 12–27 (2011)
2. Fosci, P., Psaila, G.: Finding the best source of information by means of a socially-enabled search engine. In: 16th Annual KES Conference, San Sebastian (Spain), pp. 1253–1262 (September 2012)
3. Fosci, P., Psaila, G.: Toward a product search engine based on user reviews. In: DATA 2012 Int. Conf. on Data Technologies and Applications, Rome, Italy (July 2012)
4. Hu, M., Liu, B.: Mining and summarizing customer reviews. In: Kim, W., Kohavi, R., Gehrke, J., DuMouchel, W. (eds.) KDD, pp. 168–177. ACM (2004)
5. Liu, B., Hu, M., Cheng, J.: Opinion observer: analyzing and comparing opinions on the web. In: Ellis, A., Hagino, T. (eds.) WWW, pp. 342–351. ACM (2005)
6. Robin, H., Jablonski, S.: Nosql evaluation: A use case oriented survey. In: CSC 2011 International Conference on Cloud and Service Computing, Hong Kong, China, pp. 336–341 (December 2011)
7. Strauch, C.: Nosql databases (2011),
   http://www.christof-strauch.de/nosqldbs.pdf

Database Technology: A World of Interaction

Amira Kerkad, Ladjel Bellatreche, and Dominique Geniet

LIAS/ISAE-ENSMA, Poitiers University,
Futuroscope, France
{amira.kerkad,bellatreche,dgeniet}@ensma.fr

Abstract. Interaction is a typical phenomenon in database systems. It involves several components of DBMS: the data, the queries, the optimization techniques and devices. Each component is a critical issue for the database performance. The interaction between queries is well established and recognized by the database community. The interaction among optimization techniques has been also exploited during the physical design of databases. The interaction in multi levels is usually ignored when selecting optimization techniques. In our work, we deal with the combined problem of query scheduling, buffer management and horizontal partitioning simultaneously, by proposing an interaction-aware solution. An experimental study is given to show the efficiency of our proposal.

1 Introduction

The database technology is an adequate environment for the interaction, that it may concern several components of the database: (a) the schema, (b) the queries, (c) the optimization techniques, and (d) devices. At the schema level, correlations between attributes are extremely common in the real world relational datasets [1]. This correlation has been exploited to define materialized views and indexes. At the query level, interaction has been massively studied under the problem of multi-query optimization. The data warehousing and scientific applications with their star join queries increase the rate of interaction. This interaction has been used for selecting optimization techniques such as materialized views. Recently, with the spectacular development of devices (disk, SDD, flash, etc.), several research studies exploited the interaction between devices to co-process tasks among them [2]. The interaction also touches optimization techniques. In [3,4], similarities between materialized views, indexes, data partitioning and the clustering were identified and used to facilitate their selections.

Usually, the interaction concerns only one component. In this paper, we consider the multi-component interaction, with three optimization techniques, where each one concerns one component: the query scheduling (the query level), the horizontal data partitioning (data level) and the buffer management (the device level). The query scheduling (QS) consists in defining an optimal order of executing queries to allow some queries to get benefit from already processed data. The horizontal data partitioning (HDP) is a non redundant optimization technique [5]. The buffer management (BM) consists in allocating and replacing

H. Decker et al. (Eds.): DEXA 2013, Part I, LNCS 8055, pp. 454–461, 2013.

data in the buffer pool to lower the cost of queries. Usually, these problems are treated either in isolation or pairwise such as $\mathcal{BM}$ and $\mathcal{QS}$ [6]. However, these problems are similar and complementary. But by exploring the state of art, we figure out that each problem considers the interaction on one level. For instance, the $\mathcal{HDP}$ does not consider buffer content nor query order [7,8].

To facilitate the understanding of our proposal, let us consider a motivating example. Let $\mathcal{DW}$ be a relational data warehouse with a fact table *Sales* and three dimension tables: *Time* ; *Product* ; *Customer*. On the top of this $\mathcal{DW}$, a workload of 10 star join queries is defined. The execution plans of all queries are merged in one graph called *Multi-View Processing Plan* (MVPP). MVPP is a graphical representation of a workload proposed in the context of Multi-Query Optimization [9]. Note that selection operations are pushed down after constructing the MVPP. This graph has four main levels: **(a)** leaf nodes representing the base tables, **(b)** selection nodes that may be used to partition the databases $((e.g., \{s_1, s_2 \dots s_8\}))$, **(c)** join operation nodes (e.g., $\{j_1; j_2 \dots ; j_9\}$) and **(d)** final nodes for grouping, ordering and projections (e.g., $\{gop_1; gop_2; \dots gop_{10}\}$).

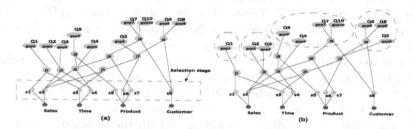

Fig. 1. Example for MVPP of 10 queries in (a) and its obtained clusters (hives) in (b)

Note that the join operation j_3 between table *Sales* and *Product* is shared by four queries Q_4, Q_7, Q_9 and Q_{10}. In order to optimize the query Q_4 involving two selections $\{s_2, s_4\}$ and a join j_3, the $\mathcal{HDP}$ may be a relevant optimization structure. The optimization of query Q_4 impacts not only query Q_4, but its benefit will propagate through all queries interacting with join node j_3: Q_7, Q_9 and Q_{10}. To spread the benefit along query plans while having the constraint of threshold W that limits the number of fragments, the choice becomes very hard to do.

If we consider the query interaction, we can group queries into four disjoint subsets depending on shared joins:

$\{Q_1\}; \{Q_2; Q_3\}; \{Q_4; Q_7; Q_9; Q_{10}\}; \{Q_5; Q_6; Q_8\}$. Note that the first join is very expensive because it involves the fact table. If the first join is optimized by the use of the $\mathcal{HDP}$ based on its selections, all queries in the same group will benefit from the partitioning. On the other hand, the intermediate results are candidates for bufferization. If we reorganize the initial structure of the MVPP by grouping queries sharing at least one node, a set of clusters is obtained as in Figure 1-b. Each cluster is called: *hive*. In each hive, we elect a query to be executed at first. This query is called *queen-bee*, and once executed, all its shared nodes are cached. Thus, queries in the same hive will be ordered and get benefit from the buffer content.

Along this paper, we exploit query interaction features to propose a new approach that deals simultaneously with the $\mathcal{HDP}$, the $\mathcal{BM}$ and $\mathcal{QS}$.

The rest of this paper is organized as follows: in section 2, we give the formalization and main challenges regarding the PBS problem. A resolution approach is proposed in Section 3 followed by an experimental study to validate our proposal in Section 4. A conclusion is given in Section 5.

2   Formalization

In [10], a formalization of the combined problem including $\mathcal{BM}$ and $\mathcal{QS}$ was given. The $\mathcal{HDP}$ selection problem involves a $\mathcal{DW}$ schema, a workload Q and threshold W. The problem aims at providing a partitioning schema, where the fact table is decomposed into N fragments ($N \leq W$) minimizing the query processing cost, where W represents a threshold fixed by the DBA. From these formalizations, we can easily identify the similarities and the complementarities between $\mathcal{BM}$, $\mathcal{QS}$ and $\mathcal{HDP}$. The combined problem including $\mathcal{HDP}$, $\mathcal{BM}$ and $\mathcal{QS}$, called "*Partitioning-Buffering-Scheduling*" problem ($\mathcal{PBM}$) is formalized as follows: Given a $\mathcal{DW}$, a workload Q, a buffer size B and a threshold W, the $\mathcal{PBM}$ consists in partitioning the $\mathcal{DW}$ such as the execution cost by considering the limited buffer size B and the threshold W will be minimized.

Beside the hardness of each problem individually, the combined problem brings new challenges regarding: (1) the resolution scenarios; (2) the nature of the new buffer objects after partitioning and (3) the global query rewriting on fragments.

1. By studying the semantic behind each optimization technique, we propose to start by the partitioning and then the two other techniques. This is due to the fact that the $\mathcal{HDP}$ alters the data component [8].

2. Theoretically, a buffer object is any subset of tuples from the universal relation UR. The UR is obtained by joining all tables in the database. Existing works concerned the non partitioned case. In these works, the objects considered are the results (or sub results) obtained from algebraic operations in the query plan. We call these objects: *elementary nodes*. In the partitioned case, the buffer objects are not the same as *elementary nodes*; because queries are rewritten to cover different sub-schemas using joins between fragments followed by unions between all partial results. In this case, the *elementary nodes* are broken into several operations inside valid subschemas. We call the inner nodes of a subschema: *subnodes*. Thus, in a horizontally partitioned schema, buffer objects are substituted by subnodes rather than elementary nodes. The number of subnodes is much larger then elementary nodes, but their sizes are often smaller.

3. Once the warehouse is partitioned, all queries will be rewritten in the fragments. This rewriting has to consider simultaneously buffer content and query order.

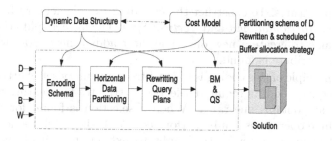

Fig. 2. Methodology of resolution of the combined problem

3   Proposal: Divide and Conquer Solution

Note that the isolated problems are very hard to solve, and existing techniques are either simple with low efficiency, or efficient with high complexity. To get a trade-off between efficiency and complexity, we propose a new divide and conquer solution for each phase guided by query interaction. In order to alleviate the algorithms, we also do an effort on the data structure that handles potential solutions. This data structure will carry out different phases of the proposed approach. Figure 2 summarizes our methodology. The entries are processed to generate an encoding schema using our dynamic data structure. This schema allows representing and handling fragments. The $\mathcal{HDP}$ algorithm runs over this encoding to find the partitioning schema with minimal cost. Once the (near) optimal partitioning schema is returned, query plans need to be traced on the new substar schemas. To do so, the dynamic data structure is exploited to detect the valid substar schemas for each query and get new plans. The new query plans are processed by the $\mathcal{BM}$ and $\mathcal{QS}$ algorithms to find the optimal order of executing queries, and the best buffer scenario. Besides the obtained partitioning schema, the algorithm returns an ordered set of queries and the associated buffer strategy. A cost model is used in both algorithms to measure the quality of a solution. The details of each phase in the proposed methodology are given along this section.

To facilitate the understanding of our proposal, we give the following assumptions: (1) no indexes are considered, (2) the workload is known in advance, and (3) the scheduling is performed offline.

3.1   Dynamic Data Structure

In this section, we describe our dynamic data structure that considers query interaction and allows : (a) representing solutions, (b) horizontal partitioning and (c) getting new query plans.

- (a) Representing solutions: one of the main problems in $\mathcal{HDP}$ is representing selection attributes and their sub-domains for the optimization process. To obtain the set of selection attributes, the MVPP is explored plan by plan. The encoding is updated by the new attributes and subdomains found in queries predicates. To construct the encoding two elementary functions are required : *Horizontal Split* and *Vertical Split*. The former splits the attribute subdomains to create a new one. The later adds a new array when a new selection attribute is found.

The principle of this coding is to start from an empty set of attributes, and for each query, add its required selection attributes by creating new arrays. Each array contains one range. When a selection is found for the current query, three operations can be performed: (**1**) If the attribute does not exist in the schema, apply a Vertical Split to extend the schema vertically by adding a new array for the attribute. The range is split into many parts to cover the new subdomains. An *else* range is added to ensure completeness. (**2**) If the attribute already exists, apply a Horizontal Split on the *else* range to add the new subdomains. (**3**) Finally, if the administrator knows the value remaining in the *else* range, it is replaced by this value. The result of this phase is an encoding schema containing the set of selection attributes and their associated sub-domains.

— (b) Partitioning schema: partitioning is ensured using the two primitives: *Merge* and *Split* applied on the encoding schema. The *Merge* function is applied on two partitions to get them in one partition. The *Split* is applied on one partition in order to be divided into two partitions if it covers at least two subdomains. Partitions can be managed by applying a series of split and merge.

— (c) Generating Profiles: the queries need to be rewritten by projecting their plans on the new schema. This rewriting phase can be performed using query profiles obtained by our dynamic data structure which allows three main functions: (1) identification of valid subschemas, (2) ordering joins inside a subschema and (3) estimating execution cost. To generate query profiles, we propose to use the obtained encoding schema and to fill each cell by 1 if the subdomain is used by the query; 0 otherwise. The cells are also indexed by selectivity of each subdomain in order to generate the selectivity factor profile. To represent subschemas, the cells values are set to 1 on the required partitions and 0 otherwise. This representation allows the cost model to identify the valid subschemas of each query by matching the query and the subschema profiles. If we consider that the best join order is obtained by minimal intermediate results, this order may be provided using the selectivity factor profile to estimate intermediate results size. In addition, these profiles allow estimating the execution cost of the workload. Consequently, the cost model proposed in [10] is extended to deal with the partitioning case.

3.2   Horizontal Partitioning : Elected Query Algorithm

To face the high complexity of the *HDP* problem, we use the interaction in the partitioning process. As it is shown in the motivating example, if some queries are elected to be optimized, the gain is spread through all queries interacting with the elected ones. Based on this feature, we propose an algorithm called

Elected Queries for $\mathcal{HDP}$ (EQHDP). The algorithm starts from the MVPP, and elects most "beneficial" queries to steer the partitioning process.

Queries are grouped into disjoint subsets, where each couple of queries sharing at least one node in the MVPP are in the same group. Queries inside each group are sorted by minimal cost. The first query (the least expensive) is elected in its group. This phase returns the set of elected queries EQ_i from each group i. The obtained set of elected queries is sorted by descending costs. That way, the most expensive queries are optimized before the threshold W is exceeded. The elected queries prune the schema of attributes depending on their requirements, i.e., only required attributes are taken. The sub-domains are tagged with the total number of elected queries using each one. Let u_{ij} be the number of EQ using this sub-domain and let k_i be the maximal value of u_{ij} in the attribute a_i. The set of attributes is sorted by maximal use (value of k_i) in order to start by partitioning on most used attributes before W is attained. After sorting the attributes, each attribute a_i is split/merged as follows: (1) The subdomains, which are not used by any elected query ($u_{ij} = 0$), are grouped in one partition P_0; (2) the most used subdomains having k_i elected queries accessing them are grouped in one partition P_k (k_i is the maximal usage value of a_i); and (3) if $N > W$ or $k_i \neq 0$ then, the remaining sub-domains are merged with P_0; otherwise, the sub-domains accessed by $k_i - 1$ elected queries are grouped in a new partition. The operation is repeated until $k_i = 0$ or $N > W$. This allows creating partitions to satisfy the most beneficial queries.

If partitioning is still possible ($N < W$), the obtained partitions are split depending on the correlation between queries accessing each sub-domain of the partition. If two subsets of elected queries require some sub-domains independently, a new partition is created to contain sub-domains used independently from the others. If $N < W$ after these merge and split operations regarding only the elected queries, then partitioning is still possible. The optimization process moves to other queries to improve their performance as well. The next set of queries is the successors of current EQs. If at least one group still has a successor, a new set of elected queries is generated by the found successors of all groups. The same process is applied by extending the encoding schema with the new set of selection attributes incrementally. The partitioning is done until $N = W$ or no more queries are left in **any** group.

3.3   Buffer Management and Query Scheduling

To reduce the complexity of the joint problem of $\mathcal{BM}$ and $\mathcal{QS}$, we propose an approach inspired from the natural life of bees. This algorithm called *Queen-Bee Algorithm* has been proposed in our previous work [10]. The basic idea behind our queen-bee algorithm is to partition the queries of the MVPP into subsets called *hives*. And for each *hive*, elect a query (*queen-bee*) to be executed first and its nodes will be cached. We associate the dynamic buffer management strategy DBM to our queen-bee algorithm. DBM mainly traverses the query plan and, for each intermediate node (operation), checks the buffer content: if the result of the current operation is already cached, it is read from the cache; else it is cached while there is enough buffer space. Once a node of a given *hive* is treated, its rank[1] value is decremented. When the rank of a node is equal 0, it will be removed from the buffer since it is useless for next queries.

Our query scheduler works on the clusters of queries (*hives*) and it shall order the queries inside each *hive* according to the buffer content. To do so, three modules extend our queen-bee algorithm [10], where global relations are replaced by local fragments: (1) generating of a query graph with connected components (QGCC), (2) sorting the components which is optional depending on whether queries have priority or not (this phase is ignored in our study) and (3) sorting queries inside each component.

4   Experimental Study

The experimental study is done using our Java Simulator which is connected to a star schema benchmark (SSB) of 100GB. The data is located on a server of 32GB of RAM and $2 \times 2.5GHz$ of CPU. The buffer pool size is set to 2 GB. We use two different workloads. The first one has 12 queries with no interaction. For the second one, we add 10 others to get 22 interacting queries. We compare our algorithm (PBS) with different cases: $\mathcal{HDP}$ using EQHDP algorithm, $\mathcal{HDP}$ using a Simulated Annealing (SA, with 500 iterations and initial temperature =300), $\mathcal{BM}$ and $\mathcal{QS}$ using the Queen-Bee algorithm and finaly, combining the three techniques using the SA. The SA is chosen in our experiments because it has been widely used in hard optimization problems, and has proven its efficiency compared to genetic and hill climbing algorithms[7].

Figure 3 shows that the EQHDP outperforms SA with interacting queries, because it is guided by interaction while the SA explores randomly the search space. The Queen-Bee optimizes workload 1, but it doesn't reduce the cost of workoad 2 because no objects are cached (no interaction). The three techniques ($\mathcal{HDP}$-$\mathcal{BM}$-$\mathcal{QS}$) are combined using a SA which gives a better performance then the isolated selection. The PBS algorithm outperforms the new SA in workload 2 contrary to workload 1. The reason is that PBS is steered by query interaction, which makes it more efficient then random search algorithms.

We also observed that $\mathcal{BM}$ after $\mathcal{HDP}$ may lower execution cost even when the initial queries are not inetracting because the partitioned schema makes new overlapping nodes. For instance, the number of nodes in the workload 1 passes from 67 elementary nodes to 1393 subnodes and in the workload 2 from 77 to 2580. As the subnodes are much smaller than elementary nodes, a large number of these nodes may fit in the buffer pool, which highers cache hits. Figure 4 shows the runtime of each algorithm (in logarithmic scale). The EQHDP and the Queen-Bee are much faster then the SA, which makes their combination in PBS more interesting then heuristics. In these experiments, we show that the PBS provides a compromise between efficiency and greediness by exploiting query interaction.

---

[1] We define the rank value of a node no_i as a counter representing the number of queries accessing no_i.

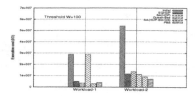 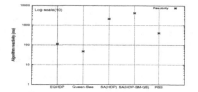

Fig. 3. Comparing performance of isolated and combined optimization techniques   **Fig. 4.** Comparing reactivity of each optimization technique

5   Conclusion

In this paper, we identify the interactions between different layers of database systems. They concern data, queries, devices and the optimization techniques. We motivate the need to develop physical design solutions by considering the interaction between these layers. To instantiate our proposal, we propose a combination of three well known problems in the literature: horizontal data partitioning, buffer management and query scheduling. Advanced solutions are given for this joint problem including algorithms, cost models and evaluations. The obtained results are encouraging.

References

1. Kimura, H., Huo, G., Rasin, A., Madden, S., Zdonik, S.B.: Coradd: Correlation aware database designer for materialized views and indexes. PVLDB 3(1), 1103–1113 (2010)
2. Breß, S., Schallehn, E., Geist, I.: Towards optimization of hybrid CPU/GPU query plans in database systems. In: Pechenizkiy, M., Wojciechowski, M. (eds.) New Trends in Databases & Inform. AISC, vol. 185, pp. 27–35. Springer, Heidelberg (2012)
3. Sanjay, A., Surajit, C., Narasayya, V.R.: Automated selection of materialized views and indexes in microsoft sql server. In: VLDB, pp. 496–505 (2000)
4. Zilio, D.C., Rao, J., Lightstone, S., Lohman, G.M., Storm, A., Garcia-Arellano, C., Fadden, S.: Db2 design advisor: Integrated automatic physical database design. In: VLDB, pp. 1087–1097 (2004)
5. Sanjay, A., Narasayya, V.R., Yang, B.: Integrating vertical and horizontal partitioning into automated physical database design. In: ACM SIGMOD, pp. 359–370 (2004)
6. Thomas, D., Diwan, A.A., Sudarshan, S.: Scheduling and caching in multiquery optimization. In: COMAD, pp. 150–153 (2006)
7. Bellatreche, L., Boukhalfa, K., Richard, P., Woameno, K.Y.: Referential horizontal partitioning selection problem in data warehouses: Hardness study and selection algorithms. IJDWM 5(4), 1–23 (2009)
8. Özsu, M.T., Valduriez, P.: Principles of Distributed Database Systems, 2nd edn. Prentice Hall (1999)
9. Sellis, T.K.: Multiple-query optimization. ACM Transactions on Database Systems 13(1), 23–52 (1988)
10. Kerkad, A., Bellatreche, L., Geniet, D.: Queen-bee: Query interaction-aware for buffer allocation and scheduling problem. In: Cuzzocrea, A., Dayal, U. (eds.) DaWaK 2012. LNCS, vol. 7448, pp. 156–167. Springer, Heidelberg (2012)

Author Index